Advances in Intelligent Systems and Computing

Volume 825

The series "Advances in Intelligent Systems and Computing" contains publications on theory, applications, and design methods of Intelligent Systems and Intelligent Computing. Virtually all disciplines such as engineering, natural sciences, computer and information science, ICT, economics, business, e-commerce, environment, healthcare, life science are covered. The list of topics spans all the areas of modern intelligent systems and computing such as: computational intelligence, soft computing including neural networks, fuzzy systems, evolutionary computing and the fusion of these paradigms, social intelligence, ambient intelligence, computational neuroscience, artificial life, virtual worlds and society, cognitive science and systems, Perception and Vision, DNA and immune based systems, self-organizing and adaptive systems, e-Learning and teaching, human-centered and human-centric computing, recommender systems, intelligent control, robotics and mechatronics including human-machine teaming, knowledge-based paradigms, learning paradigms, machine ethics, intelligent data analysis, knowledge management, intelligent agents, intelligent decision making and support, intelligent network security, trust management, interactive entertainment, Web intelligence and multimedia.

The publications within "Advances in Intelligent Systems and Computing" are primarily proceedings of important conferences, symposia and congresses. They cover significant recent developments in the field, both of a foundational and applicable character. An important characteristic feature of the series is the short publication time and world-wide distribution. This permits a rapid and broad dissemination of research results.

More information about this series at http://www.springer.com/series/11156

Sebastiano Bagnara · Riccardo Tartaglia
Sara Albolino · Thomas Alexander
Yushi Fujita
Editors

Proceedings of the 20th Congress of the International Ergonomics Association (IEA 2018)

Volume VIII: Ergonomics and Human Factors in Manufacturing, Agriculture, Building and Construction, Sustainable Development and Mining

Set 1

Editors
Sebastiano Bagnara
University of the Republic of San Marino
San Marino, San Marino

Riccardo Tartaglia
Centre for Clinical Risk Management and Patient Safety, Tuscany Region
Florence, Italy

Sara Albolino
Centre for Clinical Risk Management and Patient Safety, Tuscany Region
Florence, Italy

Thomas Alexander
Fraunhofer FKIE
Bonn, Nordrhein-Westfalen
Germany

Yushi Fujita
International Ergonomics Association
Tokyo, Japan

ISSN 2194-5357 ISSN 2194-5365 (electronic)
Advances in Intelligent Systems and Computing
ISBN 978-3-319-96067-8 ISBN 978-3-319-96068-5 (eBook)
https://doi.org/10.1007/978-3-319-96068-5

Library of Congress Control Number: 2018950646

This Springer imprint is published by the registered company Springer Nature Switzerland AG
The registered company address is: Gewerbestrasse 11, 6330 Cham, Switzerland

Preface

The Triennial Congress of the International Ergonomics Association is where and when a large community of scientists and practitioners interested in the fields of ergonomics/human factors meet to exchange research results and good practices, discuss them, raise questions about the state and the future of the community, and about the context where the community lives: the planet. The ergonomics/human factors community is concerned not only about its own conditions and perspectives, but also with those of people at large and the place we all live, as Neville Moray (Tatcher et al. 2018) taught us in a memorable address at the IEA Congress in Toronto more than twenty years, in 1994.

The Proceedings of an IEA Congress describes, then, the actual state of the art of the field of ergonomics/human factors and its context every three years.

In Florence, where the XX IEA Congress is taking place, there have been more than sixteen hundred (1643) abstract proposals from eighty countries from all the five continents. The accepted proposal has been about one thousand (1010), roughly, half from Europe and half from the other continents, being Asia the most numerous, followed by South America, North America, Oceania, and Africa. This Proceedings is indeed a very detailed and complete state of the art of human factors/ergonomics research and practice in about every place in the world.

All the accepted contributions are collected in the Congress Proceedings, distributed in ten volumes along with the themes in which ergonomics/human factors field is traditionally articulated and IEA Technical Committees are named:

I. Healthcare Ergonomics (ISBN 978-3-319-96097-5).
II. Safety and Health and Slips, Trips and Falls (ISBN 978-3-319-96088-3).
III. Musculoskeletal Disorders (ISBN 978-3-319-96082-1).
IV. Organizational Design and Management (ODAM), Professional Affairs, Forensic (ISBN 978-3-319-96079-1).
V. Human Simulation and Virtual Environments, Work with Computing Systems (WWCS), Process control (ISBN 978-3-319-96076-0).

VI. Transport Ergonomics and Human Factors (TEHF), Aerospace Human Factors and Ergonomics (ISBN 978-3-319-96073-9).
VII. Ergonomics in Design, Design for All, Activity Theories for Work Analysis and Design, Affective Design (ISBN 978-3-319-96070-8).
VIII. Ergonomics and Human Factors in Manufacturing, Agriculture, Building and Construction, Sustainable Development and Mining (ISBN 978-3-319-96067-8).
IX. Aging, Gender and Work, Anthropometry, Ergonomics for Children and Educational Environments (ISBN 978-3-319-96064-7).
X. Auditory and Vocal Ergonomics, Visual Ergonomics, Psychophysiology in Ergonomics, Ergonomics in Advanced Imaging (ISBN 978-3-319-96058-6).

Altogether, the contributions make apparent the diversities in culture and in the socioeconomic conditions the authors belong to. The notion of well-being, which the reference value for ergonomics/human factors is not monolithic, instead varies along with the cultural and societal differences each contributor share. Diversity is a necessary condition for a fruitful discussion and exchange of experiences, not to say for creativity, which is the "theme" of the congress.

In an era of profound transformation, called either digital (Zisman & Kenney, 2018) or the second machine age (Bnynjolfsson & McAfee, 2014), when the very notions of work, fatigue, and well-being are changing in depth, ergonomics/human factors need to be creative in order to meet the new, ever-encountered challenges. Not every contribution in the ten volumes of the Proceedings explicitly faces the problem: the need for creativity to be able to confront the new challenges. However, even the more traditional, classical papers are influenced by the new conditions.

The reader of whichever volume enters an atmosphere where there are not many well-established certainties, but instead an abundance of doubts and open questions: again, the conditions for creativity and innovative solutions.

We hope that, notwithstanding the titles of the volumes that mimic the IEA Technical Committees, some of them created about half a century ago, the XX Triennial IEA Congress Proceedings may bring readers into an atmosphere where doubts are more common than certainties, challenge to answer ever-heard questions is continuously present, and creative solutions can be often encountered.

Acknowledgment

A heartfelt thanks to Elena Beleffi, in charge of the organization committee. Her technical and scientific contribution to the organization of the conference was crucial to its success.

References

Brynjolfsson E., A, McAfee A. (2014) The second machine age. New York: Norton.
Tatcher A., Waterson P., Todd A., and Moray N. (2018) State of science: Ergonomics and global issues. Ergonomics, 61 (2), 197–213.
Zisman J., Kenney M. (2018) The next phase in digital revolution: Intelligent tools, platforms, growth, employment. Communications of ACM, 61 (2), 54–63.

Sebastiano Bagnara
Chair of the Scientific Committee, XX IEA Triennial World Congress
Riccardo Tartaglia
Chair XX IEA Triennial World Congress
Sara Albolino
Co-chair XX IEA Triennial World Congress

Organization

Organizing Committee

Riccardo Tartaglia (Chair IEA 2018)	Tuscany Region
Sara Albolino (Co-chair IEA 2018)	Tuscany Region
Giulio Arcangeli	University of Florence
Elena Beleffi	Tuscany Region
Tommaso Bellandi	Tuscany Region
Michele Bellani	Humanfactor^{x}
Giuliano Benelli	University of Siena
Lina Bonapace	Macadamian Technologies, Canada
Sergio Bovenga	FNOMCeO
Antonio Chialastri	Alitalia
Vasco Giannotti	Fondazione Sicurezza in Sanità
Nicola Mucci	University of Florence
Enrico Occhipinti	University of Milan
Simone Pozzi	Deep Blue
Stavros Prineas	ErrorMed
Francesco Ranzani	Tuscany Region
Alessandra Rinaldi	University of Florence
Isabella Steffan	Design for all
Fabio Strambi	Etui Advisor for Ergonomics
Michela Tanzini	Tuscany Region
Giulio Toccafondi	Tuscany Region
Antonella Toffetti	CRF, Italy
Francesca Tosi	University of Florence
Andrea Vannucci	Agenzia Regionale di Sanità Toscana
Francesco Venneri	Azienda Sanitaria Centro Firenze

Scientific Committee

Sebastiano Bagnara (President of IEA2018 Scientific Committee)	University of San Marino, San Marino
Thomas Alexander (IEA STPC Chair)	Fraunhofer-FKIE, Germany
Walter Amado	Asociación de Ergonomía Argentina (ADEA), Argentina
Massimo Bergamasco	Scuola Superiore Sant'Anna di Pisa, Italy
Nancy Black	Association of Canadian Ergonomics (ACE), Canada
Guy André Boy	Human Systems Integration Working Group (INCOSE), France
Emilio Cadavid Guzmán	Sociedad Colombiana de Ergonomia (SCE), Colombia
Pascale Carayon	University of Wisconsin-Madison, USA
Daniela Colombini	EPM, Italy
Giovanni Costa	Clinica del Lavoro "L. Devoto," University of Milan, Italy
Teresa Cotrim	Associação Portuguesa de Ergonomia (APERGO), University of Lisbon, Portugal
Marco Depolo	University of Bologna, Italy
Takeshi Ebara	Japan Ergonomics Society (JES)/Nagoya City University Graduate School of Medical Sciences, Japan
Pierre Falzon	CNAM, France
Daniel Gopher	Israel Institute of Technology, Israel
Paulina Hernandez	ULAERGO, Chile/Sud America
Sue Hignett	Loughborough University, Design School, UK
Erik Hollnagel	University of Southern Denmark and Chief Consultant at the Centre for Quality Improvement, Denmark
Sergio Iavicoli	INAIL, Italy
Chiu-Siang Joe Lin	Ergonomics Society of Taiwan (EST), Taiwan
Waldemar Karwowski	University of Central Florida, USA
Peter Lachman	CEO ISQUA, UK
Javier Llaneza Álvarez	Asociación Española de Ergonomia (AEE), Spain
Francisco Octavio Lopez Millán	Sociedad de Ergonomistas de México, Mexico

Contents

Mining

Ergonomics and Human Factors in Manufacturing

Designing a User-Centered Approach to Improve Acceptance of Innovations on the Shop Floor Using Rogers' 'Diffusion of Innovations'

Nela Murauer(✉)

BMW Group, Max-Diamand-Str. 5, 80788 Munich, Germany
nela.murauer@bmw.de

Abstract. Analyzing innovation implementation processes in theory and in the industry different approaches are observable. Whilst in Rogers' 'Diffusion of Innovations' the future user is the initiator of the implementation process, the situation is different for shop floor innovations. Mostly, it is a planner or an innovation scouting department, who decide to implement a new technology. The workers as future users will be involved later for the purpose of a proof-of-concept. This late time of user involvement often leads to a low acceptance rate. In order to enhance user acceptance in a shop floor context this paper proposes a user-centered approach of innovation implementation based on Roger's model. As a first step of the industrial application, I conduct guided interviews to involve workers into the process of my innovation use-case on smart glasses in order picking processes.

Keywords: Acceptance of innovations · Shop floor · Logistics
User-centered development · Augmented reality · Smart glasses
Order picking processes

1 Introduction

Today's challenge for manufacturing companies lies in shaping the fourth wave of the industrial revolution. In the automotive industry the use of smart wearables, robotics, and autonomous transports or for example predictive maintenance advance with the aim to create smart factories. While robotics change the assembly process, autonomous transports impact greatly on intra-logistics. An important process in intra-logistics are order picking processes. Specific to order picking processes is the degree to which automatization differs depending on the tools in use (pick-by-paper, pick-to-light, pick-by-voice, pick-by-vision, etc.). All these systems are based on human work. Especially order picking processes can benefit significantly from the picker's flexibility in task management, reaching and grasping objects with high accuracy [1]. Thus full automatization is not suitable for order picking workplaces, but the usage of assistive devices can be advantageous. For this reason, I focus in my research project on testing the use of assistive devices, in particular Augmented Reality (AR)-devices, called smart glasses.

S. Bagnara et al. (Eds.): IEA 2018, AISC 825, pp. 3–10, 2019.
https://doi.org/10.1007/978-3-319-96068-5_1

Most research projects do not sufficiently involve the shop floor-employees from the beginning of the innovation process. Comparing the frequently cited process model in Rogers' 'Diffusion of Innovations' [2] to the usual innovation process management in the automotive industry, I found the same tendency in being output- or process-focused instead of being user-centered. Innovations are thus mostly introduced in a top-down approach. For this reason, I saw the opportunity to modify Rogers' model for the context of shop floor innovations. There are two principal initiators for innovations, first the planner and second the shop floor-employee. These two roles pursue different aims, which can be ergonomic (improve working conditions and work tools) or economic (such as improvement of time, quality and costs). Regularly shop floor-initiated innovations are based on the outcome of CIP (continual improvement process). In contrast to these incremental innovations, many planners find it also necessary to pursue disruptive innovation and to shape the dynamic change towards a smart factory. In innovation projects the involvement of the workers can just be as difficult as it is important. Rogers' model includes five phases: *knowledge*, *persuasion*, *decision*, *implementation* and *confirmation* [2]. Even though planners and management pass through all phases, the workers may influence the innovation development only from the decision phase onwards – provided that they exert any influence. Tackling this lack of early involvement in future users, this paper proposes to attach a separate consultation cycle consisting of a (continual) knowledge, persuasion and decision phase for the shop floor-employees. By consulting the workers concerning their needs at the operational level, workers take the role of experts, and can subsequently influence the final product for implementation. I aim to gain higher acceptance for innovations by creating consensus in the shop floor environment.

Applying the described user-centered cycle approach to a practical example, the innovation process of my research project started with an interview series to create a knowledge basis and start the workers' persuasion phase. The interview series - including 14 employees working in the logistics department - focused on previous experiences with smart glasses, knowledge about the technology, ideas about possible changes to the process, the suspected advantages and disadvantages as well as the beneficiaries of smart glasses-assisted order picking processes.

2 Related Work

The theory of the Diffusion of Innovation was originally developed in the context of agriculture to explain the ways, reasons and speed of innovations. Rogers' theory swept as translational knowledge into many other sectors such as communication, economics or public health. Rogers analyzed in his research for example a failed diffusion, where only few people in the Peruvian village Las Molinas took up the practice of boiling water to decrease germ load [2]. If, as described in the example above, the rate of adoption is too low, few potential future users are reached. Generally, Rogers defines diffusion as a 'process by which an *innovation* is *communicated* through certain channels *over time* among the members of a *social system*' [2]. The social system, where implementation takes place, depends on the work environment. For this reason innovation should best be introduced at a workstation, which enables smooth diffusion.

The acceptance of an innovation can be maximized when future users understand its usefulness and efficiency. Hence, the perception of innovation within a social system can be a limiting factor and can slow the rate of adoption.

As mentioned before, the innovation-development process after Rogers consists of five main steps: *knowledge*, *persuasion*, *decision*, *implementation* and *confirmation* (see Fig. 1).

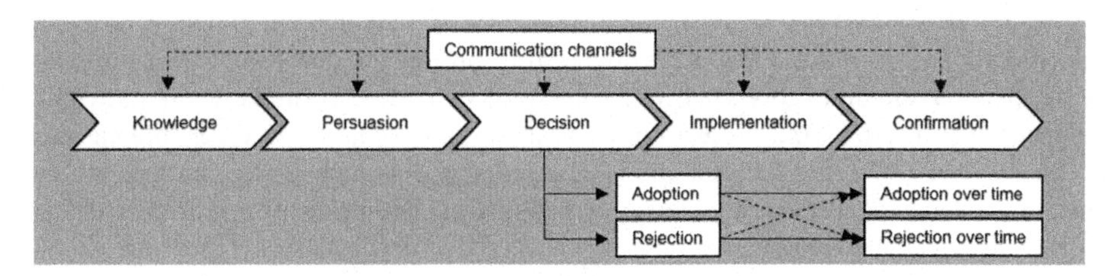

Fig. 1. Five phases of innovation development based on [2]

The knowledge phase serves to understand the functionality or purpose of an innovation. The persuasion phase gives a foundation for taking a favorable or unfavorable attitude. Building on the persuasion phase, the decision phase, which includes first tests, guides to a decision pro-adoption or pro-rejection. Once a decision is taken in favor of adopting the innovation, it can be implemented for regular use. The confirmation stage includes the review of the previous decision concerning further activities of dissemination.

3 Methodology

Comparing Rogers' innovation decision process with common innovation implementation processes in the automotive industry, I detected several differences. During my research, I observed different innovation processes in the context of bringing innovative technologies such as robotics, wearables, tablets or scanners to workstations on the shop floor. According to Rogers' innovation decision process future users hear about innovations, then obtain information about these, are subsequently convinced by the new knowledge and decide whether or not to use the innovation. In contrast, the phases at the production plant are different. Rogers concludes that early adopters are often individuals who do not have a great need for the innovation; while individuals with a higher need (based on factors such as education, socio-economic status and other determining factors) are later beneficiaries [2]. At the plant, those who are closest to the innovation - the workers – are usually not encouraged to join the group of early adopters either. At our production plant I observed two ways of improving processes. First, the continuous improvement process (CIP) collects all ideas of the shop floor workers to improve their processes. Improvements can consist of proposals to implement another technology, avoid unnecessary time spent on longer walking routes, energy or materials, or support for ergonomic work. Together with lean management planners, the employees plan, test and implement their ideas. Especially in the context of digitalization special knowledge is needed to get an overview on potential innovations, their readiness for implementation and possibilities for process improvement. Second, in production contexts innovation scouting departments analyze upcoming innovations, which are interesting in terms of implementation at the production plant and decide to test potential improvements in a so-called 'proof-of-concept'. If this test is successful, a roll-out of the innovation begins (see Fig. 2). Summing up, the planner passes through the knowledge and persuasion phase and decides - alone or in an expert team - whether an innovation will be tested on the shop floor. In contrast the worker often learns through the planner-initiated proof-of-concept about the innovation. In my research project my role is similar to an innovation planner. Yet, my research project on the usage of smart glasses in order picking processes takes a different approach and is conceptualized in such a way that workers are encouraged to become early adopters. In the following, I describe my approach to design a more user centered innovation process.

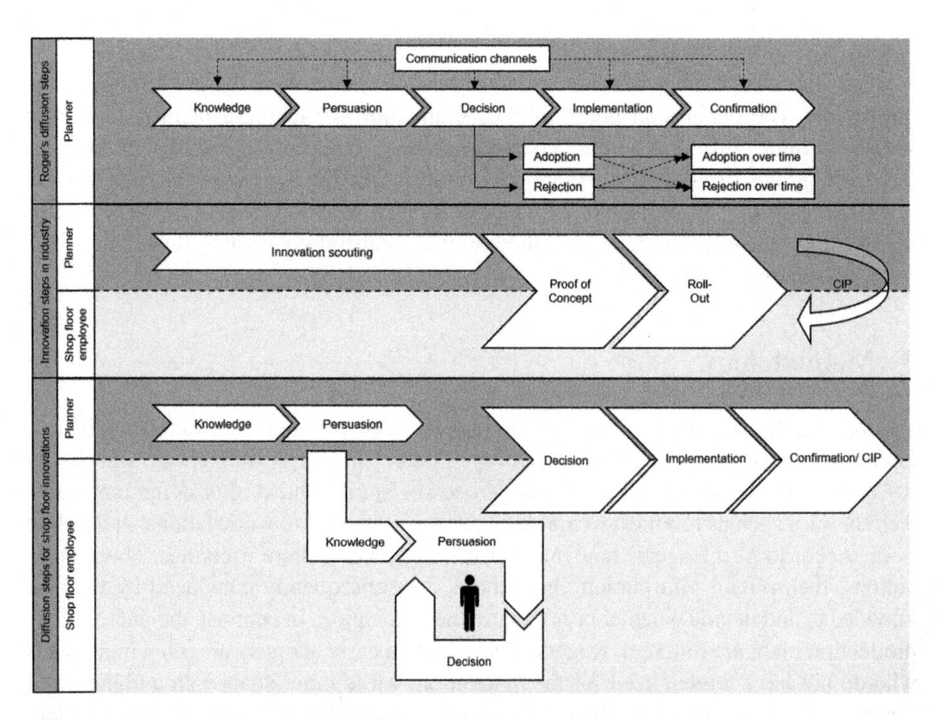

Fig. 2. Comparison of innovation processes according to [2], in the industry and our proposed process for the shop floor context.

At the beginning of my research project, I decided to work one week during the morning shift at a number of order picking workstations, which could be possible workstation for the main study. I viewed the workers as experts in order picking from the start. According to workers at the operational level, it is common that employees 'from the offices' implement a decision for a process improvement or an innovation without talking (enough) to the workers - the future users. Due to this manner of implementation, the operational employees often have a negative attitude towards innovations and must be convinced of the usefulness of an innovation after the decision for implementation has already been made. In order to avoid such negative experiences which shop floor workers connected to innovations, I tried to involve workers as much as possible during the innovation process.

Figure 2 shows the role-specific approach to an innovative project. The planner starts the process by scouting innovations (knowledge phase). He evaluates the upcoming technologies according to - for example - their potential of process improvement or economic efficiency and other characteristics, which support business strategies. During the persuasion phase he provides information to workers about the innovation, how it works, and what the differences, major risks and chances in comparison to the actual process are. The planner's job includes the creation of a common understanding of the technology. During the persuasion phase of the workers, their expert knowledge about the process can be a great input and of high value for the planner. Based on this input the planner develops a first prototype, which can be evaluated and improved and asserted several times until the worker's decision phase is ultimately completed. Shop floor employees and planer decide subsequently together which developed solution would be the best fit for the test (as part of the proof-of-concept). Of particular significance is a test for a company when it is conducted under real production conditions with the workers, the future users, as participants. Collecting the users' feedback and their preliminary assessment - whether the innovation can be permanently implemented - helps the planner and the management to decide in a user-centered way. It must be noted here, that some advantages like economic reasons can be rated more important than the workers assessment due to prioritizations in business strategies. One example where the interests of the worker and the company will not align are use cases where an innovation can replaces an employee. Shop floor employees will rarely agree with such an innovation. Nevertheless it is important to involve affected employees working at the workstation, where a technology may be introduced during the decision phase. This gives workers a chance to get acquainted with the new technology and know what changes are likely to occur to their work environment. The final decision will be made by the planner or the manager based on the worker's assessment and the business strategies. Once an innovation is implemented, the planner relies on the worker's feedback to navigate through the confirmation phase.

4 First Steps of the Industrial Application

The environment of my research project are manual man-to-goods order picking processes. Following the instructions of an order list printed on paper or visualized on a digital device, the worker collects products or components and puts them into target bins or shelves. This process is very similar to a shopping process in a supermarket with a prepared shopping list. I, in my role as innovation planner, focused on smart glasses as visualization device. I compared different hardware, advantages and disadvantages, possible use cases and related work. I was persuaded to test smart glasses at our production plant due to the benefits, which the implementation of AR promises. The stages I passed through correspond to the knowledge and the persuasion phase of the planner (see Fig. 2).

4.1 Interview Methodology

In order to involve workers into the innovation process, I first conducted guided interviews consisting of seven questions. In the beginning, I aimed to explore the respondents' knowledge and ideas on AR-technology. I asked the participants whether they had contact with Augmented or Virtual Reality (question Q_1). Then I inquired the respondents' understanding of the term 'smart glasses' in question Q_2. In question Q_3 I asked the pickers to describe how order picking workstations with AR-support look like in their imagination and how workstations will change due to the introduction of AR-technology (question Q_4). All questions served as a means to initiate the knowledge phase. Hereafter, I showed a short video prototype, which visualized an AR-design sample for the pickers. I produced this video by filming the order picking process at a workstation with a head-mounted camera GoPro® Hero 5 and overlaying this base film with a possible user interface design with the Adobe After Effects ® program. This form of prototyping helped respondents, who were unfamiliar with AR, to get an idea of the future process. For further information about prototyping methods in the AR-context see [3]. Showing the movie explained the technology to future users. This explanation is an important step to inform shop floor employees about the innovation. Thus the interviews and presentation of the movie are the starting point of the workers' knowledge phase. In this way the future user is guided to the knowledge phase by the planner. In my experience, most shop floor employees will not engage in additional technological research, hence their persuasion phase starts immediately after the planner's explanation. I supported the transition to the persuasion phase by asking the participants, what they liked and what they did not like thinking about AR-supported order picking processes (questions Q_5 and Q_6). The final question Q_7 inquired after the respondents' opinions on the potential beneficiaries of an AR-usage.

A total of 14 employees with varying roles in the logistics department participated voluntarily in the interview series. By involving mainly pickers (who are the future users of the technology), foremen, but also logistics planners, innovation experts for order picking processes and department leaders I aimed to gain a wide range of ideas and comments. The analysis of the interviews followed the step-by-step clustering process published by Meuser and Nagel [4].

4.2 Results

Only few of the interviewees already had contact with Augmented or Virtual Reality (Q_1). As examples they named a Samsung Gear, Head-up Displays in cars, Nintendo's AR-game Pokémon Go, 3D cinema and video games. One participant of an innovation department already tested some smart glasses, like for example a Google Glass. Most participants heard about the technology, but never tested it by themselves.

Summarizing the basic notions of smart glasses from the pickers' point of view, most of the descriptions were use-case rather than hardware specific. Smart glasses from their perspective are 'computerized glasses', 'like a head-up display for humans', which 'support workers' through the visualization of 'the most important information' like 'images, forms, colors, data and facts' 'in the field of view'. Processes will be 'facilitated'. 'I see [in the smart glasses], what I have to do', summarized one picker. This process transformation caused nevertheless few negative associations. 'The individual fades into the background', 'is there still a human or only half human half robot?' and 'This is a dehumanization.' Workers evaluated their employment only for their physical power and not for their cognitive abilities negatively. An interesting fact was that some participants already named criticized the smart glasses in the first definition question.

They expected that an AR-supported workstation will be faster and more flexible through the avoidance of head- and eye-movements, through the visual highlighting for minimizing the risk of confusion and especially through a lean visualization of required information. Some of the interviewees thought that the project will produce individualized user interfaces. One participant could envision to integrate actual soccer results during the evening shift as additional information or brain teasers to enhance concentration and attention during the day. Most participants assumed that the AR-support at an order picking workstation avoids errors. All answers were very goal-oriented and few process-describing. The next question addressed the changes caused by an implementation of smart glasses. The answers were quite similar to the previous question. With AR, in their opinion, the visualization will be clearer and the pickers will know faster, what they have to do. They thought that particularly the error frequency, the task completion time and the training time will decrease, which they viewed as positive aspects.

In addition to the above issues, the participants liked the expected support of the smart glasses, which could lead to 'more skilled and more confident' employees. Due to the decrease of error rates and a higher efficiency, the production plant will become 'more future-proof and competitive'. This is an important base for the employees to 'identify [themselves] with the company'. Negative comments were related to hardware limitations like battery life and wearing comfort and to the dependency on the IT-department or the supplier in case of software changings. In addition, they had doubts that refocusing the eyes will be too demanding.

With reference to the beneficiaries, the interviewees' answers were similar. The expected benefits are on the one hand more flexibility for the employees, a calmer process and a decrease of the error frequency. In their opinion the company and the management on the other hand profits from time-, cost- and error decrease, a reduction of rework, faster training times and a guided, language independent training possibility.

5 Discussion and Conclusion

The time and grade of involvement into an innovation process in a shop floor environment depends on the role of the employees. In contrast to Rogers' 'Diffusion of Innovation' the planner or an innovation scouting department rather than the worker initiate the innovation process. To support the earlier involvement of shop floor workers, a new process model based on Rogers' 'Diffusion of Innovation' for a production context was created. The active consideration of the workers' expert knowledge serves to design an optimal user-centered solution.

In the interviews workers wanted an easy, clear and comprehensible visualization of a future Augmented Reality user interface design. To develop the software according to the employees' comments and ideas, I plan to conduct a creative workshop directly on the shop floor. Due to the lack of a suitable method for creating AR-content, I will develop a tool for motivating the participants to uninhibitedly design their ideas of a possible user interface of smart glasses. The objective is to find a common ground for discussions with workers, who do not have any know how regarding Augmented Reality or smart glasses.

The idea which will be seen as most suitable, will be programmed and evaluated, continuously redesigned and evaluated until the workers decide, that it is perfect from their point of view. This describes the whole human-centered development circle shown in Fig. 2. The next step will be a test phase, which is classified under the process step 'decision phase' directly after the circle. The results of the field study, which contains user-related as well as process-related variables, supports the later decision process toward a roll-out into the production environment.

References

1. Klug F (2010) Logistikmanagement in der Automobilindustrie; Grundlagen der Logistik im Automobilbau. Springer. Heidelberg Dornrecht New York (2010)
2. Rogers EM (2003) Diffusion of innovations, 5th edn. Free Press. Simon & Schuster, New York
3. Murauer N (2018, in press) Design thinking: using photo prototyping for a user-centered interface design for pick-by-vision systems. In: Proceedings of the 11th ACM international conference on pervasive technologies related to assistive environments, New York
4. Meuser M, Nagel U (1991) ExpertInneninterviews - vielfach erprobt, wenig bedacht: ein Beitrag zur qualitativen Methodendiskussion. In: Garz D, Kraimer K (eds) Qualitativ-empirische Sozialforschung : Konzepte, Methoden, Analysen. Westdt. Verl., Opladen

Possibilities and Challenges for Proactive Manufacturing Ergonomics

Erik Brolin[1(✉)], Nafise Mahdavian[1], Dan Högberg[1], Lars Hanson[1,2], and Joakim Johansson[3]

[1] School of Engineering Science, University of Skövde, Box 408, SE-541 28 Skövde, Sweden
erik.brolin@his.se
[2] Industrial Development, Scania CV, SE-151 87 Södertälje, Sweden
[3] Bombardier Transportation Sweden AB, Banmatarvägen 1, SE-721 73 Västerås, Sweden

Abstract. This paper identifies and describes product development activities where ergonomics issues could be considered and illustrates how that could be done through a number of different approaches. The study is divided into two parts where an interview study is done to identify where in a product development process consideration of ergonomics issues are or could be done. The second part of the study includes an observation, motion capture and simulation study of current manufacturing operations to evaluate and compare three different assessment approaches; observational based ergonomics evaluation, usages of motion capture data and DHM simulation and evaluation. The results shows the importance of consideration of ergonomics in early development phases and that the ergonomics assessment process is integrated in the overall product and production development process.

Keywords: Proactive · Ergonomics · Manufacturing
Digital human modelling · Process

1 Introduction

Contemporary product and production development is typically carried out with support of computer tools where design of products and workstations are originated and evaluated within virtual environments (Pahl et al. 2007; Chandrasegaran et al. 2013). Ergonomics addresses factors important to consider in product and production development processes to ensure a good fit between humans and items being designed, whether that is a product or a workstation (Bridger 2009). Studies to evaluate the interaction between users and products, workplaces or tasks have typically been done relatively late in the development phase (Porter et al. 1993; Duffy 2012) and based on making expensive and time demanding physical mock-ups (Helander 1999; Duffy 2012). Obstacles towards more proactive ergonomics measures are found to be lack of knowledge, methods and tools for consideration of ergonomic issues together with a lack of cooperation and communication between project stakeholders (Falck and Rosenqvist 2012). Broberg (1997) reports that prerequisites for integrating ergonomics

S. Bagnara et al. (Eds.): IEA 2018, AISC 825, pp. 11–20, 2019.
https://doi.org/10.1007/978-3-319-96068-5_2

into the product development process is more time; training in ergonomics; improved contact between design, production and ergonomists; but also access to ergonomics checklists and methods. To achieve a more proactive management of ergonomic issues it is also necessary to consider the structural, technical, organizational and stakeholder-relational conditions of an organisation that enable or hinder improvement of ergonomics through the holistic systems-performance discipline of macro-ergonomics (Berlin 2011).

Different approaches for ergonomics evaluations exists and a number of factors affects the chosen approach for identifying, monitoring and controlling ergonomics problems that arise in production (Berlin 2011). A common and perhaps the most basic approach is observational based ergonomics evaluation against some kind of baseline for acceptable/unacceptable conditions, mostly related to posture analysis (Berlin 2011). Another approach for ergonomics evaluation is using motion capture data which is an objective and direct measuring technique and can be used together with observational based assessment methods (Weisner and Deuse 2014; Plantard et al. 2017). A third approach for ergonomics evaluation is using Digital human modelling (DHM) software and simulation which supports the consideration of ergonomics and human factors in virtual environments. New technology and tools such as motion capture and DHM software can enable a more proactive work in the design process when seeking feasible solutions on how the design could meet set ergonomics requirements early in the development process (Chaffin et al. 2001; Duffy 2012). However, an improved process for how to utilise advanced tools and approaches needs to be identified. The objective of this paper is to identify and describe product development activities where ergonomics issues could be considered and illustrate how that could be done through different approaches.

2 Materials and Methods

The study was divided into two parts where an interview study was done to give an understanding of the specific product realisation process at a research project partner company which develops and manufactures train propulsion and control equipment. A goal was also to identify where in the product realisation process considerations of ergonomics issues are or could be done. The second part of the study included an observation, motion capture and simulation study of current manufacturing operations to evaluate and compare three different assessment approaches; observational based ergonomics evaluation, usages of motion capture data and DHM simulation and evaluation.

2.1 Interview Study

The purpose of the interview study was to evaluate how the product realisation process of the partner company is structured and in what phases production and assembly aspects are considered, where the manufacturing ergonomics is a key part. Five qualitative, semi-structured interviews (Howitt and Cramer 2011) were conducted. The five participants' work positions varied from mechanical design engineer to manufacturing engineer, including a component engineer responsible for cables and harnesses, which gave the possibility to get a good overview of the total product realisation process at the partner company. Their work experience varied from 1 to 29 years where the interviewed manufacturing engineers had slightly less work experience. The interview questions covered the topics: competence and opinions; strategies and responsibilities; as well as working methods and procedures regarding production, assembly and ergonomics aspects in product and production development. The interviews were audio recorded and transcribed. The transcribed result from the interviews was analysed to produce an overview of the methods and procedures but also competence and opinions regarding production, assembly and ergonomics aspects in the product realisation process at the partner company.

2.2 Observation, Motion Capture and Simulation Study

The purpose of the observation, motion capture and simulation study was to evaluate with what approaches manufacturing ergonomics can be evaluated at the partner company. The study also aimed at assessing how much effort and resources different approaches would require. Three different approaches for ergonomics evaluation was included in the study: (1) observational based ergonomics evaluation using video recordings; (2) ergonomics evaluation using motion capture data; and (3) ergonomics evaluation using DHM simulations. The three approaches were used to assess one assembly station at the partner company where recordings of video and motion capture data were done with one operator. The different approaches were evaluated and rated on a number of factors, relevant to consider when integrating ergonomics assessment in the development process, such as cost of equipment, time to perform analyses, ability to assess longer assembly sequences, development phases when evaluations can be done, and necessary competence for using equipment but also for assessing manual assembly work. The comparison analysis was done based on the experience from this study when using the three different approaches but also previous experience within the research group and experiences found in literature from other researchers of the different approaches' possibilities and limitations.

3 Results

3.1 Results from Interview Study

The results from the interview study showed that the interview participants rated their competence to handle questions regarding the possibility to produce a product as high (Fig. 1). Most of them had also taken shorter company internal courses about Design for Manufacturing and Assembly (DFMA) (Ullman 2015). The competence to handle questions regarding assembly ergonomics of a product was rated as medium to high and only one manufacturing engineer had received internal education about assembly ergonomics. However, all participants rated the relevance of issues regarding production and assembly ergonomics to be high or very high for their own professional role.

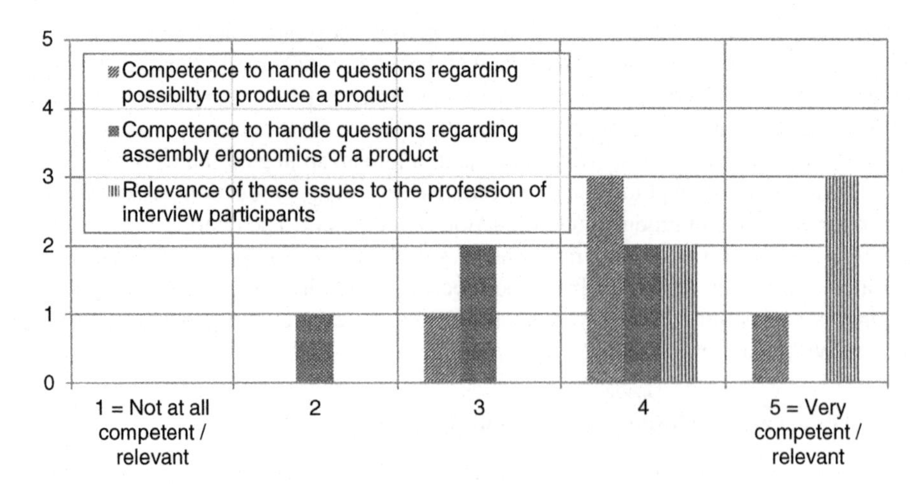

Fig. 1. Subjective ratings for competence and relevance of questions regarding production and assembly ergonomics

The mechanical design and manufacturing engineers gave a similar image of the company's product realisation process (Fig. 2). However, the mechanical design engineers put bigger emphasis on the early phases up until the detailed design phase, while the manufacturing engineers described the later phases in more detail. The component engineer responsible for cables and harnesses described a slightly different product realisation process including frequent communication with subcontractors and where detailed dimensions of harnesses were decided during the prototyping phase to be ready for the pre-series. Up until the detailed design phase the product only exist in digital formats such as CAD models or wiring diagrams for the different components.

Fig. 2. Product realisation process at the partner company

The mechanical design engineers and manufacturing engineers had slightly different views of how and in what phases production and assembly aspects were considered. The mechanical design engineers meant that they, based on experience, considered production and assembly aspects during the design phases and that they also invited manufacturing engineers and experienced assembly personnel to review the design work and give their opinions, especially in the detailed design phase but also during the earlier conceptual and preliminary design phases. The manufacturing engineers viewed the three phases: concept, preliminary and detailed design as one bigger phase focusing on DFMA. During the design phases the manufacturing engineers also considers questions such as at which production facility should the product be assembled and should additional subcontractors be involved? The manufacturing engineers have a wish and have also tried to front load the development process so that more production and assembly aspects are considered in the earlier design phases. Still, according to the manufacturing engineers, it is during the prototyping phase that most production and assembly aspects are considered and checked. The prototyping phase was introduced not too long ago to have a phase that focuses on and controls if it is possible to assemble the product or not. It is not uncommon that problems are discovered that were hard to foresee looking at digital versions of the product during the previous phases, but having a prototyping phase enables the development team to have a more finished product ready for the pre-series.

Currently the company do not use any comprehensive ergonomics evaluation method but have some basic guidelines including a lifting matrix with weight limits at different positions. In addition, production and assembly aspects are considered based on experience from previous projects. The cooperation between the design and manufacturing departments is considered to be functional and close where the design department is responsible for the early design phases up until the prototyping phase when the manufacturing department takes lead. However, the mechanical design engineers had the experience that it is not always the case that manufacturing engineers and assembly personnel show up to do the reviews that occurs during the design phases. The manufacturing engineers on the other hand felt that they have problems to really affect the design work because they do not have any clear evidence or arguments that supports their suggested design changes, or that the design has come too far when issues are discovered. The communication between the design and manufacturing departments is achieved through face-to-face meetings, phone calls and e-mails. During the design phases much communication is achieved using CAD models and this work has improved since the personnel at the manufacturing department have been given access to a 3D viewer software in which they can look at the developed product and how it is thought to be assembled. During the prototyping phase the physical prototypes are used for communication and during the pre-series the company has a system to handle design change suggestions through what is called *modification requests* which are discussed during project meetings and can result in design changes communicated through new engineering releases.

When answering a question about which assembly problems that exist in the current production operations the participants gave slightly different replies. The mechanical design engineers knew about issues with the components that they were most responsible for and mentioned tolerance issues, sharp edges and heavy components as problematic areas. The manufacturing engineers also mentioned big and heavy cables and harnesses that needs to be put deep into the product structure or handled above shoulder level. The mechanical design engineers also mentioned manufacturing errors that deviates from the drawings as possible problems while the manufacturing engineers mentioned design errors as a source of manufacturing and assembly problems. To handle and fix a production or assembly problem it is important that the solution is cost-efficient, especially if the problem or risk is expensive to eliminate. The cost needs to be covered by the project's budget or if the problem is shared between several projects it can be discussed and financed on a higher level within the company.

3.2 Results from Observation, Motion Capture and Simulation Study

The results from the observation, motion capture and simulation study showed that it is possible to use all three approaches for ergonomics evaluation included in the study (Fig. 3). However, the three approaches differ in all aspects relevant to consider when integrating ergonomics assessment in the development process such as cost of equipment, time to perform analysis, ability to assess longer assembly sequences, development phases when evaluation can be done and necessary competence for using equipment but also for assessing manual assembly work (Table 1). The analysis of the different approaches were done under the assumption that they will produce comparable results based on some relevant observational based evaluation method, e.g. RULA (McAtamney and Corlett 1993).

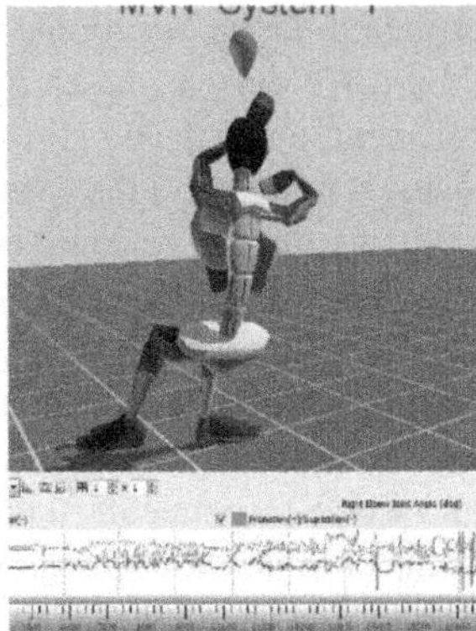

Fig. 3. Visualisations of the three approaches for ergonomics evaluation: (1) observational based evaluation, (2) motion capture recording, and (3) DHM simulation

Table 1. Comparison of relevant factors for three approaches for ergonomics evaluation

	Observational based evaluation	Evaluation using motion capture data	Evaluation using DHM simulation
Cost of equipment	Low, cost of video recording equipment	High, cost of motion capture (MC) equipment and necessary computer power	High, cost of DHM software and necessary computer power
Time to perform analysis	Low to medium, recording of work and following analysis, could be reduced if analysis is done directly on site	Medium, set up and calibration of MC equipment and recording of work, automatic analysis	Medium to high, import of CAD geometry, set up of work and simulation, automatic analysis

(*continued*)

Table 1. (*continued*)

	Observational based evaluation	Evaluation using motion capture data	Evaluation using DHM simulation
Ability to assess longer assembly sequences	Medium, longer sequences tend to give more summarized evaluations	Medium to high, longer sequences generates more data that needs to be treated	Low, each step of assembly needs to be set up
Development phases when evaluation can be done	Only when physical prototype exist, could perhaps be done earlier with rough assumptions	Could be done in early phases but might need physical prototypes or simple mock-ups to achieve an accurate interaction	Early phases when the product exist in digital format as well as later phases
Necessary competence for performing ergonomics evaluations	Expert knowledge in ergonomics assessment methods, physiology and biomechanics	Expertise in handling and use of MC equipment, some know-ledge in ergonomics assessment methods	CAD and simulation expertise, some knowledge in ergonomics assessment methods
Consideration of anthropometric diversity	Low, depends on the test participants that can be recruited	Low, depends on the test participants that can be recruited	High, any number of human models can be created and used in simulations
Accuracy of generated motion data for analysis, and subsequent analysis	Low to medium, actual motion is analysed but subsequent analysis is done subjectively	Medium to high, motions are recorded but motions can be slightly inaccurate due to poor calibration	Medium, human resembling motions are predicted with consideration to biomechanics and comfort
Fidelity and utility of illustrations and communication media	Video recordings show the actual assembly but it can be difficult to view it from specific angles due to obscuring geometries. Might be problematic due to integrity issues	Motion capture recordings makes it possible to analyse different postures in detail but it is hard to connect the recordings to the real world CAD geometry. Anonymous illustrations	Possible to analyse different postures in detail and connect the work tasks to imported CAD geometry. Anonymous illustrations

4 Discussion

The results from the interviews show that issues regarding production and assembly ergonomics are very relevant for the interview participants in their daily work. The partner company has a product realisation process, comparable to generic design processes found in literature, e.g. Cross (2008), where a big part of the process is done virtually, and physical models are not created until the prototyping phase. However, there is no clear and structured process for how to consider and check that the developed products will offer appropriate manufacturing ergonomics conditions. Instead, consideration of ergonomics is often done based on experiences and assumptions by mechanical design engineers, and when manufacturing engineers are involved to assess concepts they have problems to justify issues as they lack any agreed upon ergonomics evaluation method to refer to and lack tools to visualise the possible problems. In addition, the mechanical design engineers do not seem to be aware of all existing ergonomics issues and focuses more on apparent risks connected to mechanical structures while manufacturing engineers have knowledge of additional problems connected to assembly operations.

Therefore, it is evident that the prototyping phase is a central part of the product realisation process where many problems are discovered and solved. It seems that the prototyping phase is the main contact surface between the design and manufacturing departments where design have a final concept to communicate and manufacturing have something physical to evaluate. During this phase it is possible to do changes to the developed product as the product's design has not yet been frozen and the product has not yet entered production. However, changes during this phase may require a lot of work and time, and may require large iterations in the design work. Such changes and iterations means added costs, and hence there are clear arguments for a proactive approach where more ergonomics issues can be addressed and solved earlier in the product realisation process.

The three studied approaches for ergonomics evaluations have both advantages and drawbacks. Observational based evaluation is a low cost approach that can be done relatively fast on running production. However, to achieve acceptable analyses proper knowledge and education in ergonomics assessment methods is necessary and assessments can only be done when a physical prototype exists. Observational based assessments will by some degree be subjective and inter-observer variation will exist (Eliasson et al. 2017). In addition, longer observational based assessments tend to be more of a summary analysis of the work than actual continuous posture assessments. Motion capture based evaluation is a more costly approach but the study shows that it can be done during running production at the partner company even though set up and calibration of the system takes some time. When the system is started it can record relatively long sequences of assembly work but will also produce more data that needs to be treated and analysed. Motion capture needs some sort of physical mock-up for operators to interact with to achieve correct analysis but might be used in virtual phases through VR technology (Lawson et al. 2016). Analysing motion capture data with an algorithm that produces a result based on an ergonomics evaluation method gives an objective assessment and different postures can be studied in detail. Still, it can be hard to connect the recorded motions with CAD geometries and therefore difficult to understand what the recordings show. However, if recordings capture when an operator interacts with physical objects, e.g. in a real assembly or interacting with a mock-up, the motion capture recordings can be complemented with video recordings. Using DHM simulations to generate ergonomics evaluations is relatively costly due to high software costs and it requires much work to set up and instruct the human models to perform a complete assembly sequence. This time consuming process affects DHM tools' ability to assess longer (more than 30 s) assembly sequences. However, the approach is very suitable to use in early virtual design phases to assess different design concepts and to consider anthropometric diversity to ensure that the design fits the intended proportion of the targeted population (Brolin 2016). DHM tools are also a valuable visualisation means during early phases as it can show ergonomics issues in connection with additional CAD geometry.

It can be concluded that the partner company first and foremost need to have an ergonomics evaluation method that is agreed upon by both manufacturing and the mechanical design departments. If no such method exist, mechanical design engineers do not have any regulations to follow and fulfil and manufacturing engineers do not have defined ergonomics load level to measure and evaluate against. The current company standards that exists are too limited and do not cover all identified problem areas within the production process. If an agreed upon ergonomics evaluation method would exist it could be used with different approaches. In early development phases it can be used as a starting point for discussions and different scenarios could be assessed through these discussions. If more detailed evaluations would be necessary during early development phases, to for example give information and input to a discussion, DHM simulations would be suitable. With DHM simulations many aspect of the evaluation method could be assessed for a number of human models with varying body size. During early development phases motion capture together with VR could also be a useful approach to see and evaluate future products and assembly operations. Issues with more advanced approaches such as DHM and motion capture with VR is the additional support systems that needs to be functional to make it possible to work with CAD geometries and easily import and set up different design concepts of products and work environments. Therefore it is important to work with PDM (Product Data Management) systems and to have clever tools that automatically identifies possible interaction zones on the product. In the prototyping phase the ergonomics evaluation method should be used as a checklist to see if all requirements are met and also to assess the accuracy of DHM simulations and evaluations to further improve future simulations.

5 Conclusions

Results from the study shows that to reach proactive manufacturing ergonomics it is important to consider issues within manufacturing and assembly in very early development phases where principal design decisions are made and where product data only exist in virtual formats. It is also important that the ergonomics assessment process is integrated in the overall product and production development process and that evaluations can be done both in simulation tools but also in the physical world with real subjects to verify and validate results from virtual simulations.

Acknowledgements. This work has been made possible with the support from the research environment INFINIT at University of Skövde, supported by the Knowledge Foundation (VEAP project). Support has also been given by the participating organization in the research project. All of this support is gratefully acknowledged.

References

Berlin C (2011) Ergonomics infrastructure - an organizational roadmap to improved production ergonomics. Doctoral thesis, Chalmers University of Technology

Bridger RS (2009) Introduction to ergonomics. CRC Press, Boca Raton

Broberg O (1997) Integrating ergonomics into the product development process. Int J Ind Ergon 19(4):317–327

Brolin E (2016) Anthropometric diversity and consideration of human capabilities – methods for virtual product and production development. Doctoral thesis, Chalmers University of Technology

Chaffin DB, Thompson D, Nelson C, Ianni JD, Punte PA, Bowman D (2001) Digital human modeling for vehicle and workplace design. Society of Automotive Engineers, Warrendale

Chandrasegaran SK, Ramani K, Sriram RD, Horváth I, Bernard A, Harik RF, Gao W (2013) The evolution, challenges, and future of knowledge representation in product design systems. Comput Aided Des 45(2):204–228

Cross N (2008) Engineering design methods: strategies for product design. Wiley, Chichester

Duffy VG (2012) Human digital modeling in design. In: Salvendy G (ed) Handbook of human factors and ergonomics, 4th edn. Wiley, Hoboken

Eliasson K, Palm P, Nyman T, Forsman M (2017) Inter- and intra-observer reliability of risk assessment of repetitive work without an explicit method. Appl Ergon 62:1–8

Falck A-C, Rosenqvist M (2012) What are the obstacles and needs of proactive ergonomics measures at early product development stages? An interview study in five Swedish companies. Int J Ind Ergon 42(5):406–415

Helander MG (1999) Seven common reasons to not implement ergonomics. Int J Ind Ergon 25 (1):97–101

Howitt D, Cramer D (2011) Introduction to research methods in psychology. Pearson Education

Lawson G, Salanitri D, Waterfield B (2016) Future directions for the development of virtual reality within an automotive manufacturer. Appl Ergon 53:323–330

Mcatamney L, Corlett EN (1993) RULA: a survey method for the investigation of work-related upper limb disorders. Appl Ergon 24(2):91–99

Pahl G, Beitz W, Feldhusen J, Grote K-H (2007) Engineering design: a systematic approach. Springer, London

Plantard P, Shum HPH, Le Pierres A-S, Multon F (2017) Validation of an ergonomic assessment method using Kinect data in real workplace conditions. Appl Ergon 65:562–569

Porter J, Case K, Freer M, Bonney MC (1993) Computer aided ergonomics design of automobiles. In: Peacock B, Karwowski W (eds) Automotive ergonomics. Taylor & Francis, London

Ullman D (2015) The mechanical design process. McGraw-Hill Education, New York

Weisner K, Deuse J (2014) Assessment methodology to design an ergonomic and sustainable order picking system using motion capturing systems. Procedia CIRP 17:422–427

Human-Robot Collaboration in Manual Assembly – A Collaborative Workplace

Henning Petruck[1(✉)], Marco Faber[1], Heiner Giese[2], Marius Geibel[2], Stefan Mostert[2], Marcel Usai[1], Alexander Mertens[1], and Christopher Brandl[1]

[1] Institute of Industrial Engineering and Ergonomics, RWTH Aachen University, Bergdriesch 27, 52062 Aachen, Germany
{h.petruck,m.faber,m.usai,a.mertens, c.brandl}@iaw.rwth-aachen.de

[2] item Industrietechnik GmbH, Friedenstraße 107-109, 42699 Solingen, Germany
{h.giese,m.geibel,s.mostert}@item24.com

Abstract. The integration of humans into the assembly process in terms of human-robot collaboration (HRC) enables the flexibility of production processes with a high degree of automation, which are not flexible enough to overcome the challenges in nowadays production in all cases. However, this form of cooperation raises issues such as occupational safety or acceptance. In order to address these questions, a HRC workstation has been designed that is on the one hand characterized by traditional ergonomic design aspects with regard to conventional industrial requirements. On the other hand, data from intelligent sensors are used to adapt the system's behavior to the way the working person works. The workstation was developed on the basis of the results of a requirement analysis. This article presents the ergonomic concepts of the workplace and their implementation.

Keywords: Human-robot collaboration · Manual assembly
Ergonomic workplace design · Occupational safety

1 Introduction

Production in high-wage countries such as Germany is characterized by a high degree of task-specific automation. This results in high adaptation efforts for new product versions or variants. Considering growing numbers of variants and shorter product life cycles in many sectors, this should be evaluated critically and indicates that production systems are not sufficiently flexible in their current state to adapt to rapidly changing market conditions. One possibility to make automated production more flexible is the integration of humans in terms of human-robot collaboration (HRC). The combination of human cognitive and sensorimotor skills as well as the precision, speed and fatigue-free operation of the robot results in effective collaboration. This also applies if no direct working cooperation between humans and robots is intended, but human and robot work side by side, while carrying out different working activities.

S. Bagnara et al. (Eds.): IEA 2018, AISC 825, pp. 21–28, 2019.
https://doi.org/10.1007/978-3-319-96068-5_3

In addition, HRC workstations can be operated without safety fences and therefore with less space requirement assuming appropriately chosen robots and safety devices. This can be observed, for example, in the adhesive application at Audi AG [1]. Here, a lightweight robot applies the adhesive to the CFRP roof of the car, which cannot be repeated as accurately as the robot performs the task by humans, also due to the dimensions of the workpiece. Another indirect cooperation for applying the adhesive to the front window can be observed at BMW AG [2]. A direct cooperation between robots and humans can be found in car body construction. Here, a robot supplies the human with coolant expansion tanks, which were picked up from subjacent storage boxes manually, before the HRC workstation was set up. Due to the multiple repetitions of strenuous bending, this workplace was poorly ergonomic [3]. Another industrial application is door assembly at BMW AG, where the robot takes over the very force-intensive pressing of sound and moisture insulation [4] or lifting of the compensating housing for front axle gearboxes at the same automobile manufacturer [5].

The aforementioned examples illustrate that there are many reasons and applications for HRC. For this reason, there is not "the one" HRC workplace, but rather each workplace is developed and implemented for the individual requirements of the task. These workplaces are often only developed for the execution of a single process step. In research, there are a large number of projects dealing with the HRC, but often only individual aspects are investigated here, such as novel safety concepts for cooperation with industrial robots [6], intuitive robot control concepts for tetraplegics [7] or the adjustment of robots actions by anticipating the user's behavior based on gaze patterns [8], just to name a few.

The focus of this work is therefore on the development and design of a workstation for use in manual assembly with a wide variety of applications, which should be as versatile as possible and does not require a complete redevelopment when the production is remodeled.

In the proposed scenario, the robot assists the working person, who assembles the product on a work surface, by handing over components and tools. Thus, the working person can concentrate on the value-adding processes of production, namely the assembly. This workstation fulfils its purpose in a highly automated production process with a wide range of variants, in which workpieces and components are transferred to this workstation for the variable part of the assembly process, since this part can hardly be automated economically. The workstation is therefore an assembly island.

2 Requirements for a HRC Workplace in Manual Assembly of Multi-variant Products

Against the background of the aforementioned scenario, the work area of the workstation should be subdivided into several areas. On the one hand, there is a working area, which is shared by the working person and the robot. In this area they interact with each other and the assembly of the product is conducted. The second area of the workplace is the connection to production logistics. Both semi-finished and finished products as well as incoming components and tools must be registered and processed for transport to and from the workstation. A fundamental requirement for such a workplace is the complete accessibility of both workspaces by one or more robots. At the same time, the movement of the robot should not restrict the human working space. The workstation should be suitable for usage in various applications without great adjustment efforts, so that the workstation can be, for example, quickly converted in order to optimize it for use by left- or right-handers.

However, by involving humans and robots at the same workplace without any spatial separation, new challenges arise. Using a robot poses additional safety requirements that have to be taken into account compared to conventional workstations in manual assembly, in order to guarantee occupational safety [9]. Due to the human-centered development of the workplace we put emphasize on the ergonomic design of the entire workplace. Both principles of production ergonomics and the usage of healthy postures are important for the development. For a balanced load on the body, a height adjustment of the work surface must be implemented in order to enable the work to be carried out both in sitting and standing position.

Besides physical ergonomics, the ergonomic design of the HRC is in the foreground. It is important to make the actions of the robot comprehensible and predictable in order to avoid an increased mental load caused by supposedly arbitrary actions of the system. Both satisfactory degree of control over the system for the working person and trust in automation must be created. This requires an intuitive and user-centered design of the interaction.

3 HRC Workstation "CoWorkAs"

In consideration of the previously introduced requirements, the *Collaborative Workplace for Assembly (CoWorkAs)* was developed at the Institute of Industrial Engineering at RWTH Aachen University in cooperation with item Industrietechnik GmbH. An exemplary visualization of the workplace can be found in Fig. 1. The most important features of the workplace are described in the following.

The basic structure of the workplace is a framework consisting of four pillars, with a large workbench for the working person on the front side. The interior space of the framework represents the desired link to logistics. In addition to an interim storage for components and tools, new components that are delivered via the conveyor system are handled and sorted, bringing the workplace closer to the conveyor logistics of the production hall. The equipping with components from a parts warehouse is moved ahead of the conveyor system in the processing chain and therefore does not have to be considered separately from the equipping for the remaining manufacturing processes in the factory.

For highly flexible movements within both working areas, the robot is mounted overhead on a 2-dimensional linear guidance system, which covers the entire workstation and extends the robot's range of motion by two translational axes in the horizontal plane. This solution allows the robot to move into a pose at any time without restricting the human working space. As robot the lightweight arm SCHUNK Powerball LWA 4P was chosen. The described method of robot mounting to the workstation has not only the advantage of a large working area, but can also be

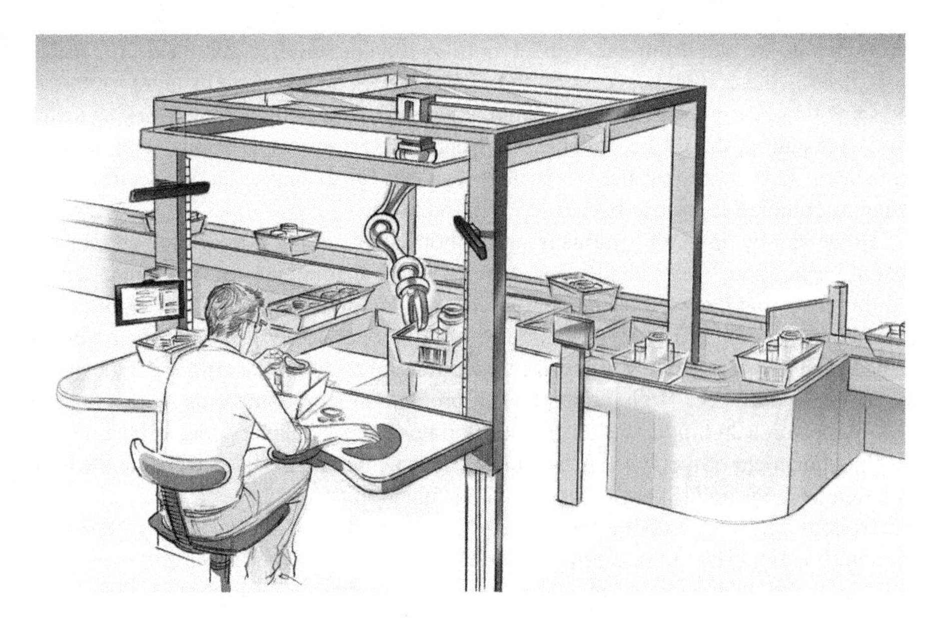

Fig. 1. Vision for human-robot collaboration in multi-variant production

advantageous from an ergonomic perspective. The hanging assembly of the robot lets the human associate the robot with the posture of the human arm hanging down from the shoulder joint. Earlier studies have shown that this anthropomorphic design of robot movements has a positive influence on the predictability of movements, mental load, and error rates in the implementation of cooperative activities [10]. Similar effects can also be assumed for the human-like hanging robot mounting.

The workstation is characterized by its modular profile construction. Therefore additional modules such as monitor, tool holders or sensors can be attached to almost any part of the workstation. This creates an almost inexhaustible design space to use the workplace in a wide variety of applications and adapt it to individual requirements.

The height adjustment of the workstation is realized by synchronized motors in each of the four pillars, whereby the entire installation including the robot and all extensions mounted directly at the workstation is height-adjustable and not only the working surface of the operator. This offers the advantage that robot movements can be carried out in the same way at any time, regardless of the height of the work surface, and that the freedom of movement is not restricted for a large height of the working surface compared to a low-adjusted working surface. The connection to a constant height conveyor system of the automated production can be solved by means of inclined conveyor belts with variable lengths. The inclination of the conveyor belts changes with the height adjustment of the workstation. For the working person, the height adjustment allows the workplace to be adapted to the individual body dimensions as well as working preferences with regard to working in a sitting or standing position.

4 Summary and Conclusion

By the implementation of the concept developed on the basis of the requirements, with CoWorkAs (see Fig. 2) a workstation for use in research and the later transfer of the validated concepts to industry was created. During the development of the workplace, special emphasis was placed on user-centered design and ergonomic layout of the workplace. With the workstation, which was designed to satisfy all design requirements, groundwork was laid for the investigation of questions concerning HRC in manual assembly. However, the interaction concepts between human and robot have only been implemented partially so far. After their implementation, they have to be validated empirically in order to answer questions about acceptance, trust in automation, or effectivity of the selected functions.

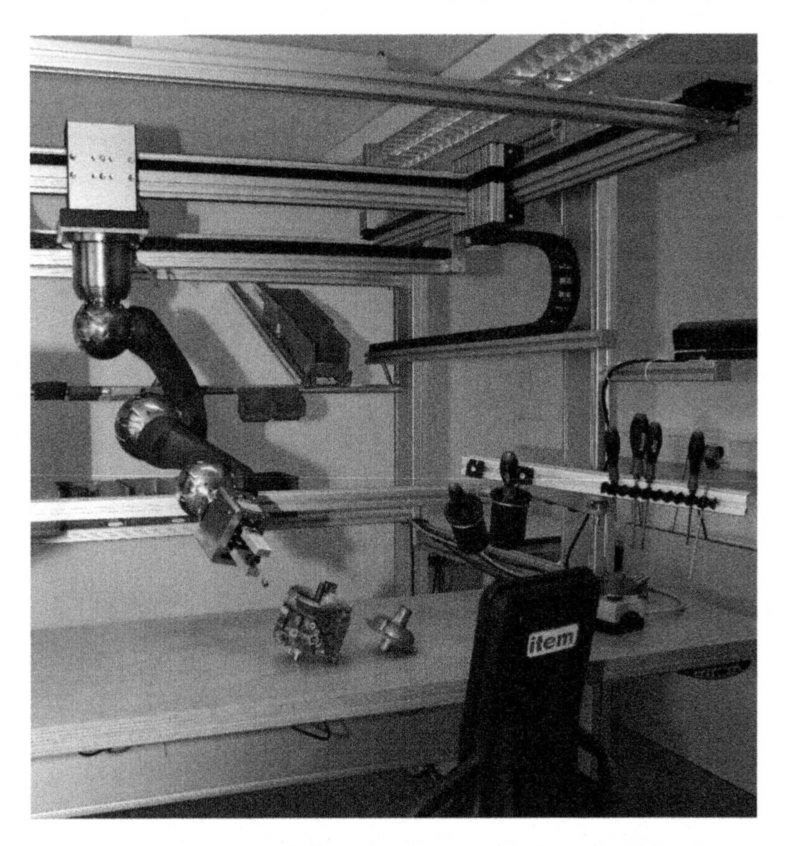

Fig. 2. Physical installation of the HRC workplace "CoWorkAs"

Among other things, the adaption of the system's behavior to the working person is intended, in order to take individual characteristics, behavioral patterns and work preferences into account. This adaptation, one of the core concepts of this workplace, takes place in various dimensions. On the one hand, there is the spatial adaptation that takes place by adapting the handover positions between human and robot. When the robot delivers a component or tool, it is guided by the position and handiness of the working person and carries out a bringing movement towards the working person's hand instead of approaching a fixed hard-coded position. The robot does not move at maximum speed, but rather slows down the movement with increasing proximity to the working person and stops completely at a very close proximity, so that the working person has to carry out the last step in the direction of the robot. This deceleration is realized by a distance-dependent speed, in this case the distance between the gripper of the robot and the hand of the working person. The further the robot moves away from humans, the faster it is allowed to move. The positions of the body parts required for this speed control and for adapting the transfer position are provided from the merged data of several depth imaging cameras to increase the robustness against occlusion.

Fig. 3. Visualization of alternative configuration of HRC-workplace "CoWorkAs"

On the other hand, the system adapts to a certain extent to the cycle times of the working person instead of fixed cycle times. Under the hypothesis that a reduction of mental load can be achieved by this approach, the component should not be brought to the working person until it is actually needed. A handover process, which is initiated too early, can lead to stress for the worker and a period of time in which the robot cannot be used for other activities. On the other hand, too late initiation leads to a forced pause for the worker and delays in the production process. To adapt the cycle time, it is necessary to predict the required assembly time. This can be achieved, for cxample, by comparing the target times according to MTM and the actual times required for previous work steps [11].

Furthermore, the workstation must be adapted for use in productive operation. However, it has been pointed out that this can be done without considerable effort due to the modular design. In addition, other fields of application are conceivable for such a workplace. In the current state, almost every collaboration in which the working person works on a work surface can be implemented with this working station. Another example with more component boxes and mobile side tables is visualized in Fig. 3. However, this work surface can also be replaced or redesigned in order to transfer the concepts of the work place to other areas of application such as assembly line production.

Acknowledgements. The authors would like to thank the German Research Foundation DFG for the kind support within the Cluster of Excellence "Integrative Production Technology for High-Wage Countries".

References

1. Audi AG (2017) Human robot cooperation: KLARA facilitates greater diversity of versions in production at Audi. https://www.audi-mediacenter.com/en/press-releases/human-robot-cooperation-klara-facilitates-greater-diversity-of-versions-in-production-at-audi-9179
2. Robotics & Automation News. BMW shows of its smart factory technologies at its plants worldwide. https://roboticsandautomationnews.com/2017/03/04/bmw-shows-off-its-smart-factory-technologies-at-its-plants-worldwide/11696/
3. beAutomotive. First-time cooperation between robots and humans on Audi assembly line. Retrieved from http://beautomotive.be/first-time-cooperation-robots-humans-audi-assembly-line/
4. FourByThree. Robots and humans team up at BMW to digitally disrupt auto industry. http://fourbythree.eu/robots-and-humans-team-up-at-bmw-to-digitally-disrupt-auto-industry/
5. KUKA AG (2017) HRC system in production at BMW Dingolfing. https://www.kuka.com/en-de/press/news/2017/06/bmw-dingolfing
6. Vogel C, Fritzsche M, Elkmann N (2016) Safe human-robot cooperation with high-payload robots in industrial applications. In: The eleventh ACM/IEEE international conference on human robot interaction, pp 529–530. IEEE Press (2016)
7. Nelles J, Kohns S, Spies J, Brandl C, Mertens A, Schlick CM (2016) Analysis of stress and strain in head based control of collaborative robots—a literature review. In: Goonetilleke R, Karwowski W (eds) Advances in physical ergonomics and human factors. Advances in Intelligent Systems and Computing, vol 489, pp 727–737. Springer, Cham
8. Huang CM, Mutlu B (2016) Anticipatory robot control for efficient human-robot collaboration. In: The eleventh ACM/IEEE international conference on human robot interaction, pp 83–90. IEEE Press

9. Faber M, Bützler J, Schlick CM (2015) Human-robot cooperation in future production systems: analysis of requirements for designing an ergonomic work system. Procedia Manuf 3:510–517
10. Kuz S, Schlick C, Lindgaard G, Moore D (2015) Anthropomorphic motion control for safe and efficient human-robot cooperation in assembly system. In: Proceedings of the 19th triennial congress of the IEA, vol 9, p 14
11. Petruck H, Mertens A (2017) Predicting human cycle times in robot assisted assembly. In: Advances in ergonomics of manufacturing: managing the enterprise of the future: proceedings of the AHFE 2017 international conference on human aspects of advanced manufacturing, 17–21 July 2017, The Westin Bonaventure Hotel, Los Angeles, California, USA

Human Work Design: Modern Approaches for Designing Ergonomic and Productive Work in Times of Digital Transformation – An International Perspective

Peter Kuhlang(✉), Manuela Ostermeier, and Martin Benter

MTM-Institut, Deutsche MTM-Vereinigung e. V., Eichenallee 11, 15738 Zeuthen, Germany
{Peter.Kuhlang,Manuela.Ostermeier, Martin.Benter}@dmtm.com

Abstract. Today, enabling productive and ergonomic work processes, work methods and work systems plays a significant role and in the future, it is going to gain even more significance. The "Ergonomic Assessment Worksheet" (EAWS) is a screening tool to assess the physical workload on the human body in different workplaces. It was developed for ongoing production and for production planning in the automotive industry and similar industries. With EAWS, physical stress can be assessed in a very detailed way without requiring a lot of effort. Aspects of successive stress superposition can be greatly simplified for short cycle tasks. The results of the evaluation are the basis for the communication between management and workers councils. The new process building block systems Human Work Design (MTM-HWD®) describes motions of people in conjunction with an ergonomics assessments procedure - in this case EAWS - in one step. This way, it allows a direct correlation in designing productive and ergonomic work. This contribution presents principles, practical application cases and the standardized education concept of EAWS and MTM-HWD® in the light of their international application.

For both methodologies, capturing or tracking of motions is essential in order to collect information about body postures of human beings, forces and loads for manual handling as well as information of the frequency of the repetitiveness of the upper limbs. An automatic collection of these kind of data by a motion capturing suit (AXS) and the connected gloves as well as the evaluation of the collected data by an EAWS and MTM-HWD® analysis will also be presented in this contribution from a theoretical and practical (business case) perspective.

Keywords: Human Work Design · International standards
Ergonomic Assessment Worksheet · Motion capture

S. Bagnara et al. (Eds.): IEA 2018, AISC 825, pp. 29–37, 2019.
https://doi.org/10.1007/978-3-319-96068-5_4

1 EAWS Training - A International Standard

Well-designed work helps to prevent stress, maintain health and performance, and drive employee motivation along the value chain. Planning and evaluation as well as the holistic design of workplaces under ergonomic aspects are therefore becoming more important in companies [1].

The Ergonomic Assessment Worksheet (EAWS) is mainly used in industrial work environment for the prospective, preventive and corrective ergonomic design of work systems [2, 3]. The use of EAWS provides a comprehensive assessment of the physical burden on the whole body and on the upper limb. The goal is the holistic design of the work system and processes in combination with an ergonomic risk analysis. A foresighted ergonomics achieves the realization of ergonomic demands in the design process.

For the worldwide uniform (standardized) application and the training of a method – in particular a method for the evaluation of ergonomic (biomechanical) risks – a substantial challenge results from the increasing internationalization of the production. The reality is that a large proportion of companies using methods such as MTM (Methods-Time Management) or EAWS maintain or plan production sites abroad. Nevertheless or just because of that all future users of this method have to be trained identically as to scope and content.

A high quality training throughout the world will be achieved and guarantee by globally uniformed admission requirements, training materials and syllabi. A worldwide acknowledged training standard is guaranteed by the German MTM Association (Deutsche MTM-Vereinigung e. V. in particular by [4],

- ensuring the availability and multilingualism of the qualification offer,
- clearly defined quality requirements for instructors (training to an instructor and in the implementation of courses) and ensuring these requirements,
- clearly defined and worldwide comparable training programs (scope, duration of the individual training, the training materials, didactic supports, examinations, and so on),
- the international reputation of the MTM methodology and the related connections and degrees as well as a high distribution and number of users.

The training to an EAWS-Practitioner, for example, provides in two 5-day courses the basic for carrying out a safe and quantified Risk analysis with EAWS. The MTM software TiCon4 EAWS is used to support the calculations for the analysis of practical examples. Upon successful completion of the training, the EAWS-practitioner must refresh and update his knowledge at an interval of three years [5].

2 EAWS in Practice Use

EAWS was developed between 2006 and 2008 in cooperation of international experts from work safety, work medicine, physiologic/biomechanics, work system designing for use in industrial companies. The development was coordinated from IMD [6]. It is used for the prospective, preventive and corrective ergonomic design of workplaces. Its focus are typical biomechanical loads that impair the human body on an industrial environment [7].

The Assessment basics of the EAWS are international (e.g. ISO) and national (e.g. DIN) standards, acknowledged assessment tools and relevant literature [5].

Using the EAWS means recognizing ergonomic deficits already in the early phase of product development process, i.e. already in design and planning, and therefore minimizing health risks from the start. EAWS is unique because it offers the chance to calculate different types of biomechanical loads and aggregate that into a total risk score. With EAWS, it is possible to assess biomechanical loads of postures, action forces and manual material handling. These load situations are then aggregated in the whole body score. Loads onto the upper limbs by highly repetitive activities can also be assessed with EAWS. Furthermore, it is possible to assess biomechanical loads of postures, action forces and manual material handling. These stress situations are then aggregated in the whole body score [5].

Figure 1 shows the structure of the EAWS Worksheet in refer to the types of the biomechanical loads [6].

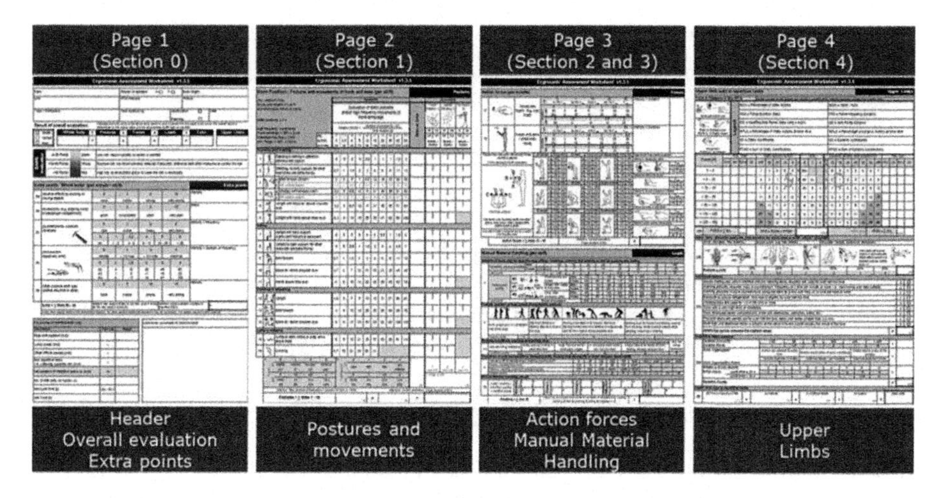

Fig. 1. Form structure of EAWS [6]

Additional to the types of biomechanical loads listed above, the user has to assess general, organizational, and technical data (e.g. date of analysis, assessed workplace, cycle time) for the assessment of a work task [6].

To describe the ergonomic risk of the workplace for the employee, EAWS uses a risk score measured in points. A higher point value is equal to a higher risk. The overall evaluation of the risk is divided in two values – the risk of the whole body (Section 0–3) and the upper limbs (Section 4). The risk score of the whole body and the upper limb is classified in three colours according to the traffic light scheme (green, yellow, red). See Fig. 2 for the details to the colours and their meanings.

A low risk is shown with green, the workspace is designed as "recommended" and "no action is needed". Red means "high risk", this workplace configuration is "to be avoided" and an "action to lower the risk is necessary" [6].

0-25 Points	Green	Low risk	recommended; no action is needed
>25-50 Points	Yellow	Possible risk	not recommended; redesign if possible, otherwise take other measures to control the risk
>50 Points	Red	High risk	to be avoided; action to lower the risk is necessary

Fig. 2. Overall evaluation of EAWS [6] (Color figure online)

3 Automate EAWS Assessment

The use of EAWS requires expertise, detailed information about the work system and time. To leverage the application and spread the method further, systems are tested that support users and generate EAWS analysis partially or completely automatically. One possibility is the detection of movements of the human body, the so-called motion capturing.

3.1 AXS Suit for Motion Capture

AXS Motionsystem Korlátolt Felelősségű Társaság developed the AXS suit. In collaboration with the German MTM Association the AXS suit is developed further. The AXS suit is equipped with sensors, which provide information about occurring forces, existing body and hand-arm postures. Due to the automatically generated data, the AXS suit is a good way to partially automate the EAWS analysis [7].

3.2 Use of AXS in Combination with EAWS

In order to be able to collect the relevant data automatically, an employee wears the suit. The employee performs his work tasks with the suit (see Fig. 3a). The software belonging to the suit picks up the information from the sensors and converts them into

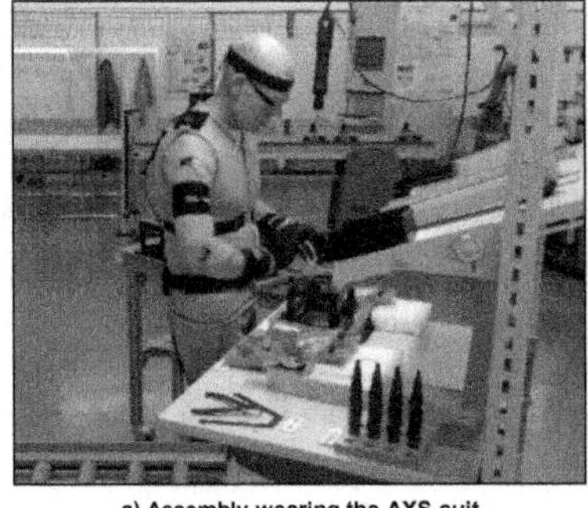

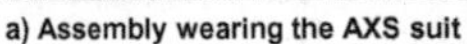

a) Assembly wearing the AXS suit

b) Manikin in the AXS software

Fig. 3. Visualisation of motion capturing data with the AXS suit positions, forces and direction of force. Figure 3b shows the 3D simulation of the recorded data.

By using the AXS suit, a large part of data required for the EAWS analysis is automatically recorded. This results in an important advantage. Additional general information must be entered manually in the software. General information are e.g. the work organization and technical data (see Fig. 4) [7].

Work organization	
Real shift duration	480 min
Lunch break	30 min
Other official pauses	40 min
Number of breaks	3
Cycle	60 s
Technical data	
Weight pump housing	3 kg
Force for screw driver	20 N
Other weights/forces	≤ 0,5 kg/5 N

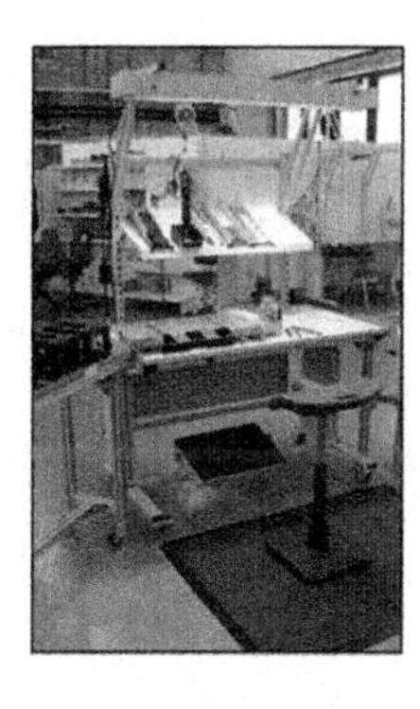

No.	Work task
1	Place housing into mounting device
2	Place gasket on housing
3	Assemble housing lid to housing
4	Turn on 4 screws by hand
5	Tighten 4 screws with cordless power screwdriver
6	Turn housing
7	Place shaft into mounting device
8	Place 2 gear wheels onto shaft
9	Assemble 2 sealing rings with pliers
10	Place 2 ball bearings onto shaft
11	Place AU "shaft" into housing
12	Aside housing

Fig. 4. General information of the work system

3.3 Evaluation AXS with EAWS

To test the quality of the partially automatically generated EAWS analysis of the AXS suit, the suit was used at various workstations. An EAWS expert checked the results of the suit. He carried out manual EAWS analyses for these work stations and compared them with the results from the AXS suit.

One exemplary workstation for the testing is located at the Institute of Production Systems of the Technical University of Dortmund. Figure 5 shows the workplace for the assembly of a pump housing.

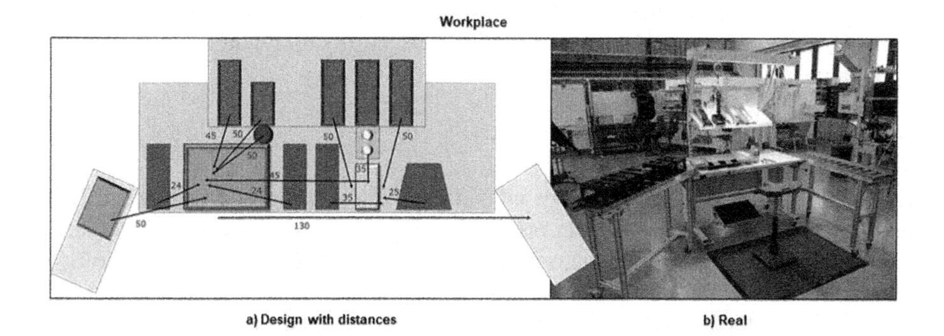

Fig. 5. Workplace describing (Color figure online)

First, the EAWS expert recorded the general information. Based on the influencing factors (postures, forces, weights etc.) he calculated the risk score and determined the risk area (green, yellow, red) for this work task.

To generate the analysis with the AXS suit, the suit recorded one work cycle. Then the user had to add manual the missing information e.g. shift duration, cycle time and breaks. This additional information is identical. This way, these aspects cannot cause any deviations in the analyses.

Table 1 presents the comparison of both results [7]. It shows that the automatically generated analysis of the AXS suit is more conservative then the manual analysis. However, this is an advantage in an assistance system, since no risks can be underestimated. In this example, the deviation is 26% (19 points (manual) vs 24 points (AXS)).

Table 1. Comparison of the EAWS analyses (manual and automated)

Risk area	Point value of the EAWS analysis			
	Manual analysis	Automated analysis	Deviation	
Total for whole body	19.0	24.0	5.0	26%
Upper limbs	17.0	17.3	0.3	2%

The tests and validations have shown that the AXS suit including his software, despite the discrepancies found, is a suitable support for carrying out EAWS analyses.

Another modern approach to design productive and ergonomic work systems is introduced subsequently.

4 MTM-HWD® for Assessment of Ergonomics and Time

In the industrial work environment, the ergonomic design of work tasks and work places is becoming increasingly important. Nonetheless, in order to design processes, the planners need a temporal evaluation of the tasks. The new MTM method Human Work Design (MTM-HWD®) is a method that captures and assesses both time and ergonomics [8–11]. Thus, the use of MTM-HWD® allows designing of productive and ergonomic work systems.

4.1 MTM-HWD®

MTM-HWD® captures all necessary influencing factors using one worksheet. Using the sheet to create the analysis can prevent errors and ensures that the user systematically records all factors. For an easier handling, this method is available with software support.

The MTM-HWD® form is divided into the following subareas:

- Description
- General settings
- Lower limbs

- Trunk
- Head/neck
- Upper limbs
- Quantity and time values

Each subarea has further subdivisions to describe the activities as completely as possible. A significant difference to most analysis methods is the description in pictograms. The individual subareas are explain in more detail below.

4.2 Describing and General Settings

The user has to fill the describing and the general settings first. This is the first step in characterizing a work task with MTM-HWD®. Describing is the part, where the user describes the process step in his own words. General settings are the object (part, tool, actuator, no object and means of transport 1–4), the action (e.g. obtain, deposit, retract), the active limb (hand, finger, foot) and the passive limb(s) (carrying/inactive, carrying with object, hold an object or not a).

4.3 Lower Limbs, Trunk and Head/Neck

The lower limb describes the posture and the movements of the leg. An additional information is the distance, the working conditions and the stability placed (see Fig. 6a). The trunk has three possible postures (flexion, rotation und lateral bending). Every posture is divided in different ranges (see Fig. 6b). The head/neck has two possible postures (head and eye travel). Head posture means the head is neutral position or bends to left/right/forward. Eye travel describes the distance the eyes have to cover (including or not the neck) in cm (see Fig. 6c).

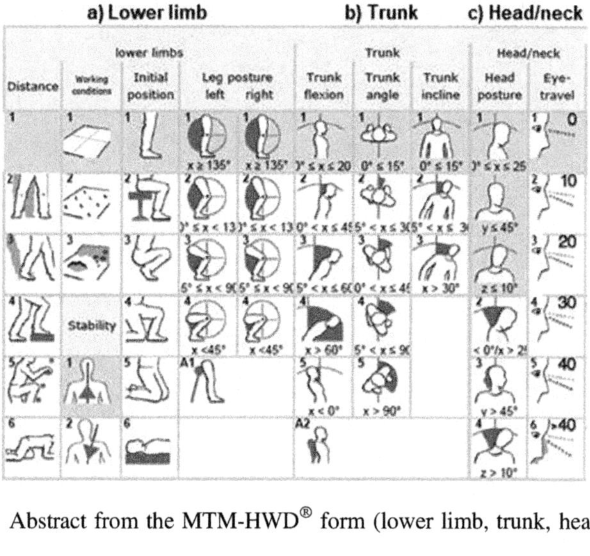

Fig. 6. Abstract from the MTM-HWD® form (lower limb, trunk, head/neck)

4.4 Upper Limbs

The part upper limbs describes many different influencing variables. One reason for this diversity is, that most work tasks are performed with the hand-arm system. Another reason for this is the aim to assess work tasks ergonomically [10, 11].

The different postures are (see Fig. 7):

(a) The Upper arm posture describes the deflection of the arm relative to the neutral position (hanging to the side).
(b) Hand position describes the height of the worker's hand relative to her/his shoulder.
(c) Arm extension refers to the distance between hands and shoulder joint.
(d) Wrist posture describes, analogous to trunk posture, whether the worker rotates, bends, or inclines the wrist.
(e) Weight/force describes the load to the employee during his/her task. The direction of force belongs to this area.
(f) Distance class describes the travelled distance. There are different distance ranges.
(g) Provision refers to the arrangement or position of the object during the OBTAIN (describes the situation to get an object under control).
(h) Positing accuracy, assembly position and positing condition describe the accuracy required in placing an object, i.e. whether the worker has to pay attention to the symmetry, shape, characteristics, or weight of the object.
(i) Grasping movement describe the finger posture necessary to gain control of the object, or adjust the control.
(j) Mode of grasping is divided in three categories (contact, grasp and embrace) and shows some examples.
(k) The describing of vibrations would be needed if a tool like an electric or pneumatic screwdriver is used.

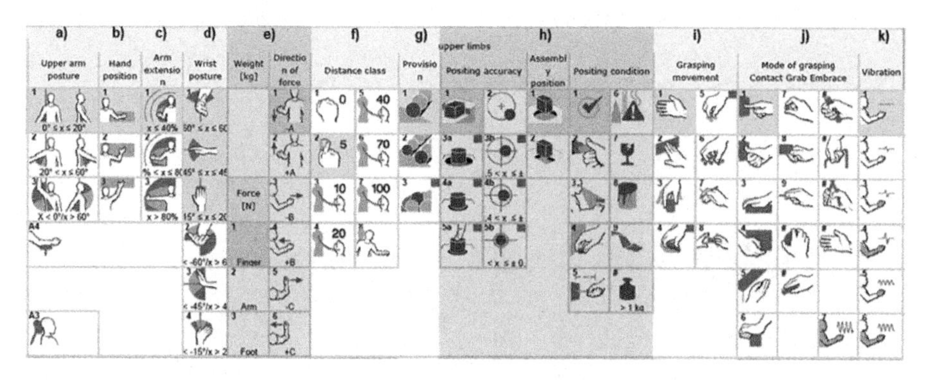

Fig. 7. Abstract from the MTM-HWD® form (upper limbs)

4.5 Result and Outlook

Once the user has recorded all influencing factors in the system, both a temporal and an ergonomic evaluation are carried out. The basis for the temporal evaluation is provided by MTM-1, the MTM basic system. This guarantees a reliable and comparable time. Based on the information collected in the system, an EAWS score is automatically generated. Here the advantage of an MTM-HWD® analysis becomes visible – one analysis leads to several results. This means that the time spent on different types of workplace analyses can be reduced massively.

Currently the AXS suit and MTM-HWD® are subject of further validations in the field and the results will be presented in future publications.

References

1. Schlick C, Bruder R, Luczak H, Mayer M, Abendroth B (2010) Arbeitswissenschaft, 3rd edn. Springer, Heidelberg
2. Lavatelli I, Schaub K, Caragnano G (2012) Correlations in between EAWS and OCRA index concerning the repetitive loads of the upper limbs in automobile manufacturing industries. Work 41:4436–4444
3. Mühlstedt J, et al (2012) Ergonomic assessment of automotive assembly tasks with digital human modelling and the ergonomics assessment worksheet (EAWS). Inderscience Enterprise Ltd.
4. Ostermeier M, Kuhlang P (2017) Internationale Methoden- und Ausbildungsstandards zur Bewertung ergonomischer Risiken. In: TBI2017
5. Schaub K, Caragnano G, Britzke B, Bruder R (2013) The European assembly worksheet. Theoret Issues Ergon Sci 14(Issue 6):616–639
6. Deutsche MTM-Vereinigung e.V.: Lehrgangsunterlage – EAWS-Praktiker. Deutsche MTM-Vereinigung e.V., Hamburg [2914]
7. Benter M, Rast S, Ostermeier M, Kuhlang P (2018) Automatisierung von Ergonomiebewertungen durch Bewegungserfassung am Beispiel des Ergonomic Assessment Worksheet (EAWS). In: Gesellschaft für Arbeitswissenschaft e.V. (ed) Gestaltung der Arbeitswelt der Zukunft - 64. Kongress der Gesellschaft für Arbeitswissenschaft, Dortmund
8. Finsterbusch T, Wagner T, Mayer M, Kille K, Bruder R, Schlick C, Jasker K, Hantke U, Härtel J (2014) Human work design - Ganzheitliche Arbeitsgestaltung mit MTM. In: Gesellschaft für Arbeitswissenschaft e.V. (ed) Gestaltung der Arbeitswelt der Zukunft - 60. Kongress der Gesellschaft für Arbeitswissenschaft, pp 324–326. GfA-Press, Dortmund
9. Finsterbusch T, Kuhlang P (2015) A new methodology for modelling human work – evolution of the process language MTM towards the description and evaluation of productive and ergonomic work processes. In: Proceedings of 19th triennial congress of the IEA, Melbourne, 9–14 August 2015
10. Finsterbusch T, Petz A, Faber M, Härtel J, Kuhlang P, Schlick CM (2016) A comparative empirical evaluation of the accuracy of the novel process language MTM-human work design. In: Schlick C, Trzcieliński S (eds) Advances in ergonomics of manufacturing: managing the enterprise of the future. Advances in Intelligent Systems and Computing, vol 490. Springer, Cham
11. Kuhlang P, Finsterbusch T, Rast S, Härtel J, Neumann M, Ostermeier M, Schumann H, Mühlbradt T, Jasker K, Laier M (2017) Internationale Standards zur Gestaltung produktiver und ergonomiegerechter Arbeit. In: Dombrowski U, Kuhlang P (eds) Mensch – Organisation – Technik im Lean Enterprise 4.0, pp 91–154. Shaker, Aachen

Fukushima-Daiichi Accident Analysis from Good Practice Viewpoint

Hiroshi Ujita(✉)

Institute for Environmental & Safety Studies, 38-7 Takamatsu 2-chome, Toshima-ku, Tokyo 171-0042, Japan
kanan@insess.com

Abstract. Fukushima-Daiichi Accident has been analyzed from the viewpoint of good practice, that is from resilience engineering viewpoint. The good cases of resilience response are observed in individual base and organizational base as below: The effectiveness of insight on accident cases (inundation in Madras, 9.11 terrorism and B.5.b. order for the countermeasures) and of the risk evaluation, Decision of continuation of sea water infusion (individual base), Reflection of the experience on Chuetsu-Oki Earthquake, Improvement of seismic building which is equipped emergency power source system and air conditioning system (organizational base), Deployment of fire engines (organizational base), The effectiveness of command system in ordinal time (on-site of organizational base), and Support by cooperation companies and manufacturers (designers and site workers of organizational base). It is important to 'establish the feedback system on organization learning in ordinal time', that is to establish the system admitting violation of order. The decision at on-site are given priority than other ones. The representative example is the decision of sea water infusion continuation which was given priority at on-site, even though the official residence and the main office of Tokyo Electric Power Company had ordered to stop the infusion.

Keywords: Resilience engineering · High reliability organization
Risk literacy · Fukushima accident · Bounded rationality
Information limitation · Time constraint · Context

1 Introduction

Whereas the direction which discusses the safety from the accident analysis, a new trend of analytical methods such as resilience engineering, high reliability organization, or risk literacy research [1–3], which analyze the various events by focusing on the good practices, are becoming popular.

The results of Fukushima Daiichi Accident investigation with diversified characteristic have been released until now. Based upon the analyses of the investigation, the success cases (good practice) and failure cases for emergency responses were analyzed concerning to personal- response capability, organizational-response capability, and communication ability with external organizations, and then the problems of responses were extracted. The action of sea-water infusion on Fukushima Daiichi nuclear power

S. Bagnara et al. (Eds.): IEA 2018, AISC 825, pp. 38–50, 2019.
https://doi.org/10.1007/978-3-319-96068-5_5

plant No.1 was paid attention and analyzed based on the 'Accident Analysis Report of Fukushima Nuclear Accident' [4] (Accident Report) by TEPCO, Tokyo Electric Power Company, especially focus on the decision making of continuation of pouring sea water.

2 History of Accident and Human Error Type

The history of accident and human error type trend is shown in Fig. 1 [5]. In the era when the plant system was not so complicated as in the present age, it was thought that technical defects are the source of the problem and accidents can be prevented by technical correspondence. As the system became more complex, it came to the limit of human ability to operate it, accidents caused by human error occurred. Its typical accident happened at Three Mile Island (TMI) nuclear power plant in 1979. For this reason, individuals committing errors are considered to be the source of the problem, improvement of personnel capacity by appropriate selection and training of personnel, and proper design of interface design are considered effective for error prevention.

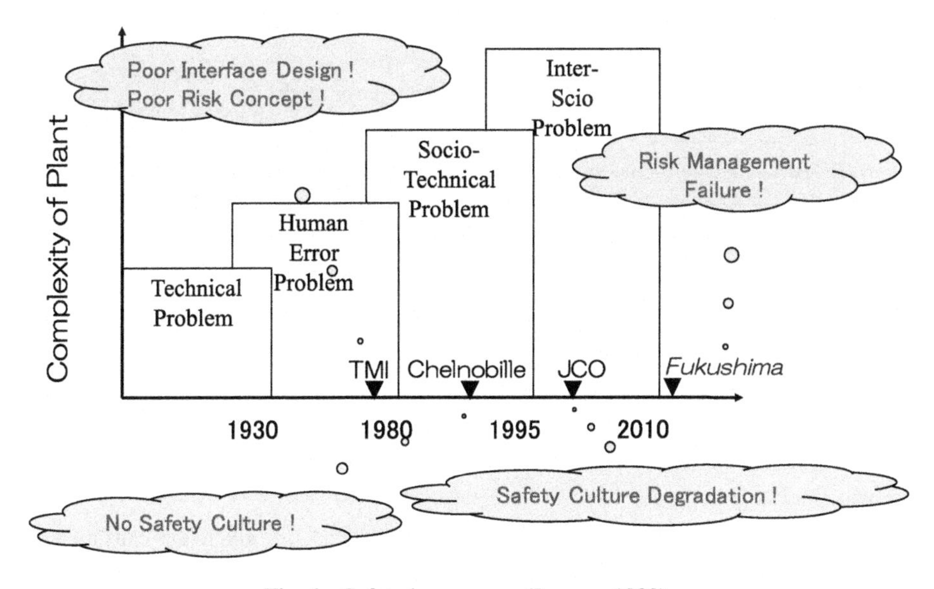

Fig. 1. Safety issue scope (Reason, 1993).

Thereafter, accidents caused by complicated interrelationships of elements such as technology, human, society, management, organization and so on occurred, and then the problem was interaction between society and technology. Furthermore, not only within the plant and enterprises but also accidents where the relationship failure with external stakeholders and organizations is a source of the problem becomes noticeable, and a framework for comprehensive problem solving including inter-organizational relationship has become to be necessary. A recent accident is a so-called organizational accident in which the form of an accident is caused by a complex factor and its influence reaches a social scale [5].

For this reason, the analytical methods, as well as type and social perceptions of error or accident, are changing with the times also. Human error and Domino accident model had initially appeared, are then has been changing to system error and Swiss cheese accident model, and recently move to safety culture degradation and the organizational accident.

3 Accident Model and Error Model

As a result that the technology systems become huge, complex and sophisticated, safety issues are shifted to the problem of organization from human, and further from hardware, such socialization is occurring in every technical field. For this reason, the analytical methods, as well as type and social perceptions of error or accident, are changing with the times also. Table 1 shows trends of the accident model and error model [6]. Human error and Domino accident model had initially appeared, are then changing to system error and Swiss cheese accident model, and recently move to safety culture degradation and the organizational accident.

Table 1. Accident model and error model.

Accident Model	Error Model	Analysis Method	Management
Domino (failure chain)	Component Failure and human Error	Cause-Consequence Link	Encapsulation, Seek & Destroy
Swiss Cheese (Loss of Diversity)	System Error (Common mode Failure)	Risk Evaluation	Defense & Barrier
Organizational Accident (Fallacy of Defense in Depth)	Safety Culture Degradation	Behavioral Science Safety Culture Check List	Monitor & Control (Organizational Learning)

Design ↕ Management

A conventional accident model is the domino model, in which the causation of trouble and the error is analyzed and measures are taken. In the model, slip, lapse, and mistake are used which are the classification of the unsafe act to occur by on-site work. These are categorized as the basic error type, while violation which is intentional act violating rule has become increased recently and considered as cause of social accident.

Design philosophy of the defense in depths has been established, and the accident to occur recently is caused by the excellence of the error of a variety of systems. The analysis of the organization blunder is necessary for the analysis by this Swiss cheese accident model in addition to conventional error analysis.

An organization accident is a problem inside the organizations, which reaches earthshaking event for the organization as a result by the accumulation of the best intentions basically. It is an act of the good will, but becomes the error. As for the organization accident, the interdependence inside of the organization or between the organizations is accumulated by fallacy in the defense in depths, and it becomes a problem of the deterioration of the safety culture in its turn. The organizational management based on the organization analyses such as behavioral sciences will be necessary for these measures.

4 Bounded Rationality in Context vs. Judge by God

In the field of cognitive science and the cognitive system engineering, the human being is considered as to think and judge something reasonably along context while there are information limitation and time constraint. Sometimes the decision may be judged as an error by the outside later. It is called "bounded rationality in the context" vs. "judge by God". The absurd action of the organization had been often explained in human illogicality conventionally, while the approach has recently come out to think that the human being rationality is the cause [7].

There are three approaches proposed from Organizational (Behavioral) Economics, business cost theory (reluctant to do), agency theory (information gap), and proprietary rights theory (selfishness) as shown in Table 2 [8]. Business cost theory analyses action of opportunity principles and sunk cost, agency theory, moral hazard and adverse selection (lemon market), and proprietary rights theory, cost externality. The common supposition is "the bounded rationality and the utility maximization".

Table 2. Three approaches by behavioral economics (The organization fails rationally: Kenshu Kikusawa 2009).

	Transaction cost theory (Bothersome)	Agency theory (Information gaps)	Property rights theory (Selfish)
Object	Business relationship	Agency relationship (Principle and Agency)	Ownership relationship
Inefficiency	• Opportunistic behavior • Sunk cost	• Moral Hazard • Adverse Selection (Lemon Market)	Externalities
Institutional resolution	Transaction cost saving system (Organization style change: fellow- centralized-decentralized)	Agency cost reduction (Symmetrical information)	Internalization of externalities (Ownership allocation)
Example	• One-man management - external monitoring	• Work sharing	• Fellow consciousness and organizational concealment

Common assumptions: bounded rationality and utility maximization

It is necessary to find the social context that the error is easy to occur, in the engineering for human beings hereafter. In other words, a way of thinking has changed in the direction to analyzing the social context that is easy to cause an error, from analyzing direct cause of the error. Because this direction is beyond the range of conventional ergonomic treating the contents of the error, it is very difficult. However, we should recognize it now, if we do not analyze an error in the viewpoint the relationship between safety and the environmental element surrounding human being, we can not to lead to measures. The measures should be matched with human rational characteristics.

Countermeasure on Business cost theory is business cost saving system which changes organizational style from group organization, via. centralization of power type organization, and to decentralization of power type organization, agency theory, agency cost reduction system based on mutual exchange of the information, and proprietary rights theory, internalization of the system externality based on proprietary rights distribution.

5 The Methodology on Resilience Engineering, High Reliability Organization, and Risk Literacy

Whereas the direction which discusses the safety from the accident analysis, a new trend of analytical methods such as resilience engineering, high reliability organization, or risk literacy research, which analyze the various events by focusing on the good practices, are becoming popular.

The resilience is the intrinsic ability of a system to adjust its functioning prior to, during, or following changes and disturbances, so that it can sustain required operations under both expected and unexpected conditions. A practice of Resilience Engineering/ Proactive Safety Management requires that all levels of the organization are able to:

- Monitor
- Learn from past events
- Respond
- Anticipate

Organizational process defined by the High Reliability Organization is shown below. There are 5 powers in 2 situations.

- Preparedness for Emergency Situation in Ordinal Time:
  - Carefulness (Confirmation),
  - Honesty (Report),
  - Sensitivity (Observation),
- Emergency Response in Emergency Situation:
  - Alert (Concentration),
  - Flexibility (Response),

Ability of Risk Literacy is also defined by followings, which is largely divided to 3 powers and further classified to 8 sub-powers.

- Analysis power
  - Collection power
  - Understanding power
  - Predictive power
- Communication power
  - Network power
  - Influence power
- Practical power
  - Crisis Response Power
  - Radical Measures Power

6 The Analysis Based on Resilience Engineering, High Reliability Organization, and Risk Literacy

6.1 Chronological Analysis

The analysis for the detail of sea-water infusion to Fukushima-Daiichi No. 1 unit has been performed. Chronological analysis was drawn up from "The main chronological analysis of Fukushima Daiichi nuclear power plant No. 1 unit from earthquake occurrence to the next day" and "The status response relating to pouring water to No. 1 unit", which came from the 'Accident Report'.

The chronological analysis shows that the preparation of sea-water infusion is decided and ordered concurrently with pouring freshwater, and also shows that continuation of sea-water infusion is decided as on-site judgment although an official residence and the main office of TEPCO directed to stop the infusion, in taking into consideration of the intention of official residence. The necessity of continuation of sea-water infusion was recognized consistently by on- site judgment, and these measures were taken. This means that the main office of TEPCO and the official residence violate the fundamental principle which on-site judgment should be preceded in emergency situation.

6.2 Organizational Factors Analysis

The process of sea-water infusion on Fukushima Daiichi No. 1 unit is analyzed from the point of resilience capability, high reliability organization capability, and risk-literacy capability [1–9]. The analysis example by risk-literacy capability is shown in Table 3 from the viewpoint of risk-management, which is described in 'Introduction of Risk Literacy- Lessons Learned from Incidents' [3]. The definition of risk literacy capability which can extract communication power that is important both for ordinary time and for emergency situation is the most appropriate for analysis. The horizontal axis shows response capabilities which are suggested in each study, and the vertical axis shows each level of individual, organization, which is further divided management

Table 3. Evaluation of response capability from the viewpoint of risk literacy: the analysis of sea water infusion process on Fukushima Daiichi No. 1.

Risk literacy / Analysis level		In Ordinal Time				In Emergency Situation		
		Analysis Power			Communication Power		Practical Power	
		Collection Power	Comprehensive Power	Predictive Power	Information Transmission Power	Influence power	Crisis Response Power	Radical Measures Power
Individual		•Damage of Tsunami	•Risk recognition of Tsunami damage	•Risk recognition of Power Loss	—	—	• Continuation of Sea water infusion	• Training for emergency
organization	on-site	•Collection of accidents: •Jyogan-Tsunami	•Earthquake •Tsunami •Evaluation of influence range by PSA	Recognition of accidents damage	•Information sharing at on-site	•Command system (on-site) •Centralized at seismic building *•Contact between Control Room & Emergency Response Room*	•Infusion of fresh water and sea water •Vent •Prevention of damage expansion	•Preparation of seismic building and fire engines •Command system •AM measures •Tsunami protection measures
	Management	•Collection of accident: Jyogan-Tsunami, JNES Tsunami PSA, Infusion at Le Blayais & Madras	*•Risk misrecognition of Tsunami damage*	*•Risk misrecognition of Power Loss*	*•Information sharing between main office and on-site* Failure event	•TV conference system (2F site) *•Confusion in command system between main office and on-site*		•Review the education and training system Success event
External correspondence (official residence, etc.)		•Anti- terrorism in overseas *Collection of example : 9.11 terrorism- B.5.b.*	•Classification of importance on accidents *•Risk misrecognition of earthquake and Tsunami*	•Importance of external events *•Risk misrecognition of infrastructure damage*		*•Media, local government, publicity to overseas* *•Confusion of command system among official residence, main office, & on-site*	*•Delay of initial response* *•Governmental command system*	•Support by vendor and cooperation company •Support by external organizations *•Drastic measures : Structure reform (Regulation/ Electric power company*

and on-site, and correspondence to outside such as official residence. The gothic font in green means success case or good practice, while the italic font in red means failure case.

The analysis of emergency correspondence like this kind of huge accident cannot be analyzed enough using conventional framework. As a whole of one organization, the classifications were reviewed and revised from two points. One is that the differences in correspondence and the problems in cooperation between on-site and administration department cannot be clarified. And the other one is that communication power has two sides which are the information cooperation in ordinal time and collaboration in case of emergency situation. The analysis power and information transmission power correspond to ordinal time, and the influence power and normal time skill correspond to the case of emergency situation. In this analysis, the contact with official residence and the cooperation inside government are also included in correspond to external organizations.

7 Success and Failure Cases

From the viewpoint of Resilience Engineering, the cases of success and failure are listed and analyzed.

7.1 Good Practice of Resilience Response

The good practices of resilience response are observed in individual base and organizational base as below.

- The effectiveness of insight on accident cases (inundation in Madras, 9.11 terrorism and B.5.b. order for the countermeasures) and of the risk evaluation.
  - Decision of continuation of sea water infusion (individual base)
- Reflection of the experience on Chuetsu-Oki Earthquake.
  - Improvement of seismic building which is equipped emergency power source system and air conditioning system (organizational base)
  - Deployment of fire engines (organizational base)
- The effectiveness of command system in ordinal time (on-site of organizational base).
- Support by cooperation companies and manufacturers (designers and site workers of organizational base).

For Tsunami countermeasures in other plants, good practices were also observed in organizational base:

- Onagawa plant level was enough high to prevent Tsunami disaster due to clear understanding of the effect and sharp decision by top management,
- Tokai-Daini plant installed water stuck for sea-water pump to protect infusion due to good communication and corporation among prefecture officers and utility managers.

The reason why the good cases are occurred in on-site, the officers and workers always felt that their mission is to carry out with the sense of ownership and also with critical mind. They had trained the accident management in ordinal time, which works effectively in emergency situation, which is the just significant frame derived from the development of safety culture. It is important to 'establish the feedback system on organization learning in ordinal time'. But there were a little lack of information between control room and emergency response room, they will be able to solve by taking physical measures to clarify the circumstances at on-site. The TV conference system of Fukushima Daini nuclear power plant had worked effectively to communicate among on-site, the main office of TEPCO and the outside organizations. Furthermore using the white board for information sharing, which is the good practice that the resilience works well, could prevent the confusion at on-site.

7.2 The Failure of Comprehensive Power in Organization, the Fallacy of Composition of Risk Awareness

The many failure cases are defined under national government level and nuclear industry level which are the problems of rare event awareness and of organization culture. Although everyone had same recognition for the risk of power loss and Tsunami, the accurate decision had not been made by national government level, just only made by individual level. The ordinal time training at on-site also work in emergency situation at the accident, while the level of administration department and government didn't work well.

- Risk misrecognition of Loss of offsite power and damage by Tsunami (national government level, industry level)
- Confusion of command system (organization base- between on-site and the main office of TEPCO)
- Confusion of command system (external correspondence base- national government level, and organization base-among official residence, regulation, and the main office of TEPCO)

The continuations of emergency training in ordinal time with assuming the severe accident progression is considered to be the effective way. As many lacks in emergency correspondence in management department and in national government level are observed, and then the emergency training is necessary in management level, in which responsibility assignment is regularly taken, the incident seriousness is evaluated, and the mode is switched from ordinal time to emergency situation.

7.3 Points in Common and Difference in Fukushima Daiichi and Daini, Toukai Daini, and Onagawa Plants

Cold Shutdown could be successfully achieved due to survive if power supply, where good practices of corporation among operators and between on-site and head office were observed based on correct information. Because Air Cooling EDG (Emergency Diesel Generator) continued to operate in Fukushima Daiichi plant 5 and 6 units, and 3 EDGs in Fukushima Daini plant. Especially, TV conference system was effectively functioned in Fukushima Daini plant, because of enough information by survival indicators energized from survived batteries.

TEPCO put the priority onto countermeasures on Chuetuoki Earthquake and only considered without making countermeasures on Tsunami affect. Important Anti-seismic Building and Portable Fire Engine had been installed for seismic countermeasures in every plant of TEPCO before huge earth quake and Tsunami, and those apparatuses were used in the accident effectively in Fukushima Daiichi and Daini plants. Good practices of Tsunami countermeasures of site height decision process of problem institution by professionals and acceptance by administration performed in both Toukai Daini and Onagawa Plants.

One point we should understand is that Tsunami estimated height has drastically changed as time passed, which means it is not reliable information. At the information limitation circumstance, four plants showed good resilience examples. The difference of the correspondence is due to difference in priority policy. Fukushima made efforts on earthquake measures, on the other hand, Onagawa and Tokai Daini, efforts on Tsunami measures. It seems that the differences in suffered effect and in its correspondence between Fukushima Daiichi 1–4 and 5–6 & Daini both owned by TEPCO are due to relative relationships between actual tsunami height and site height.

7.4 Consideration on Organizational Problem

The true nature of the problem in Japanese organization that doesn't change from when the 'Substance of Failures' [9], in which Japanese military operation failures in the World War II were analyzed, was written by Tobe, Nonaka, et al. Failure cause was described as standpoint of irrationality in Japanese on this book. But the problems in organization are not able to be resolved by irrationality in Japanese. It should be explained using by bounded rationality which Kikusawa advocate in 'Absurdity of Organization' [8]. His idea is that decision making which are made under limited circumstances based on limited information will end in failure from the eye of God. He also advocates destroying the bounded rationality for failure measures. It means that it is important to 'establish the system admitting violation of order in emergency situation'. The decision at on-site are given priority than other ones. The representative example is observed in the decision of sea water infusion continuation, the decision at on-site were given priority even though the official residence and the main office TEPCO had ordered to stop the infusion. Otherwise it is the failure case that occur delay of PCV vent, for time loss to get the permission of national government and local government.

It is important to 'establish the feedback system on organization learning in ordinal time'. The continuations of emergency training in ordinal time with assuming the severe accident progression is considered to be the effective way. The emergency training is necessary in management level, in which responsibility assignment is regularly taken, the incident seriousness is evaluated, and the mode is switched from ordinal time to emergency situation.

The problems as above can be explained by "Homogeneous way of thinking" and "Concentric Camaraderie", as shown in Fig. 2, which are the hindrance on safety pursuit in Japan. 'Bottom-up decision making structure' connects to 'Absence of top management', and then becomes to 'Delay of decision making and Lack of understanding on valuing safety'. Due to the Japanese are excellent as noncommissioned officer (Soldier: Russian, Junior officer: French, Chief of staff: German, General: American), they often show their ability at emergency situation. But Japanese are short of management abilities, they often make heavy intervention or omission.

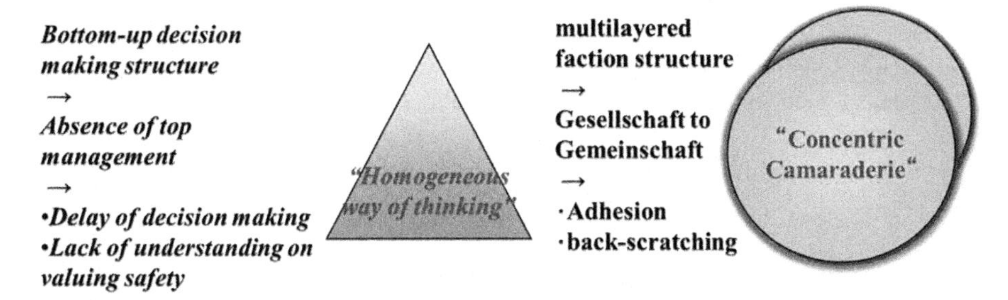

Fig. 2. "Homogeneous way of thinking" and "Concentric camaraderie", obstruction factors on safety pursuit in Japan.

'Multilayered faction structure' appeared in "Concentric Camaraderie" makes 'Organization from Gesellschaft to Gemeinschaft', and then 'Adhesion and back-scratching' are spread in the organization. For the "Concentric Camaraderie", the feedback system in organization learning leads to the failure due to be preceded to internal logic than social common sense even in national government level or nuclear industry level.

8 Discussion

Rare Event is high consequence with low frequency. Low consequence with high frequency event is easy to treat by commercial reason, while it is very difficult to handle the rare event even the risk is just the same. "Unexpected event" has been used frequently, but it is the risk-benefit issues to assume or not. Tsunami Probabilistic Risk Analysis has been carried out, and safety related personnel knew the magnitude of the effect well.

Regardless of the initiating event, lack of measures to "Station blackout" is to be asked. According to the "Defense in Depth" concept reflecting Fukushima accident, we should consider three level safety functions; usual normal system, usual safety system, and newly installed emergency system including external support functions. Anyway the diversity is significantly required for not only future reactor concept but also existing plant back-fit activities.

Swiss Cheese Model proposed by Reason, J indicates operational problem other than design problem [5]. Fallacy of the defense in depth has frequently occurred recently because plant system is safe enough as operators becomes easily not to consider system safety. And then safety culture degradation would be happened, whose incident will easily become organizational accident. Such situation requires final barrier that is Crisis Management.

Concept of "Soft Barrier" has been proposed [6]. There are two types of safety barriers, one is Hard Barrier that is simply represented by Defense in Depth. The other is Soft Barrier, which maintains the hard barrier as expected condition, makes it perform as expected function. Even when the Hard Barrier does not perform its function, human activity to prevent hazardous effect and its support functions, such as manuals, rules, laws, organization, social system, etc. Soft Barrier can be further divided to two measures; one is "Software for design", such as Common mode failure treatment, Safety logic, Usability, etc. The other is "Humanware for operation", such as operator or maintenance personnel actions, Emergency Procedure, organization, management, Safety Culture, etc.

9 Conclusion

The good practices or success cases of resilience response are observed in individual base and organizational base as below.

- The effectiveness of insight on accident cases (inundation in Madras, 9.11 terrorism and B.5.b. order for the countermeasures) and of the risk evaluation.
  - Decision of continuation of sea water infusion (individual base)
- Reflection of the experience on Chuetsu-Oki Earthquake.
  - Improvement of seismic building which is equipped emergency power source system and air conditioning system (organizational base)
  - Deployment of fire engines (organizational base)
- The effectiveness of command system in ordinal time (on-site of organizational base).
- Support by cooperation companies and manufacturers (designers and site workers of organizational base).

The many failure cases are defined under national government level and nuclear industry level which are the problems of rare event awareness and of organization culture. The ordinal time training at on-site also work in emergency situation at the accident, while the level of administration department and government didn't work well.

- Risk misrecognition of Loss of offsite power and damage by Tsunami (national government level, industry level)
- Confusion of command system (organization base- between on-site and the main office of TEPCO)
- Confusion of command system (external correspondence base- national government level, and organization base- among official residence, regulation, and the main office of TEPCO)

It is important to 'establish the feedback system on organization learning in ordinal time', that is to establish the system admitting violation of order. The decision at on-site are given priority than other ones. The representative example is the decision of sea water infusion continuation which was given priority at on-site, even though the official residence and the main office TEPCO had ordered to stop the infusion.

References

1. Hollnagel E (2009) Safety culture, safety management, and resilience engineering. In: ATEC aviation safety forum
2. Weick KE, Sutcliffe KM (2001) Managing the unexpected. Jossey-Bass, San Francisco
3. Lin S (2005, in Japanese) Introduction of risk literacy-lessons learned from incidents. NIKKEI-BP, Tokyo
4. TEPCO (2012) Accident analysis report of fukushima nuclear accident
5. Reason J (1997) Managing the risks of organizational accidents. Ashgate, Brookfield
6. Ujita H, Yuhara N (2015, in Japanese) Systems safety, Kaibundo, Japan

7. Ujita H (2017) Requirement on personnel and organization for safety and security improvement by accident and error model. In: Human Computer Interaction 2017, Vancouver, 12 July 2017
8. Kikusawa K (2000, in Japanese) Absurdity of organization. In: DIAMOND 2000
9. Tobe N, et al (1984, in Japanese) Substance of failures. In: DIAMOND 1984

The Ergonomics of the "Seated Worker": Comparison Between Postures Adopted in Conventional and Sit-Stand Chairs in Slaughterhouses

Natália Fonseca Dias[1(✉)], Adriana Seára Tirloni[1], Diogo Cunha dos Reis[1,2], and Antônio Renato Pereira Moro[1,2]

[1] Technological Center, Federal University of Santa Catarina, Florianópolis, SC, Brazil
ergonomia.nd@gmail.com

[2] Biomechanics Laboratory, CDS, Federal University at Santa Catarina, Florianópolis, SC, Brazil

Abstract. This case study aims to compare conventional and sit-stand chairs regarding Brazilian Regulatory Standard 36 (NR-36) requirements. Two types of chairs present in a slaughterhouse in southern Brazil were compared (sitting and sit-stand posture). The seven NR-36 requirements of seat characteristics were used: (1) to have height adjustable to the worker's height and the nature of the tasks performed; (2) little or no conformation at the base of the seat; (3) rounded front edge; (4) backrest adapted to the body for protection of the lumbar region; (5) easy-to-use adjustment systems; (6) constructed with material that prioritizes thermal comfort; and (7) footrest that adapts to the length of the worker's legs in cases where the operator's feet do not reach the floor. The sit-stand chair met 6 requirements of the NR-36, lacking the attribute that requires backrest (4), in contrast, the conventional chair met all requirements. The function of the backrest in the sitting posture is linked to the reduction of load on the spine. Although the sit-stand chair has no backrest, the posture adopted when using this chair is more natural and distributes 60% of the body weight in the lower limbs, resulting in less intervertebral disc compression. Based on this principle, the sit-stand posture may be an alternative to be tested in slaughterhouses. It was concluded that both conventional and sit-stand chairs met most of NR-36 requirements, being options for alternating postures.

Keywords: Sitting posture · Sit-stand chair · Slaughterhouse

1 Introduction

In Brazil, the poultry industry employs around 3.5 million people, both directly and indirectly. And an average of 350,000 workers work directly in slaughterhouses [1]. The high number of workers is justified by the country's production. Brazil is the world´s leader in export activity and the second country with the largest production [2].

S. Bagnara et al. (Eds.): IEA 2018, AISC 825, pp. 51–59, 2019.
https://doi.org/10.1007/978-3-319-96068-5_6

The development of work-related musculoskeletal disorders (WMSDs) often occurs in industrial workers [3]. The main risk factors when it comes to the tasks performed at slaughtering tasks are the intense pace of work, inadequate postures and the application of high forces when handling loads [4], insufficient recovery time [5] vibrations and low temperatures [6], favoring the development of WMSDs.

The Brazilian Regulatory Standard 36 (NR-36) on health and safety at work in meat processing industries [7] is a mandatory compliance requirement for Brazilian companies [8] and comprises many items. In the item on furniture and workstation, the NR-36 mentions requirements to be met in relation to posture, including general items established in NR-17 - Ergonomics [9] and adds some specific items to the sector. One of them is that all furniture must be adapted or designed to allow the alternation between standing and sitting postures, meeting the minimum condition of one seat for every three workers [7].

As a way to reduce the occupational hazards related to the ergonomic aspects, there is a great need of improvement of the workstations in order to reduce the biomechanical requirements, providing adequate working conditions and good posture, allowing postural alternations [10]. The most adequate posture for the development of labor activities is the one that prevents compensatory movements, properly distributes loads and conserves energy according to physiological and biomechanical needs [11]. The human body assumes three basic postures during everyday activities: lying, sitting and standing [10]. In industrialized countries, about ¾ of the workstations are suitable for sitting posture [12]. However, it is highly recommended that the workstations allow postural alternation between sitting and standing work, meeting the orthopedic and physiological requirements [12].

When compared to the standing posture, the sitting posture requires less energy expenditure, reduces joint overload in the lower limbs, decreases hydrostatic pressure in lower limb circulation, and provides stability for the development of tasks that require visual and motor control [13]. Therefore, in the standing posture, the overload in the intervertebral discs is smaller, due to the little moment of displacement of the trunk [14].

According to other studies, an ideal posture is when the trunk is kept upright, keeping the curvatures of the spine, and where the angle between the trunk and thighs remains around 128° (±7°) [15, 16]. This posture, classified as sit-stand posture, is considered neutral with reduction of musculoskeletal overloads and preservation of vertebral curvatures within ideal physiological and biomechanical values [17]. Therefore, this study aims to compare conventional and sit-stand chairs regarding Brazilian Regulatory Standard 36 (NR-36) requirements.

2 Method

This study used an exploratory method, a quantitative approach, with a case study design. Two types of existing chairs were compared in a slaughterhouse in the south of Brazil (a conventional chair and a sit-stand chair). For this, the seven requirements of NR-17 and NR-36 were used, which describe the specific characteristics of the seats: to have height adjustable to the height of the worker and the nature of the tasks performed; little or no conformation at the base of the seat; rounded front edge; backrest adapted to the body for protection of the lumbar region; easy-to-use adjustment systems; built from material that prioritizes thermal comfort; and footrest that adapts to the length of the worker's legs in cases where the operator's feet do not reach the floor. The last requirement was included in this study, considering that in Brazilian slaughterhouses, the chairs used often have the footrest. The data were presented by means of simple frequency (Fig. 1).

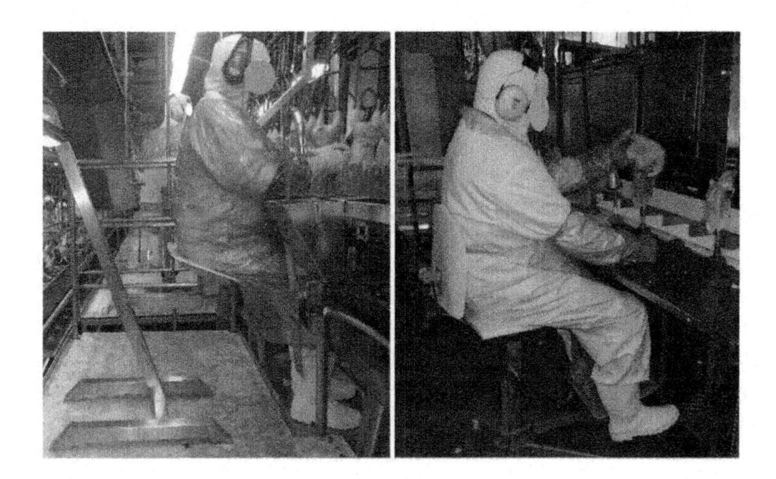

Fig. 1. Sit-stand posture (left) and sitting posture (right) used in the analyzed slaughterhouses.

3 Results

The results of the analysis of the chairs regarding compliance with the requirements of NR-36 are presented in Table 1.

The sit-stand chair met 6 of the 7 requirements demanded by the NR-36, except for the required backrest adapted to the body for protection of the lumbar region. On the other hand, the conventional chair met all of the requirements.

4 Discussion

There are few studies with quantitative data to define the relationship between beneficial and harmful postures to the human body [11]. Chaffin, Andersson and Martin [13] state that workstations should be adjusted according to the individual characteristics of workers, offering a suitable chair and work surface that is consistent with the tasks to be performed, reducing the ergonomic risks. Regarding the sitting posture, there is still no common ground when it comes to classifying the best angulation of the lumbar spine [11, 18, 19].

In industrialized countries, about ¾ of the jobs are suitable for the sitting posture [12]. According to Iida [10], among the advantages of performing tasks in the sitting

Table 1. Comparison of two types of chairs regarding seat requirements.

Seat Requirements - NR-36		Conventional chair	Sit-stand chair
1	Height adjustable to the height of the worker and the nature of the tasks performed	x	x
2	Features little or no conformation at the base of the seat	x	x
3	Rounded front edge	x	x
4	Backrest with shape slightly adapted to the body for protection of the lower back	x	–
5	Easy-to-use adjustment systems	x	x
6	Built from material that prioritizes thermal comfort	x	x
7	Footrest that adapts to the length of the worker's legs, in cases where the operator's feet do not reach the floor	x	x
Requirements that have been met		7	6

posture are: reduced energy consumption compared to the standing posture, fatigue reduction, lower mechanical and hydrostatic pressure in the lower limbs and facilitation of the development of precision activities. On the other hand, it has disadvantages such as: increased pressure in the ischial tuberosity, limitation of reach and reduction of blood circulation in the lower limbs.

Mörl and Bradl [20] have analyzed office worker´s tasks, and found that the sitting posture is predominant in 82% of the cases over the standing posture. A relationship was established between lumbar pain and the sitting posture due to the low activation of the lumbar muscles, which transmit the overload to the ligaments and intervertebral discs. In another study, no significant differences were found between the activation of the lumbar muscles in the posture while sitting on an upholstered office chair, on an exercise ball and on a wooden bench [21].

In Brazil, a study with 995 slaughterhouse workers found that, in relation to the positions adopted during task performance, 84.4% of the workers claimed to work standing up, 40.4% stated that there were not enough chairs for all of the workers and 59.4% reported not having enough work space to perform their tasks properly [22].

Industries do not make posture adaptations a priority, since they are considered intangible ergonomic improvements, that is, they cannot quantify the financial return such improvements provide [10]. However, with the implementation of NR-36, slaughterhouses must meet the minimum requirements to reduce risks in the activities developed, one of the requirements being the furniture and workstations, which interfere in the posture adopted by the workers.

Messing, Tissot and Stock [23] interviewed 7,757 workers (4,534 men and 3,223 women) and found that workstations that do not allow posture management are strongly associated with pain in the legs or calves as well as in the ankles or feet.

According to NR-17 – Ergonomics [9], whenever the task can be performed in the sitting posture, the workstation must be designed or adapted to this position. On the other hand, a more current standard recommends that whenever the work can be performed alternating the standing position with the sitting posture, the workstation must be designed or adapted to favor the alternation of positions [7].

Other comparative studies on the condition of postural alternation between the standing and sitting positions showed the reduction of the discomfort in the upper limbs and in the trunk of workers who performed the postural alternation [24–26]. Seo et al. [24] presented effective pain reduction results in lower limbs during sit-stand posture when compared to the sitting and standing postures.

An alternative to postural adaptation during the tasks is the sit-stand posture, allowing better mobility and allowing greater reach, besides facilitating rapid transitions between the standing and sitting postures [13, 27]. Although, few laboratory studies have analyzed this postural condition and yet there were limitations regarding seat and backrest angulations [24–26].

Chester, Rys and Konz [25] carried out a comparative study analyzing the standing posture, sitting and using a sit-stand chair, with 18 subjects, where the sit-stand chair had characteristics of 12° seat-back inclination and without a lumbar support device, concluding that the sitting posture was comfortable for those who were surveyed, while the standing posture was the most uncomfortable. The respondents noticed discomfort in the thoracic spine, hips and thighs while remaining in the sit-stand posture. The sitting posture is a frequent condition in the daily life of human beings, however, it causes a considerably greater overload on the vertebral discs than the standing posture [28, 29].

Ebara et al. [30] conducted a study with 24 students, comparing three working conditions in a typing task: (1) sitting at standard workstations (standard), (2) sitting on a chair with the work surface elevated to standing position (high-chair) and (3) a combination of 10-min sitting and 5-min standing with the same setting as that in the high-chair condition (sit-stand). An analysis was conducted for subjective musculoskeletal discomfort, subjective sleepiness, sympathetic nerve activity and work performance. The results showed greater discomfort in the forearm/wrist/right hand, thighs and hips during the high-chair condition. Nevertheless, the authors point out the difficulty in finding suitable furniture for this condition. In relation to sleepiness during the task, the standard condition indicated the highest levels, but without statistical significance. In the sympathetic nerve activity criterion, the sit-stand condition demonstrated the highest levels, followed by the high-chair condition. In the work performance, the sit-stand condition showed better results compared to the other situations.

During the execution of a task performed on a bench/table on a horizontal surface the tendency is for the worker to lean forward, resulting in a damaging position for the spine, increasing the internal pressure on the intervertebral discs by 35% when compared to the standing posture [31]. Chaffin, Anderson and Martin [13] show that the change from standing posture to sitting changes the curvature of the lumbar spine, consequently increasing the pressure in the intervertebral discs.

Moro [32] found that furniture that provides angulations greater than 120° in the hip joint, sit-stand posture, allow a better distribution of body weight, reducing discomfort and preventing occupational diseases. Bendix, Winkel and Jessen [33] studied three seat modulations: fixed inclination at +10°, fixed inclination at −5° and mobile inclination between −8° and +19°, which resulted in increased foot edema and seat pressure in the −5° condition.

One study analyzed the posture in three different furniture situations (chair and table) with 37 students [32]. In situation "A", the chair had a 30° (sit-stand posture) angle and the table had 15°. In condition "B", the chair maintained a 30° angle and the table was horizontal. While in the "C", situation the chair and table were set horizontally, and a backrest was added to the spine. The study verified the concentration of weight discharges in the seat support with the use of a load-monitoring prototype in its several support and resting points. The results indicated a greater concentration of load in the seat support during situation "C", resulting in 73.2% of body weight, while situations "A" and "B" presented 51.5% and 50.5% respectively.

ISO 14738 [34] recommends the sit-stand posture in workstations where adequacy to the sitting posture is not available, avoiding the prolonged time in the standing posture. Although the sit-stand chair has no backrest, the posture adopted when using this chair is more natural and distributes 60% of the body weight in the lower limbs [34].

According to Vergara and Page [35], chairs that allow postural mobility of the spine aid in reducing discomfort, and the presence of the backrest contributes to this reduction. But, Moro [32] found a minimum load, ± 1% of the body weight deposited on the backrest of the chair, noting that the subjects of the study leaned forward and moved away from the backrest. The author suggests that this result may have been a consequence of the type of task performed during the experiment – reading.

Oliver and Middledicth [36] claim that the backrest lowers pressure on the intervertebral discs, but will only have an effect when properly supported, avoiding discomfort in the lower back. Therefore, it is important to emphasize the importance of raising awareness among workers about the correct use of the chair and its adjustments, reducing back strain on the lumbar spine. The adaptation of the man-machine-environment system must take place in an efficient manner, reducing the harmful consequences on the worker [10], optimizing the well-being and overall performance of systems [37].

To ensure that the seat and backrest suit the characteristics of each worker, the chair should have adjustment systems that are easy to handle. These characteristics should ensure comfort and mobility, avoiding improper compensation and posture [38].

Neufert [39] highlights the importance of adjusting the height of the backrest, which when positioned at the top, will limit the movement of the scapulae, and when it is put in a much lower position, it will impair the movement of the buttocks. This support should be placed in the lumbar region to achieve a more lordotic curvature, which allows the reduction of stress on the spine [13] but it does not guarantee posture stability.

Regarding the other requirements of the NR-36, the wide variation in height of the Brazilian population implies the need to provide furniture that meets at least 95% of this general population, including the determination of the adjustable height of ergonomic chairs [40]. Regarding the seat of the chair, the requirement of little or no conformation in the base is due to the fact that seats with very defined shapes, in which the buttocks fit, makes posture changes difficult [10].

In addition, another requirement of the seats is that they contain the anterior rounded edge, avoiding compression of the tissues and, consequently, decreasing the venous return in the lower limbs, besides causing discomfort for the workers [41, 42].

To avoid thermal discomfort for the workers in cold environments, the work seats must be made of material that prioritizes thermal comfort. Use of metallic seats should therefore not be allowed [38].

In cases where the worker's feet do not reach the floor, a footrest must be provided. This measure aims to eliminate muscular overloads caused by situations in which the lower limbs hang loose, without any support [38]. Chaffin, Andersson and Martin [13] indicate that the worker should sit with knees bent, feet resting on the floor, transferring body weight to the seat and floor.

Slaughterhouse jobs must comply with the NR-36 requirements, providing workers with adequate conditions to perform their tasks. In relation to the sitting posture and the use of chairs, the interference of the furniture in the human-machine interaction should be highlighted. Thus, it is recommended that companies should provide furniture to enable posture alternation (standing, sitting and/or stand-posture) according to the individual characteristics of workers, in order to reduce ergonomic risks and avoid the onset of musculoskeletal disorders.

5 Conclusion

It is clear from the results of this study that both the conventional and sit-stand chairs met the majority of the NR-36 requirements. They are therefore options for alternating postures. However, it is important to point out the need for workstations to be suitable for sitting and standing postures, allowing the alternation between them, avoiding prolonged time in static postures. In addition, further studies need to be carried out in order to investigate the satisfaction of slaughterhouse workers regarding the use of conventional and sit-stand chairs, to verify their usability, the impact of the backrest on the lumbar region and, also, to analyze the models of chairs used in several slaughterhouses.

References

1. Brazilian Association of Animal Protein (ABPA). História da Avicultura no Brasil. http://abpa-br.com.br/setores/avicultura. Accessed 10 Feb 2018
2. Brazilian Association of Animal Protein (ABPA) (2017) Annual Report 2017. ABPA, São Paulo. http://abpa-br.com.br/storage/files/final_abpa_relatorio_anual_2017_ingles_web.pdf. Accessed 10 Feb 2018
3. Rasotto C, Bergamin M, Simonetti A, Maso S, Bartolucci GB, Ermolao A, Zaccaria M (2015) Tailored exercise program reduces symptoms of upper limb work-related musculoskeletal disorders in a group of metalworkers: a randomized controlled trial. Man. Ther. 20 (1):56–62
4. Caso MA, Ravaioli M, Veneri L (2007) Esposizione a sovraccarico biomeccanico degli arti superiori: la valutazione del rischio lavorativo nei macelli avicoli. Prevenzione Oggi 3(4):9–21

5. Punnett L, Wegman DH (2004) Work-related musculoskeletal disorders: the epidemiologic evidence and the debate. J Electromyogr Kinesiol 14(1):13–23
6. OSHA (2013) Occupational Safety and Health Administration. Prevention of musculoskeletal injuries in poultry processing. https://www.osha.gov/Publications/OSHA3213.pdf. Accessed 22 Feb 2018
7. Brasil (2013) Ministério do Trabalho. Norma Regulamentadora NR-36 - Segurança e saúde no trabalho em empresas de abate e processamento de carnes e derivados. Portaria MTE nº 555, de 18 de abril de 2013, Brasil
8. Brasil (1978) Ministério do Trabalho. Norma Regulamentadora NR 01 – Disposições Gerais. Portaria GM n.º 3.214, de 08 de junho de 1978, Brasil
9. Brasil (1978) Ministério do Trabalho. Norma Regulamentadora NR 17 – Ergonomia. Portaria GM nº 3.214, de 08 de junho de 1978, Brasil
10. Lida L (2005) Ergonomia. Projeto e Produção. 2nd ed. Edgard Blücher, São Paulo
11. Claus AP, Hides JA, Moseley GL, Hodges PW (2009) Is 'ideal' sitting posture real? Measurement of spinal curves in four sitting postures. Man. Ther. 14(4):404–408
12. Kroemer KH, Grandjean E (2005) Manual de ergonomia: adaptando o trabalho ao homem, 5th edn. Bookman Editora, Porto Alegre
13. Chaffin DB, Andersson GBJ, Martin BJ (2006) Occupational biomechanics, 4th edn. Wiley, Hoboken
14. Corlett EN (2006) Background to sitting at work: research-based requirements for the design of work seats. Ergonomics 49(14):1538–1546
15. Keegan JJ (1953) Alterations of the lumbar curve related to posture and seating. J Bone Joint Surg Am 35(3):589–603
16. Grandjean E (1973) Ergonomics of the home. Taylor & Francis, London
17. Grandjean E, Hünting W (1977) Ergonomics of posture — review of various problems of standing and sitting posture. Appl Ergon 8(3):135–140
18. Moro ARP (2000) Análise biomecânica da postura sentada: uma abordagem ergonômica do mobiliário escolar. Doctoral dissertation, Tese (Doutorado em Educação Física) UFSM, Universidade Federal de Santa Maria. Santa Maria (2000)
19. Pynt J, Higgs J, Mackey M (2001) Seeking the optimal posture of the seated lumbar spine. Physiother Theory Pract 17(1):5–21
20. Mörl F, Bradl I (2013) Lumbar posture and muscular activity while sitting during office work. J Electromyogr Kinesiol 23(2):362–368
21. McGill SM, Kavcic NS, Harvey E (2006) Sitting on a chair or an exercise ball: various perspectives to guide decision making. Clin Biomech 21(4):353–360
22. Ruiz RC, Kupek E, Menegon FA, Ruiz PGM (2017) Roteiro para a ação. Revista Proteção 311:60–67
23. Messing K, Tissot F, Stock S (2008) Distal lower-extremity pain and work postures in the Quebec population. Am J Public Health 98(4):705–713
24. Seo A, Kakehashi M, Tsuru S, Yoshinaga F (1996) Leg swelling during continuous standing and sitting work without restricting leg movement. J Occup Health 38(4):186–189
25. Chester MR, Rys MJ, Konz SA (2002) Leg swelling, comfort and fatigue when sitting, standing, and sit/standing. Int J Ind Ergon 29(5):289–296
26. Taillefer F, Boucher JP, Comtois AS, Zummo M, Savard R (2011) Physiological and biomechanical responses of different sit-to-stand stool types on women with and without varicose veins. Le Travail Humain 74(1):31–58
27. Antle DM, Vézina N, Côté JN (2015) Comparing standing posture and use of a sit-stand stool: analysis of vascular, muscular and discomfort outcomes during simulated industrial work. Int J Ind Ergon 45:98–106

28. Harrison DD, Harrison SO, Croft AC, Harrison DE, Troyanovich SJ (1999) Sitting biomechanics part I: review of the literature. J Manipulative Physiol Ther 22(9):594–609
29. Claus AP, Hides JA, Moseley GL, Hodges PW (2016) Thoracic and lumbar posture behaviour in sitting tasks and standing: progressing the biomechanics from observations to measurements. Appl Ergon 53:161–168
30. Ebara T, Kubo T, Inoue T, Murasaki GI, Takeyama H, Sato T, Suzumura H, Niwa S, Takanishi T, Tachi N, Itani T (2008) Effects of adjustable sit-stand VDT workstations on workers' musculoskeletal discomfort, alertness and performance. Ind Health 46(5):497–505
31. Zapater AR, Silveira DM, Vitta AD, Padovani CR, Silva JCPD (2004) Postura sentada: a eficácia de um programa de educação para escolares. Ciência Saúde Coletiva 9:191–199
32. Moro AR (2000) Análise do sujeito na postura sentada em três diferentes situações de mobiliário cadeira-mesa simulado em um protótipo. Revista Eletrônica do Estudo do Movimento Humano, Kinein 1(1)
33. Bendix T, Winkel J, Jessen F (1985) Comparison of office chairs with fixed forwards or backwards inclining, or tiltable seats. Eur J Appl Physiol 54(4):378–385
34. International Standards Organization (2002) Safety of machinery-anthropometric requirements for the design, ISO 14738. International Standards Organization, Geneva
35. Vergara M, Page A (2002) Relationship between comfort and back posture and mobility in sitting-posture. Appl Ergon 33(1):1–8
36. Oliver J, Middleditch A (1998) Anatomia funcional da coluna vertebral. Revinter, Rio de Janeiro
37. IEA - International Ergonomics Association, http://www.iea.cc/whats/index.html. Accessed 14 Feb 2018
38. Brasil (2017) Ministério do Trabalho. Secretaria de Inspeção do Trabalho. Manual de Auxílio na Interpretação e Aplicação da Norma Regulamentadora nº 36: Segurança e Saúde no Trabalho em Empresas de Abate e Processamento de Carnes e Derivados. Brasília - DF
39. Neufert E (2013) Arte de projetar em arquitetura: princípios, normas e prescrições sobre construção, instalações, distribuição e programa de necessidades dimensões de edifícios, locais e utensílios; consultor para arquitetos, engenheiros, aparelhadores, estudantes, construtores e proprietários, 18th edn. Gustavo Gili, São Paulo
40. Brasil (2002) Ministério do Trabalho. Secretaria de Inspeção do Trabalho. Manual de Aplicação da Norma Regulamentadora nº 17: Ergonomia. Brasília - DF
41. Brandimiller PA (1999) Corpo no Trabalho, Guia de Conforto e Saúde para Quem Trabalha em Microcomputadores. Editora Senac, São Paulo
42. Panero J, Zelnik M (2014) Dimensionamento humano para espaços interiores: um livro de consulta e referência para projetos, 1st edn. Gustavo Gili, São Paulo

Simple and Low-Cost Ergonomics Interventions in Isfahan's Handicraft Workshops

Mohammad Sadegh Sohrabi(✉)

Department of Industrial Design, School of Architecture and Urban Design, Art University of Isfahan, Isfahan, Iran
ms.sohrabi@aui.ac.ir

Abstract. Introduction: In developing countries (IDCs), Handicraft is one of the industries in which a major part of the workforce is still contained with high prevalence of WMSDs complaints. Major problems associated with handicraft producing operations, awkward postures in different parts of body (i.e. neck, shoulders, elbows, wrists/hands, upper back, lower back, thigh, knees, and ankles). It is concluded that the high rate of absenteeism has an adverse effect on quality and quantity of production, efficiency of workers and organization. The aim of this study was to implantation of simple and cost-intensive ergonomic interventions based on the ILO ergonomic checkpoints in Isfahan Meenakari pots handicraft workshops.

Methods: This is an interventional study by using ILO Ergonomic checkpoints (Second edition) for assessment ergonomic risk factors in Isfahan handcraft workshops around Imam square. This assessment performed base on checkpoint guideline.

Results: After workplace assessment Base on adjusted checklist four intervention designed and prototyped. Those interventional projects were (1) Bending machine redesigning, (2) Resting facilities designing, (3) Raw material trolley designing, and (4) Hand-tools stand designing.

Conclusion: With practical Checklist guidelines, simple and cost-effective interventions designed to improve the level of ergonomics and the health of workers at handicraft workshops.

Keywords: Handicraft · Ergonomics · Intervention · ILO checkpoints

1 Introduction

Handicrafts are defined as those activities through which some goods are manufactured by hand with certain skills and individual creation as well as enough experience is considered as a perquisite for most of them. Handicraft is mostly carried out individually in small workshops. In this countries, Handicraft is one of the industries in which a major part of the workforce is still contained with high prevalence of WMSDs complaints [1–3].

Unfortunately, small-scale integration (SSI) despite many studies of musculoskeletal disorders in the Iranian labor and industry has been less studied. In general,

S. Bagnara et al. (Eds.): IEA 2018, AISC 825, pp. 60–70, 2019.
https://doi.org/10.1007/978-3-319-96068-5_7

the upper limbs of the body, such as arms and hands, are in many of the main handicrafts that are involved with Upper extremity disorders [4, 5].

A review of the literature shows that there is limited research on the prevalence and occurrence of musculoskeletal symptoms and their contributing risk factors among handicraft workers in Iran. The summary of the results of these studies includes these matters; According to Khandan et al. [4], musculoskeletal disorders are most important occupational illnesses among aged workers in Isfahan handicrafts. 92% of the workers in handy craft workshops are right-handed, and most of the musculoskeletal injuries respectively were observed in lower back 37% and 34% neck regions [4]. Also based on results of another research in 2013 on 100 craftsman working in Isfahan, Iran. The highest percentages of complaints related to severe musculoskeletal discomfort were reported in right shoulder (%36), right wrist (%26), neck (%25), and upper right arm (%24), respectively [6, 7]. Results of research on handicraft workers in Tabriz, Iran showed that overall prevalence of musculoskeletal complaints, particularly in the neck (57.9%), lower back (51.6%), and shoulders (40.5%) were relatively high [8]. In other research on hand-woven shoe-sole making workshops in Iran, the mean severity of low back, shoulder, and upper back pain was higher than in other parts of the body. Therefore, handicraft work can be considered a job with a high risk of MSDs [9].

Major problems associated with handicraft producing workers, awkward postures in multiple parts of body (i.e. neck, lower back, shoulders, elbows, wrists/hands, upper back, knees, and ankles). It is concluded that the high rate of absenteeism has an adverse effect on quality and quantity of production, efficiency of workers and organization and many work-related health problems [2, 3, 7].

Nowadays using ergonomic checkpoints, checklists and guidelines in small enterprises construction sites, handicraft workshops and agriculture in industrially developing countries is very common to improve workers health and increase productivity [10]. One of these beneficial checkpoints is International Labour Office (ILO) ergonomic checkpoints that included practical and easy-to-implement solutions for improving safety, health and working conditions. Using its action-checklist and obvious illustrations and information, the ILO undertook participatory ergonomics training to small enterprises, home workers and farmers [11].

"Ergonomic checkpoint based on numerous examples of practical ergonomic improvements achieved at low cost. There are many such examples worldwide, including ergonomically designed tools, carts, materials-handling techniques, workstation arrangements, work environments, worksite welfare facilities and group work methods. This guideline presents 132 ergonomic interventions aimed at creating positive effects without relying on costly or highly sophisticated solutions. The emphasis is on realistic solutions that can be applied flexibly manner and contribute to improving working conditions and productivity". This action-oriented ergonomics tools can be used in varied large and small industries in nine section include [12];

1. Materials storage and handling (checkpoints 1–17)
2. Hand tools (checkpoints 18–31)
3. Machine safety (checkpoints 32–50)
4. Workstation design (checkpoints 51–63)

5. Lighting (checkpoints 64–72)
6. Premises (checkpoints 73–84)
7. Hazardous substances and agents (checkpoints 85–94)
8. Welfare facilities (checkpoints 95–105)
9. Work organization (checkpoints 106–132)

An ergonomic checkpoints have been used in previous interventions and have been approved as a useful and effective tool. This checklist and its guideline used in Helali and Shahnavaz [13], Helali [14], Motamedzade [15], Ahmadi et al. [16], Ishkandar and Shamsul [17] studies which has largely produced useful results on the objectives of ergonomic interventions through the provision of practical and easy-to-use ergonomic solutions for improving working conditions and productivity for making workplace changes particularly for use in Industrially Developing Countries (IDCs). It is a valuable tool for identifying simple, practical and inexpensive solutions to ergonomic problems, especially in small and medium-sized companies.

To improvement of the ergonomic situation in handicraft workshops researchers recommended to optimize workstations, automation of some activities, design appropriate tools, pay attention to working hours and resting times, training and raise awareness about how the work is done correctly [4, 7, 9]. This study aimed to implantation of simple and cost-intensive ergonomic solutions based on the ILO ergonomic checkpoints in Isfahan Meenakari pots handicraft workshops.

2 Methods

2.1 Study Design

This interventional study performed in Feb–Jun 2017 in Isfahan, Iran. Most of the handicraft workshops in Isfahan are around the Naqsh-e Jahan square. First, all Meenakari pots handicraft workshops around Naqsh-e Jahan Square Observed and then by using a short version of the ergonomic checklist [12], high risk workshops were selected. Due to the risk factors of the ergonomic checklist book, workshops with high force or repetitive motions were selected to continue the study. In these workshops, the main body of Mina's pot was made from a copper sheet with a rotating hand-held process. Then, in the selected workshops, Ergonomic evaluation was performed based on the full version of ergonomic checklist [12] and the main risk factors were determined. Finally, considering the risk factors identified in different areas, four intervention designed and prototyped. Extended information of study process showed in Fig. 1.

2.2 Study Population and Sample

The number of active Meenakari pots made handicraft workshops around the Naqsh-e Jahan Square are about 63 workshops. After the preliminary review by observations and using short version of ergonomics checklist, 14 workshops were selected for more extended evaluation and intervention. These 14 workshops selected based on the priority specified in Ergonomic checkpoints manual, severity of ergonomic risks evaluated

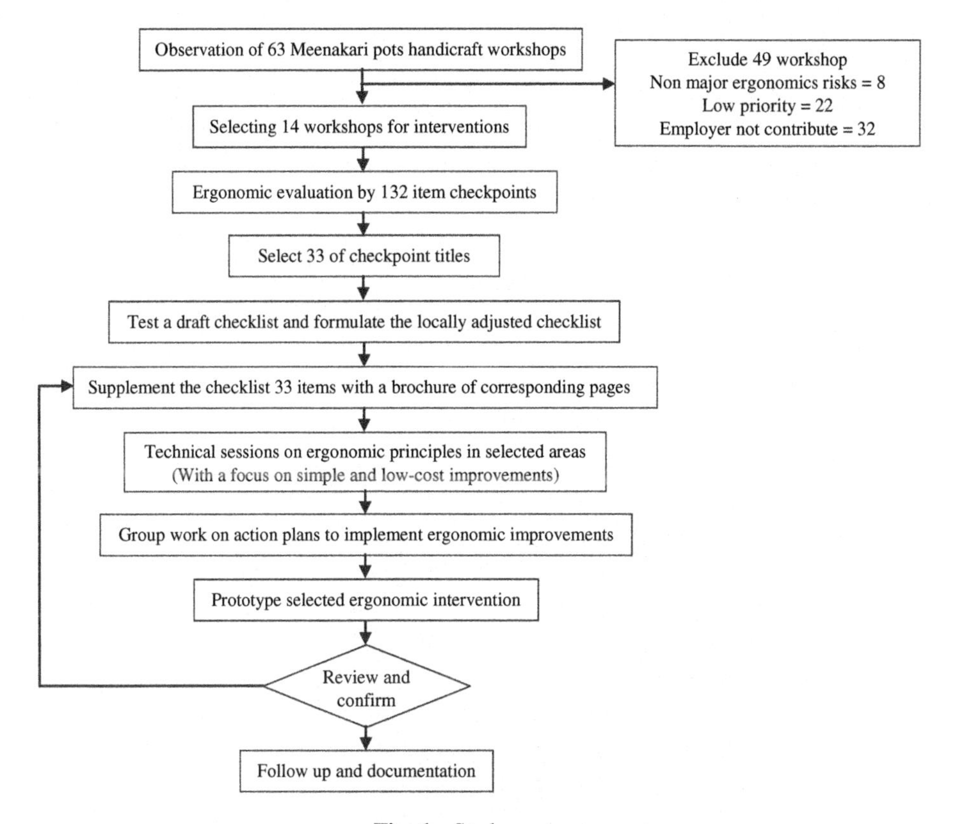

Fig. 1. Study process

and participation of workshop employers in ergonomic intervention program. Excluded workshops contains low intervention priority, non-major ergonomics risks or employer not contribute in research program.

2.3 Materials and Tools

The main tool used for collecting ergonomics information of workshops and conducting an evaluation was ILO Ergonomic checkpoints (Second edition) published in 2010 [12]. The Persian version of this book has been used for evaluations. Applying selected checkpoints to the workplace done based on ergonomics guidelines and illustrations [11, 12]. Also design of ergonomic interventions developed based on the book guidance, brochure-type information sheets and Individual Consciousness. Other tools used in this study include a camera for recording work environment images and non-structured interviews to understanding the status of workshops and workers health problems.

3 Results

After observations and quick ergonomic assessment of workshop to find major risk that treat health, safety and productivity draft checklist build and formulate. This 33 items checklist based on brochure of corresponding pages in ILO ergonomic checkpoints book. According to workshops condition main topic of ergonomics risk factors was related to materials storage and handling, Hand tools, Machine safety, Workstation design and Welfare facilities issues. As for Table 1, selected 33 items checklist guided to four intervention projects; (1) Bending machine redesigning, (2) Resting facilities designing, (3) Raw material trolley designing, and (4) Hand-tools stand designing.

This intervention topics came from selected checklists outcome and technical sessions on ergonomic principles in selected areas in workshops locations. Current view of common handicraft workshop showed in Fig. 2.

Fig. 2. Meenakari pots handicraft workshops, current workstation

Technical sessions done with a focus on simple and low-cost improvements in next step. In this sessions participants of stockholders argue about problems and solutions. The limitations of the solutions included practicability, low cost, and preserve nature of the handicrafts workshop. The nature of handicrafts in these workshops is very important and eliminates the possibility of industrial interventions such as automation and so on.

Fig. 3. Interventional projects. Up left; Bending machine redesign. Up right; Resting facilities. Bottom left; Raw material trolley. Bottom right; Hand tools stand

3.1 Bending Machine Redesigning

The objectives of the project include modifying the bending machine's on/off button, guarding rotating parts of machine, adding a local lighting system, adding an oil vapor absorption system, and adding an auto-absorbing system to the copper swarfs. At the time of running the study, details of the redesign and changes were made on the machine, and a technical plan and guide for making modification were provided to the workshop employers. The 3D view of the device is shown in the Fig. 3A.

3.2 Resting Facilities Designing

The purpose of this intervention was to create a suitable area for rest and food and drink in the workplace. The limitations of the solutions were the lack of space at the workshop and the need for the closest resting place to work and design with the least cost. In this project, a table and portable chair system were designed and implemented to set up a resting place when needed. The execution of this system is shown in the Fig. 3B.

3.3 Raw Material Trolley Designing

The main goal of this project was to facilitate the transportation of raw materials in the workshop and storage of manufactured pots. Using ILO checkpoints guides and sympathy workers, a trolley was designed to carry copper sheets in the workshops and

Table 1. Adjusted checklist for ergonomic evaluation and selecting related projects

Sections	Selected items	Related to project
Materials storage and handling		
	1.Improve the layout of the work area so that the need to move materials is minimized	Raw material trolley designing
	2.Use carts, hand-trucks and other wheeled devices, or rollers, when moving materials	
	3.Use mobile storage racks to avoid unnecessary loading and unloading	
	4.Use multi-level shelves or racks near the work area to minimize manual transport of materials	
	5.Reduce manual handling of materials by using conveyers, hoists and other mechanical means of transport	
	6.Instead of carrying heavy weights, divide them into smaller lightweight packages, containers or trays	
	7.Provide handholds, grips or good holding points for all packages and containers	
	8.Move materials horizontally at the same working height	
	9.Eliminate tasks that require bending or twisting while handling materials	
	10.Keep objects close to the body when manually handling materials	
Hand tools		
	11.Select tools designed for the specific task requirements	Hand-tools stand designing
	12.Provide safe power tools and make sure that safety guards are used	
	13.Use hanging tools for operations repeated in the same place	
	14.Minimize the weight of tools	
	15.For hand tools, provide the tool with a grip of the proper thickness, length, shape and size for easy handling	
	16.Provide hand tools with grips that have adequate friction or with guards or stoppers to avoid slips and pinches	
	17.Provide a "home" for each tool	
	18.Provide enough space for stable postures and stable footing during power tool operation	
Machine safety		
	19.Design controls to prevent unintentional operation	Bending machine redesigning
	20.Make emergency controls visible and easily accessible from the natural position of the operator	
	21.Make different controls easy to distinguish from each other	

(*continued*)

Table 1. *(continued)*

Sections	Selected items	Related to project
	22.Make sure that the worker can see and reach all controls comfortably	
	23.Use natural expectations for control movements	
	24.Purchase machines that meet safety criteria	
	25.Use properly fixed guards or barriers to prevent contact with moving parts of machines	
Workstation design		
	26.Place frequently used materials, tools and controls within easy reach	Hand-tools stand designing
Lighting		
	27.Provide local lights for precision or inspection work	Bending machine redesigning
	28.Relocate light sources or provide shields to eliminate direct and indirect glare	
Premises		
	29.Install effective local exhaust systems that allow efficient and safe work	
	30.Recycle wastes to make better use of resources and protect the environment	
Hazardous substances and agents		
	31.Isolate or cover noisy machines or parts of machines	
	32.Ensure safe wiring connections for equipment and lights	
Welfare facilities		Resting facilities designing
	33.Provide rest facilities for recovery from fatigue	

temporarily store the pots. The dimensions of this trawl are 55 × 45 × 91 cm. In the Fig. 3C 3D view and prototype of trolley showed.

3.4 Hand-Tools Stand Designing

There is main design requirements of hand tools stand; Arranging tools for working with bending machine are available in such a way that the manual tools were in most accessible to the worker with minimal bending and twisting of the worker to the place of the tools. This stand was designed according to the location of the worker related to bending machine, anthropometric data of the employed workers, and the shape and size of the tools. It is shown in the Fig. 3D in 3D view and a prototype constructed.

4 Conclusion

In this study, the purpose is using a simple and useful text tools to create ergonomic interventions in the workshops of Isfahan's Naqsh-e Jahan Square by focusing interventions on simple and low-cost solutions. These interventions were conducted by sharing the principles and solutions of ergonomics with the stakeholders and workers of the handicraft workshops. Using checklist tips, all related items and ergonomics risk factors were recorded. In meetings selected items with high priority grouped together, afterward they became problematic issues for creating intervention projects.

In the present study due to the time constraints and the lack of implementation of all interventions until deadline of research, the feasibility of the effects of health, comfort and productivity of workers is not evaluated. Notwithstanding, in the initial observations that were carried out after the evaluation of prototypes in handicraft workshops, the employee satisfaction has increased and the lags and lost time have decreased. In another similar study on the redesign of handicraft tools in India, the level of comfort was 42.69% and the productivity increased by 29.62%. This suggests that it could directly improve the level of comfort and productivity after the implementation of ergonomics design interventions in handicrafts enterprises [18]. In this study, ILO ergonomic checkpoints were used as a guide and an action-oriented ergonomics tools qualitative method. This tool also allows for quantitative measurement, and has also been developed as a weighted Checkpoints by Ahmadi et al. that can be used in studies using a quantitative study [16].

Several previous interventions have been used successfully of ILO ergonomic checkpoints. In this studies, this guideline has been used to lead the process of making solutions and managing the project. Motamedzade's study in Iranian Tire Manufacturing Industry [15] and Ishkandar and Shamsul's study in Malaysia Melaka Oil Palm Plantation [17] and Helali's study in Gilan, Iran [14] can be noted.

In small-scale integration (SSI) or small and medium-sized enterprises (SMEs) the most important issue is the need for adopting participatory approach as part of resolving ergonomics problems at work place. Also, focusing on low-cost and cost-effective solutions is very effective in ergonomic programs [5, 19]. In this study, despite the simplicity of the research process or cost of interventions, qualitative results have been effective on the results of the work.

Finally, the positive aspects of using checkpoints are simplicity of concepts, ease of implementation, and the span of evaluation aspects. In addition to the guidance given in this checkpoints, can plan and implementation ergonomic interventions according to the condition of the workplace. Also, this method didn't need very high level science and will be train with minimal instruction and use of illustrations for each content. Eventually, conducting participative ergonomic studies in small industries or handicraft workshops requires an ergonomist with leadership and facilitation skills. The role of ergonomist facilitators in the project is to ensure that interventions are speeded up and that targets at workers' health and productivity.

Acknowledgments. This article was extracted from the ergonomics class work of Industrial Design Bachelors in the Art University of Isfahan in the second semester of 2016–2017 that conducted by the author.

Conflicts of Interest. None declared.

References

1. Durlov S, Chakrabarty S, Chatterjee A, Das T, Dev S, Gangopadhyay S et al (2014) Prevalence of low back pain among handloom weavers in West Bengal, India. Int J Occup Environ Health 20(4):333–339
2. Meena M, Dangayach G, Bhardwaj A (2012) Occupational risk factor of workers in the handicraft industry: a short review. Int J Res Eng Technol 1(3):194–196
3. Meena M, Dangayach G, Bharadwaj A (eds) (2011) Impact of ergonomic factors in Handicraft Industries. In: Proceedings of the international conference of mechanical, production and automobile engineering
4. Khandan M, Koohpaei A, Talebi F, Hosseinzadeh Z (2015) Assessment of ergonomic risk factors of repetitive tasks among handicraft workers in Isfahan, Iran, using the assessment of repetitive tasks tool. Health Syst Res 11(4):713–718
5. Sain MK, Meena M (2016) Occupational health and ergonomic intervention in Indian small scale industries: a review. Int J Recent Adv Mech Eng 5(1):13–24
6. Shakerian M, Rismanchian M, Khalili P, Torki A (2016) Effect of physical activity on musculoskeletal discomforts among handicraft workers. J Edu Health Promot 5:8
7. Shakerian M, Rismanchian M, Torki A, Fadaee P, Alian M, Saeedi M (2014) The evaluation of the effect of physical activity on musculoskeletal discomforts among one of the handicraft industries. Health Syst Res 10(3):587–598
8. Dianat I, Karimi MA (2016) Musculoskeletal symptoms among handicraft workers engaged in hand sewing tasks. J Occup Health 58(6):644–652
9. Veisi H, Choobineh A, Ghaem H (2016) Musculoskeletal problems in iranian hand-woven shoe-sole making operation and developing guidelines for workstation design. Int J Occup Environ Med 7(2):87–97
10. Kogi K (2007) Action-oriented use of ergonomic checkpoints for healthy work design in different settings. J Hum Ergol (Tokyo) 36(2):37–43
11. Kawakami T, Kogi K (2005) Ergonomics support for local initiative in improving safety and health at work: international labour organization experiences in industrially developing countries. Ergonomics 48(5):581–590
12. International Labour Office (2010) Ergonomic checkpoints: practical and easy-to-implement solutions for improving safety, health and working conditions. Second edition, Geneva
13. Helali F, Shahnavaz H (1996) Ergonomics intervention in industries of the industrially developing countries. Case study: Glucosan—Iran. In: Proceedings of human factors in organizational design and management (ODAM)—V Amsterdam, The Netherlands, North-Holland, vol 141, no 6
14. Helali F (2009) Using ergonomics checkpoints to support a participatory ergonomics intervention in an industrially developing country (IDC)-a case study. Int J Occup Saf Ergon 15(3):325–337
15. Motamedzade M (2013) Ergonomics intervention in an Iranian tire manufacturing industry. Int J Occup Saf Ergon 19(3):475–484

16. Ahmadi M, Zakerian SA, Salmanzadeh H (2017) Prioritizing the ILO/IEA ergonomic checkpoints' measures; a study in an assembly and packaging industry. Int J Ind Ergon 59:54–63
17. Ishkandar M, Shamsul B (2016) Developing ergonomics intervention for improving safety & health among smallholders in melaka oil palm plantation: a participatory action oriented approach. Malays J Hum Factors Ergon 1(1):36–42
18. Meena M, Dangayach G, Bhardwaj A (2015) An ergonomic approach to design hand block tool for textile printing handicraft industry. Int J Recent Adv Mech Eng 4(4):19–30
19. Hermawati S, Lawson G, Sutarto AP (2014) Mapping ergonomics application to improve SMEs working condition in industrially developing countries: a critical review. Ergonomics 57(12):1771–1794

Low Back Biomechanics of Keg Handling Using Inertial Measurement Units

Colleen Brents[1(✉)], Molly Hischke[1], Raoul Reiser[2], and John Rosecrance[1]

[1] Department of Environmental and Radiological Health Sciences, Colorado State University, Fort Collins, CO 80521, USA
Cbrents1@rams.colsotate.edu

[2] Department of Health and Exercise Science, Colorado State University, Fort Collins, CO 80521, USA

Abstract. Workers handling beer kegs experience risk factors associated with occupational low back pain (heavy loads, awkward trunk postures). Many breweries are small and lack resources for materials handling equipment, causing much work to be done manually (including keg handling).

Measuring worker motions (kinematics) may aid job design to reduce risk factors associated with lifting. Low back motions may be evaluated using observation techniques and devices which are oftentimes inconvenient in the field. Application of wireless inertial measurement units (IMUs) for human motion provides whole body kinematics information, including low back.

The present study used a 3-dimensional motion capture system to investigate low back kinematics during keg handling at a Colorado brewery. Specifically, five workers lifted spent kegs onto a clean and fill line. Workers wore 17 IMUs as they handled kegs. Low back angular displacements were assessed during keg handling at two heights (low, high). Repeated measures analyses were performed with each trunk angular displacement variable as a function of lift condition.

Differences in low back kinematics between lift conditions were identified. During low lifts, torso flexion was significantly greater than high lifts. A broader range of angular displacements was observed in high lifts. Data collection was feasible during operational hours due to IMU's small design. Data collected from experienced workers provided researchers with information directly applicable to keg handling in small breweries. Results from the study can help improve workplace design in a craft brewery, reduce risk, and create safer work.

Keywords: Craft beer industry · Manual materials handling · Low back pain Inertial measurement units

1 Introduction

Disability from low back pain (LBP) accounts for one third of disabilities from work related risk factors, according to the Global Burden of Disease 2010 (Driscoll et al. 2014). In 2015, occupational low back injuries in the U.S. had an overall incidence rate of 17.3 per 10,000 full time workers (Bureau of Labor Statistics 2016). Furthermore,

S. Bagnara et al. (Eds.): IEA 2018, AISC 825, pp. 71–81, 2019.
https://doi.org/10.1007/978-3-319-96068-5_8

the overall economic impact of LBP in the U.S. (direct costs such as lost wages and rehabilitation plus indirect costs such as retraining, and reduced efficiency and quality) exceeded $200 billion annually (Dagenais et al. 2008; Marras et al. 2009).

Low back pain is the most common work-related illness in the U.S., and its causation has been extensively researched. Despite the abundance of these investigations, the causes of occupational LBP in terms of exposure are still poorly understood in many occupational research groups (Amell et al. 2001; Gatchel and Schultz 2014; Jones and Kumar 2001; Putz-Anderson et al. 1997). Numerous studies have identified an association between low back injuries and manual materials handing (MMH) tasks (Faber et al. 2007; Lavender et al. 2012; Magora 1975; Marras 1993; van Dieën et al. 2010; Zurada et al. 2004). Awkward postures, long durations, and high frequency of MMH are occupational risk factors associated with LBP (Basahel 2015; Coenen et al. 2014; de Looze et al. 1994; Potvin 2008).

The craft brewing industry is a growing area that relies on manual materials handling, especially small breweries. Within the last ten years, the number of breweries in the U.S. has more than doubled, with over 6,000 breweries in 2017 (Brewers Association 2017). Colorado has the second most craft breweries in the U.S. with over 300 in 2016 (Brewers Association 2016). Craft breweries start as small businesses without resources to invest in expensive automated materials handling equipment, thus, much of materials handling is performed manually by workers. To date, there have been few studies related MMH or LBP in the brewing industry (Abaraogu et al. 2015; Jones et al. 2005). As of 2015, brewery injuries averaged 4.2 injuries per 100,000 h (Brewers Association 2016). The lack of studies and presence of occupational injuries in brewing industry presents a research opportunity on MMH and LBP.

Kegs, a container for beer, are one example of commonly handled materials in craft breweries. Half-barrel kegs were the most commonly used kegs at the craft brewery in the present study. With a capacity of 58.7 L (15.5 U.S. gallons), half-barrel kegs are 59.0 cm high, 41.0 cm wide, and weigh between 13.6 to 72.6 kg (60 to 160 lbs.) when empty or filled with beer. Kegs must be cleaned before they can be refilled with beer.

One approach to study LBP is to investigate the postures and motions (kinematics) workers experience during actual tasks. A way to quantify kinematics in an occupational setting is with 'wearable technology.' Lab studies of occupational tasks are limited in their ability to represent the actual work practices, demands, conditions and environment. Inertial measurement units allow collection of continuous human motion data during work tasks in the actual work environment. Inertial measurement units have been validated to measure human motion, with an emphasis on MMH (Faber et al. 2016; Kim and Nussbaum 2013; Schall et al. 2015). Characterizing trunk motion could be used to improve workplace design, reduce risk, and create a safer workplace in a craft brewery during tasks involving keg handling. The purpose of this research was to assess trunk posture using sternum and pelvic sensor data when craft brewery workers transferred spent kegs from incoming pallets onto a cleaning and filling conveyor.

2 Methods

2.1 Subjects

Five male employees at a craft brewery who regularly operated the kegging line voluntarily participated in this institutional review board-approved cross-sectional study after providing written informed consent. Participants self-reported that they were free of LBP at the time of study.

2.2 Keg Line Process

At the craft brewery used in the present study, kegs were manually lifted and flipped from incoming pallets onto a conveyor roller. Due to the design of the conveyor roller, kegs were placed upside down. Kegs were washed externally, then internally, and finally filled with beer. During an eight-hour shift, a worker may handle as many as 600 or as few as 100 kegs depending on the packaging demands of the scheduled beer. Kegs are transported around the brewery arranged eight to a pallet, often stacked two high (the height of the keg handle at 69.8 or 139.5 cm).

2.3 Instrumentation and Procedure

Study participants were fitted with seventeen IMU sensors (Xsens, Enschede, NT). Worker setup and calibration was done in accordance with the Xsens MVN system manual in an area with minimal ferromagnetic interference. The IMU suit was worn for approximately 90 min while subjects operated the kegging line during operational hours of the craft brewery. Data for was recorded in MVN Studio BIOMECH software

Fig. 1. Workstation setup of the kegging line includes the conveyor roller (A) and incoming spent kegs on a pallet. Kegs are stacked two high on the pallet, high (B) and low (C).

on an HP laptop. Video was simultaneously recorded for the 90 min during keg handling tasks. Due to the variability in the number of kegs handled, present study participants were recorded while they handled 64 kegs, or 32 kegs from each lift condition, which took approximately 90 min. A keg lift was defined as the point of initial grasp to final release of the worker's hands when transferring the keg from the pallet the conveyor. Keg origin height was tagged as a high or low, as shown in Fig. 1.

2.4 Data Processing and Analysis

2.4.1 Peak and Average Trunk Angular Displacement

Lift height condition and initiation/termination time points were identified using the video recordings and sensor fusion data for each subject. In Xsens MVN Studio, data was recorded in Euler angles and exported in quaternions form, which was converted into rotation angles using Matlab 2017b (Roetenberg et al. 2013). Using a custom R code in RStudio Version 1.0.136, 2016, lift time points and rotation angle data were combined to identify descriptive statistics of the trunk angular displacement during keg lifts (R Core Team 2016). Trunk angles were calculated from pelvic and sternum sensor data. Maximum values corresponded to flexion in the sagittal plane. Minimum values corresponded to extension (or minimal flexion) in the sagittal plane. Averages of trunk angular displacement in each plane were identified for all lifts under both height conditions. Evaluating trunk positions of workers assists researchers, clinicians, and other stakeholders to characterize the spectrum of work-related postures for the different lift conditions.

2.5 Statistical Analysis

The data were determined to be normally distributed, therefore no data transformation took place prior to statistical analysis. The present study had equal sample sizes between high and low lifts.

A repeated measures design with replicates and univariate analysis of variance (ANOVA) test was performed for each torso angular displacement as a function of lifting height origin. The lift height was the condition (low and high) and the within-subject factor. The interaction between subject and lift condition was investigated. Post hoc analysis consisted of least square means (*Lsmeans*) to examine significant effects. Significance was determined if $p \leq \alpha = 0.05$.

3 Results

All five male workers at the craft brewery who regularly operated the keg line participated and were free of LBP at the time of the study. Anthropometric information inputted into the biomechanical model is shown in Table 1. Two keg lift height conditions were analyzed in the present study and sample graphs of trunk angular displacement are shown in Figs. 3 and 4. Average trunk flexion was significantly greater during low lifts compared to high lifts by an average magnitude of 4.2° ($p = 0.03$).

Minimum flexion was significantly greater during low lifts compared to high lifts, with participants experiencing an average of 3.34° of trunk extension ($p = 0.04$). While not significant, overall magnitudes of maximum flexion were larger while workers handled kegs from low heights. While the differences between maximum flexion were not significant, the confidence interval for high lifts was larger, suggesting there is more variability in trunk motion when workers are transferring kegs from high levels. Estimated trunk angular displacement is shown in Table 2. The difference between high and low keg lift durations was not significant ($p = 0.46$).

Table 1. Anthropometric data of study participants. (All variables except age and body mass, were included in the sensor fusion algorithm for the IMU system)

Variable	Units	Mean	SD	Minimum	Maximum
Age*	*years*	31.0	4.5	28.0	39.0
Body mass*	*kg*	5.9	74.8	78.8	88.9
Body height	*cm*	179.0	5.6	173	186
Foot size	*cm*	31.6	1.3	30.3	33.0
Arm span	*cm*	174.5	4.8	168	181.0
Ankle height	*cm*	10.7	1.0	9.5	12.0
Hip height	*cm*	93.4	3.4	90	99.0
Hip width	*cm*	27.5	1.9	26	30.5
Knee height	*cm*	51.2	2.6	47.5	54.5
Shoulder width	*cm*	36.4	5.3	27	40.0
Shoe sole height	*cm*	3.7	1.1	2.0	4.8

Table 2. Estimated angular trunk displacement in both keg lift height conditions. Values reported in angles as mean (lower, upper confidence interval). Significant values have *

Motion	High	Low	p-value
Average flexion	15 (9, 21)	19 (12, 27)	0.03*
Maximum flexion	35 (23, 47)	40 (32, 47)	0.19
Minimum flexion	−1 (−7, 5)	−3 (−10, 4)	0.04*

4 Discussion

The investigators utilized IMUs to characterize trunk motion in keg handling from high and low lifts to improve workplace design and reduce injury risk of the workers. Lift height impacted the amount of flexion experienced by the worker during keg handling.

4.1 Trunk Postures

Workers' average trunk posture in the sagittal plane was significantly greater during low lifts compared to high lifts ($p > 0.05$). When lifting kegs from a low height, workers experienced an average of four degrees of additional forward flexion. Kegs originating from low positions are 69.8 cm lower than their high counterparts (the height of a keg on a pallet), therefore requiring the worker to bend farther forward when initiating the lift.

Magnitudes of trunk extension were observed to be significantly different between high and low keg lifts ($p > 0.05$). Workers lifting kegs from low heights on average experienced trunk extension three degrees greater than when lifting from high locations. Momentum could be factor when workers are lifting kegs, and trunk extension is a way to maintain this energy.

While the differences were not significant ($p > 0.05$), the confidence interval of trunk maximum flexion was broader during high lifts than low lifts, suggesting workers can experience a more diverse range of motion in the sagittal plane.

Available research on trunk posture exposure during brewing tasks is limited, but measurements collected during keg line operation can be compared to other studies. Trunk motion has been investigated in other industries with known associations to work related musculoskeletal disorders (including LBP). Keg line operators lifting kegs from both height conditions experienced similar peak flexion when compared to a database of 17 distribution centers (37°, Marras, Lavender, Ferguson, Splittstoesser, and Yang, Marras et al. 2010). Keg line operators experienced less peak sagittal flexion than Southeastern reforestation planters (75°, Granzow et al. 2018). Peak sagittal flexion in brewery workers was greater than registered nurses (36°, Schall et al. 2016).

4.2 Limitations

Only five young healthy men participated in the present study. While the participants represented all the keg line operators at the craft brewery when the study was conducted, the results may not be generalizable to MMH workforce. Data was collected at a craft brewery that utilized both manual and automated processes. The keg line required kegs to be manually loaded then were automatically cleaned, filled, and palletized. Therefore, postural patterns may not be generalizable to the full range of kegging demands regarding craft brewery equipment variability.

To minimize task interference, the order in which the worker lifted kegs from the pallet was not controlled. Lift strategy could vary with a keg's horizontal location on the pallet, but only vertical orientation was considered in the present study.

Magnetic immunity due to the IMU algorithm enabled data to be collected in an industrial setting (Roetenberg et al. 2013). Since the completion of this study, updated algorithms have been released with greater levels of magnetic immunity. While operating the kegging line operators occasionally had to check tank pressure (located across the brewery). When the worker walked to the tanks, they would exceed the range of the antennae. The hoses and tanks presented additional magnetitic interference that compromised the accuracy of the IMU system. Due to these circumstances, recording was paused and resumed when the worker returned to the kegging line. Future data collection under these conditions, with the updated algorithm, would be able to follow the worker throughout the brewery. The IMU system and sensor fusion algorithm approximate body segment kinematic data based on sensor data and a biomechanical model. While validation studies have been conducted and established the accuracy of the IMU system, there are inherent limitations in the sensors and biomechanical models (Banos et al. 2014; Schall et al. 2016; Schall et al. 2015). Despite these limitations, the portability and small design of IMUs enables researchers to conduct studies during operational hours in the field, thus increasing ecologic validity (Allread et al. 2000).

Although the present study had limitations, the results provided valuable insight into torso kinematic demands of a keg handling task in a craft brewery. Given the positive reception to the IMUs use in breweries, further research is hopeful.

4.3 Future Directions

Results of the present study introduced baseline information on the trunk motion calculated from IMU sensor data for the sternum and pelvis during keg handling tasks. The data generated during this study can be used for multiple studies, from trunk kinematics as a function of lift weight to a characterization of other brewery tasks to other anatomical biomechanics studies.

Other studies of trunk motion in industries quantify the amount of time spent in 'extreme' (at or greater than 45° of flexion) postures. A study on hand planters in reforestation projects observed sagittal trunk motion and calculated percent time extreme flexion (Granzow et al. 2018). While the mean maximum sagittal flexion of brewery workers conducting kegging tasks was less than 45°, the values of the upper confidence intervals could be considered 'extreme'. This lack of 'extreme' forward flexion could be due to the design of kegs (the handle is 59.0 cm high which enables workers to grasp the keg without reaching all the way to the floor). Other tasks in breweries (from transporting pallets to handling raw ingredients) might require greater levels of trunk flexion. Future studies could investigate trunk posture across a variety of brewing-related tasks.

Future studies could also address additional factors related to occupational LBP other than trunk posture. Characteristics of the load (such as weight) influence the magnitude of compressive and shear forces generated about the low back. Compressive and shear forces at the lumbosacral joint have been correlated to LBP, and NIOSH has recommended action limits to reduce the risk of developing LBP (Marras 1993; Marras and Granata 1995; Waters et al. 1993). Keg weight was assumed to be constant at 13.6 kg (empty keg weight) for the present study. Lifting kegs with residual beer were up to the worker's discretion. If they determined that they could safely complete the lift with some amount of beer residue, the data were included in the analysis. Because of this, the true weight of the keg could have exceeded the assumption of 13.6 kg and workers' trunk motions could have changed to accommodate heavier loads. Lifting unexpectedly heavy loads has been suggested to be associated with LBP, but research is ongoing (Van Der Burg et al. 2000; Van der Burg and Van Dieën 2001). Future studies could consider how different keg weights impact the torso joint angles and peak loading.

Previous studies have estimated compressive and shear resulting forces during occupational task simulations in lab settings. Portable force plates (van Dieën et al. 2010) and in-shoe pressure sensors (Khurelbaatar et al. 2015) have been used to achieve this. Recently, an inverse dynamics model was developed that combines IMU kinematic data and ground reaction force and moment model to predict internal resulting forces (Faber et al. 2016; Karatsidis et al. 2018). The ability to estimate trunk and spinal kinetics in the field will increase our understanding of how work factors can contribute to occupational LBP.

There are several practical factors to consider regarding the feasibility of using IMUs and a biomechanical model in the field. The cost of the full body IMU system can be a significant barrier to implementation. The wireless sensors and biomechanical model are significantly more expensive than single IMU worker devices (Lee et al. 2017; Schall et al. 2015). Future studies should investigate reliability of simplified multiple IMU systems to measure specific body regions (Kim and Nussbaum 2013; van Dieën et al. 2010; Vignais et al. 2013). Alternatively, the full body suit generates large, vast amounts of data. Task-specific information captured with full body IMU systems in the field could generate databases that can be used to characterize multiple components of the body during a work task.

The scope of this project was to use IMUs to measure trunk motion in the sagittal plane. Given the available developing technology, the continuation of measuring worker exposures in the natural work environment is hopeful.

5 Conclusions

The craft brewery used in the present study is in the process of redesigning the kegging line to reduce the musculoskeletal demands. Results from this study were shared to assist in that design process. Since completion of this study, the brewery installed additional safety mirrors above the pallet delivery area and increased the forklift traffic for delivering pallets of used kegs.

Craft breweries are a rapidly growing, yet under-studied industry. Using IMUs, we effectively characterized low back angular displacement of kegging tasks. Specifically, we compared trunk motion when workers handled kegs from different pallet heights. While the overall duration of the handling task is low compared to a complete eight-hour shift, the workers are exposed to angular displacements and estimated forces associated with a risk of developing LBP. Due to the portability and simplicity of the IMU system, investigators will be able to return and measure the impact of the intervention in the future. Body and joint kinematic information can be used to improve workplace design in craft breweries, reduce risk, and create safer work.

Many risk factors associated with occupational LBP exist in MMH jobs. Effectively quantifying trunk postures during task performance helps researchers understand how work design influences worker movement. Inertial measurement units are lightweight, noninvasive instruments that, when secured to a worker, record movements without disrupting their natural movements. When combined with a biomechanical model, IMUs' coordinate data can be used to estimate whole body kinematics but the reliability of sensor fusion algorithms and filters for IMUs continues to be a challenge. Current validation studies, algorithm developments, and high portability suggest that IMUs continue to be applied in future studies. While the scope of this study was trunk flexion, many other approaches and levels of analysis are possible.

Acknowledgements. The authors declare no conflict of interest. This study was funded under Mountains and Plains Education Research Center, NIOSH Grant T42OH009229.

References

Abaraogu U, Okafor U, Ezeukwu A, Igwe S (2015) Prevalence of work related musculoskeletal discomfort and its impact on activity: a survey of beverage factory workers in Eastern Nigeria. Work 52:627–634. https://doi.org/10.3233/WOR-152100

Allread WG, Marras WS, Burr DL (2000) Measuring trunk motions in industry: variability due to task factors, individual differences, and the amount of data collected. Ergonomics 43:691–701. https://doi.org/10.1080/001401300404670

Amell TK, Kumar S, Rosser BWJ (2001) Ergonomics, loss management, and occupational injury and illness surveillance. Part 1: elements of loss management and surveillance: A review. Int J Ind Ergon 28:69–84. https://doi.org/10.1016/S0169-8141(01)00013-0

Banos O, Toth M, Damas M, Pomares H, Rojas I (2014) Dealing with the effects of sensor displacement in wearable activity recognition. Sensors 14:9995–10023. https://doi.org/10.3390/s140609995

Basahel AM (2015) ScienceDirect investigation of work-related musculoskeletal disorders (MSDs) in warehouse workers in Saudi Arabia. Procedia Manuf 3:4643–4649. https://doi.org/10.1016/j.promfg.2015.07.551

Brewers Association (2016) Industry Updates: U.S. Bureau of Labor Statistics Data Suggests Improved Brewery Safety - Brewers Association (WWW Document). https://www.brewersassociation.org/industry-updates/u-s-bureau-labor-statistics-data-suggests-improved-brewery-safety/. Accessed 2 Apr 2017

Brewers Association (2017) U.S. hits record-breaking 6,000 breweries in operation. Brewers Association Press Release (WWW Document). https://www.brewersassociation.org/press-releases/2017-craft-beer-review/. Accessed 18 Apr 2018

Bureau of Labor Statistics (2016) Nonfatal occupational injuries and illnesses with days away from work 2015. https://www.doi.org/USDL-15-2205

Coenen P, Kingma I, Boot CRL, Bongers PM, van Dieën JH (2014) Cumulative mechanical low-back load at work is a determinant of low-back pain. Occup Environ Med 71:332–337. https://doi.org/10.1136/oemed-2013-101862

Dagenais S, Caro J, Haldeman S (2008) A systematic review of low back pain cost of illness studies in the United States and internationally. Spine J. https://doi.org/10.1016/j.spinee.2007.10.005

de Looze MP, Kingma I, Thunnissen W, van Wijk MJ, Toussain HM (1994) The evaluation of a practical biomechanical model estimating lumbar moments in occupational activities. Ergonomics 37:1495–1502

Driscoll T, Jacklyn G, Orchard J, Passmore E, Vos T, Freedman G, Lim S, Punnett L (2014) The global burden of occupationally related low back pain: estimates from the global burden of disease 2010 study. Ann Rheum Dis 73:975–981. https://doi.org/10.1136/annrheumdis-2013-204631

Faber GS, Chang CC, Dennerlein JT, van Dieën JH (2016) Estimating 3D L5/S1 moments and ground reaction forces during trunk bending using a full-body ambulatory inertial motion capture system. J Biomech 49:904–912. https://doi.org/10.1016/j.jbiomech.2015.11.042

Faber GS, Kingma I, van Dieën JH (2007) The effects of ergonomic interventions on low back moments are attenuated by changes in lifting behaviour. Ergonomics 50:1377–1391. https://doi.org/10.1080/00140130701324622

Gatchel RJ, Schultz IZ (2014) Handbook of musculoskeletal pain and disability disorders in the workplace. https://www.doi.org/10.1007/978-1-4939-0612-3

Granzow RF, Schall MC, Smidt MF, Chen H, Fethke NB, Huangfu R (2018) Characterizing exposure to physical risk factors among reforestation hand planters in the Southeastern United States. Appl Ergon 66:1–8. https://doi.org/10.1016/j.apergo.2017.07.013

Jones T, Kumar S (2001) Physical ergonomics in low-back pain prevention. J Occup Rehabil 11:309–319. https://doi.org/10.1023/a:1013304826873

Jones T, Strickfaden M, Kumar S (2005) Physical demands analysis of occupational tasks in neighborhood pubs. Appl Ergon 36:535–545. https://doi.org/10.1016/j.apergo.2005.03.002

Karatsidis A, Jung M, Schepers HM, Bellusci G, de Zee M, Veltink PH, Andersen MS (2018) Predicting kinetics using musculoskeletal modeling and inertial motion capture

Khurelbaatar T, Kim K, Lee SK, Kim YH (2015) Consistent accuracy in whole-body joint kinetics during gait using wearable inertial motion sensors and in-shoe pressure sensors. Gait Posture 42:65–69. https://doi.org/10.1016/j.gaitpost.2015.04.007

Kim S, Nussbaum MA (2013) Performance evaluation of a wearable inertial motion capture system for capturing physical exposures during manual material handling tasks. Ergonomics 56:314–326. https://doi.org/10.1080/00140139.2012.742932

Lavender S, Marras W, Ferguson S, Splittsteosser R, Yang G (2012) Developing physical exposure-based back injury risk models applicable to manual handling jobs in distribution centers. J Occup Environ Hyg 9:450–459

Lee W, Seto E, Lin K-Y, Migliaccio GC (2017) An evaluation of wearable sensors and their placements for analyzing construction worker's trunk posture in laboratory conditions. Appl Ergon. https://doi.org/10.1016/j.apergo.2017.03.016

Magora A (1975) Investigation of the relation between low back pain and occupation. Scand J Rehabil Med 7:146–151

Marras WS (1993) Dynamic measures of low back performance. American Industrial Hygiene Association

Marras WS, Knapik G, Ferguson S (2009) Loading along the lumbar spine as influence by speed, control, load magnitude, and handle height during pushing. Clin Biomech 24:155–163. https://doi.org/10.1016/j.clinbiomech.2008.10.007

Marras WS, Lavender SA, Ferguson SA, Splittstoesser RE, Yang G (2010) Quantitative biomechanical workplace exposure measures: Distribution centers. J Electromyogr Kinesiol 20:813–822. https://doi.org/10.1016/j.jelekin.2010.03.006

Potvin JR (2008) Occupational spine biomechanics: A journey to the spinal frontier. J Electromyogr Kinesiol 18:891–899. https://doi.org/10.1016/j.jelekin.2008.07.004

Putz-Anderson V, Bernard B, Burt S (1997) Musculoskeletal disorders and workplace factors: a critical review of epidemiologic evidence for work-related musculoskeletal disorders of the neck, upper extremity, and low back, Second edn. U.S. Department of Health and Human Services, Public Health Service, Centers for Disease Control and Prevention, National Institute for Occupational Safety and Health, Cincinnati

R Core Team (2016) R: a language and environment for statistical computing

Roetenberg D, Luinge H, Slycke P (2013) Xsens MVN: full 6DOF human motion tracking using miniature inertial sensors 3

Schall MC, Fethke NB, Chen H (2016) Evaluation of four sensor locations for physical activity assessment. Appl Ergon 53:103–109. https://doi.org/10.1016/j.apergo.2015.09.007

Schall MC, Fethke NB, Chen H, Gerr F (2015) A comparison of instrumentation methods to estimate thoracolumbar motion in field-based occupational studies. Appl Ergon 48:224–231. https://doi.org/10.1016/j.apergo.2014.12.005

Van der Burg JCE, Van Dieën JH (2001) Underestimation of object mass in lifting does not increase the load on the low back. J Biomech 34:1447–1453. https://doi.org/10.1016/S0021-9290(01)00118-X

Van Der Burg JCE, Van Dieën JH, Toussaint HM (2000) Lifting an unexpectedly heavy object: The effects on low-back loading and balance loss. Clin Biomech 15:469–477. https://doi.org/10.1016/S0268-0033(99)00084-4

van Dieën JH, Faber GS, Loos RCC, Paul P, Kuijer FM, Kingma I, Van Der Molen HF, Frings-Dresen MHW (2010) Validity of estimates of spinal compression forces obtained from worksite measurements. Ergonomics 53:792–800. https://doi.org/10.1080/00140131003675091

Vignais N, Miezal M, Bleser G, Mura K, Gorecky D, Marin F (2013) Innovative system for real-time ergonomic feedback in industrial manufacturing. Appl Ergon 44:566–574. https://doi.org/10.1016/j.apergo.2012.11.008

Zurada J, Karwowski W, Marras W (2004) Classification of jobs with risk of low back disorders by applying data mining techniques. Occup Ergon 4:291–305

Workload Estimation System of Sequential Manual Tasks by Using Muscle Fatigue Model

Akihiko Seo[1(✉)], Maki Sakaguchi[1], Kazuki Hiranai[1], Atsushi Sugama[2], and Takanori Chihara[3]

[1] Tokyo Metropolitan University, Hino, Japan
aseo@tmu.ac.jp
[2] National Institute of Occupational Safety and Health, Kiyose, Japan
[3] Kanazawa University, Kanazawa, Japan

Abstract. In this study, we sought to develop a system to evaluate the workload of multiple sequential tasks using a digital human and muscle fatigue model, as well as test its validity using a sequential task experiment. The muscle fatigue model is the three-component model introduced by Xia et al. The model assumes that the muscle motor unit consists of resting, activated, and fatigued components. We used a temporal smoothed value of the active component ratio to the non-fatigued component to estimate workload. A system was developed using this model to evaluate workload of any combination of sequential tasks of the single manual handling task. A sequential task consisting of three kinds of material handling task performed by a digital human and real environment was prepared as a validity test. We found that the estimated workload using the simulation and the subjective scores showed a similar pattern with the load of the sequential tasks and repetitions.

Keywords: Physical workload · Sequential task · Muscle fatigue

1 Objectives

Estimating the physical workload of multiple sequential tasks is required to create and improve production systems. The multiple task analysis of the National Institute for Occupational Safety and Health (NIOSH) lifting equation [1] is one of the methods used for this purpose. The sequential lifting index [2] is used for more complex multiple task analyses. The muscle fatigue model has been used to evaluate the workload of each body parts. Xia and Fley-Law [3] and Ma et al. [4] proposed a simple muscle fatigue model that was implemented in a digital human software [5]. However, using the muscle fatigue model in a digital human system and its validity have not been firmly established.

The purpose of this study was to develop a system to evaluate the workload of multiple sequential tasks using a digital human and muscle fatigue model and test its validity using a sequential task experiment.

S. Bagnara et al. (Eds.): IEA 2018, AISC 825, pp. 82–86, 2019.
https://doi.org/10.1007/978-3-319-96068-5_9

2 System Outline

2.1 Muscle Fatigue Model

The muscle fatigue model is a three-component model introduced by Xia and Fley-Law [3]. The model assumes that the muscle motor unit consists of resting ($\boldsymbol{M_R}$), activated ($\boldsymbol{M_A}$), and fatigued ($\boldsymbol{M_F}$) components. The proportion of the three components changes with muscle activity by the target load (***TL***) or joint torque ratio and resting time. The component ratio of $\boldsymbol{M_F}$ or the ratio of target load to non-fatigued component (brain effort: ***BE***) are used to evaluate workload.

$$\boldsymbol{BE} = \frac{\boldsymbol{TL}}{100\% - \boldsymbol{M_F}} \times 100[\%] \quad (1)$$

Based on this Eq. (1), ***BE*** is zero when the muscle is relaxed (i.e. ***TL*** = 0). However, the subjective feeling of fatigue requires a certain recovery time after the muscle is deactivated. We, therefore, used the exponential smoothing for smoothing ***BE*** to fit the exponential response of the subjective complaint of muscle fatigue. The temporal smoothed ***BE*** (***SBE***) by exponential smoothing is as follows:

$$\boldsymbol{SBE_i} = (1 - \boldsymbol{\alpha})\boldsymbol{SBE_{i-1}} + \boldsymbol{\alpha BE_i} \quad (2)$$

Where, $\boldsymbol{SBE_i}$ and $\boldsymbol{SBE_{i-1}}$ are ***SBE*** at time $\boldsymbol{i}$ and $\boldsymbol{i} - 1$, $\boldsymbol{BE_i}$ is ***BE*** at time $\boldsymbol{i}$, $\boldsymbol{\alpha}$ is smoothing coefficient ($0 \leq \boldsymbol{\alpha} \leq 1$).

We used ***SBE*** to properly estimate the low muscle activity period with the remaining fatigued component.

2.2 Workload Estimation for Multiple Sequential Tasks

Before starting the estimation of multiple sequential task, we make temporal data file of the joint torque ratio to the maximum joint torque of a single manual task using a biomechanical model of the digital human. Any combinations of sequential tasks for a single manual task can be designed with this system. The workload of the sequential tasks can then be calculated using the muscle fatigue model. Following steps demonstrates how the data are processed. Step-6 compares the subjective score of workload as part of the validity test.

1. Make a single task motion of material handled by the digital human
2. Calculate the time series data of joint torque ratio as the ***TL*** during a single task using a biomechanical model
3. Make a set of sequential task list consisting of any combination of a single manual task, any repetition, and any resting duration.
4. Calculate the time series data of the ***BE*** using the muscle fatigue model
5. Obtain the smoothed ***BE*** (***SBE***) as the evaluation values of workload using the exponential smoothing method
6. Resampling the ***SBE*** at the time when the subjective workload is asked

3 Validity Test

We prepared a sequential task consisting of three kinds of material handling task in the digital human and real environment. The tasks included sitting, standing, and squatting postures holding 10-kg baggage with symmetric lifting posture. The participants carried out one baggage holding task within 25 s and repeated the 25-s task for 5 times. Short breaks of 25 or 50 s were also included during the task to control the total workload and test the effect of the break. The torque ratios of the shoulder, lumbar, and hip joint during baggage lifting tasks were calculated at the frame rate of 10 Hz using a biomechanical model. We then calculated the workload estimation values of $\boldsymbol{M_F}$, $\boldsymbol{BE}$, and $\boldsymbol{SBE}$ from the torque ratio data using our system. The parameters of the fatigue model were referred from Flay-Law et al. [6]. The real experimental data consisted of a subjective score of the workload based on Borg CR10 scaling that was recorded every 25 s during the same sequential task procedures.

Five male students participated in this experiment. The average and standard deviation of their age, height, and weight were 22.6 ± 1.0 y, 171.8 ± 3.7 cm, and 65.9 ± 3.6 kg, respectively.

The preliminary study included five types of combinations of a single task. The result of a task set of sitting (5 times)-standing (5 times)-squatting (5 times)-resting (50 s) are shown in Figs. 1 (shoulder joint), 2 (lumbar joint), and 3 (hip joint). The estimated workload using the simulation and the average subjective scores of the five participants showed a similar pattern of increasing and decreasing with the sequential tasks and repetitions.

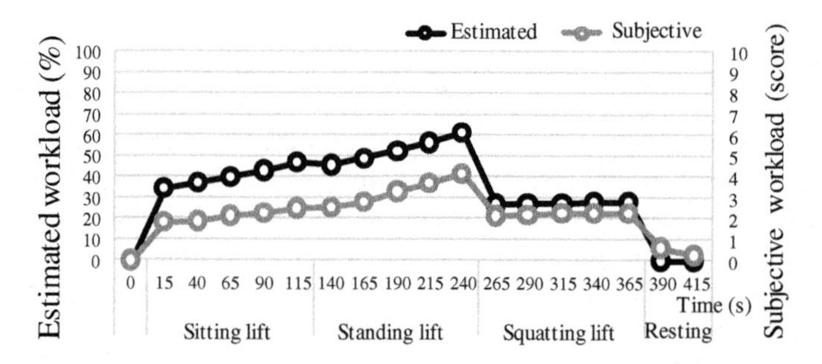

Fig. 1. Comparison of estimated and subjective workload of the shoulder joint ($\alpha = 0.069$).

4 Discussion

This study showed the procedure of the workload estimation of any numbers of repetitive tasks from one cycle task data of joint torque ratios using the muscle fatigue model. As shown in Figs. 1, 2 and 3, the procedure can also estimate workload of each joint for any combination of multiple tasks. Some differences remain between the estimated values and the subjective score of workload. It may be brought by the over/under estimation of joint torque ratio, which is not related to the fatigue model.

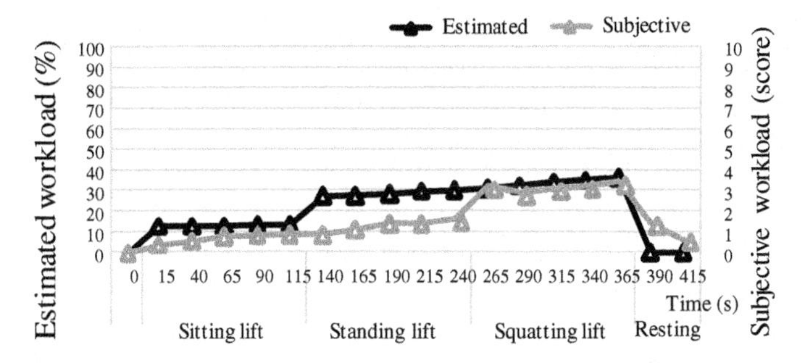

Fig. 2. Comparison of estimated and subjective workload of the lumbar joint ($\alpha = 0.074$).

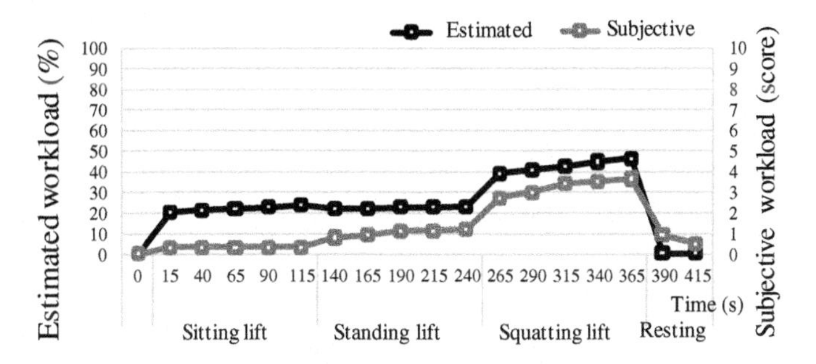

Fig. 3. Comparison of estimated and subjective workload of the hip joint ($\alpha = 0.058$).

Refining the parameters of the exponential smoothing may be necessary to fit the data during resting period.

5 Conclusion

The muscle fatigue model can evaluate sequential tasks using exponential smoothing.

This work was supported by JSPS KAKENHI Grant Number JP16K01247.

References

1. Waters TR, Puts-Anderson V, Garg A, Fine LJ (1993) Revised NIOSH equation for the design and evaluation of manual lifting tasks. Ergonomics 36(7):749–776
2. Waters TR, Lu ML, Occhipinti E (2007) New procedure for assessing sequential manual lifting jobs using the revised NIOSH lifting equation. Ergonomics 50(11):1761–1770
3. Xia T, Frey-Law LA (2008) A theoretical approach for modeling peripheral muscle fatigue and recovery. J Biomech 41:3046–3052

4. Ma L, Chablat D, Bennis F, Zhang W (2009) A new simple dynamic muscle fatigue model and its validation. Int J Ind Ergon 39(1):211–220
5. Abdel-Malek K et al (2008) Santos: a digital human in the making. University of Iowa, Center for Computer Aided Design
6. Frey-Law LA, Looft JM, Heitsman J (2012) A three-compartment muscle fatigue model accurately predicts joint-specific maximum endurance times for sustained isometric tasks. J Biomec 45:1803–1808

Epidemiological Survey of Occupational Accidents: A Case Study in the Flour and Animal Feed Business

Lucas Provin[1(✉)] and Cristiane Nonemacher Cantele[2]

[1] Vicato Alimentos, Rua Dr. João Neto 106, Sanaduva, RS, Brazil
provin@vicato.com.br
[2] Cavaletti Professional Seating S.A, 99706-54 Erechim, RS, Brazil
ergonomia@cavaletti.com.br

Abstract. Occupational accidents constitute a serious problem that can affect different organizational environments, ergonomic and safety-oriented actions can minimize risk factors. The objective of this study is to carry out an epidemiological-logical survey of occupational accidents at a company in the flour and animal feed segment, located in a small city, in the northern portion of Rio Grande do Sul. The research was characterized as a case study, exploratory and quantitative approach, developed through bibliographic and documentary analysis. The analysis occurred from January 2010 until September 2017, considering the cases of work accidents based on the following variables: number of accidents, gender, age, type of accident, function performed by the injured employee, days of absence, costs per accident, places of highest incidence of accidents in the company, part of the body affected in the work accident. The results showed the occurrence of 28 work accidents in the period, being 89% typical and most occurring in 2012 (29%). The highest incidence is between employees from 26 to 45 years old, mainly in the functions of full and junior machine operator, being traumas that affect the hands the main occurrence (48%). It was verified that 36% of the accidents let the employee away from his duty from a period between10 and 15 days, generating the direct cost of U$ 12,355.00 for the company in the period. The conclusion is that epidemiological surveys of occupational accidents can generate useful information for companies, enabling the development of ergonomic interventions capable of improving occupational safety, preventing new accidents, which can positively reflect the productive process, as well as reduce costs and improve health and quality of life in the workplace.

Keywords: Work accidents · Epidemiology · Ergonomics · Safety
Occupational health · Work hazard

1 Introduction

Workplace accidents are events that affect workers' health and/or life. They can occur in a large number of organizations, especially in the industry, generating damages to the own worker, to the social security system and to the companies. The work accident

S. Bagnara et al. (Eds.): IEA 2018, AISC 825, pp. 87–104, 2019.
https://doi.org/10.1007/978-3-319-96068-5_10

is a matter present in the Federal Constitution, being also ruled by complementary legislation and specified labor area. However, despite the existence of laws that seek to regulate aspects of labor relations, it is important to consider that in many cases occupational accidents are also related to occupational safety, ergonomic issues and planning and control the production processes.

In this perspective, it has become an important management tool, aimed at improving the environment and quality of life at work, which reflects in the development of employees' activities and, consequently, in the dynamics of relations and results of companies, acting directly or indirectly in the prevention of occupational accidents, especially in relation to osteomio articular injuries. In addition, it studies the relationship between man and the work environment and "contributes to the design and modification of work environments maximizing production, while pointing out the best health and well-being conditions for those working in these environments" (Marques et al. 2010, p. 1).

By focusing on the ergonomic approach, the company tends to be concerned with people, the physical aspects of workplaces (environment and equipment) and the relationship between them and their function. Thus, ergonomics "tends to develop jobs that reduce the biomechanical and cognitive stocks, trying to put the operator in a good work position" (Silveira and Salustiano 2012, p. 75).

In this sense, this study aimed to understand the correlation between the work activity, the occurrence of work accidents and the ergonomic issues involved and the impact on the production process as a way to broaden the discussion of these issues in the scope of Production Engineering.

The relevance of this study is justified in order to generate epidemiological information useful for making decisions about work accidents, possible ergonomic risks and their relationship with productive environment.

The research problem to be answered was the following: what are the characteristics of work accidents that occurred in the company under study, from 2010 to 2017. Thus, the general objective was to carry out an epidemiological survey of occupational accidents, considering the cases occurred in the period from 2010 to 2017. The specific objectives involved: to contextualize and deepen theoretically aspects of work accidents and their relationship with ergonomic aspects; to collect information about the occurrences of work accidents in the company and the expenses with these accidents; to identify possible ergonomic and accident risks that must be improved, aiming at a more ergonomic and healthy work space and, consequently, a more productive environment.

Thus, we have as a context of analysis a company that operates in the field of flour and animal feed, located in a small city in the north of the State of Rio Grande do Sul. The purpose of the survey was to contribute to the company in the evaluation of work accident cases and possible relations with ergonomic aspects, highlighting the financial variables involved, problems generated by accidents in the dynamics of production, among others.

2 Work Accidents

2.1 Historical Evolution

The occurrence of occupational accidents and concern for workers' health goes back to the first manifestations of work in society. However, was after the Industrial Revolution that the concern with the health and safety of the worker occurred with greater intensity. "The misfortune of labor is correlated with the Industrial Revolution and the machinery" (Fernandes 1995, p. 33). Thus, the replacement of manual labor by the machines and the low experience of the workers contributed to the beginning of the occurrence of accidents during the work practice. According to Martins (2002, p. 411), "the loom and the steam engine were the cause of work-related accidents. From that moment on, there begins to be a concern for the injured. It was verified that the injured at work could not get new placement in other companies, being totally unprotected ". Today the International Labor Organization is a respected and fundamental organization to solve issues related to work in the world and to the struggle for social justice and balance in labor relations. In Brazil, in 1904, a special legislation was developed for labor misfortunes, and in 1919 the first labor accident law (Law No. 3.724) was adopted, considering the Occupational Risk Theory, based on European doctrine, don't arguing who was guilty of the act. The Civil Code of 1916 brought some rules of protection. It was sought to compensate damages caused by award accident, from aquiline guilt. Based on this thought, there is reparation for the damage done to the things of others due to the demonstration of guilt, and the damage must be proven, who committed it and the nexus between the damage and the fault - causal nexus. Thus, following the trend that was indicated in the context of the country, the Federal Constitution of 1934, ensured in cases of work accident the right of the worker to resort to the social security system. Thus, the transformations in the labor legislation, related to the work accident, combine three elements: "one of historical order, with free or salaried labor; with the extensive use of machinery; and, economical, with the work carried out under a company regime, being the venture risk assumed by the employer, which lives up to the product or profit "(Fernandes 1995, p. 26). The national legislation on occupational accidents evolved from the demands of society itself and from established labor relations, following the transformation of these relations and giving a new perspective of protection to the worker.

2.2 Work Accidents and Statistics in Brazil

According to the Ministry of Social Security, the work accident or the transit accident, is the accident occurred in the exercise of the professional activity at the service of the company or in the transit, residence/work/residence, and that causes bodily injury or functional disturbance that causes the loss or reduction (permanent or temporary) capacity for work or, ultimately, death (Brazil 2017a, b, c). This type of accident is characterized as such only when there is "a link between work and the effect of the accident" (Martins 2002, p. 399). It should be noted that they were compare to accident at work, as determined by the Ministry of Social Security: I - the accident linked to work; II - the accident suffered by the insured in the place and time of the work; III - the disease arising from accidental contamination of the employee in the exercise of his activity; IV - the accident suffered by the insured, even if outside the place and working hours, in the execution of order or in the performance of service under the authority of the company.

2.3 Relation Between Work Accidents and Drop in Productivity

The work accident automatically generates a break in the production cycle, reducing productivity in the workplace. The removal of an employee because of an accident creates a network of consequences. According to Zocchio (2002), work accidents and occupational diseases affect the company by generating damages to the patrimony, reduction of productivity, increase of the amount of costs including operational by requiring the hiring of new employees, relocations or overtime payment.

Therefore, it is relevant to encourage and improve safety at work, as a way to minimize the occurrence of accidents and positively influence on labor productivity. Marras (2000) emphasizes the importance of prevention to eliminate the causes of work accidents, and the company must maintain a hygiene and safety system, prioritizing the ergonomics of the environment and functions, and making workers aware of actions to protect his own life and the life of his coworkers.

Therefore, behavioral change and investment in ergonomics and safety are fundamental in this process.

2.4 Coasts for Work Accidents for Companies

Workplace accidents are unexpected events that generate costs for companies. Zocchio (2002) points out that the work accident is an abnormal occurrence and undesirable in the exercise of work that interrupt the activity where they occur; interfere negatively in other activities as well; which can cause physical and mental damage, and may even cause death, causing damage to companies and also to the socioeconomic system of the country, as they directly affect Social Security.

Especially for companies, the consequences of awork accident are perceived in the bureaucratic difficulties with the official entities and damage to the image of the company before the market; expenses with first aid and transportation of the injured to the place of service, loss of productive time of other employees to assist the injured or production stops to comment on the subject; damage or loss of material, tools, equipment or machines (Marras 2000).

Accidents generate direct and indirect costs for the company. Ferreira and Peixoto (2012) point out that the direct cost, also known as insured cost, is one that is not directly related to the accident and represents a permanent cost to the employer, the monthly contribution of the companies' being called "Work Accident Insurance (WAI)" and is calculated on the company's basis at three levels of risk (light, medium and severe) and a percentage on the company's contribution payroll (1%, 2% and 3%, respectively).

Production Engineering originated in the 19th Century in the Industrial Revolution, with the emergence of the first industries and the capitalist system, at which time a mass production of various products was established through the modernization and discovery of new technologies. From this, the industries began to develop rapidly and with this emerged several challenges in the organization of the productive process (Abrepo 2017).

Production Engineering operates in the integrated systems of men, machines and equipment, with the purpose of designing, installing and improving such systems, as well as evaluating the results of these systems (2002, p. 19). Among the areas of production engineering, can be highlighted product engineering, factory design, production processes, method and process engineering, production planning and control, production costs, quality, organization and maintenance planning, reliability engineering, ergonomics, occupational hygiene and safety, logistics and distribution, and operational research (Cunha 2002). Given its specificity, Production Engineering acts directly on issues that interfere with productivity, such as lack of ergonomic conditions, occupational diseases and work accidents.

According to Gonçalves Filho and Ramos (2015), occupational accidents are public health problems, with a great burden for society as a whole. Besides affecting the social issue, with deaths and mutilations of workers, it causes economic and social security losses, as well as affecting the productive forces. In this way, Production Engineering can become an instrument that extends prevention issues, restructuring production systems, contributing to the learning and analysis of accidents, understanding risks, solving problems and protecting people.

3 Ergonomics

3.1 Ergonomic

Working conditions have undergone rapid changes over the years, with a large number of workers whopper form their activities with the help of machines. In addition, many activities generate repetitive stress or are being developed in an unsuitable environment. It is in this sense that ergonomics emerges as a way to avoid harmful effects and also to promote efficient and pleasant health conditions (Kroemer and Grandjean 2008).

Ergonomics is the study of the adaptation of work to the human being, and "to transform work is the primary purpose of ergonomic action" (Guérin et al. 2001, p. 1). One of the intentions of ergonomics is to reduce the harmful consequences over the worker, including reduce of fatigue, stress, errors and accidents, providing health, safety and satisfaction to workers during their interaction with the production system (Iida and Guimarães 2012).

Ergonomics studies various aspects: posture and body movements (seating, standing, pushing, pulling and lifting loads), environmental factors (noise, vibration, lighting, climate, and chemical agents), information (information collected by looking, hearing and other senses), relations between displays and controls, as well as positions and tasks (appropriate, interesting tasks) (Dul and Weerdmeester 2012).

In Brazil, the Regulatory Standards - NR 17 aims to establish parameters that allow the adaptation of working conditions, to the psychophysiological characteristics of the work, to give more comfort, security and allow people to have a better result at work.

These conditions of work include transportation and unload of materials, to the furniture, to equipment and to conditions of work place and organization (Brazil 1990).

3.2 Environmental Factors

Environmental factors of a physical and chemical nature involve noise, vibrations, illumination, climate, and chemicals substances that can affect workers' health, safety, and comfort (Dul and Weerdmeester 2012).

Unfavorable environmental conditions can cause discomfort, increasing the risk of accidents and causing considerable damage to health. (Iida and Guimarães 2012). The presence of loud noises in the work environment can disturb and, over time, eventually cause deafness. Vibration can affect the whole body or only part of the body. The intensity of the light that falls on the work surface should be sufficient to ensure a good visibility, not harming the worker in its activity. In the same way, the climate must meet the various conditions, including air temperature, radiant heat, air velocity and relative humidity (Dul and Weerdmeester 2012).

According to NR 17, the environmental conditions of work must be adequate to the psychophysiological characteristics of the workers and the nature of the work to be performed.

3.3 Occupational Biomechanics

Biomechanics is part of Ergonomics, since posture and movement are part of the work activity. Occupational biomechanics is "a part of general biomechanics, which deals with body movements and work-related forces" (Ida and Guimarães 2012, p. 149).

Thus, ergonomics seeks to minimize the consequences for workers, since work can influence functional capacity, generating effects on the body and health (Guérin et al. 2001). Based on postural analysis and movement in the work environment, it is possible to perceive its consequences, since the physical interaction of the worker in his activity, whether with the machines, tools and other equipment, as well as the movements and forces that he performs, is related to the development of musculoskeletal disorders related to work - DORT (Iida and Guimarães 2012).

That is why the posture and body movement aspects in the environment are elements of Ergonomics study and should, from this analysis, be corrected or avoided. It is advisable to create favorable conditions for tasks that require constant movement, respecting the rhythm of each worker, granting pauses during the execution of the task and not exceeding both weight and repetition (Moraes and Silva 2015).

3.4 Occupational Health and Quality of Life at Work

Elements such as hygiene and safety in the work environment are fundamental for the promotion of health and quality of life. According to France and Rodrigues (1996), values on health and illness are built in the company under the focus of productivity, under the principles that are adopted of social responsibility and the value that is given to the preservation of people, work accident histories and organizational culture. According to these authors, companies are organized in such a way to respond efficiently to the needs of customers, within the characteristics of the product of its history, the characteristics of the market and its competitors and suppliers. According to Paraguay (1990), the sources of stress and contribute to the reduction of workers' quality of life and well-being are: environmental factors (noise, illumination, temperature, ventilation at inappropriate levels or limits), organizational factors (involvement and participation in the work, organizational support, supervision style, managerial support, organizational schemes, career plans) and the organization of work based on the mental aspects of the same, such as monotony or work overload, production and work rhythm, temporal pressures, meaning of work and nature of tasks. A clean, organized, ergonomic, airy and illuminated work environment, as well as the orientation and use of safety equipment, contribute to the promotion of workers' health and well-being (Paraguay 1990).

According to Souza (2017), it is relevant to investigate workers' health related to the identification of risk situations and the evaluation of health impacts caused by the work activity and its environment, placing health problems in the different production contexts, mapping morbidity and mortality, identifying temporal trends and groups of workers with higher risks, among other surveys that generate information that is fundamental for improving the environment and working conditions. The worker's health risks approach allows the control of causes of accidents, physical, chemical and biological agents that cause injuries, physical stress and mental overload.

Attention to regulations is very important to reduce accidents that greatly affect worker's health and quality of life, when they do not suppress their lives, leaving deep marks in families and in the companies themselves and other workers. In addition, occupational diseases and occupational accidents amplify problems in the productive system, as they generate absenteeism, turnover and production decrease.

4 Methodology

This research was characterized as exploratory, with a quantitative approach, based on a bibliographic and documentary case study. According to Gil (2009), the exploratory research is one that aims to provide greater familiarity with the problem, with a view to making it more explicit. It is a type of flexible research both in its planning and in its execution, being that usually involves bibliographical survey; interviews with people who have had practical experiences with the problem researched and analysis of examples that stimulate understanding.

On the other hand, quantitative research is characterized by the use of quantification in the collection and treatment of information through statistical techniques, aiming at guaranteeing results and avoiding distortions of analysis and interpretation, allowing a greater safety margin for inferences (Diehl and Tatim 2004).

As for the techniques and procedures, the study involved the development of bibliographical and documentary research, based on a case study. For Yin (2001, p. 19), case studies involve a single analysis and become an important research strategy "when 'how' and 'why' questions are posed, when the researcher has little control over the events and when the focus is on contemporary phenomena inserted in some context of real life."

In this study, the context concerns a company that operates in the production of flour and animal feed located in a small city in the north of the State of Rio Grande do Sul. The bibliographic research was developed from the analysis of bibliographic material already published on the subject.

The documentary research was carried out from the analysis of the company's reports, from January 2010 to September 2017. The following variables were investigated: number of accidents, gender, age, type of accident, function performed by the injured employee, days of removal, costs per accident, places of highest incidence of accidents in the company, part of the body affected in the work accident.

The data were analyzed based on descriptive statistics, considering frequency and/or percentage number, and were organized in graphs with the objective of carrying out an epidemiological survey of the company's work accidents. Data analysis involved the interpretation of the results obtained, in order to meet the general and specific objectives of the research.

5 Results and Discussion

The data was collected with the Department of Human Resources Management of the company, considering the period from January 2011 to June 2017, in the sectors inherent to the production process. Currently, the company has 215 employees in total, of which 166 are located in subsidiary 1 of the parent company, 19 in a subsidiary in Sananduva/RS and 30 in a subsidiary in Marcelino Ramos/RS (Table 1) of relevant attributes enabling to verify how the person's image is composed in relation to productivity issues and spatial issues.

Table 1. Number of employees by sector Vslocation Source: Research data, 2017.

Company units	Sectors	Number of employees	Away
Parent unit - subsidiary 1	Wheatmill	25	1 (INSS)
	Corn Mill	6	1 (Union)
	Factory of animal food	24	3 (INSS)
	Workshop	6	
	Qualitycontrol	7	
	Warehouse	2	
	Return	3	1 (INSS)
	Expedition Wheat Mill	11	1 (INSS)
	Expedition animal food	27	2 (INSS)
	Transportation	21	
	Administrative	31	
	Direction	3	
Parent unit - subsidiary 2	Silos	19	1 (INSS)
Unit 2	Unit 2 Mill/Production	30	1 (INSS)
Total	**215**	**11**	

Especially in the wheat and corn flour production sector in the parent plant, there are 31 employees, 29 active and 2 away from the job. In the feed production sector of the parent plant, 24 employees are employed, 21 of which are active and 3 are away from the job. It should be noted that the company provides personal protective equipment (PPE) such as: boot, cap, ear protector and mask, according to what establishes the legislation and labor regulations.

5.1 List of Accidents

It should be noted that during the period of analysis there were 28 work accidents, of which 25 were typical (89%) and 03 (11%) of path (Fig. 1).

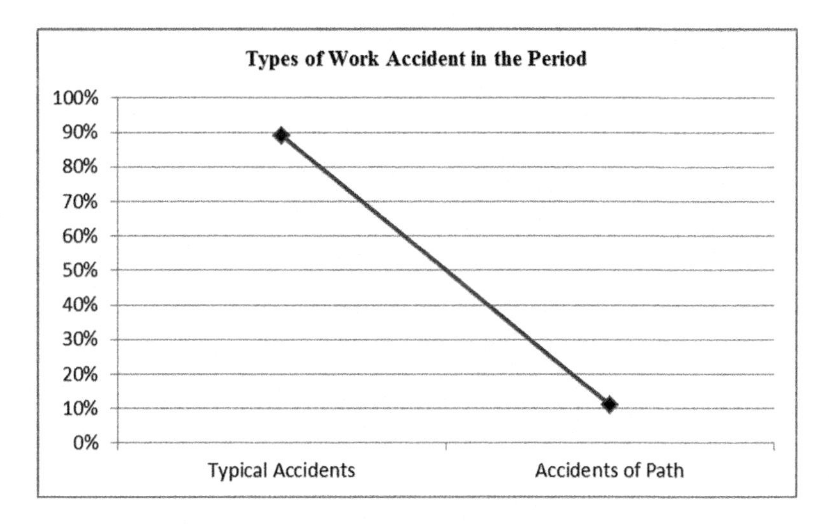

Fig. 1. Types of work accident in the period. Source: Research Data, 2017

It is verified that in the studied period the highest incidence was typical work accidents, occurring in the company's site (Brazil 2017a, b, c). In agreement with the data found in the literature, these types of work accidents are the most evident in the whole country (Brazil 2015). On the other hand, road accidents, that means, those occurring between the residence and the place of work of the insured and vice versa (Brazil 2017a, b, c) represented a much lower percentage in the period of analysis.

5.2 List of Incidents and Their Respective Years

According to Fig. 2, the highest incidence year was 2012, with eight cases (29%), followed by 2011 with five (18%) and 2014 and 2015 with four accidents each (14%).

Due to the lack of records in the company, it was not possible to estimate the reasons for the higher incidence of accidents in 2012. However, can be consider that this data may be related to the production rhythm, insufficient training or causes linked to the organizational climate, factors generally described in the literature.

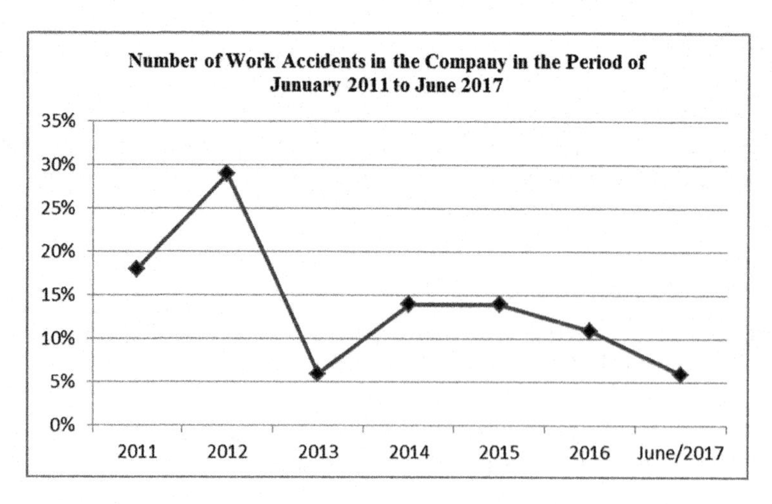

Fig. 2. Number of work accidents in the company from January 2011 to June 2017. Source: Research Data, 2017

5.3 Accident Reporting by Gender and Age Group

All employees who suffered accidents are male, with the age groups observed in Fig. 3, the majority (58%) are between 26 and 45 years of age.

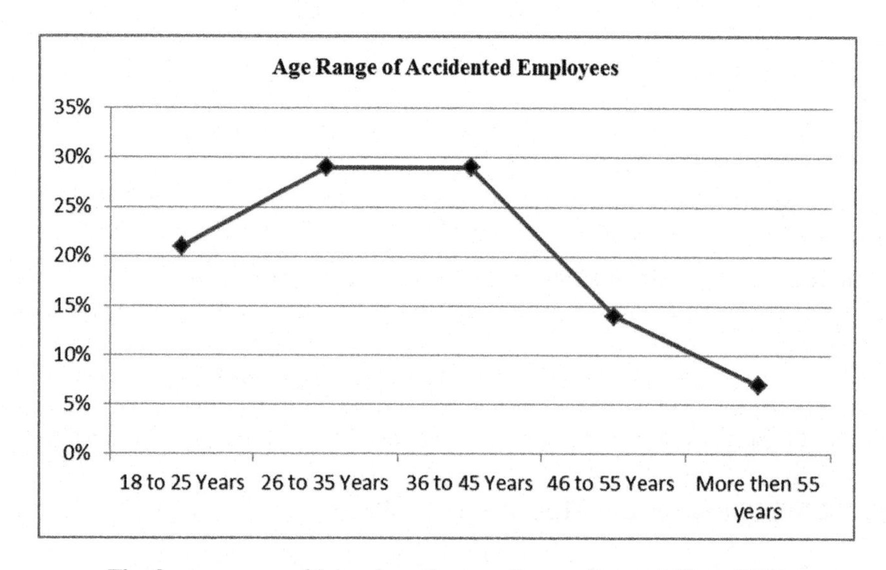

Fig. 3. Age range of injured employees. Source: Research Data, 2017

It should be consider that all employees who have suffered accidents are male, because the study company, the production sector, object of this study, has a strictly male labor force, due to the characteristics of this process. However, the literature points to a higher incidence of work accidents among men, and they are the ones that suffer the most according to Federal Government statistics (Brazil 2015).

Regarding the age group, it should be considered that data from the literature point to a higher occurrence in the age group with or over 40 years (Almeida and Branco 2011). In addition, older workers, due to their greater experience, may be less likely to crash, especially when they already operate machinery or other activities in the production sector for a longer time, meaning more experience in the area.

5.4 List of Accidents and Their Respective Functions

Among the cases of accidents, the main functions and respective places where the employees worked were highlighted. It can be seen that the functions of full and junior machine operator count for 32% of accidents, followed by functions and auxiliary of full production and conferring of junior merchandise with 11% each. Agreeing that the accident rate and the relationships with the task of operating machinery may indicate that a specific intervention should be carried out on site (Fig. 4).

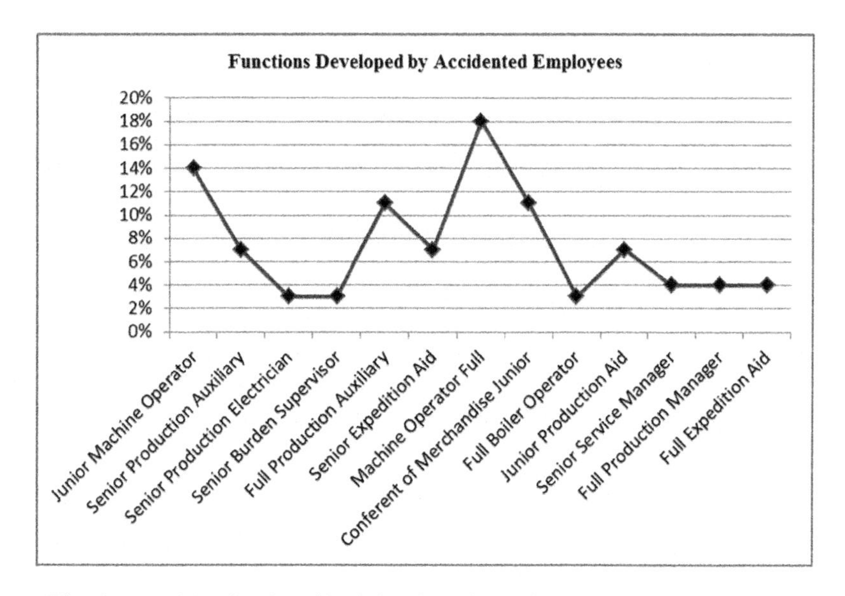

Fig. 4. Functions developed by injured workers. Source: Research Data, 2017

All position nomenclatures are based exactly on what the employees will perform. They are basically split by hiring time, meaning a newly hired employee up to two years of experience in the function is characterized as junior in function, moving to full and usually after five years in the senior role.

5.5 List of Accidents and Their Respective Sectors

Regarding the sectors where labor accidents occurred, it can be observed in Fig. 5 that the majority occurred in the feed mill (31%), followed by the wheat mill sector (23%) and the wheat expedition sector (19%). It should be noted that in these sectors the main risk factors are the intense use of machines, involving belts, gears along the packaging process, separation of packages, among others. In addition, there is a repetition of movements on the part of the employees, which can be an accident causing factor.

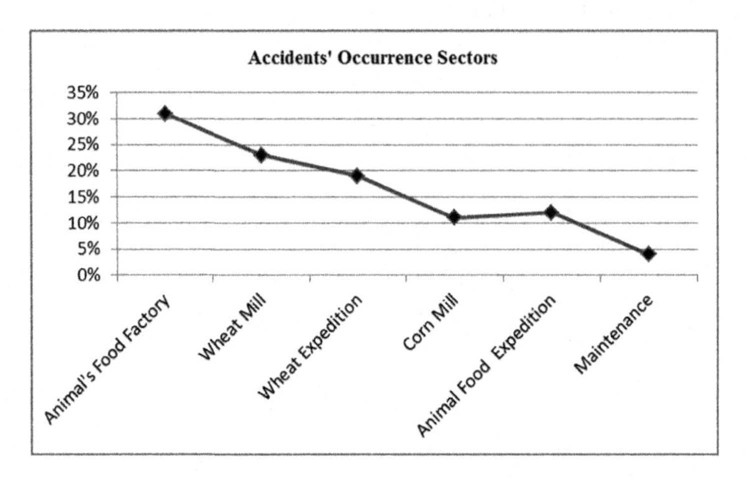

Fig. 5. Accidents' Occurrence Sectors.

The repetition of movements is a factor causing occupational diseases, being recognized as elements that generate sickness of the workers, contributing to the occurrence of accidents in the exercise of the work, either for characterizing corporal injury or functional disturbance that causes death or the reduction of the capacity to work (Fernandes 1995).

5.6 List of the Traumas that Generate Work Accidents

Figure 6 shows the types of trauma suffered by employees that characterized workplace accidents in the company. The largest of them affected the hand of the employees (48%), 28% of which were cuts, 12% of finger fractures, 4% of finger amputations and another 4% of the scaphoid fracture.

It is found that most of the traumas suffered by employees relate to acute events generated by the mechanical risk to which they are exposed in the industry throughout the manufacturing process. It should be noted that the mechanical risk in the industry represents potential risks and requires special manipulation by the manipulator, and the avoidance or minimization of these risks requires that the machine and the operator be in perfect condition (Iida and Guimarães 2016).

In addition, these indicators corroborate with the statistics of occupational accidents (Brazil 2015), pointing out that among the five main registered occupational accidents

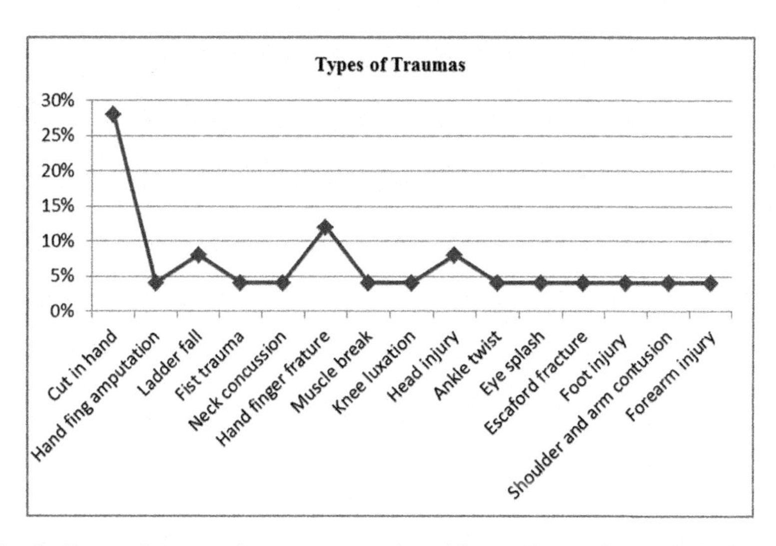

Fig. 6. Types of traumas that generate work accidents. Source: Research Data, 2017

situations, three involve the hands as a wrist and hand injury, a fracture at the level of the wrist and hand injuries and superficial trauma of the wrist and hand. None of the mentioned amputations or cuts are recurrent to the use of cutting tools, such as saws, guillotines, etc. The company does not have any of these tools, the great majority of accidents involving cuts and amputations originated in belts, and pulleys of the production and packaging machines, however in this context an intervention must be performed.

5.7 List of Absence Days

Figure 7 shows that 36% of the accidents involved leave between 10 and 15 days, and until October 5, 2017, four employees who suffered accidents in 2012 (fall), 2013 (fall), 2016 (arm fracture) and 2017 (shoulder injury) are still away from the job, all included in the INSS under benefit B91*.

* According to Brazilian legislation, the benefit B91 refers to the payment of monetary amounts to workers who are away of due to work-related injuries.

5.8 List of Work Accidents and Costs

Figure 8 shows the range of work accidents cost for the company in the period of analysis. It was verified that the majority of the accidents (52%) were in the range of U $ 150.00 to U$ 300.00 these values only correspond to the salary costs in the period of absence of each employee, it means that they do not include costs with lost productivity, production delays related to the absence of the employee or relocation of personnel. It should be noted that the costs of work accidents involve direct and indirect costs, some easily quantifiable, and others that can only be expressed in qualitative terms (European Agency for Safety and Health at Work 2002).

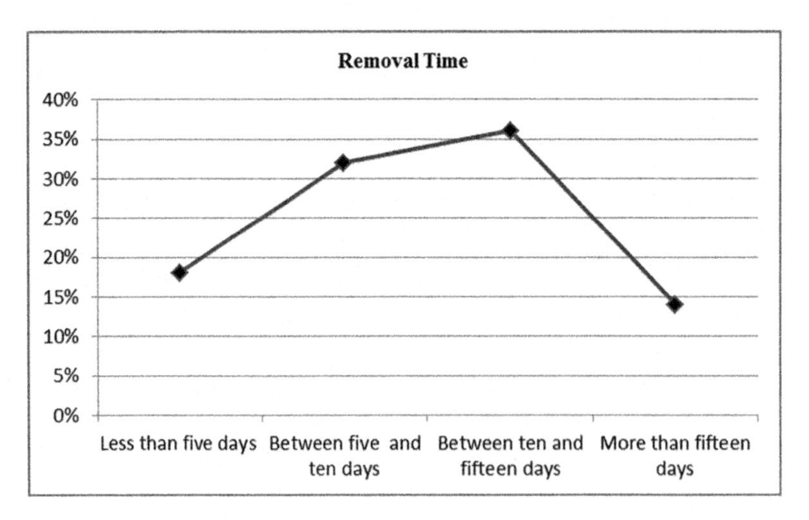

Fig. 7. Removal time of workers who suffered an accident in the company. Source: Research Data, 2017

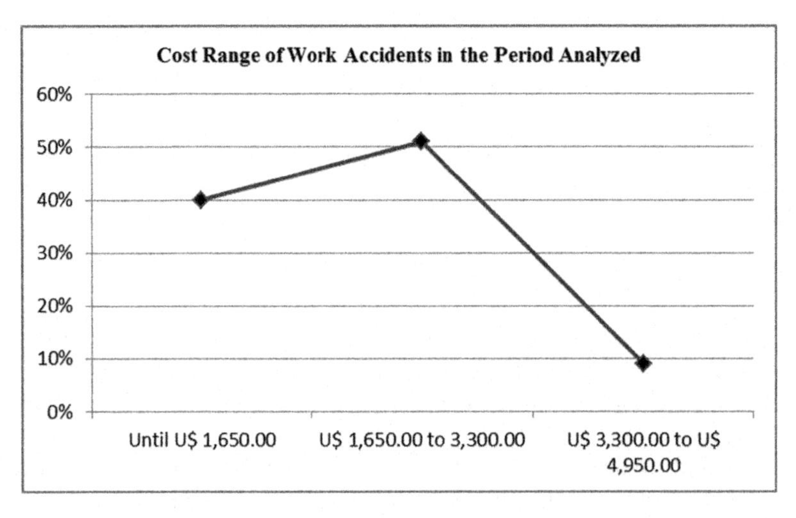

Fig. 8. Cost range of work accidents for the company in the period. Source: Research Data, 2017

5.9 Ratio of Amount Spent on Work Accidents

Figure 9 refers to amounts spent per year on work accidents for the company, and the year 2012, as it was the one that had the most accidents, was also the one that generated the most cost, in a total of U$ 3,749.00. During the study period, adding up all the expenses, were spent U$ 12,355.00.

It has been realized that in addition to the costs of dismissals, when we talk about work accidents, we can also mention the problems they can cause such as: legal

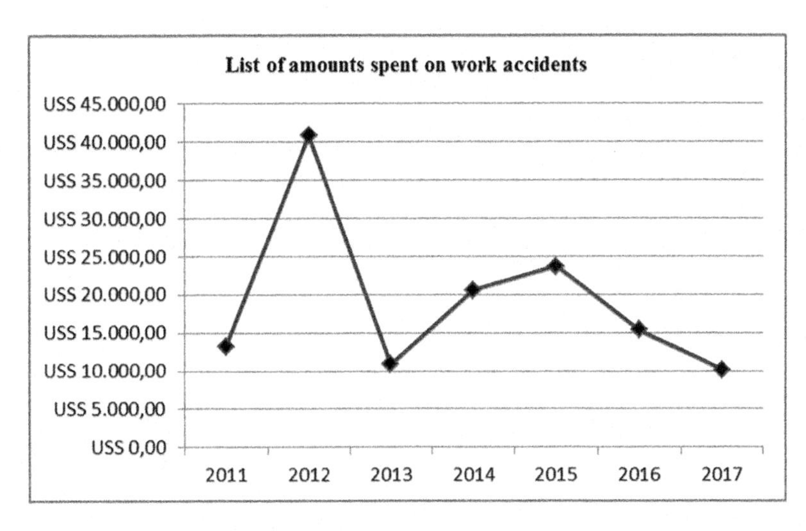

Fig. 9. List of amounts spent on work accidents. Source: Research Data, 2017

proceedings, audit and intervention of the Ministry of Labor. Legal processes are the most common, and they lead many companies to unburden large amounts of money to pay for something that could often be avoided by using or enforcing ergonomics programs and work injury prevention programs. Especially on legal proceedings, it should be noted that the country's legislation confers the civil liability of the employer, in the event of an accident at work, being disciplined in the Federal Constitution itself, according to article 7, item XXVIII.

In this perspective, the employer responds not only to public insurance against work accidents, but also to the damages caused to the worker by reason of fraud or guilt (Martins 2002).

The data collected from this study can serve as a guide for the company to improve the management of the social, economic and financial impacts of occupational accidents. From the data collection it is possible to produce indicators capable of making combinations about the frequency and severity of accidents, facilitating the company's suitability to legal and normative issues, as well as developing methods to follow the different sectors and functions, creating preventive measures that may include training, lectures, demonstrations and attention to the use of safety equipment for the protection and quality of life in the workplace.

6 Conclusion

The production system as a whole, from the factory floor to the management positions, are in a way linked to safety at work, accident prevention and employee health promotion should be a priority.

The objective of this study was to carry out an epidemiological survey on work accidents in a flour and animal feed manufacturer. It was verified that this objective was reached, since several data were collected, being the most relevant aspects the age group, the relevant age group was between 26 and 45 years, in the functions of full and junior machine operator, the absence time varied between 10 to 15 days and that generated an approximate direct cost of US$ 12,355.00 for the company in the period, not counting the amounts spent related to the legal costs of possible labor claims, since the separation nexus verified was an accident. In this sense, the need for the accrued amount of expenditure could have been invested in various actions, such as: ergonomic interventions, work safety programs to prevent new accidents and their consequences and to comply with current legislation, with emphasis on Regulatory Standards 17 and 12.

It is also valid to point out that the amounts spent on such facts could have been invested in process improvements and optimization of the productive flow, not counting the psychological effect on those workers with limb loss due to amputation. It was concluded, therefore, the importance of the Production Engineer to be attentive to the human factors that compose the productive process, with the objective of promoting better conditions of work, health and safety. As limitations of this study can be emphasized the fact that it is a case study, and generalizations cannot be made to other organizational environments. In addition, data on the impact and costs of accidents directly incident to the productivity and production sector of the company could not be ascertained.

This survey did not end the discussion. In this sense, we suggest the continuity of the study with the development of a control and formation of a more specific database on work accidents, facilitating the planning of actions of environmental improvement and training, as well as the application of ergonomic intervention programs and work safety, in order to minimize possible causes.

References

ABEPRO: Brazilian Association of Production Engineering. Production Engineering: large area and curricular guidelines. http://www.abepro.org.br/arquivos/websites/1/Ref_curriculares_ABEPRO.pdf. Accessed 25 Mar 25

Almeida PCA, Branco AB (2011) Work accidents in Brazil: prevalence, duration and social security expenditure. Braz J Occup Health 36(124):195–207

Brazil (1988) Constitution of the Federative Republic of Brazil

Brazil (1991) Law No. 8,213, of July 24, 1991. Provides for the Plans of Benefits of Social Security and provides other measures. http://www.planalto.gov.br/ccivil_03/leis/L8213cons.htm. Accessed 5 Oct 2017

Brazil (2017a) Ministry of Labor. NR 12 - Safety at work in machinery and equipment. http://trabalho.gov.br/seguranca-e-saude-no-trabalho/normatizacao/normas-regulamentadoras/norma-regulamentadora-n-12-seguranca-no-trabalho-em-maquinas-e-. Accessed 25 Mar 2017

Brazil (2017b) Ministry of Labor. NR 17 - Ergonomics. http://trabalho.gov.br/images/Documents/SST/NR/NR17.pdf. Accessed 25 Mar 2017

Brazil (2017c) Ministry of Social Security and Ministry of Finance. Social security benefits. http://www1.previdencia.gov.br/aeps2007/16_01_01_01.asp. Accessed 27 Oct 2017

Brazil (2015) Ministry of Social Security and Ministry of Finance. Statistical Yearbook of Work Accidents, vol 1, Brasília, MF

Canton JA, Fontes ARM, Torres I (2016) Post-implantation discussion of the possible contributions of ergonomics in the automation project of production lines. GEPROS. In: Production, Operations and Systems Management, Bauru, a. 11, n. 4, pp 267–283

Coelho FU (2004) Course of civil law, vol 2. Saraiva, São Paulo

Cordeiro R (1995) Epidemiological surveillance of occupational diseases: some ideas. Health Soc 4(1–2):107–110

Cunha GD (2002) A current overview of production engineering. http://www.abepro.org.br/arquivos/websites/1/PanoramaAtualEP4.pdf. Accessed 25 Mar 2017

Diehl AA, Tatim DC (2004) Research in applied social sciences: methods and techniques. Prentice Hall, São Paulo

Dul J, Weerdmeester B (2012) Practical ergonomics, 3rd edn. Blucher, São Paulo

European Agency for Safety and Health at Work (2002) Socioeconomic costs resulting from occupational accidents. www.fiocruz.br/…/safety%20e%20saude%20no%20work/FACTS%2027%20. Accessed 20 Nov 2017

Fernandes A (1995) Work accidents: from the sacrifice of work to prevention and reparation - legislative evolution, current events and perspectives. LTr, São Paulo

Ferreira LL (2015) On the ergonomic analysis of work or AET. Revista Brasileira de Saúde Ocupacional 40(131):8–11

Ferreira LS, Peixoto NH (2012) Worksafety I. Santa Maria: UFSM, CTISM, Sistema Técnica Aberta do Brasil

França ACL, Rodrigues AL (1996) Stress and work: basic guide with a psychosomatic approach. Atlas, São Paulo

Gil AC (2009) How to elaborate research projects, 4th edn. Atlas, São Paulo

Goncalves Filho AP, Ramos MF (2015) Accident at work in production systems: approach and prevention. Manage Prod 22(2):431–442

Guérin F, Laville A, Danillo F, Duraffourg J, Kerguelen A (2001) Buy-ender the work to transform it: the practice of ergonomics. Blucher: Vanzolini Foundation, São Paulo

Iida I, Guimarães LBM (2016) Ergonomics: design and production, 3rd edn. Blucher, São Paulo

Kroemer KHE, Grandjean E (2008) Ergonomics manual: adapting work to man, 5th edn. Bookman, São Paulo

Marques A, Tavares E, Souza J, Magalhães JA (2010) Ergonomics as a determining factor in the good progress of production: a case study. Revista Ana-grama Interdisciplinary Sci. J. Graduation a. 4, ed

Marras JP (2000) Administration of human resources: from operational to strategic, 3rd edn. Futura, São Paulo

Martins SP (2002) Social securitylaw, 18th edn. Atlas, São Paulo

Moraes AO, Silva RL (2015) Ergonomics in the work environment: workers' perception of the relationship between working conditions and turnover. REGIT, Fatec-Itaquaquecetuba, SP, vol 2, no 4, pp 107–125

Nascimento AM (2005) Course on labor law: history and general theory of labor law - individual and collective labor relations, 20th edn. Saraiva, São Paulo

Paraguay AIBB (1990) Stress, content and work organization: ergonomic contributions to improve working conditions. RevistaBrasileira de SaúdeOcupacional 70:40–43

Prates CCO. Historical evolution of accident legislation in Brazil. http://www.revistapersona.com.ar/Persona10/10Prates.htm. Accessed 03 Oct 2017

Silveira LBR, Salustiano EO (2012) The importance of ergonomics in the study of times and movements. R & D Prod Eng 10(1):71–80

Souza PRR. Epidemiology and health of the worker. www.imiarj.com.br/Epidemiologia.pdf. Accessed 05 Oct 2017

Yin RK (2001) Case study: planning and methods, 2nd edn. Bookman, Porto Alegre

Zocchio A (2002) Practice of accident prevention: ABC of work safety. 7. They are worth it: LTr

Driving the Company's Players to Take Ownership of Ergonomics

J.-P. Zana(✉)

Ergonomist ZConcept, 15 rue de Lagny, 75020 Paris, France
jeanpierrezana@gmail.com

Abstract. To ensure the success of this ownership initiative, companies have to adhere to 5 major key performance points:

1. show genuine commitment by the company's senior management,
2. mobilize all the necessary skills and make them work together,
3. identify the priorities,
4. integrate and organize the sustainability of the initiative,
5. devote sufficient time to planning and to exploiting feedback.

The first stage starts with a screening-diagnosis campaign, representative of the company's activities. The analyses of the work situations observed must enable companies to undertake the first preventive actions. The approach must be suited to the company, and the recommendations must promote preventive actions.

The second stage consists of a transfer of ergonomics skills by training people to become 'officers' in ergonomics as applied to the company's activities. This entails training-action in the field with workshops organized by implementing the methods and tools proposed for the analysis of the activity as well as a method for appraising the physical workload or physical demands of the job.

The third stage to be put in place involves sharing actions between plants, between factories (for manufacturing groups) and between companies within the same business sector, that could be coordinated by professional organizations.

It is fair to say today that those companies which have undertaken diagnoses, pursue their actions with the same desire to continue making progress in evaluating risks liable to cause MSDs (musculoskeletal disorders) and, above all, in rolling out a culture of prevention of risk exposure factors.

It is still too early to assert the role of this initiative in the decline in the number of MSDs, but companies committed to the issue are implementing long-term actions for the ongoing improvement of work situations and seeing the emergence of a prevention culture and a desire to involve all stakeholders.

Keywords: MSD · Applied ergonomics · Physical workload · Prevention

1 Intoduction

In 2007, at the request of the Technical Committee of the Chemical Industries of the Occupational Risks Department [*Comité Technique des Industries de la Chimie de la Direction des Risques Professionnels*] (DRP) of the French National Health Insurance

S. Bagnara et al. (Eds.): IEA 2018, AISC 825, pp. 105–109, 2019.
https://doi.org/10.1007/978-3-319-96068-5_11

Fund for Employees [*Caisse Nationale d'assurance Maladie des travailleurs salariés*] (CNAMTS), I was required to conduct ergonomic diagnoses in the perfumery, plastics processing and rubber industrial sectors. The idea was to figure out why these sectors were seeing more musculoskeletal disorders (MSDs) than the other sectors of the chemical industry.

The initial diagnoses, bolstered by the desire of both employers and employees to press ahead, have helped us design together, step by step, a prevention initiative which is now rolled out across other sectors of activity. For the past 10 years, companies that have undertaken to improve work situations are still pursuing the initiative, thereby fulfilling occupational health demands requiring iterative action plans.

2 The Applied Ergonomics Initiative

The initiative presented herein is the result of a co-construction with volunteer companies and technicians from the French Pension and Occupational Health Insurance Funds [*Caisses d'Assurance Retraite et de Santé au Travail*] (CARSAT) which monitor these companies on a daily basis to assist them in their prevention initiatives regarding occupational risks in general and MSDs in particular.

2.1 Condition of Implementation

In order to ensure that the initiative is a success, the technical committee has recommended 5 points which should be adhered to by companies wishing to engage in the initiative:

- show genuine commitment by the company's senior management,
- mobilize all the necessary skills and make them work together,
- identify the priorities,
- integrate and organize the sustainability of the initiative,
- devote sufficient time to planning and to exploiting feedback. Only two levels of headings should be numbered. Lower level headings remain unnumbered; they are formatted as run-in headings.

2.2 The Various Stages of the Global Prevention Initiative

Once the various stakeholders have confirmed their commitment, a degree of latitude should be given to those employees tasked with the operational capability of the initiative. In fact, every company should draw up an action plan taking account of the seasonality of its business, shutdowns, production peaks, etc. The introduction of an action plan and its follow-up at health and safety meetings are vital if the initiative is to be long-lasting. It is essential that the company's various players are trained in applied ergonomics soon after the diagnosis of work situations, so that the employees can see for themselves the initial achievements, which ensures their commitment from the onset. This training should involve employees from various sectors (human resources (HR), hygiene, safety and environment (HSE), production, maintenance, etc.).

The program of this essentially practical training–action is centered on the activities of the company's various sectors. The aim of the remainder of the initiative is a local rollout first, followed by a rollout across the professional sector, before reaching out to the outside.

The other stages of the initiative should be put in place gradually, involving the various departments based on the priorities that came to light following the diagnosis, albeit taking account of the employees' ability to get involved in the initiative.

Table 1. The various phases of the global prevention initiative

1.	SCREENING / DIAGNOSIS OF WORK SITUATIONS
•	Compilation of occupational injuries and diseases reports over the last 3 years,
•	Observation of work situations (workstations) and risk exposure factors,
•	Analysis of determinants and constraints of factors of exposure to the physical workload and MSD risks,
•	Reporting to the company and introduction of preventive action and follow-up plan.
2.	TRANSFER OF SKILLS
•	Training as ergonomics officers of employees from various departments (HR, HSE, maintenance, member of CHSCT [health, safety and working conditions committee], method, process based on the diagnosis and the requests of sectors, plants, departments…
•	Application of method to analyze the activity by involving employees,
•	Integration of social dialog/labor relations with the method for assessing the physical workload,
•	Identification of approaches to technical, organizational and human solutions by taking account of professional knowledge, expertise and know-how,
•	Iterative preventive actions by involving employees.
3.	ROLLOUT ACROSS THE COMPANY'S OTHER SECTORS OR DEPARTMENTS
•	Creation of a library of work situations, accessible to all sectors of activity,
•	Internal communication on the actions conducted,
•	Information or training-action for learning the methods used by the ergonomics officers,
•	Integration within the approach of the support departments (occupational health, technical design, purchasing, etc.) in order to raise their awareness of the company's work situation ergonomics.
4.	ROLLOUT REACHING OUT TO PROFESSIONAL BRANCH – EXTERNAL COMMUNICATION
•	Feedback to the other companies in the sector of activity and professional branches for sharing and a more global improvement of working conditions,
•	Sharing experiences at professional events, on social networks…
5.	SUSTAINABILITY OF THE PREVENTION INITIATIVE
•	Training of all new recruits and raising awareness of assessment of their activity,
•	Regular assessment of work situations using the same methods,
•	Follow-up of action plans by all company bodies.

Some companies organize meetings for their ergonomics officers twice a year in order to review improvements made, problems encountered, and training needs. The meetings are also an opportunity to add to the work situations library set up by one of them. This has become an essential sharing tool to avoid having to start from scratch in similar work situations or to continue improving certain workstations thanks to the progress made in some equipment or to modifications made to processes.

For companies concerned by the initiative described herein, a working meeting of ergonomics officers is planned by year-end or early 2019 so they can share their results/findings, queries and needs…

Sample Heading (Forth Level). The contribution should contain no more than four levels of headings. The following Table 1 gives a summary of all heading levels.

3 Results

This organization of the initiative is one that accommodates optimum ownership by the company's various players, of the occupational health and safety issue, by building on ergonomics principles. The kind of applied ergonomics principles that we implement should be adapted to the company's activities, the prevention already achieved, any existing prevention culture, and the ability of all the players to commit to the prevention initiative to aim for a better quality of life in the workplace.

The diagnosis, when it is pragmatic and close to the actual activity, is the starting point to the establishment of a social dialog on health and safety issues, with a systemic, global and shared approach.

The methodologies developed in the training of ergonomics officers are deployed to the employees, personnel representatives, the technology division or the maintenance methods department, and the management committee. There are two methods-tools developed by the prevention institution in France. Their implementation is straightforward, but requires much teaching expertise and pragmatism from the trainer. They involve the players concerned who are given analysis grids which they can share, in order to seek a consensus together on the operating methods and possible solutions.

These analysis matrices represent an objective foundation for identifying and assessing risk exposure factors, they encourage shared orientations for searching for prevention pathways, and they help inform the action plan and the risk assessment document. The first level of analysis through the risk identification grid is complemented by an in-depth analysis based on ergonomics standards. It should be noted that these are state of the art for the workload assessment methods or for the design of work situations that best protect the operators.

With is approach, it is fair to say that companies which have taken up the initiative establish their actions over the long term by a growing commitment of stakeholders who can see year on year the improvements made to working conditions. Prevention is an issue that is raised practically every day in some factories. An identical approach should be rolled out to encourage return to work. Another issue that we cover but is at a fledgling stage is the involvement of medical players, those involved in collective and individual occupational health and those in charge of the individual health of patients who have consulted them, but who know nothing (or next to nothing) about any links between their patients' jobs and their pathologies/conditions. To this should be added the fact that the other therapists treating these employees-patients are no more familiar with the issue.

To approach this topic, we draw on the recommendations of the International Social Security Association (ISSA) which, in a series of documents entitled Vision Zero, proposes guidelines for prevention, promotion of health, and return to work.

4 Conclusion

The initiative being put forward is a co-construction, deployed since 2007, first at professional organizations in the perfumery, rubber and plastics processing sectors. It was later extended to the chemical sector as well as in other service and industrial sectors with volunteer companies which, a few years later, are continuing the actions initiated. The installation of what can be regarded as a prevention culture must continue with the international recommendations also suggesting return to work and job retention. By way of conclusion, let me quote two brief precepts chosen by two groups of ergonomics officers anxious to rally round their company's stakeholders even more vigorously:

- "People" at the heart of the company
- Aware of risks and responsible for risks – For a better life together.

References

1. Atain-Kouadio J-J, Claudon L, Maziere P, Meyer J-P, Navier F, Turpin-Legendre E, Verdebout J-J, Zana J-P (2014) Analysis method of physical workload for the prevention of MSDs, INRS, ed6161
2. Zana JP, Victor B, Gravier E, Bulquin Y, Katlama Rouzet V, Burgaud M (2009) Démarche de prevention durable des TMS – engagement de trois organisations professionnelles
3. Lollia P-A (2016) Les compétences à l'épreuve du dialogue, FMT Mag, N° 118
4. Zana JP, Meyer JP, Turpin-Legendre E (2012) Application of European and international ergonomics standards for the manual handling activities of lifting, moving and pushing/pulling, poster in 18th World Congress on Ergonomics, Recife, Brazil
5. Daniellou F, Simard M, Boissières Y (2016) Human and Organizational Factors in industrial safety, les specifications of industrial safety, 2009-04, ICSI editor LNCS. http://www.springer.com/lncs. Accessed 21 Nov 2016

An Ergonomic Program in a Chemical Plant of Rhodia/Solvay in Brazil

Valmir Azevedo[1,2(✉)]

[1] Rhodia Solvay, Paulínia, Brazil
valmir.azevedo@solvay.com
[2] Unicamp University, Campinas, São Paulo, Brazil
vazevedo@fcm.unicamp.br

Abstract. Since 2002, an ergonomic program, assumed by the management of the industrial site, including the participation of employees, being done comes to identify, analyze, solve and control ergonomic risk factors at work. The steps of the ergonomic process are: 1. training of the employees in ergonomics (main concepts, risk factors identification, simplified and some advanced methods of ergonomic analysis, record activities/tasks; discussion and choice of solutions, action plan, implementation of solution, follow-up); 2. formation of the working group by plant; 3. identification of activities with potential risk to cause damage to health; 4. ergonomic analysis of activities/tasks at workplace and photo/filming of them; 5. discution of collected data, seeing photos/films and classification of risk degree, in group; 6. discussion and choice of solutions in group; 7. elaboration of technical report; 8. fill indicator «ergostatus» ; 9. elaboration of action plan by the team of the plant; 10. implementation of solution; 11. follow-up after implementation of solution; 12. update the indicator. The main achievements were: office furniture adequacy; labs furniture and practices adequacy; elimination of manual lifting to loads >20 kg; mounting of several mechanical lifting load devices; installation of force reduction devices in valves; automation of processes (ex.: filling and charging barrels at logistic; silica amorphous packing); improvement of management of equipments (fork-lifts in logistic); the initiative of each area or plant to identify and eliminate ergonomic risk factors (nowadays each area or plant has an employee in charge of ergonomics issues).

Keywords: Ergonomics · Chemical Plant · Worker's participation

1 Introduction

In Brazil, companies that hire employees following the CLT - Consolidation of Labor Laws - have the legal obligation, among many, to apply the recommendations contained in the Ergonomics standard, to NR17. This standard presents the general lines of approach to ergonomics related to work and refers to methods of ergonomic analysis according to the specificity of each work activity. Many multinational companies, following their specific HSE policies, also apply ergonomic risk assessment procedures to their units. The company Rhodia/Solvay, a chemical industry, based in Belgium, but with sites installed in several countries applies, in the case of HSE subjects, including

S. Bagnara et al. (Eds.): IEA 2018, AISC 825, pp. 110–115, 2019.
https://doi.org/10.1007/978-3-319-96068-5_12

Ergonomics, at least the legislation of the country where the site is installed. An Ergonomics Program assumed by the management of the industrial site, based on the NR 17 standard and on the rigor of the updated scientific technical knowledge of this discipline, added by employee participation has been underway at the Paulínia plant, in Brazil, since 2002. In the following topics we present the details of this program (Photo 1).

Photo 1. Paulínia Chemical Plant of Rhodia/Solvay in Brazil.

2 The Steps of an Ergonomic Program

The steps of the ergonomic process are: 1. training of the employees in Ergonomics (main concepts, risk factors identification, simplified and some advanced methods of ergonomic analysis, record activities/tasks; discussion and choice of solutions, action plan, implementation of solution, follow-up); 2. formation of the working group by plant; 3. identification of activities with potential risk to cause damage to health; 4. ergonomic analysis of activities/tasks at workplace and photo/filming of them; 5. discution of collected data, seeing photos/films and classification of risk degree, in group; 6. discussion and choice of solutions in group; 7. elaboration of technical report; 8. fill indicator «ergostatus» ; 9. elaboration of action plan by the team of the plant; 10. implementation of solution; 11. follow-up after implementation of solution; 12. update the indicator (Photo 2).

Photo 2. Work group in ergonomics – Paulinia Chemical Plant of Rhodia/Solvay in Brazil.

3 Main Achievements

The main achievements were: office furniture adequacy; labs furniture and practices adequacy; elimination of manual lifting to loads >20 kg; mounting of several mechanical lifting load devices; installation of force reduction devices in valves; automation of processes (ex.: filling and charging barrels at logistic; amorphous silica packing); improvement of management of equipments (forklifts in logistic); the initiative of each area or plant to identify and eliminate ergonomic risk factors (nowadays each area or plant has an employee in charge of Ergonomics). The solution to eliminate or reduce the intensity of ergonomic risk factors range from simple adjustments of the positioning of a computer monitor on a desk to complex modifications that involve automating the packaging of chemical substances in powder or in the packing of liquid products. In parallel, gymnastics activities in the workplace are performed daily by the teams (Photo 3).

4 Main Results

In 2009, in a critical analysis system (CAS) covering all HSE practices, there were 91 tasks with high risk to cause injury in the upper limbs or in the vertebral column. By 2011 all these tasks were resolved and, as a consequence, the risk of illness was reduced or eliminated. In the last critical analysis system (CAS), in February 2018, there were only 05 tasks, with moderated risk degree to solve. As a result of the ergonomics program, the occurrence of WRMD or a more severe Low Back Pain cases are rare events (Photo 4).

Photo 3. Before: manual moving barrels in Paulínia Chemical Plant of Rhodia/Solvay in Brazil.

Photo 4. Mechanical moving barrels and automatic filling barrels in Paulínia Chemical Plant of Rhodia/Solvay in Brazil.

Photo 5. Signaling about ergonomics in Paulínia Chemical Plant of Rhodia/Solvay in Brazil.

Photo 6. Gymnastic at work in Paulínia Chemical Plant of Rhodia/Solvay in Brazil.

5 Conclusion

In conclusion, the political will of the industrial site management of a Chemical Plant of Rhodia Solvay, in Paulínia, Brazil, associated with employee participation and technical scientific support were determinant to reduce or eliminate the ergonomic risk factors of many work activities (Photos 5 and 6).

References

1. Taylor FW (1995) Princípios da Administração Científica. Atlas, São Paulo
2. Guerin F (2008) Compreender o trabalho para transformá-lo. Edgar Blucher, São Paulo
3. Daniellou FA (2004) Ergonomia em Busca de Seus Princípios. Edgar Bluccher, São Paulo
4. Wisner A (2003) A Inteligência no Trabalho. Fundacentro, São Paulo
5. NR 17 (1990) Ministério do Trabalho, Brazil
6. www.iea.cc
7. Couto HA (1995) Ergonomia Aplicada ao Trabalho, vol I. Ergo Editora, Belo Horizonte
8. Couto HA (1996) Ergonomia Aplicada ao Trabalho, vol II. Ergo Editora, Belo Horizonte
9. Couto HA (2000) Novas Perspectivas na Abordagem Preventiva das LER/DORT. Ergo, Belo Horizonte
10. Chaffin AM (2001) Biomecânica ocupacional. Ergo Editora, Belo Horizonte
11. Wilson JR (2000) Fundamentals of ergonomics in theory and practice. Appl Ergonomics 31:557–567
12. Moore JS, Garg A (1995) The Strain Index: a proposed method to analyze jobs for risk of distal upper extremity disorders. Am. Ind. Hyg. Assoc. J. 56:443–458
13. Colombini D, Occhipinti E (1996) Proposta di un indice sintetico per la valutazione dell`esposizione a movimenti repetitive degli arti superiori (OCRA Index). Medicina del Lavoro. Med Lav 87(6):526–548
14. Waters TR, Putz-Anderson V, Garg A, Fine LJ (1993) Revised NIOSH equation for the design and evaluation of manual lifting tasks. Ergonomics 36(7):749–776
15. Waters TR, Lu ML, Piacitelli A, Werren D, Deddens JA (2011) Efficacy of the revised NIOSH lifting equation to predict risk of low back pain due manual lifting: expanded cross-sectional analysis. J Occup Environ Med 53(9):1061–1067
16. Blehm C, Vishnu S, Khattak A, Mitra S, Yee RW (2005) Computer vision syndrome: a rewiew. Surv Ophthalmol 50(3):253–262
17. Reason J (2000) Human error: models and management. BMJ 320(7237):768–770

Ergonomic Analysis on the Assembly Line of Home Appliance Company

Isabel Tacão Wagner[1], Jessica Nogueira Gomes e Silva[1], Vanessa Rezende Alencar[1], Nilo Antonio de Souza Sampaio[1,2], Antonio Henriques de Araujo Junior[1], Jose Glenio Medeiros de Barros[1], and Bernardo Bastos da Fonseca[1(✉)]

[1] Rio de Janeiro State University - FAT, Resende, RJ, Brazil
bernardobastosf@gmail.com
[2] Educational Association Dom Bosco, Resende, RJ, Brazil

Abstract. Benefits provided by ergonomics application in assembly systems design are linked to the reduction in occupational injury risks and to the improvement of physical and psychosocial conditions of the workforce with a drastic reduction in all costs linked to absence, medical insurance, and rehabilitation. Moreover, ergonomics improvements improve quality and operators productivity. An activity analysis was performed in a fan assembly production line. The study was developed in a company located in the Rio de Janeiro State that produces home appliances. The activity analysis consisted of systematic observations and interviews with the assembly line operators to understand the activities that are developed in the assembly of a fan, besides identifying possible limitations and existing occupational hazards in some stage of production. In the fans assembly line it was identified that operators does repetitive up and down movements in order to open a bag to packed the fan, which can cause pain, injury and muscle fatigue. As a solution, it was proposed to install a blower at the workstation that allows the opening of the bag and it prevents the operator performs repetitive movements and reduces activity time. This study shows the need for applications of ergonomic improvements that enable positive changes in the fans assembly line. The study is going to present the company and to expose the solution with future gains in relation to production efficiency and operator health.

Keywords: Ergonomics · Assembly systems design · Production process

1 Introduction

Assembly lines are production systems that include serially located workstations in which operations (tasks) are continuously processed. They are employed in various industries like the automobile, electronics, home appliances, etc., where the objectives is to produce large series of the same or similar products [1]. Several activities performed in assembly systems; in particular those associated with repetitive movements and with considerable level of stress or with extended assumption of uncomfortable

S. Bagnara et al. (Eds.): IEA 2018, AISC 825, pp. 116–123, 2019.
https://doi.org/10.1007/978-3-319-96068-5_13

postures, might be correlated to the insurgence of work related musculoskeletal disorders - WMSDs [2]. It is possible to notice a strong link between assembly systems and ergonomics, both in theory and in practice.

Benefits provided by ergonomics application in assembly systems design are linked to the reduction in occupational injury risks and to the improvement of physical and psychosocial conditions of the workforce with a drastic reduction in all costs linked to absence, medical insurance, and rehabilitation [3]. Furthermore, ergonomics improvements improve quality and operators productivity [4].

A healthy work environment adapted to the worker has become a reality. The appropriate comfort and accident prevention methods and specific pathologies for each specific activity are offered to the individual to avoid poor posture and repetitive strain injuries. When applied in an organization, the company provides a favorable environment in the workday, improving posture, preventing diseases and reducing sedentary lifestyle. In addition, it contributes to reducing out-of-pocket costs and increasing productivity.

The ergonomic analysis is an intervention stage and addresses the task (prescribed work) and activity (actual work) [5]. This analysis is done through observation and notes, which should contain information about the activity being carried out, the worker's behaviors as well as the working conditions to which he is subjected. To do so, it is necessary to interview with the operators so that they can clarify about the activities that execute as well as clarify doubts that may appear during the observation.

The study was developed in a company located in the Rio de Janeiro State, Brazil, that produces home appliances; the activity analysis was performed in a fan assembly production line and the analysis was performed in a fan assembly production line. The purpose of this paper is to analyze the physical ergonomics in a workstation on the assembly line of ventilators. With the purpose of exposing solutions and improvements to the employees, facilitating and helping in their performance, and to the company, in view of the increase in production.

2 The Assembly Line System Studied

The fans assembly line process consists of three phases: assembly, testing and packaging (see Fig. 1). The assembly area is composed of eight cells, where each cell is composed of an assembly workstation (station 1) and a test station (station 2). For every four cells, there is a packaging workstation (station 4) and dispatch (station 3). The focus of the study was on the packing station (see Fig. 2). The blue parts represent where all parts and supplies are positioned to power the assembly line.

In the packaging process (station 4), the operator disassembles part of the fan and picks up a plastic bag to pack the fan body. Then the operator puts the fan components in a box, where an operator is responsible for closing the box and placing it in a space to be taken to the company's stock. The cycle of the stage where the fan body is packed with the plastic bag was determined by the company to be carried out in 5.6 s. However, it has been observed that this activity is developed in 12.5 s due to two identified aspects: the disposal of the bags and static electricity. Just to open the bag,

Fig. 1. Fan assembly line.

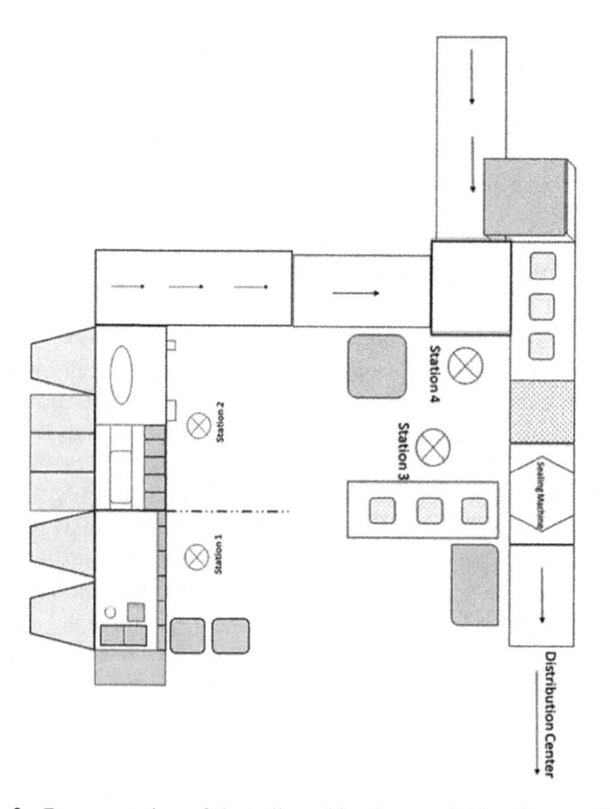

Fig. 2. Representation of the cell working layout. (Color figure online)

it takes 8.4 s. This situation increases the activity time of opening the bag and causes the operator to perform repetitive up and down movements in order to open it, which can cause pain, injury and muscle fatigue.

3 Methods

The methodology used to perform the work is based on activity analysis through direct and indirect observation, interview, as well as a documentary analysis [6, 7]. The direct observation used was intensive, consisting of two phases: (1) the first, based on observation and (2) the second one, based on interview. In this first stage, we obtained information about reality through the analysis of the systematic process of the facts or phenomena we wish to study.

Data collection was performed as follows:

- Follow-up of the operator's work routine through visits.
- Analysis of the activities carried out, observing the time spent by each one.
- Verification of posture, movements and furniture used to perform tasks.

In the interview phase, the researcher was able to talk to the operator to get more information and to know their feelings about the work.

In indirect observation, however, the process was studied through other means without being in person, through photos and videos. This method allows you to view, review and pause several times each phase of activity, being able to check details that were sometimes imperceptible in face-to-face visits.

Documentary analysis was done both before and after the visits. Before, it was necessary to do a field research, analyzing the company, its concept, what is produced, its areas, etc. At the end of the visits it was necessary to check some data to validate some information between the prescribed work (task) and the actual (activity).

4 Results and Discussion

In the activity analysis of the station 4 at the assembly line it was observed that operator took on average 12.5 s to perform the activity to open the plastic bag and put it in the base. This task is scheduled to take place in approximately 5.6 s according to with the procedure (prescribed work) of the company. In addition, the operator performs repetitive movements with the arms that do not match those that would be ideal, according to the principles of ergonomics.

It was observed that the time is due to the difficulty of the operator to open the bag of packaging which can occur for the following reasons: bags ready so disorganized and the static plastic bags.

In relation to handling, the operator performs repetitive movements that may become uncomfortable and may cause muscle fatigue and pain, such as: move the arms repeatedly upwards and downwards, in order to open the bag, and twisting of the body, to pick up the bags (see Fig. 3).

Fig. 3. Repetitive movement up and down to open the bag.

4.1 Proposed Design Submitted to the Company

Based on a detailed analysis made through videos of the activities, it was concluded that a faster, ergonomic and organized open the bags would be necessary and would bring significant improvements to the operator productivity and reduction in occupational injury risks. Below, the two proposed solutions.

Automatic Blower
An automatic blower reduces significantly operator effort in trying to open the bag. Only be done the movement to carry the bag to the nozzle and trigger the blower. In addition, reduce the time required for the activity of opening of bags, which, as noted, has difficulties to be performed.

By timing the videos obtained in the interviews, it was observed that the operator present at the moment, it took 8.4 s on average to pick and open the bag, totaling approximately 29 s for the total package of the product (3 and 4). In addition, it was also measured the time of assembly and test 1 and 2, to determine the total time of the fan assembly and packaging, which was estimated at 192 s.

According to the Table 1 that refers to the calculations performed, it is possible to notice that with this total time of assembly, would be produced at the end of a month (24 days) 72000 fans from the model of 40 cm, considering the production of 8 cells in three shifts existing in the company, being that operators use 6 h and 40 min of the 8 h of work, since it is deducted 1 h of lunch and two pauses of 10 min in the shift.

Due to the difficulty of obtaining the time trial that the blower would open the bags, it was assumed that the opening time of the bags with automatic blower would be for 5 s. With this, retracing the same calculations arrives at the conclusion that would be produced 73152 fans, so 1152 fans more per month. In the Table 2 below is shown the calculations performed.

Table 1. Assembly of fans according to the data collected.

Time and quantity of assembled fans without the blower - actual situation	
Time do assembly a fan (in seconds)	192 s
Time available for assembling fan in one shift	24000 s
Quantity assembled in one shift per day	125
Quantity assembled in the three shifts	375
Quantity assembled in the three shifts X 24 work day	9000
Total amount (fans) per month in 8 cells	72000

Therefore, such a measure would be effective in increasing the productivity of the company because it would have the time of assembling, testing and packaging the product reduced by 2%, so that the company produces a larger quantity at the same time available for assembling.

Table 2. Assembly of fans according to proposed design

Time and quantity of assembled fans with the blower – future situation	
Time do assembly a fan (in seconds)	189 s
Time available for assembling fan in one shift	24000 s
Quantity assembled in one shift per day	125
Quantity assembled in the three shifts	375
Quantity assembled in the three shifts X 24 work day	9000
Total amount (fans) per month in 8 cells	72000

Comparatively, a decrease of approximately 3 s in the production of the fan would bring a great increase in the volume of production.

However, it should be noted that the company's production is seasonal and the analysis was performed in the month of October, next to the summer season when the sale of fans increases. If the data were obtained in other months, the results could be different.

Plastic Bag Holder

The use of a "plastic bags holder", complementary to the blower auto helps in the organization of bags, causing the work environment also become more organized. These plastic bags holders may be purchased with ease and have low cost and in addition, facilitate the operator in the open bag activity.

Another alternative for the better disposition of the bags would be the use of a coil of plastic bags, already with the marked cuts. Similar to the plastic bag reels on the market, the reel would be resting on a carrier facing the operator, which would cause fewer movements to operator (see Fig. 4). In addition, the coil holder has a blade that cuts each bag, which facilitates the opening.

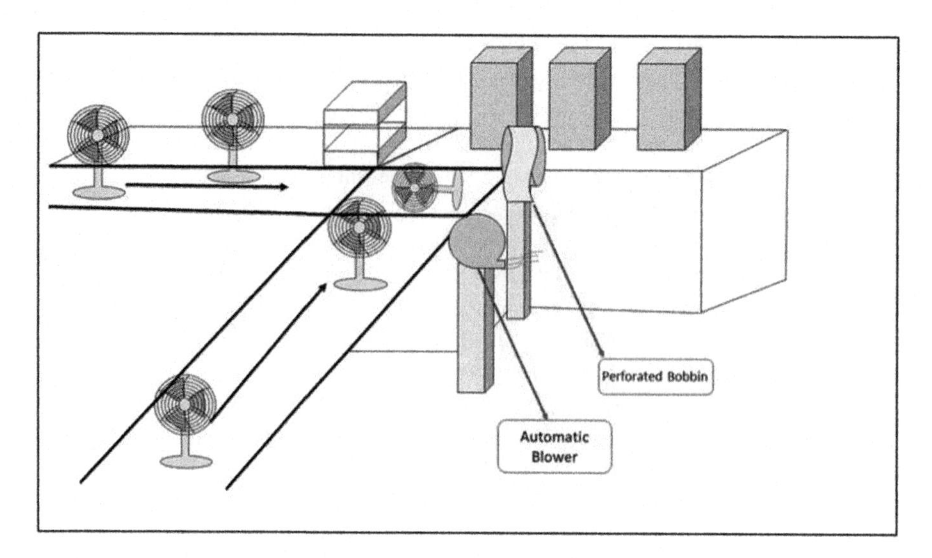

Fig. 4. Workstation with the proposed solutions.

The adoption of these presented solutions can bring greater movement savings to the worker because they would be better positioned, thus reducing the possibility of stress, fatigue, pain among other related pathologies and, therefore, producing greater health and satisfaction to the operator. For this, it is advisable to develop a prototype with the solutions to perform tests and make the necessary adjustments.

5 Results and Discussion

In a time when technology is present in many ways in assembly lines, there are still several activities where there is physical effort on the part of the operator.

In this way, the ergonomic study is extremely important. Ergonomics is based on the study of human beings and their environments, trying to adapt the environment to human work taking into account its limitations and variation of size. The application of this scientific study can bring several benefits, such as improvement in the quality of life of the worker, reduction in the number of absences and absences, reduction of waste, professional valorization and productivity.

Through the case studied in this paper we obtained results from the application of ergonomic improvements that indicated the need to implement changes in the process of packaging production of fans, as the deployment of a blower and a plastic bag holder. From these results and other studies carried out, it can be noted that the ergonomics allied technology results in significant advances, not only for the employee and the company.

References

1. Dolgui A, Gafarov E (2017) Some new ideas for assembly line balancing research. IFAC-PapersOnline 50(1):2255–2259
2. Wick JL, McKinnis M (1998) The effects of using a structured ergonomics design review process in the development of an assembly line. In: Kumar S (ed) Advances in Occupational Ergonomics and Safety. IOS Press
3. Carey EJ, Gallwey TJ (2002) Evaluation of human postures with computer aids and virtual workplace design. Int J Prod Res 40(4):825–843
4. Drury CG (2000) Global quality: linking ergonomics and production. Int J Prod Res 38 (17):4007–4018
5. Montmollin M (2011) A Ergonomia. Instituto Piaget, Brazil
6. Kirwan B, Ainsworth LK (1992) A guide to task analysis. Taylor & Francis, London
7. Stanton N et al (2005) Handbook of human factors and ergonomics methods. CRC Press, Boca Raton

Evaluation Metrics Regarding Human Well-Being and System Performance in Human-Robot Interaction – A Literature Review

Jochen Nelles(✉), Sonja Th. Kwee-Meier, and Alexander Mertens

Institute of Industrial Engineering and Ergonomics, RWTH Aachen University, Bergdriesch 27, 52062 Aachen, Germany
j.nelles@iaw.rwth-aachen.de

Abstract. This literature review provides an overview on evaluation metrics regarding human well-being and system performance in human-robot interaction. In this context a systematic literature search in the Web of Science and IEEE Xplore databases was carried out. Thus, 30 relevant contributions out of 3854 studies were analyzed by multistage filtering. To gain an overview on and compare the different approaches and results, the studies are summarized in tables according to the following criteria: author, year, title, task, study design, measurement methods, population, and results. The evaluation metrics presented in this contribution in principle can be divided into questionnaire-based surveys and psychophysiological measurement methods. In addition, the studies are classified into evaluation metrics for measuring well-being and performance.

The research was carried out with regard to an industrial engineering and ergonomics context, but is independent of specific application areas.

Keywords: Evaluation · Human-robot interaction · Literature review Metrics · Performance · Taxonomy · Well-Being

1 Introduction

The different taxonomies of human-robot interaction and robots are complex and can be divided into communication channel, robot task, physical and temporal proximity, kind of interaction, field of application, robot morphology, human interaction role, degree of robot autonomy, and team composition [1]. Evaluation metrics of human-machine-systems can be divided into theory-based, expert-based and user-based evaluation approaches. Expert evaluations are qualitative (e.g. they are carried out using a cognitive walkthrough) and are based on ISO 9241-110 [2], usability heuristics, style guides or other guides. User-based evaluations can be both qualitative (e.g. based on questioning, method of thinking aloud or self-confrontation) or quantitative (e.g. by means of video/audio analysis of user handling or eye tracking analysis). Furthermore, the evaluation can be sorted according to the evaluation date. In a summative evaluation, an already finished product is assessed globally and collectively. On the other hand, during a formative evaluation the assessment takes place during system design

S. Bagnara et al. (Eds.): IEA 2018, AISC 825, pp. 124–135, 2019.
https://doi.org/10.1007/978-3-319-96068-5_14

and the assessment procedure is integrated into the development process of an emerging product. A mixed form is the formative-summative evaluation, where a partially finished product is evaluated during the product development process, for example with regard to the ergonomic quality of individual system functions. User testing may be performed at different degrees of completion of the product, for example at early stages using models, scenarios or sketches, or at later stages using prototypes, laboratory tests or field validations [2–5].

Due to the heterogeneous taxonomies of human-robot interaction and a wide variety of evaluation metrics, a general method for the evaluation of human-robot interaction does not exist. The approach presented in this article focuses on the ergonomic design and evaluation of human-robot interaction. In "The Discipline of Ergonomics" issued by the International Ergonomics Association (IEA) from 2000, Ergonomics (or human factors) is defined as "the scientific discipline concerned with the understanding of the interactions among humans and other elements of a system, and the profession that applies theoretical principles, data and methods to design in order to optimize well-being and overall performance" [6].

This leads to the following research questions: (1) Which are the factors that influence human well-being and system performance of human-robot interaction?, and (2) which evaluation metrics can be used to determine human well-being and system performance of human-robot interaction?

In this paper, a literature review with regard to evaluation metrics of human well-being and system performance of human-robot interaction is presented. In order to conduct a scientific survey with regard to human well-being and system performance in the context of human-robot interaction, a systematic literature search in Web of Science and IEEE Xplore databases was carried out. Thus, 30 relevant contributions out of 3854 studies were analyzed by multistage filtering. In principle, the evaluations used in the studies can be divided into questionnaire-based surveys and psychophysiological measurement methods. In some cases, several questionnaires were combined and supplemented with a psychophysiological measurement method. Furthermore, the evaluation metrics differ on whether human well-being or system performance is assessed.

2 Methods

2.1 Search Strategy

A systematic search in two online databases was conducted to select papers published in peer-reviewed journals and conferences since 2000. In Web of Science, 1823 papers until 30th January 2017 and in IEEE, 2031 papers until 28th February 2017 were considered.

2.2 Inclusion Criteria

Studies were limited to those that were published in English. The full paper selection was based on the following steps: first, papers were excluded based on title; second, papers were excluded based on abstract; and last, papers were excluded based on full text. Additionally, five papers were included based on forward search. The following search keywords were combined: human* OR employee OR worker AND robot* AND interact* OR collaborat* OR cooperat*. Further groups of search keywords were supplemented with the refine results function on the basis of the intersection stated above: design approach OR design stud*, acceptance OR well-being OR human factor, workspace OR workshop OR working room OR working space, evaluat OR interview OR survey OR quest*, usability, framework, ergonom*, version*, workload, trust.

2.3 Methodological Quality Assessment

All full papers were assessed according to the following criteria: (1) 'title': does the title fit the searching context?; (2) 'task': what was the task of the participants?; (3) 'study design': did the authors describe the operationalization of the study in a comprehensible way, is the study replicable?; (4) 'measurements/method': what kind of measurement technology or questionnaire was used in the study? Is the method described sufficiently?; (5) 'population': does the study differentiate the subjects, e.g. according to age, gender or further characteristics? (6) 'results': does the study contain relevant results?

3 Results

The initial search identified 3854 references. After title and abstract screening, 53 references were retrieved for full text appraisal. After full text appraisal, 25 studies were identified to meet the inclusion criteria. In addition, 5 papers based on forward search were integrated (Tan et al. 2008–Tan et al. 2010). Altogether, 30 studies were included in this review. Figure 1 shows a flow diagram of the study selection (see Fig. 1). The extracted details of the included studies can be found in Table 1.

4 Discussion

It should be noted that the studies in Table 1 and Fig. 2, which are based on the inclusion criteria of the literature review, only represent a part of the metrics for the evaluation of well-being and performance. Finally, it is noticeable that the experimental design with the dependent and independent variables as well as the questionnaires and measures used are very heterogeneous. Thus, there is no general answer to the research questions: (1) Which are the factors that influence human well-being and system performance of human-robot interaction?, and (2) which evaluation metrics can be used to determine human well-being and system performance of human-robot interaction?

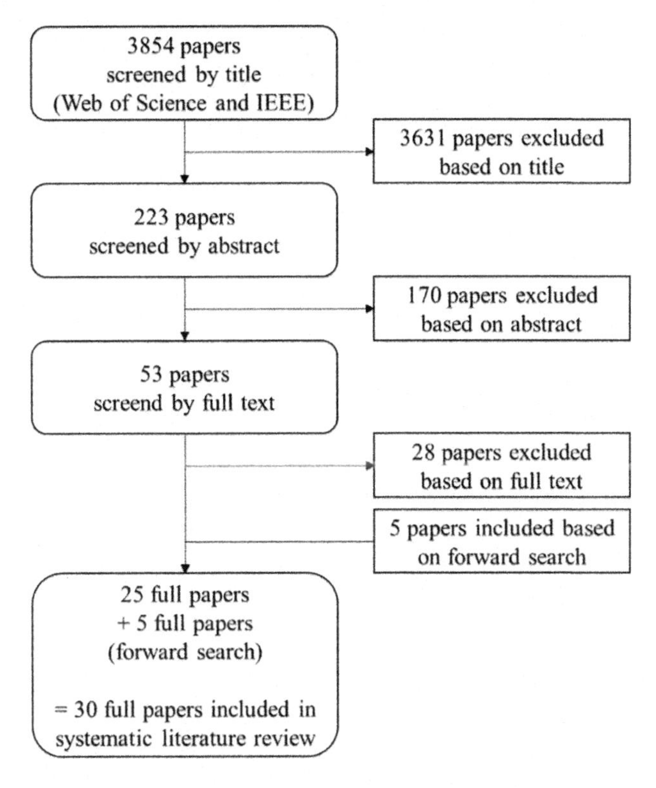

Fig. 1. Flow diagram of the paper selection

However, the evaluation metrics presented in this article can help to select suitable instruments for the evaluation of human-robot interaction. Basically, the literature review of the studies presented in Table 1 can be divided into 24 evaluation metrics for well-being and performance. Furthermore, the metrics can be divided into questionnaire-based surveys and psychophysical measurement methods. Figure 2 is based on Table 1 and shows an overview of the evaluation metrics, with the primary sources added to the figure (see Fig. 2) [37–42].

The selection of suitable evaluation methods depends, inter alia, on the individual application context, the resulting experimental design, and the associated dependent and independent variables. Using the example of the described application context – the support of emergency personnel under increased psychological stress during detection of improvised explosive devices and determination of the hazard potential with a remote-controlled robotic manipulator using sensor data – further questionnaires and measurement methods like eye tracking or electrooculography should be considered [43].

Table 1. Extracted details of included studies, in alphabetical order (n = 30)

Author	Title	Task	Study design	Measurements	Sample	Results
Bortot et al. [7]	Directly or on detours? How should industrial robots approximate humans?	Investigation of the influence of the trajectory in the movement of the true center point on the personal feelings of the subjects	Scenario 1: visual interaction, scenario 2: audio task	Subjective evaluation, state trait anxiety inventory (STAI-S), NASA TLX	19 participants	A variable trajectory has a negative influence on the well-being of the participants
Charalambous et al. [8]	The Development of a scale to evaluate trust in industrial human-robot collaboration	-Questionnaire design	Literature review. Questionnaire design and test phase. Elimination of questions with regard to Cronbach's Alpha	Trust scale questionnaire, depending on human (safety, experience), robot (performance and physiological properties) and task	176 students	"Development trust scale" questionnaire, 10 questions, five-point Likert scale
Daniel et al. [9]	Simplified human-robot interaction: modeling and evaluation	Robot control with two control concepts (old vs. new)	Robot control with old and new control concept.	Efficiency (time to complete), usability, general trust in automation, video analysis	16 participants	New robot control concept is faster
Dehais et al. [10]	Physiological and subjective evaluation of a human-robot object hand-over task	Manual handover task between robot and participant	Three handover tasks	Subjective self-evaluation; psychophysiological measurement: skin conductance, muscle deltoid activity	12 participants	Subjective evaluation is consistent with physiological measurement
Hanrock et al. [11]	A meta-analysis of factors affecting trust in human-robot interaction	-Meta-analysis	-Meta-analysis: 10 studies with correlating factors and 11 studies with experimental effects	Trust dependent upon human (12 factors), robot (13 factors) and environment (8 factors)	–	Human and environmental factors hardly affect trust. Biggest influences are factors related to the robot, especially performance and characteristics of the robots
Kessler et al. [12]	A Comparison of trust measures in human-robot interaction scenarios	Navigation with Lego Mindstorm robot	Two studies with different information density	Trust in automation scale vs. human robot trust scale, Negative attitudes towards robotics (NARS)	58 students (15m, 43f)	Scale provides different results
Liu et al. [13]	A projection-based making-human-feel-safe system for human-robot cooperation	Robot trajectories and workspace monitoring	Workspace monitoring with/without projection	In-house development questionnaire with regard to feeling of safety	6 participants	Without projection the feeling of safety was lower

(continued)

Table 1. *(continued)*

Author	Title	Task	Study design	Measurements	Sample	Results
Maurtua et al. [14]	FourByThree: imagine humans and robots working hand in hand	–	Prototype of human-robot interaction	Real-time calculation of rapid upper limb assessment (RULA) score	–	Integration of RULA score in robot and software design was shown to be beneficial for a more ergonomic body posture
Murata et al. [15]	Ergonomics and cognitive Engineering for robot-human cooperation	–	Study on the basis of men-are-better-at/machines-are-better-at principle	–	–	Various evaluation methods: workload/NASA TLX, psychophysiological methods: pupillary fluctuation, electroencephalography
Murphy et al. [16]	Survey of Metrics for Human-Robot Interaction	- Meta-analysis	Meta-analysis: 29 studies with regard to metrics of human-robot interaction	–	–	Taxonomy of human-robot interaction metrics: human-robot system (productivity, efficiency, reliability, safety, coactivity)
Nomura et al. [17]	Prediction of human behavior in HRI using psychological scales for anxiety and negative attitudes toward robots	Human-robot interaction and communication	Development of questionnaires NARS and RAS, answering questionnaires before and after an experiment	Negative attitudes towards robotics (NARS), robot anxiety scale (RAS), state trait anxiety inventory	38 students (22m, 16f)	NARS and RAS scores are gender independent. Men interact differently with robots than women at a same score.
Novak et al. [18]	Psychophysiological responses to different levels of cognitive and physical workload in haptic interaction	Haptic interaction with manipulator	Three levels of cognitive and physical workload	Psychological state, complexity of the tasks, psychophysiological methods: heart rate, respiratory rate, skin conductance, skin temperature	30 participants (age: 19–46, 23m, 7f)	Use of psychophysiological methods is possible with light and heavy physiological work load
Ogorodnikova et al. [19]	Methodology of safety for a human robot interaction designing stage	- Model	Summary of risk assessment methods and relevant factors for safe human-robot interaction	Fault tree analysis, product failure mode and effects analysis (FMEA)	–	Risk assessment method for human robot interaction
Pini et al. [20]	Evaluation of operator relief for an effective design of HRC workcells	-Simulation of partly automated human robot workplace	Comparison between simulated partly automated human robot workplace and simulated manual assembly workplace	Modified rapid upper limb assessment score, energy consumption (kcal), efficiency (time to complete)	–	At partly automated workplaces workload is reduced, task execution is faster and energy consumption is lower

(continued)

Table 1. (*continued*)

Author	Title	Task	Study design	Measurements	Sample	Results
Profanter et al. [21]	Analysis and semantic modeling of modality preferences in industrial human-robot interaction	Four assembly tasks, e.g. Pick & Place	Four different input modalities: touch, gesture, 3D pen, and speech	Questionnaires regarding Usability, age and experience, video analysis	30 participants (23m, 7f)	Gesture control is the favored robot control concept
Rahman et al. [22]	Trust-based compliant robot-human handovers of payloads in collaborative assembly in flexible manufacturing	Payload handover between human and robot	Six subtasks with handover between human-robot and robot-human. Handover strategy/grade of cautions robot to human handover depending on robot's trust in human co-worker	Measurement of robot's trust in human co-worker depending on human hand speed/efficiency/performance, handover success rate/human errors; questionnaires regarding workload with NASA TLX, human's trust in robot with Likert scale	20 students	Measurement of robot's trust and handover adjustment increases success rate of payload handover
Sadrfaridpour et al. [23]	An integrated framework for human-robot collaborative assembly in hybrid manufacturing cells	Bringing a set of two fitting parts and installing them on the top of the main part brought by the robot-process is repeated 7 times	Pilot study with manual control and advanced control depending on human performance	Trust in automation scale, workload NASA TLX, performance	5 participants	Trust is higher in advanced robot mode
Sanders et al. [24]	The influence of modality and transparency on trust in human-robot interaction	Two monitors with primary task (robot control) and secondary task (camera observation)	Three levels of communication (constant, context-sensitive, low), three modalities (text, acoustic, graphical)	Human robot trust, Trust in automation, workload NASA TLX	71 participants (age: 18–22)	Graphical modality leads to greatest trust, constant flow of communication is preferred, information overflow had no negative impact
Sim et al. [25]	Extensive assessment and evaluation methodologies on assistive social robots for modelling human–robot interaction – A review	-Review	Review: 110 studies with regard to assessment and evaluation methodologies of assistive social robots	–	–	Overview of strengths, weaknesses and uniqueness of assessment and evaluation methodologies

(*continued*)

Table 1. *(continued)*

Author	Title	Task	Study design	Measurements	Sample	Results
Tan et al. [26–31]	Six (forward search) publications regarding human-robot interaction are based on each other	Cable assembly, distribution of subtasks between human and robot	Varying	Task and collaboration analysis, mental workload depending on motion speed and human-robot distance, system performance	5 experts, 5 participants, 7 participants	Mental workload increased as the robot motion speed increased. Mental workload increased as distance was reduced
Weiss et al. [32]	The USUS Evaluation Framework for Human-Robot Interaction	-Model	Evaluation framework for human-robot interaction consisting of expert evaluation, user studies, questionnaires, physiological measures, focus groups and interviews	Usability, social acceptance, user experience and societal impact	–	Evaluation framework/methodological mix
Weiss et al. [33]	A methodological adaptation for heuristic evaluation of HRI	Video-based heuristic evaluation of usability	Expert evaluation of three human-robot interaction scenarios	Usability heuristics	10 experts	Video-based heuristic evaluation valuable instrument for getting information about problems
Weiss et al. [34]	First application of robot teaching in an existing industry 4.0 environment: does it really work?	Three scenarios with/without robot: manufacturing/teaching, laboratory/teaching, assembly/collaboration	Three human-robot interaction scenarios	Negative attitudes towards robots, workload NASA TLX, system usability scale, acceptance questionnaire, video and audio analysis	5 participants for each scenario	Higher acceptance after tests with human-robot interaction
Weistroffer [35]	Assessing the acceptability of human-robot co-presence on assembly lines: a comparison between actual situations and their virtual reality counterparts	Human-robot interaction during assembly of a car door: manual assembly vs. virtual environment	Real-world vs. virtual reality assembly	Six point Likert scale questionnaire: acceptance, efficiency, safety, usability; performance: timing, errors; psychophysiology: heart rate, electrodermal response	7 participants (6m, 1f)	VR is well suited for assembly activity simulation. In the questionnaire, close cooperation with the robot is not considered negative, but electrodermal activity increases.
Yagoda and Gillan 2012 [36]	You want me to trust a robot? The development of a human–robot interaction trust scale	Handling of unmanned aeroplane and unmanned vehicle	Study 1 expert interview: assessment of human-robot factors, study 2 simulation: assessment of unmanned aeroplane and unmanned vehicle	–	11 experts (more than 5 years experience), 100 participants	Human robot interaction (HRI) trust scale

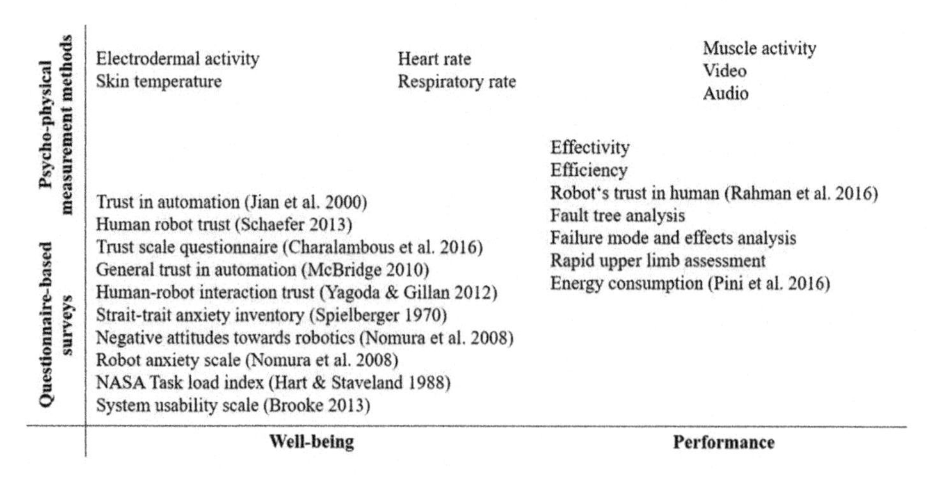

Fig. 2. Evaluation metrics of well-being and performance, divided into questionnaire-based surveys and psychophysical measurement methods [37–42].

Acknowledgments. The literature review was performed within the research project "DUCH-BLICK" (Grant No. 13N14329), funded by the German Federal Ministry of Education and Research (BMBF) in the context of the national program "Research for Civil Security" and the call "Civil Security – Aspects and Measures of Coping with Terrorism". In addition, this publication is part of the research project "MeRoSy" (Grant No. 16SV7190), which is also funded by the German Federal Ministry of Education and Research (BMBF).

References

1. Onnasch L, Maier X, Jürgensohn T (2016) Mensch-Roboter-Interaktion – Eine Taxonomie für alle Anwendungsfälle. 1. Edition. Dortmund: Bundesanstand für Arbeitsschutz und Arbeitsmedizin (baua Fokus), project number: F 2369, PDF file. https://doi.org/10.21934/baua:fokus20160630 (in german)
2. ISO 9241-110 (2006) Ergonomics of human-system interaction – Part 110: Dialogue principles
3. ISO 9241-210 (2010) Ergonomics of human-system interaction – Part 210: Human-centered design for interactive systems
4. Nelles J, Brandl C, Mertens A (2018, in German) Regelkreismodell für die menschzentrierte Gestaltung und Evaluierung einer Mensch-Roboter-Interaktion am Beispiel eines Mensch-Roboter-Arbeitsplatzes. In: Arbeit(s)Wissenschaf(f)t – Grundlage für Management und Kompetenzentwicklung. 64. Kongress der Gesellschaft für Arbeitswissenschaft
5. Schlick CM, Bruder R, Luczak H (2010, in German) Arbeitswissenschaft. Springer, Heidelberg
6. Dul J, Bruder R, Buckle P, Carayon P, Falzon P, Marras WS et al (2012) A strategy for human factors/ergonomics: developing the discipline and profession. Ergonomics 55 (4):377–395. https://doi.org/10.1080/00140139.2012.661087

7. Bortot D, Born M, Bengler K (2013) Directly or on detours? How should industrial robots approximate humans? In: Kuzuoka H (ed) 2013 8th ACM/IEEE International Conference on Human-Robot Interaction (HRI), 3–6 March 2013, Tokyo, Japan (including workshop papers). IEEE, Piscataway, pp 89–90
8. Charalambous G, Fletcher S, Webb P (2016) The development of a scale to evaluate trust in industrial human-robot collaboration. Int J Soc Robot 8(2):193–209
9. Daniel B, Thomessen T, Korondi P (2013) Simplified human-robot interaction: modeling and evaluation. Model. Identif. Control Norw. Res. Bull. 34(4):199–211
10. Dehais F, Sisbot EA, Alami R, Causse M (2011) Physiological and subjective evaluation of a human-robot object hand-over task. Appl Ergon 42(6):785–791
11. Hancock PA, Billings DR, Schaefer KE, Chen JYC, de Visser EJ, Parasuraman R (2011) A meta-analysis of factors affecting trust in human-robot interaction. Hum Factors 53(5):517–527
12. Kessler TT, Larios C, Walker T, Yerdon V, Hancock PA (2017) A comparison of trust measures in human–robot interaction scenarios. In: Savage-Knepshield P, Chen J (eds) Advances in human factors in robots and un-manned systems, Proceedings of the AHFE 2016 international conference on human factors in robots and unmanned systems, 27–30 July 2016, Walt Disney World®, Florida, USA. (Advances in Intelligent Systems and Computing, vol 499, pp 353–364. Springer, Cham
13. Liu D, Kinugawa J, Kosuge K (2016) A projection-based making-human-feel-safe system for human-robot cooperation. In: Proceedings of 2016 IEEE international conference on mechatronics and automation, 7–10 August 2016, Harbin, China, pp 1101–1106. IEEE, Piscataway
14. Maurtua I, Pedrocchi N, Orlandini A, Fernandez JG, Vogel C, Geenen A, Althoefer K, Shafti A (2016) FourByThree: imagine humans and robots working hand in hand. In: 2016 IEEE 21st international conference on emerging technologies and factory automation (ETFA), 6–9 September 2016, Berlin, Germany, pp 1–8. IEEE, Piscataway
15. Murata A (2000) Ergonomics and cognitive engineering for robot-human cooperation. In: Proceedings of the 9th IEEE international workshop on robot and human interactive communication, IEEE RO-MAN 2000, 27–29 September 2000, pp 206–211. Osaka Institute of Technology, Osaka, IEEE Service Center, Piscataway
16. Murphy RR, Schreckenghost D (2013) Survey of metrics for human-robot interaction. In: Kuzuoka H (ed) 2013 8th ACM/IEEE international conference on human-robot interaction (HRI) (including workshop papers), 3–6 March 2013, Tokyo, Japan, pp 197–198. IEEE, Piscataway
17. Nomura T, Kanda T, Suzuki T, Kato K (2004) Psychology in human-robot communication: an attempt through investigation of negative attitudes and anxiety toward robots. In: Proceedings of the 13th IEEE international workshop on robot and human interactive communication, RO-MAN 2004, 20–22 September 2004, Kurashiki, Okayama, Japan, pp 35–40. IEEE Operations Center, Piscataway
18. Novak D, Mihelj M, Munih M (2011) Psychophysiological responses to different levels of cognitive and physical workload in haptic interaction. Robotica 29(03):367–374
19. Ogorodnikova O (2008) Methodology of safety for a human robot interaction designing stage. In: Conference on human system interactions, HSI 2008, 25–27 May 2008, Krakow, Poland, pp 452–457. IEEE, Piscataway
20. Pini F, Ansaloni M, Leali F (2016) Evaluation of operator relief for an effective design of HRC workcells. In: 2016 IEEE 21st international conference on emerging technologies and factory automation (ETFA), 6–9 September 2016, Berlin, Germany, pp 1–6. IEEE, Piscataway

21. Profanter S, Perzylo A, Somani N, Rickert M, Knoll A (2015) Analysis and semantic modeling of modality preferences in industrial human-robot interaction. In: Burgard W (ed) 2015 IEEE/RSJ international conference on intelligent robots and systems (IROS), 28 September 2015–2 October 2015, Hamburg, Germany, pp 1812–1818. IEEE, Piscataway
22. Rahman SMM, Sadrfaridpour B, Wang Y (2016) Trust-based optimal subtask allocation and model predictive control for human-robot collaborative assembly in manufacturing. In: Proceedings of the ASME 8th annual dynamic systems and control conference – 2015, presented at ASME 2015 8th annual dynamic systems and control conference, 28–30 October 2015, Columbus, OH, USA. The American Society of Mechanical Engineers, New York
23. Sadrfaridpour B, Saeidi H, Wang Y (2016) An integrated framework for human-robot collaborative assembly in hybrid manufacturing cells. In: 2016 IEEE international conference on automation science and engineering (CASE), pp 462–467. IEEE
24. Sanders TL, Wixon T, Schafer KE, Chen JYC, Hancock PA (2014) The influence of modality and transparency on trust in human-robot interaction. In: 2014 IEEE international inter-disciplinary conference on cognitive methods in situation awareness and decision support (CogSIMA), 3–6 March 2014, San Antonio, TX, USA, pp 156–159. IEEE, Piscataway
25. Sim DYY, Loo CK (2015) Extensive assessment and evaluation methodologies on assistive social robots for modelling human–robot interaction – a review. Inf Sci 301:305–344
26. Tan JTC, Duan F, Zhang Y, Arai T (2008) Task decomposition of cell production assembly operation for man-machine collaboration by HTA. In: IEEE international conference on automation and logistics, ICAL 2008, 1–3 September 2008, Qingdao, China, pp 1066–1071. IEEE, Piscataway
27. Tan J, Duan F, Zhang Y, Arai T (2009) Extending task analysis in HTA to model man-machine collaboration in cell production. In: IEEE international conference on robotics and biomimetics, ROBIO 2008, 22–25 February 2009, Bangkok, Thailand (rescheduled from December 2008, original conference date: 14–17 December 2008), pp 542–547. IEEE, Piscataway
28. Tan JTC, Duan F, Zhang Y, Watanabe K, Kato R, Arai T (2009) Human-robot collaboration in cellular manufacturing: design and development. In: 2009 IEEE/RSJ international conference on intelligent robots and systems, IROS 2009, St. Louis, Missouri, USA, 10–15 (i.e. 11–15) October 2009, pp 29–34. IEEE, Piscataway
29. Tan JTC, Zhang Y, Duan F, Watanabe K, Kato R, Arai T (2009) Human factors studies in information support development for human-robot collaborative cellular manufacturing system. In: The 18th IEEE international symposium on robot and human interactive communication, RO-MAN 2009, 27 September 2009–2 October 2009, Toyama, Japan, pp 334–339. IEEE, Piscataway
30. Tan JTC, Arai T (2010) Analytic evaluation of human-robot system for collaboration in cellular manufacturing system. In: IEEE/ASME international conference on advanced intelligent mechatronics (AIM), 6–9 July 2010, Montreal, QC, Canada, pp 515–520. IEEE, Piscataway
31. Tan JTC, Arai T (2010) Information support development with human-centered approach for human-robot collaboration in cellular manufacturing. In: IEEE RO-MAN, 19th IEEE international symposium on robot and human interactive communication, 13–15 September 2010, Viareggio, Italy, pp 767–772. IEEE, Piscataway
32. Weiss A, Bernhaupt R, Lankes M (eds) (2009) The USUS evaluation framework for human-robot interaction AISB2009: proceedings of the symposium on new frontiers in human-robot interaction

33. Weiss A, Wurhofer D, Bernhaupt R, Altmaninger M, Tscheligi M (2010) A methodological adaptation for heuristic evaluation of HRI. In: IEEE RO-MAN, 19th IEEE international symposium on robot and human interactive communication, 13–15 September 2010, Viareggio, Italy, pp 1–6. IEEE, Piscataway
34. Weiss A, Huber A, Minichberger J, Ikeda M (2016) First application of robot teaching in an existing industry 4.0 environment. Does it really work? Societies 6(3):20
35. Weistroffer V, Paljic A, Fuchs P, Hugues O, Chodacki J-P, Ligot P, Morais A (2014) Assessing the acceptability of human-robot co-presence on assembly lines: a comparison between actual situations and their virtual reality counterparts. In: RO-MAN, The 23rd IEEE international symposium on robot and human interactive communication, 25–29 August 2014, Edinburgh, Scotland, UK, pp 377–384. IEEE, Piscataway
36. Yagoda RE, Gillan DJ (2012) You want me to trust a ROBOT? The development of a human–robot interaction trust scale. Int J Soc Robot 4(3):235–248
37. Brooke J (2013) SUS: a retrospective. J Usability Stud 8(2):29–40
38. Hart SG, Staveland LE (1988) Development of NASA-TLX (Task Load Index): results of empirical and theoretical research. Adv Psychol 52:139–183
39. Jian J-Y, Bisantz AM, Drury CG (2000) Foundations for an empirically deter-mined scale of trust in automated systems. Int J Cognit Ergon 4(1):53–71
40. McBridge SE (2010) The effect of workload and age on compliance with and reliance on an automated system. Georgia Institute of Technology, Ph.D. thesis
41. Schaefer KE (2013) The perception and measurement of human-robot trust. Ph.D. thesis, University of Central Florida, Florida
42. Spielberger C, Gorusch R (1970) State-trait anxiety inventory, manual for the state-trait anxiety inventory. Consulting Psychologist Press
43. Kwee-Meier S, Nelles J, Mertens A, Schlick C (2017, in German) Integrierte Informationsvisualisierungen bei der Mensch-Roboter-Kooperation zur Unterstützung von Einsatzkräften unter erhöhter psychischer Belastung. In: Soziotechnische Gestaltung des digitalen Wandels – kreativ, innovative, sinnhaft. 63. Kongress der Gesellschaft für Arbeitswissenschaft

Towards an Engineering Process to Design Usable Tangible Human-Machine Interfaces

Michael Wächter[1(✉)], Holger Hoffmann[2], and Angelika C. Bullinger[1]

[1] Chemnitz University of Technology, 09111 Chemnitz, Germany
michael.waechter@mb.tu-chemnitz.de
[2] innosabi GmbH, Widenmayerstraße 50, 80538 Munich, Germany

Abstract. Internet of Things (IoT) technologies gave rise to a multitude of new opportunities and challenges for creating human-machine-interfaces. Especially for mobile devices like tablets, tangible user interfaces (TUI), providing the haptic controls for manipulating the GUI, and in extension tangible human-machine interfaces (tHMI), also including hardware elements for handling and operating the computing device, are to be designed. Currently only rudimentary design guidelines exist and state-of-the-art engineering methods are not comprehensive enough to cover the elaborate design process. In this paper, we follow a design science based approach to develop and evaluate an engineering process for usable tHMI. We empirically evaluate the engineering process by creating a tHMI for a mobile assistance system for maintenance workers, focusing on the subjective quality measures for the created artifacts. Measuring electrical activity in involved muscles as an objective benchmark, we show that tHMIs created using our method are significantly better than current available commercial alternatives. The prototype resulting from our instantiation provides a basis for useable mobile devices in an industrial context and is evaluated as highly suitable for use by domain experts. Hence our engineering process is one of the first comprehensive approaches that covers all aspects of designing, building and evaluating tHMI for IoT devices in production settings and is supplemented with a choice of rigorous methods for evaluating usability in the field.

Keywords: Usability · tangible Human Machine Interface · Internet of Things

1 Introduction

In the wake of Internet of Things (IoT) technologies and the ubiquitous availability of computing devices in all areas of human work, new opportunities and challenges for creating human-machine-interfaces arose [1, 2]. For those types of ubiquitous computing devices not only new software interfaces (GUI) are to be designed, but also tangible user interfaces (TUI) providing the haptic controls for manipulating the GUI and in extension tangible human-machine interfaces (tHMI), also including hardware elements for handling and operating the computing device [3, 4]. One prime example are mobile devices like tablets, which initially developed for consumer use now find

S. Bagnara et al. (Eds.): IEA 2018, AISC 825, pp. 136–147, 2019.
https://doi.org/10.1007/978-3-319-96068-5_15

their way into production settings, e.g. for machine control. When designing usable tHMI for these devices to be used on the shopfloor special requirements from industry, e.g. one-handed versus two-handed operation, must be considered. However, only rudimentary design guidelines exist and current engineering methods are not comprehensive enough to cover the design process. In this paper, we follow a design science based approach [5] to develop and evaluate an engineering method for usable tHMI.

2 Research Method

Our research design is based on the framework of design science research (DSR) according to [5]. The systematic realization of solutions for a problem relevant in practice – in our case the need for a holistic approach to the user-centered design of usable tangible human-machine interfaces – through the iterative and scientifically rigorous design and evaluation of artifacts.

We focus on the detailed analysis of the requirements of for an engineering process and its instantiation. Focus group of our research are planners, i.e. the managers responsible for planning industrial assistant systems, and developers of shop floor maintenance systems. This ensures the practicality of the engineering process. For the development of the engineering process, we apply product ergonomics, system engineering, interaction design, user-centered design, human-computer interaction and (rapid) prototyping as our foundation from the scientific knowledge base. The evaluation of the engineering process is realized iteratively in the context of case studies according to [6]. To facilitate this, the individual phases of the engineering method are applied in field studies. Figure 1 gives an overview of our research method as an instantiation of the research design according to [5].

3 Development of the Engineering-Process

3.1 Requirement Analysis

Purpose of Requirement Analysis. An artifact of DSR is considered successfully implemented if it fulfills previously collected requirements. This includes requirements elicited from the application domain (i.e., planners and developers in the production environment) as well as requirements derived from the solution domain (e.g., by limitations of existing solutions) [5]. In addition, requirements can be deduced from procedural models in the scientific literature for a user-centered design of human-machine interfaces, which represent the status quo of the knowledge base.

Systematic Approach. In order to capture the requirements for an engineering method for the design of usable tHMI, we conducted a structured literature review as an initial first step to assess the existing knowledge base [7]. Additionally, we conducted a focus group (n = 5) to analyze the requirements of planners and developers, i.e. users of user-centered development processes in industrial practice.

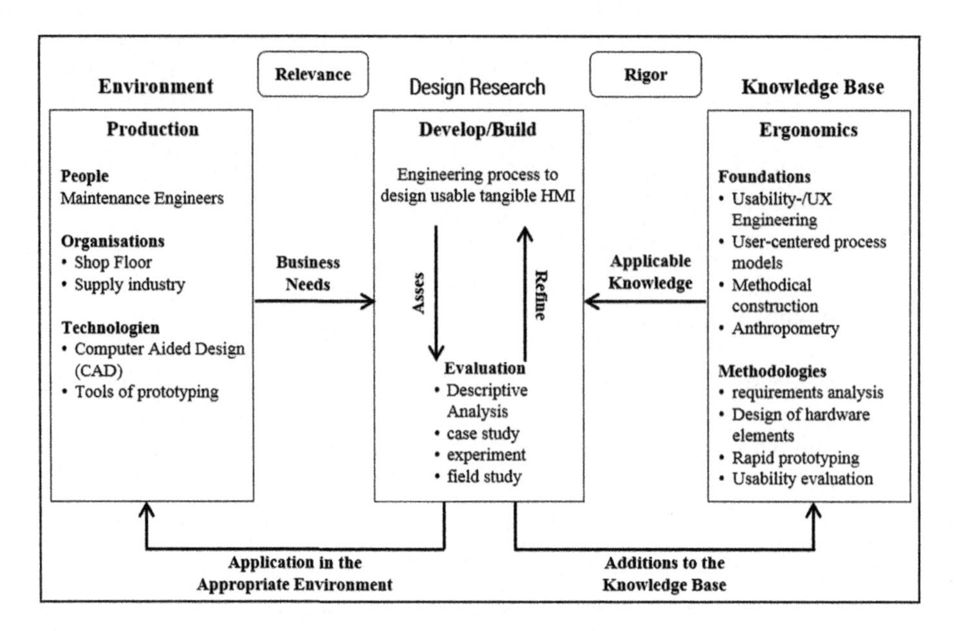

Fig. 1. Instantiation of Hevner et al.'s design science research framework

Identified Requirements. The requirements we identified can be classified as requirements for user-centered process models, general requirements for methods used and specific requirements for the design of tangible HMI. Table 1 details the identified requirements for user-centered process models from the perspective of planners and developers.

Table 1. Requirements for user-centered process models of planners and developers.

Requirement	Explanation	Source
Specific methods	Context-specific methods for user-centered development for the individual phases	[8–11]
Users' evaluation	Current involvement of users in the evaluation of product models	[8, 10–12]
Early prototypes	Early design of prototypes to integrate user feedback into further development	[8, 12–14]

The listed requirements complement the basic rules of user-centered design. Furthermore, the following general requirements for the given procedures and tools arise (Table 2):

Table 2. General requirements for methods used.

Requirement	Explanation	Source
Standardized methods	Use of standardized methods for a comparable quality of the results	[8, 10]
Low application time	Use resource-efficient methods to minimize time and costs	[8, 11, 13–17]
Simple evaluation	For the evaluation of the used methods no previous knowledge should be necessary	[8, 10, 14, 17]

Existing methods and tools for the user-centered development of software interfaces can only partially be transferred to the design of tangible human-machine interfaces. Table 3 shows the specific requirements for the development of useable tHMI:

Table 3. Specific requirements for the design of tangible HMI.

Requirement	Explanation	Source
Design methods	Support of planners and developers by methods for designing tangible HMI	Focus group with planners and developers (n = 5) of assistance systems for production
Consideration of anthropometry	Anthropometric variables are an integral part of tangible HMI design	
Evaluation methods	Support of planners and developers by methods for evaluating tangible HMI	

User-centered process models described in the literature only meet these requirements to a limited extent, especially with regard to the proposed methods. The identified requirements for methods are an essential basis for creating the engineering process.

3.2 Engineering Process

The basic structure of our proposed engineering process results from the analysis of existing models of usability engineering, UX engineering and methodological design and consists of four phases, highlighted in Fig. 2. Each process phase contains the basic steps of a user-centered approach - analysis, design, prototyping and evaluation [18]. For this purpose, the users of the engineering method are provided with context-specific methods that correspond to the identified requirements of the planners and developers. By designing a tangible HMI, methodological prototyping has special significance within the engineering process.

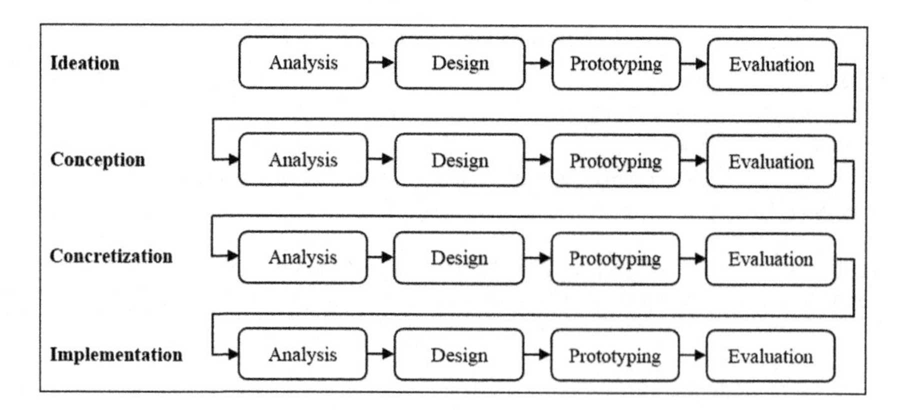

Fig. 2. Structure of the engineering process

- Ideation, to study the user requirements for an assistance system as well as the context of use and the identification of central functions. In addition, various design drafts for the individual functions are the basis for the following phases.
- Conception, to first develop an ergonomic design of the handles and to analyze existing solutions in literature and practice. Together with the results of ideation, different handle shapes are created, which are evaluated by future users and provide suggestions for a final handle shape.
- Concretization, to design a final handle shape based on the findings in the conception phase. The identified functions of ideation are also analyzed, designed, prototyped and evaluated by future users.
- Implementation, to design a functional model that includes all identified functions prototypically and is evaluated in terms of its usability within the context of a real-world application scenario. The results of this final study demonstrate the usability of the developed tHMI.

4 Evaluation of the Engineering Process

4.1 Use Case

Digitized maintenance as part of industrial production provides maintenance workers with all relevant information such as machine condition, maintenance history and repair instructions, in-situ on the shopfloor via mobile devices [19, 20]. Currently available mobile devices for industrial use lack significantly concerning systemic requirements resulting from changes in the domain due to digitization and user requirements. Although hardware manufacturers offer mobile solutions for the production and logistics sector, they only consider industrial features such as robust construction, dust and splash water protection [1]. For our evaluation we designed the artifact iteratively over four case/field studies – to test the suitability of the identified methods for analysis, design, prototyping and evaluation. As part of the successive studies, the tangible man-machine interface was developed for use with a mobile assistance system for maintenance personnel. Every phase of the process was carried out together with planners, developers and users from the maintenance department of three major German car manufacturers/supplier. Figure 3 shows the designed prototypes as final results of the respective phases.

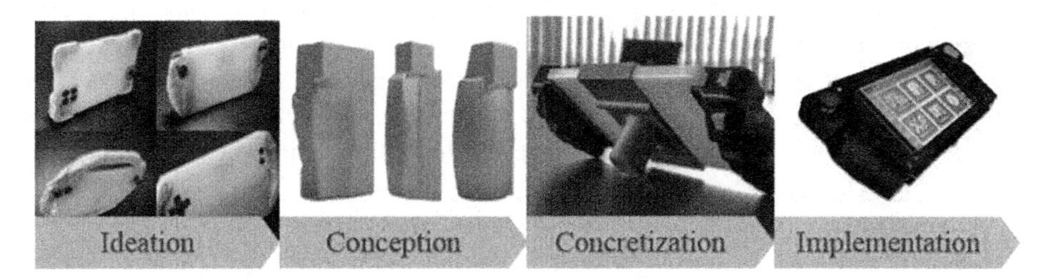

Fig. 3. Prototypes in the different phases.

4.2 Subjective Evaluation – Field Study

Ideation. Following the first phase, non-involved maintenance workers (n = 6) evaluated the prototyped handle shapes out of clay. Their positive feedback shows that the process-affiliated methods enable an efficient and practicable identification of the requirements. The resulting requirements have been confirmed by maintenance staff and demonstrate the successful use of the proposed methods.

Conception. The evaluation of the three resulting handles was implemented in three focus groups with maintenance workers (n = 15). While the design for handle form 1 is based on the dimensions for actuators [21], handle 2 is based on a mobile panel currently used in production settings. The base for handle 3 is the combined draft design from the initial Ideation phase. The evaluation task was to use the three different concept models with one hand and two hands, as the users would do in their daily jobs. In settings randomized following the Latin square [22], the maintenance staff evaluated the comfort of each handle with the CQH questionnaire [23] after each round, resulting in detailed data on ease of use with both-handed operation via physical controls and for one-handed touchscreen operation.

The analysis of the evaluation data shows that handle shapes 1 and 3 are preferred by our evaluation panel. On a 7-step Likert scale from A (very uncomfortable) to B (very comfortable), the maintenance workers rated the overall comfort of handle 1 with M = 5.00 (SD = 1.41) and handle 3 with M = 4.40 (SD = 1.54) best, while handle 2 with M = 2.27 (SD = 1.10) is rated significantly worse. The qualitative content analysis [24] for the collected feedback and the potential for improvement of the handles supports the quantitative results of the questionnaire. During conception it emerged that users would prefer a combination of the size of handle 3 and the ergonomic design of handle 1 through finger indentation.

Concretization. Due to the high influence of the handle shape on the handling of an assistance system, we evaluated the results of the concretization phase – the 3 handle shapes and the newly designed final handle shape – following the same evaluation approach as in the previous phase.

The descriptive analysis of the questionnaires (n = 15) confirms a better rating for the overall comfort for the grip forms 1, 3 and the final grip compared to handle 2. This is rated significantly worse again with M = 2.21 (SD = 0.58). The data on the perceived comfort of one- and two-handed gripping show interesting differences: The final grip is rated best with one-handed gripping with M = 5.5 (SD = 1.40), handle 3 with two-handed gripping with M = 6.21 (SD = 0.70). This result can be attributed to the ergonomic design of handle 4, which give the user more grip when handling one-handed (Fig. 4).

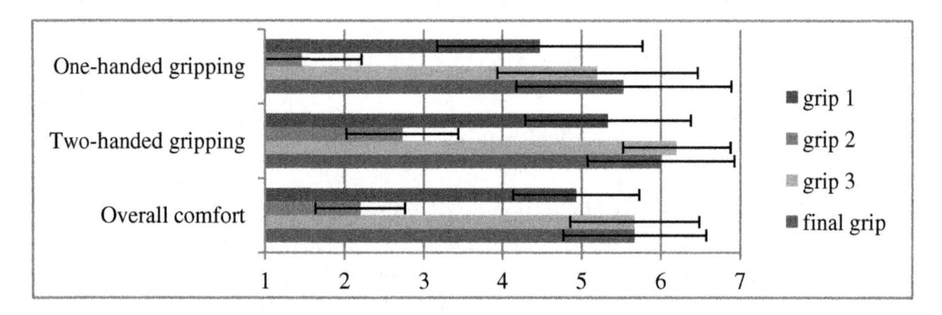

Fig. 4. Evaluation of the overall comfort of the various handles

While the quantitative data initially didn't show a clear improvement in the final grip (concretization phase) compared to handle 3 (conception phase), the qualitative data show that the final handle shape, as a result of the existing finger indentations and adjusted handle diameter as a function of anthropometry, felt significantly better becomes.

Implementation. The final evaluation is conducted using a fully functional model implemented as part of a usability test in a real maintenance scenario with potential future users (n = 20). To ensure the comparability of the reviews, the subjects went through the usability test twice. In a Within-Subject design, the subjects started randomized with a conventional tablet or the designed functional model. After each run, the maintenance workers evaluated their mobile device using the System Usability Scale (SUS) questionnaire. The results of the SUS show that the maintenance workers significantly prefer ($p < 0.01$) the tHMI designed following our engineering process with M = 88.75 (SD = 7.27) compared to the conventional tablet with M = 68.25 (SD = 11.73).

The subjective evaluation of the designed artifact shows that the engineering process with the identified procedures and tools meets the requirements of the application domain. Our user-centered approach makes it possible to design and test a usable tangible HMI of assistance systems in coordination with the users. In order to confirm the results of the subjective evaluation, an objective evaluation of the four grips was implemented.

4.3 Objective Evaluation – Laboratory Experiment

Approach. To confirm the users' positive subjective perception of the handle shape designed following our approach we conducted an additional laboratory experiment to gather objective measurement data using electromyography (EMG) to measure electrical activity of selected muscles. We measure those physiological markers for the different handle shapes during one-handed operation of the software interface of a standard tablet. During the evaluation, participants (n = 32, male, mean age: 31.6 years, all right-handed) hold the tablet using their left hand to interact with the software interface using their dominant right hand. After starting the evaluation software, ten randomized numbers are displayed in two rounds, which the test person enters via a keypad on the touch surface [3]. A randomization of the order following the Latin square [22] takes into account possible sequence and learning effects. During the experiment, we measure the five muscles suspected to be the most stressed in such a usage scenario using EMG (frequency: 1500 Hz): biceps brachii (BB), flexor carpi ulnaris (FCU), flexor carpi radialis (FCR), brachioradialis (BR), and M. flexor pollicis brevis (FPB). Before the subjects started the first trial run, the maximum electrical activation of each muscle at maximum contraction (MVC) was determined by three static measurements over three seconds each. We use this MVC as a baseline to normalize EMG data recorded during the experiment to have a basis for comparing the measurement data between different subjects. The normalized values represent the percentage of electrical activation of the muscles in relation to the maximum activation and provide an indicator for the present strain of the individual muscles as a function of the handle shape considered [25]. After a three-minute recovery period, the subject began with the first handle shape. For each handle, three measurements were taken with two-minute rest periods.

Statistical Analysis. For the statistical analysis of the recorded data with SPSS, the data was cleaned in the first step. All outliers identified by boxplot (triple interquartile range) were extracted, we examined the resulting clean dataset using a RMANOVA followed by Bonferroni post-hoc testing. The results of the RMANOVA show significant differences between the different handle shapes. Their degrees of freedom were corrected depending on the sphericity according to Greenhouse Geisser. Depending on the muscles examined, the Bonferroni Post-Hoc Test shows significant differences in muscle strain between the different grip types.

Findings. The differences in the percent strain of subjects' biceps brachii show interesting results. Handle 2, rated worst by subjectively perceived comfort, causes a significantly ($p = 0.01$) lower strain in this muscle than the other grips with $M = 12.93$ percent of MVC ($SD = 7.21$). The BB is a large muscle of the upper arm, which is responsible for holding the tablet due to the flexion. Because the tested devices including handles have the same weight, this factor is excluded. However, with the measured load range in the lower third (between 12–18%), it can be assumed that there is only small influence on the perceived comfort of the handles (Fig. 5, lower values are better).

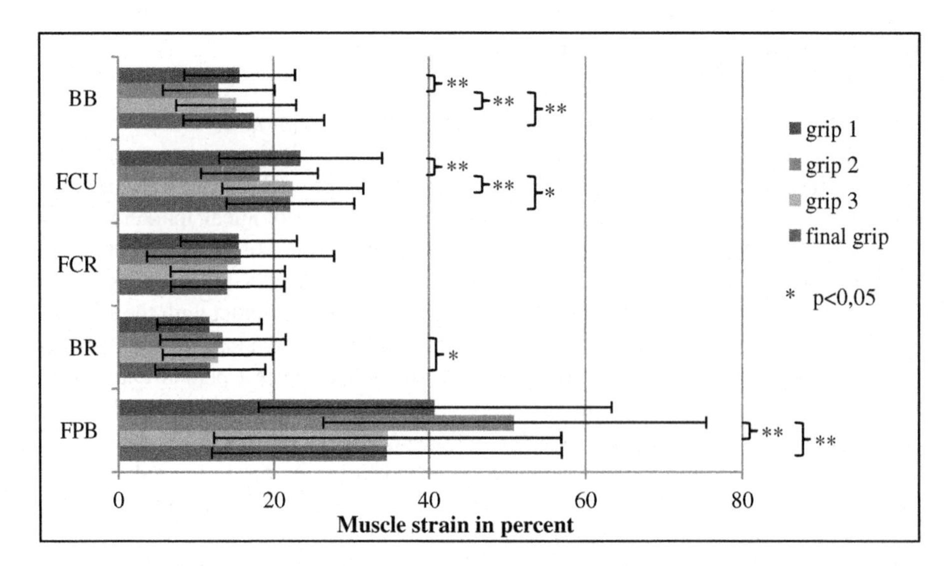

Fig. 5. Strain on the muscles during one-handed handling of the grip variants

The results for the flexor carpi ulnaris (FCU) show similar results. With M = 18.22 (SD = 7.5), grip 2 causes significantly lower stress (p = 0.01, p = 0.05) of the muscle, which indicates a comparably higher ulnar abduction when using the other handle variants [26].

There are no significant differences between the handles in evaluation for the flexor carpi radialis (FCR). The significant (p = 0.05) lower stress of the brachioradialis (BR) for grip variant 4 (M = 11.82, SD = 7.09) compared to grip variant 2 (M = 13,48, SD = 8,09) may indicate a changed semi pronation position [26], which is a consequence the different handle shape, but does not indicate a significant difference in the subjective comfort rating.

In the results of the M. flexor pollicis brevis (FPB), which is responsible for the gripping movement of the hand, subjects using grip 2 showed significantly higher stress (M = 50.84, SD = 24.45) compared to handles 3 and 4 (p = 0.001), which were rates best in the subjective comfort perception. The direct influence of this muscle on holding the handles and the high differences in the strain of about 15%, which deems to make further research in the causes necessary, suggest a direct correlation to the subjective comfort assessment via the questionnaire CQH and to the qualitative feedback of the users.

The results of our laboratory experiment show that the subjective results can also be objectively reproduced and that the methods and tools used in the engineering process allow valid statements about the design of handles. In summary, the process mix used shows that the inclusion of qualitative feedback for a user-centered product design is absolutely necessary, since users' suggestions for improvement via a questionnaire are difficult to determine and would otherwise not reach the product designers and thus go unnoticed.

5 Conclusion and Future Work

The in-situ application of the engineering process for developing an appropriate tHMI to make a commodity tablet usable on the shopfloor delivered three key insights. By applying the process along with the appropriate context-specific methods for each of the phases we show:

- that the engineering method provides various methods for analyzing the requirements and design of the functions, which are suitable for comprehensively and efficiently capturing the needs and requirements of the users as well as those requirements stemming from the application context of a shopfloor assistance system,
- that the design of the handles for an assistance system has a significant impact on the perception of user comfort and thus has an impact on the acceptance of the entire system,
- that our proposed evaluation methods can be successfully put into practice within the scope of the four phases of the development process. Furthermore, they provide important insights from future users for the subsequent design activities.

The evaluation of the designed artifact shows that the methods used in the engineering process fulfill the requirements of the planners and developers. Using the example of the mobile assistance system for maintenance workers we could demonstrate, that our user-centered approach enables designing and testing a subjectively and objectively usable tangible HMI of assistance systems in collaboration with potential users.

The functional model as a design artefact resulting from our process is suitable for the task it was designed for, outperforming existing commercial systems in key aspects. Low-cost prototyping technologies such as 3D printing enable rapid design and evaluation cycles within the individual process phases. This allows planners and developers as well as users to check their requirements for the tangible HMI interface at early stages and often during the whole design process. Consequently, planners and developers are given the opportunity to integrate user requirements in a resource-saving way throughout the entire design process, thereby ensuring that the tangible HMI is highly suitable for use when applying our process.

To evaluate our design process empirically we instantiate it to create a tHMI for a mobile assistance system for maintenance workers, detailing the individual iterative steps and focusing on the subjective quality measures for the created artifacts. We demonstrate that tHMI created using our method are significantly better than current available commercial alternatives by comparing objective measurements, the users' electrical activity in involved muscles. Additionally, we show our tHMI to be subjectively preferable to a conventional tablet by evaluating feedback from potential future users, employing methods like the CQH and SUS. The prototype resulting from our instantiation provides a basis for useable mobile devices in an industrial context and is rated as highly suitable for use. Hence our proposed engineering process can be considered one of the first comprehensive approaches that covers all aspects of designing, building and evaluating tHMI – especially for IoT devices in production settings – and is supplemented with a choice of rigorous methods for both subjectively and objectively measuring its resulting artefacts' usability in the field.

The attitude of the product designer regarding the importance of usability can have a significant impact on the implementation of the results of the evaluation with users of the tangible HMI. This results in additional research needs in the continuation of the evaluation of our engineering process with further examples. With a changed set of actors and stakeholders in the development team, the priority of usability among users of the engineering method should be further investigated. Additional potential lies in the application of our engineering method in other application domains, for example the design of tangible human-machine interfaces in assistance systems used for special interest groups like senior citizens and people with disabilities as well as tHMI in certain usage scenarios, e.g. in the vehicle or in sports. Given the suitability of the methods used, even outside of the domain production, the cost-effective use of 3D prototyping methods make it possible to establish the engineering method in the emerging start-up and maker scene and to effectively confirm the importance of product ergonomics.

References

1. Gorecky D, Schmitt M, Loskyll M (2017) Mensch-Maschine-Interaktion im Industrie 4.0-Zeitalter. In: Vogel-Heuser B, Bauernhansl T, Hompel Mt (eds) Handbuch Industrie 4.0 Allgemeine Grundlagen: Allgemeine Grundlagen, pp 219–236. Springer, Wiesbaden
2. Schmitt M, Meixner G, Gorecky D et al (2013) Mobile interaction technologies in the factory of the future. IFAC Proc. Vol. 46(15):536–542
3. Pereira AL, Miller T, Huang Y et al (2013) Holding a tablet computer with one hand: effect of tablet design features on biomechanics and subjective usability among users with small hands. Ergonomics 56(9):1363–1375
4. Wächter M, Bullinger AC (2016) Gestaltung gebrauchstauglicher tangibler MMS für Industrie 4.0 – ein Leitfaden für Planer und Entwickler von mobilen Produktionsassistenzsystemen. Zeitschrift für Arbeitswissenschaft 70(2):82–88. https://doi.org/10.1007/s41449-016-0020-0
5. Hevner AR, March ST, Park J et al (2004) Design science in information systems research. MIS Q 28(1):75–105
6. Venable J, Pries-Heje J, Baskerville R (2012) A comprehensive framework for evaluation in design science research. In: Peffers K, Rothenberger M, Kuechler B (eds) DESRIST 2012, vol 7286. LNCS. Springer, Heidelberg, pp 423–438
7. Wächter M (2018) Engineering-Methode zur Gestaltung gebrauchstauglicher tangibler Mensch-Maschine-Schnittstellen für Planer und Entwickler von Produtionsassistenzsystemen. Wissenschaftliche Schriftenreihe des Institutes für Betriebswissenschaften und Fabriksysteme, Heft 127, Chemnitz
8. van Kuijk J, Kanis H, Christiaans H et al (2015) Barriers to and enablers of usability in electronic consumer product development: a multiple case study. Hum Comput Interact 32 (1):1–71
9. van Eijk D, van Kuijk J, Hoolhorst F et al (2012) Design for usability; practice-oriented research for user-centered product design. Work 41(Suppl 1):1008–1015
10. Glende S (2010) Entwicklung eines Konzepts zur nutzergerechten Produktentwicklung - mit Fokus auf die "Generation Plus". Dissertation, Berlin
11. Bruno V, Dick M (2007) Making usability work in industry. In: Thomas B (ed) Conference of the computer-human interaction special interest group (CHISIG) of Australia, pp 261–270

12. Hemmerling S (2002) Evaluation in frühen Phasen des Entwicklungsprozesses am Beispiel von Gebrauchsgütern. In: Timpe K, Baggen R (eds) Mensch-Maschine-Systemtechnik: Konzepte, Modellierung, Gestaltung, Evaluation, 2. Aufl., Stand: Februar 2002. Symposion, Düsseldorf, pp 299–317
13. Boivie I, Gulliksen J, Göransson B (2006) The lonesome cowboy: a study of the usability designer role in systems development. Interact Comput 18(4):601–634
14. Vredenburg K, Mao J, Smith PW et al (2002) A survey of user-centered design practice. In: Wixon D (ed) The SIGCHI conference, p 471
15. Chilana PK, Ko AJ, Wobbrock JO et al (2011) Post-deployment usability. In: Tan D, Fitzpatrick G, Gutwin C et al (eds) The 2011 annual conference, p 2243
16. Gulliksen J, Boivie I, Göransson B (2006) Usability professionals—current practices and future development. Interact Comput 18(4):568–600
17. Rosenbaum S, Rohn JA, Humburg J (2000) A toolkit for strategic usability. In: Turner T, Szwillus G (eds) The SIGCHI conference, pp 337–344
18. Norman DA (2013) The design of everyday things: revised and expanded edition. Basic Books, New York
19. Scheer A (2013) Industrie 4.0: Wie sehen Produktionsprozesse im Jahr 2020 aus? IMC AG
20. Schröder C (2016) Herausforderungen von Industrie 4.0 für den Mittelstand. Gute Gesellschaft - soziale Demokratie #2017plus
21. Deutsches Institut für Normung (2010) Ergonomische Anforderungen an die Gestaltung von Stellteilen(894-3)
22. Sedlmeier P, Renkewitz F (2008) Forschungsmethoden und Statistik in der Psychologie. PS Psychologie. Pearson Studium, München [u.a.]
23. Kuijt-Evers LFM, Vink P, de Looze MP (2007) Comfort predictors for different kinds of hand tools: differences and similarities. Int J Industr Ergon 37(1):73–84
24. Glaser BG, Strauss AL (1967) The discovery of grounded theory: strategies for qualitative research Observations. Aldine Pub. Co., Chicago
25. Bischoff C, Schulte-Mattler WJ, Conrad B (2009) Das EMG-Buch: EMG und periphere Neurologie in Frage und Antwort, 2nd edn. Thieme e-book library, Thieme, Stuttgart
26. Schünke M, Schulte E, Schumacher U et al (2007) Prometheus - Lernatlas der Anatomie: Allgemeine Anatomie und Bewegungsystem, 2., überarb. und erw. Aufl. Thieme, Stuttgart [u.a.]

Thumb Plastic Guard Effect on the Insertion of Push Pins Using Psychophysical Methodology

Alejandro Iván Coronado Ríos[2(✉)], Delcia Teresita Gamiño Acevedo[1], Enrique Javier De la Vega Bustillos[1], and Francisco Octavio Lopez Millan[1]

[1] Industrial Engineer Department, Instituto Tecnológico de Hermosillo, Hermosillo, Sonora, Mexico
[2] Ford Stamping and Assembly Plant, Hermosillo, Sonora, Mexico
acoronl4@ford.com

Abstract. The present study was made through the usage of the psychophysical methodology applied to 14 women previously being trained for 4 days making insertion tests with the usage of their thumb finger on time frames of 8 h for five days to obtain knowledge. Maximum strength level in the insertion of 1 to 4 Push-pins using a thumb splint. During the investigation 14 workstations with push pins involved on the labor operation were selected, more than 90% were above the recommended acceptance limit of strength, which is of 10 lb from 1–7 push pins and 7 lb from 8–12 (Longo et al. 2004). This research evaluates the usage of a thumb plastic guard customized to plant population to implement new protective personal equipment in order to increase the strength of this muscular group in workstation with push pins.

Test was divided into 4 weekly sessions; First is an inclusion per minute with dominant hand, non-dominant and both thumbs per minute, Second: two insertions per minute with dominant hand, non-dominant and both thumbs, Session 3: three insertions per minute with dominant hand, non-dominant and both thumbs, Session 4: four insertions per minute with dominant hand and non-dominant. It showed that performance with splint increase since without this device applied force does not exceed 11 lb.

Keywords: Psychophysical methodology · Maximum strength level

1 Introduction

This research evaluates work procedures which are modified to prevent work-related injuries, at the level of the upper limbs, mentioning three of the most common caused by repetitive De Quervain's Tenosynovitis, Rizartrosis or osteoarthritis movements and finger in spring, which manifest themselves among the employees of the automotive industry, procedures such as the placement of push pins where force is exerted during an 8 h shift brings incidences of skeletal muscle injury due to traumatic disorders

S. Bagnara et al. (Eds.): IEA 2018, AISC 825, pp. 148–152, 2019.
https://doi.org/10.1007/978-3-319-96068-5_16

Cumulative (DTA's). The use of splint reduces this type of lesions increasing efficiency and productivity within the plant.

Reduce the incidence of injuries in thumb to make efforts of pressure, it will benefit to entire population of workers, in addition this analysis also would provide important information to companies requiring avoid or reduce expenses for injuries.

Other specific objectives related to this research are the enhancement of acceptable effort of push pins insertion, increase productivity and improve installation component feasibility.

2 Methodology

Lesions caused by a repetitive stress also known as Repetitive Stress Injuries (RSIs) are injuries that occur when too much pressure is exerted on a part of the body, resulting in inflammation (pain and swelling), injured muscles or tissue damage.

The RSIs are common injuries related to physical exertion and usually affect people in this case that they remain long time using their thumb on the inclusion of Push-pins generating diseases or disorders in ligaments, muscles, tendons or joints located in upper limbs in this case the hand (Fig. 1).

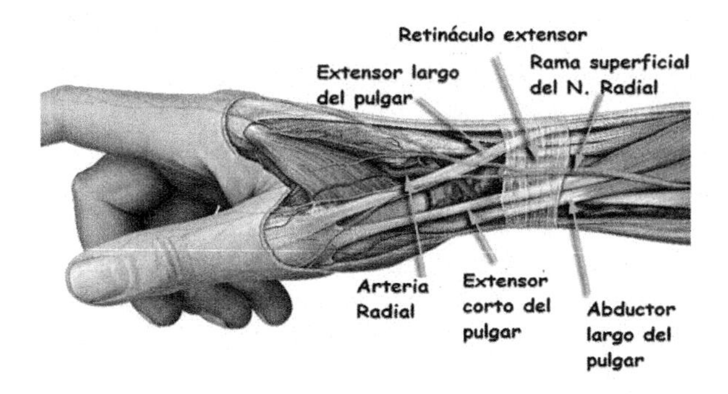

Fig. 1. Anatomy of the hand with synovial sheaths

This analysis provides elements that act as criteria and indicators for preventive and corrective programs. Similarly, this analysis recognizes the environmental factors and their influence on the worker, identifies the factors of risk, its location, measurement and quantification. Some of these activities come from a problem that can be the recognition of a risk of an accident at work, an occupational disease which will lead us to find the solution in accordance with the objective pursued in terms of occupational health.

This study involves the analysis of 14 workstations with push pins, 92.85% of these workstations are above of the acceptable limit of push pins insertions and to prove a splint thumb in order to enhance the acceptable insertion effort of 10 lb of 1 to 4 push pins.

Research took place at Ford Motor Company Hermosillo involving women employees of final assembly lines focused on workstation that contains push pins insertions (Fig. 2).

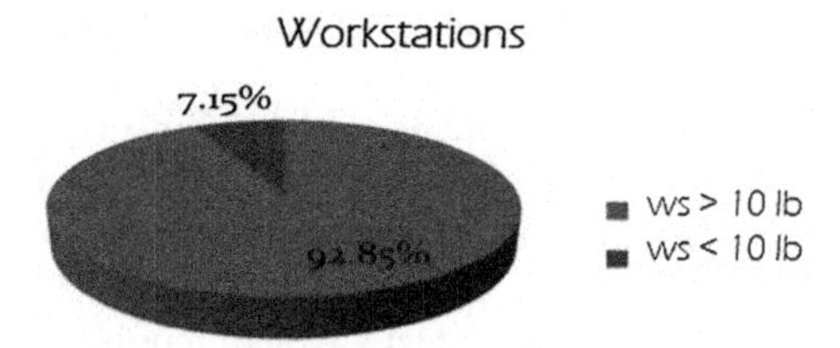

Fig. 2. Thumb insertion effort workstations at Ford Hermosillo

In our case, it was necessary to first identify the risk, for the operator at the workplace, so an analysis to achieve a good design for a splint that will help to reduce the risk of injury.

First training testing was performed using the thumb of the dominant hand, the second test thumb of the non-dominant hand and the third with both thumbs, on three different tests used the same type of push-pin. In each session the inclusion of Push-pins was increasing respectively:

session 1-1 push-pins
session 2-2 push-pins
session 3-3 push-pins
session 4-4 push-pins

These training sessions were divided into 4 sessions, data was collected at finish of each one, adding together the two were taken 5 sessions of data collection; In addition, these periods were constituted in work and rest.

Already completed the training period was test session which consisted on diverse types of activity per session, using the thumb of the dominant hand, the second test of the thumb of the non-dominant hand and the third with both thumbs.

These test sessions were divided into 5, where subsequently after each test data were taken, is worth mentioning that part of the training was to experiment with which mentally may be familiar to perform each test instructions. In it was explained that had

Fig. 3. Dynamometer test

to apply his best without feeling any discomfort or pain, all this pretending a working day of 8 h, in addition, these instructions were also used in previous experiments where psychophysical methodology was used to determine maximum strength. The following equipment was used for the experiment already placed in base and dynamometer Push-pin fastener (Fig. 3).

3 Results

The following table shows results of forces applied in pounds of a statistical analysis: media, fashion and deviation standard of each session and maximum/minimum force in hand dominant, non-dominant and both, in addition to the calculation of percentiles (Table 1).

Table 1. Result of research

Session 1								
	Mode	Media	Max	Min	Std Dev.	Percentile		
						5	50	75
Dominant	12.2	14.72	26.7	7	3.72	9.7	14.4	17.4
Non Dominant	12.5	12.91	27.5	6.3	2.93	8.7	12.5	15.3
Both Thumbs	20	22.03	41.2	10.8	6.17	14.4	20.2	26.8

Session 2								
	Mode	Media	Max	Min	Std Dev.	Percentile		
						5	50	75
Dominant	12.3	14.06	27.7	6.2	3.56	8.9	13.6	16.4
Non Dominant	10	11.99	25.2	5	2.88	7.6	11.8	14
Both Thumbs	20.5	21.02	41.3	8.9	6.11	12.4	20.1	24.7

Session 3								
	Mode	Media	Max	Min	Std Dev.	Percentile		
						5	50	75
Dominant	12	12.16	25.1	4.1	3.62	6.8	12	14.4
Non Dominant	10	11.18	24	4.3	3.11	6.6	10.8	13
Both Thumbs	15	17.83	34.1	6.1	5.05	9.9	17.4	20.8

Session 4								
	Mode	Media	Max	Min	Std Dev.	Percentile		
						5	50	75
Dominant	10	10.9	21.5	3.9	2.83	6.3	10.8	12.8
Non Dominant	10	10.01	21.1	3.9	2.44	6	10	11.6

The summary of each test was divided into 4 sessions, where subsequently after each test data were taken in the following manner:

Session 1: an inclusion per minute with dominant hand, non-dominant and thumbs together per minute. Session 2: two insertions per minute with dominant hand, non-dominant and thumbs together per minute. Session 3: three insertions per minute with dominant hand, non-dominant and thumbs together per minute. Session 4: four insertions per minute with dominant hand and non-dominant. It showed that performance with splint increase since without ferrule applied force does not exceed 11 lb.

4 Conclusions

We can conclude that the 13 of evaluated women (14 participants) exceed the 10 lb without causing them to fatigue and guarantees greater strength insertion using the splint. They showed improvement with the use of the splint by applying one higher than 10 lb, exceeding the acceptable standard of insertion in push pins (thumb), Demonstrating the use of splint enhance the applied force, reducing injuries caused by this type of activities carried out in the automotive industry.

References

Chapelot D, Pichon A (2013) Improvement of energy expenditure prediction from heart rate during running. Physiol Meas 35:253–266. IOP Science

Chaves A (2008) Tenosinovitis estenosante, SCielo

De Diego F (2008) Lesiones Tendinosas de mano y muñeca en el ámbito laboral, pp 30–35

Kroemer K, Kroemer H, Kroemer-Elbert K (2001) Ergonomics how to design for ease and efficiency. Prentice Hall, Englewood Cliffs

LaDou J (2007) Diagnóstico y tratamiento en medicina laboral y ambiental. McGraw-Hill, México

Longo N, Potvin J, Stephens A (2004) A psychophysical analysis to determine aceptable forces for repetitive thumb insertions

Muñoz JE (2011) Ergonomía. Universidad de Antioquia, Colombia

Niebel B, Freivalds A (2009) Ingeniería Industrial Métodos, estándares y diseño del trabajo, vol 12. Editorial McGraw Hill

http://espanol.arthritis.org/espanol/la-artritis/preguntas-frecuentes/pf-oa-pulgar/. (s.f.)

Estimation of Lifting and Carrying Load During Manual Material Handling

Mitja Trkov[1,2(✉)] and Andrew S. Merryweather[1,2]

[1] Department of Mechanical Engineering, University of Utah, Salt Lake City, UT, USA
{m.trkov, a.merryweather}@utah.edu
[2] Rocky Mountain Center for Occupational and Environmental Health, University of Utah, Salt Lake City, UT, USA

Abstract. Low back injuries and low back pain are often caused by improper task execution, overuse or lack of guidance and training. Our current understanding of dose-response relationships between risk factors that contribute to these injuries remains unclear. Enhanced monitoring of risk factors contributing to injuries could provide more complete exposure-response information. It is difficult to continuously monitor workers and their exposures to ergonomic risk factors using existing technologies. This paper presents a practical approach to advance continuous measurements of common risk factors by quantifying the weight of an object during lifting and carrying, lift frequency and lift duration during manual material handling (MMH). We estimate these parameters based on the ground reaction forces (GRF) and considering trunk dynamics. The results show that by considering trunk dynamics and applying simple signal processing techniques, we can precisely estimate these risk parameters. These parameters can then be used to estimate injury risk of workers. The developed methodology is designed for real-time continuous monitoring applications and sets the foundation for future development of in-field monitoring of workers with wearable sensors.

Keywords: Manual material handling · Injury risk estimation
Ground reaction force · Low back pain · Wearable sensors

1 Introduction

Low back injuries (LBI) and especially low back pain (LBP) rank among the most prevalent sources of occupational musculoskeletal disorders. Low back injuries (LBI) and especially low back pain (LBP) are often caused by improper task execution, overuse and lack of guidance and training [1, 2]. Our current understanding of the dose-response relationships between risk factors that contribute to these injuries remains unclear. Part of the reason for our incomplete understanding of how risk factors contribute to injuries is due to difficulties obtaining comprehensive exposure information for individuals working complex tasks. Often, exposures are estimated from snap-shots of work tasks, interviews with workers, periodic observations or "representative" video measurements.

S. Bagnara et al. (Eds.): IEA 2018, AISC 825, pp. 153–161, 2019.
https://doi.org/10.1007/978-3-319-96068-5_17

The goal of this study was to develop a practical approach to advance continuous measurements of the risk factors of LBP as defined in the revised NIOSH lifting equation (RNLE) [3, 4]. A recent study monitored hand forces during MMH using an instrumented force shoes and an inertial motion capture suit [5]. However, full-body motion capture system is inconvenient for in-field measurements. In this paper, we focus on quantifying the weight of an object, lifting frequency and lift duration during manual material handling (MMH) using limited number of body segment acceleration measurements. The proposed methodology is designed to enable continuous real-time monitoring and data collection, with the intent to establish robust methods for using wearable sensor technologies for collecting ergonomic risk factor measurements in the workplace.

2 Methods

2.1 Testing Protocol

We recruited a young healthy male participant with no previous musculoskeletal disorders or back pain history to perform the MMH tasks, see Fig. 1. These tasks included lifting, lowering and side-to-side positioning/transferring of the 5.25 kg (11.6 lbs) box with proper handles. The subject lifted the box from the ground on one side of the platform and put it on the table in the center and lifted it again and lowered it to the other side, see Fig. 1. The process was then repeated in the reverse direction. The total vertical lifting distance was 60 cm (23.5 in.) and the horizontal distance between the lifts was 1 m (39 in.). The box handles were positioned 25 cm (10 in.) from its bottom. To add larger variability in task executions, the subject was asked to perform lifts at a slow and fast self-selected speeds for controlled and quick, jerky movements performed in squatting and stooping postures. While performing lifting tasks, the subject was constantly standing over two force plates to continuously collect the complete ground reaction force (GRF) data.

Fig. 1. Subject performing (a) lifting and (b) carrying task during the experiments.

In addition, the subject was asked to walk on a custom walkway (0.9 m × 9.6 m) with and without carrying a box with a mass of 18.25 kg (40.2 lbs). The walkway contained a total of six force plates (three Bertec, Columbus, OH, USA and three AMTI, Watertown, MA, USA) measuring GRF data.

Forty-seven reflective markers were placed on a subject to capture whole body kinematics. Three additional markers were placed on the box to capture the kinematics of the handling object. Kinematic data were collected using an optical motion capture system (OptiTrak, NaturalPoint Inc, Corvalis, OR, USA) with a 120 Hz sampling frequency. The force plate data and motion capture data were synchronized using an external analog trigger signal. The subject was informed about the testing protocol and signed the informed consent form approved by the Institutional Review Board at the University of Utah.

2.2 Data Processing

All six force plate data were recorded at 1200 Hz, filtered using a fourth-order Butterworth filter with a cut-off frequency of 10 Hz and down sampled to 120 Hz to match the kinematic data. Kinematic data were filtered using a fourth-order Butterworth filter with a 5 Hz cut-off frequency [6].

2.3 Load Estimation Algorithm

Load mass was estimated by subtracting the subject's body mass and limb dynamics from the GRF for the lifting and carrying trials. We estimate load mass as:

$$m_{\text{Load}} = \left(F_{zGRF} - m_{\text{Subject}} * g - \sum\nolimits_{i=1}^{n} m_i * a_i\right)/g \qquad (1)$$

where F_{zGRF} is the z-axis component of GRF (normal force), $m_{Subject}$ is the body mass of the subject, g is the gravitational constant, and m_i and a_i are the mass and acceleration of the individual body parts, respectively. We separately subtracted dynamics of either the whole body (n = 9) or trunk only dynamics (n = 1) for comparison purposes. In the whole body dynamics, we included the dynamics of nine body parts; left and right thighs, shanks with feet, upper arms, forearms with hands and trunk-neck-head part. The segment masses and center of mass locations were determined as in [6]. All calculated load estimations were averaged over the duration of the lifting event and compared to the known weight of the load/box.

Lifting Events. The lifting events were detected based on the estimated lifting object mass signal (m_{Object}) that was filtered using a second-order Butterworth filter with a 1 Hz cut-off frequency and averaged with a moving time window interval. The time window was 1.5 s for the data considering trunk only dynamics and 0.75 s for data considering whole body dynamics. Fixed thresholds of 3 kg and 1.5 kg, were set to detect the start and end of the lifting events, respectively. Kinematics of the box (i.e. the absolute velocity of a single marker attached to the box) were used as a reference to determine the actual start and end of the lifting events. Automatic detection of the start and end of the event was determined when the object's velocity exceeded and dropped below the threshold velocity value of 0.03 m/s, respectively.

The time duration of the lifting events was computed as the difference between the start and end of the lifting event. We computed this separately for the events determined using box marker data and from the lifting load estimation data considering whole body dynamics and trunk dynamics only.

Information of detected lifting events was used to continuously compute the lifting frequency $\omega_{Lift}(t)$ defined as

$$\omega_{Lift}(t) = \frac{N_{Lift}(t)}{t - t_0} \tag{2}$$

where $N_{Lift}(t)$ is the number of detected lifting events at time t, and the time difference $t - t_0$ is the total trial duration expressed in minutes that started at time t_0. In addition, we compute instantaneous frequencies ω_i of the individual lifting events defined as

$$\omega_i = \frac{1}{t_i}, \quad for\ i = 1, \ldots, N_{Lift} \tag{3}$$

where t_i is the duration of the ith lift defined as the duration between the starts of the two consecutive lifts and N_{Lift} is the total number of lifts.

Carried Load Estimation. The load carrying events during walking were analyzed between the heel strike on the second consecutive force plate and the toe off at the second to last force plate. Such timing configuration guaranteed capturing GRF fully supported by the force plates. A GRF normal force threshold of 50 N was set to detect the heel strike and toe off events on those two force plates that defined the time interval used for estimating the carrying load. Trials where the subject did not step on all six force plates consecutively were not considered in the analysis.

We summed the normal GRF measurements of all the force plates and applied second-order Butterworth filter with a 1 Hz cut-off frequency to suppress the effects of body and load dynamics in the GRF signal. Additionally, we computed a continuous average over the last 0.5 s time window of the filtered signal to even further suppress the GRF oscillations due to walking dynamics.

3 Results

3.1 Lifting Event Detection

Estimations of object weight, frequency and duration of the lifting tasks were possible using only the normal component of the GRF measurements and trunk dynamics. Figure 2 shows the results of the estimated lifting load. Compared to the actual load, the average estimated load mass was within 3.9% (5.1 $\pm$ 0.5 kg) considering whole body dynamics and within 9.7% (4.8 $\pm$ 0.5 kg) considering trunk only dynamics.

Figure 3 shows the results of the lifting frequencies. The calculated average continuous lifting frequencies from the GRF data were both 9.7 $\pm$ 1.4 lifts/min

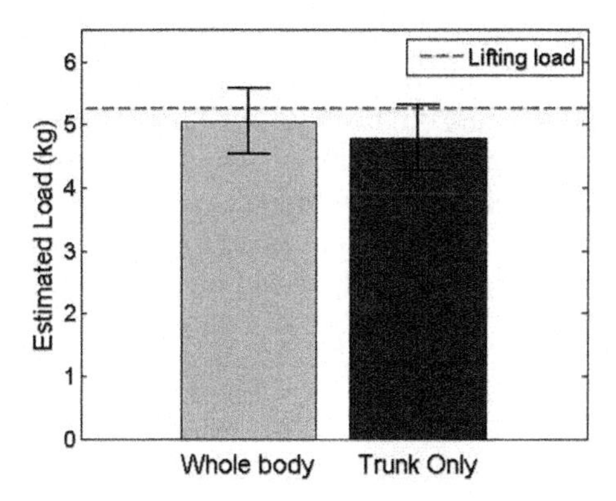

Fig. 2. Estimated lifting load considering whole body dynamics and trunk only dynamics.

considering whole body or trunk only dynamics. Compared to the frequency determined from the box marker velocity data they both deviated less than 0.2%. In addition, we present the instantaneous frequency results ω_i presented as discrete events after each lift. The instantaneous frequency results show greater variations compared to the continuous frequency measurements for both methods. In Fig. 3, we intentionally plot the average lifting frequency over the whole trial as computed in the RNLE (average over 15 min) to show the discrepancies compared to the instantaneous frequency estimation.

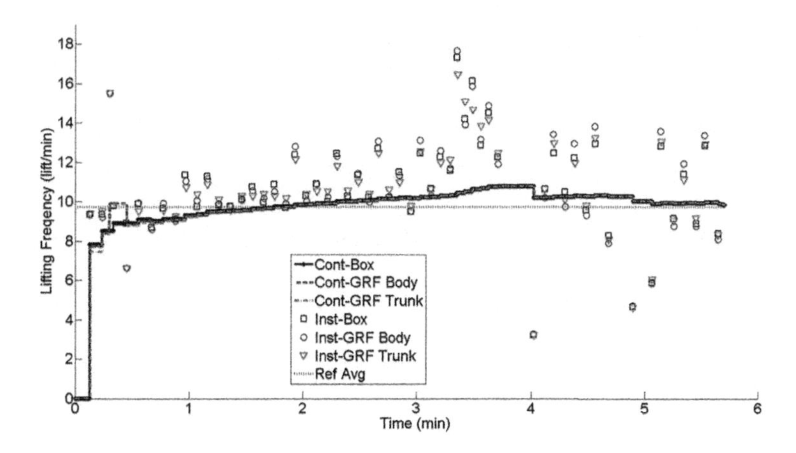

Fig. 3. Estimated lifting frequency results as continuous measurement (lines) and discrete instantaneous values (dots) determined from the box marker position data and GRF considering whole body dynamics or trunk only dynamics. A reference value represents the average lifting frequency during the whole trial.

Figure 4 shows the results of the lifting/lowering durations for the motion capture and GRF data. The average lifting duration estimated from the box marker data was

2.2 ± 0.4 s and was taken as a reference. The lifting durations from the GRF considering whole body and trunk only dynamics were 2.3 ± 0.6 s and 2.5 ± 0.7, respectively, deviating 1.4% and 13.3% from the lifting duration for the box marker data. The results obtained from both methods during fast squat or stoop lifting events show a high correlation, while the lifting durations from GRF data during squatting slow lifts overestimated those obtained from the box marker velocity data. These lifting duration estimates considering trunk only dynamics show greater discrepancies compared to the data considering whole body dynamics.

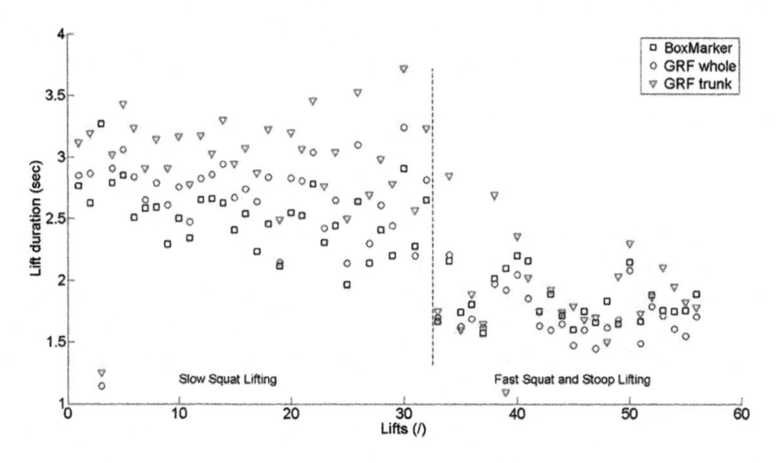

Fig. 4. Estimated lifting duration of an individual lift determined from the box marker position data and GRF considering whole body dynamics or trunk only dynamics.

3.2 Load Carrying

Figure 5(a) shows the results of three individual signals of continuous carried load estimations for a representative trial considering the subject's body mass and trunk dynamics. The subject mass and trunk dynamics subtracted from the initial signal shows large oscillations indicating a larger contribution of limb and heavy carrying load dynamics. After applying filtering and averaging the signal, our estimation closely matches the actual load mass. The average load estimation results of the initial, filtered and averaged filtered signals during the carrying event are 18.4 ± 14.6 kg, 18.7 ± 1.2 kg and 18.3 ± 0.6 kg, respectively, see Fig. 5(b). The mean load estimations of all three signals accurately estimated the actual load, however, the large standard deviation of the initial signal makes it impossible to use for instantaneous and continuous load estimation. The standard deviation decreases with the applied signal filtering and averaging which provides instantaneous and accurate information about the carrying load during the entire carrying duration (1.6 s–4 steps).

We further confirmed the results by comparing the estimations considering trunk only and whole body dynamics. Figure 6 shows average load estimation results during load carrying. The mass estimation of the carried load during walking was overestimated by 0.3% (18.3 ± 0.6 kg) considering whole body dynamics and underestimated

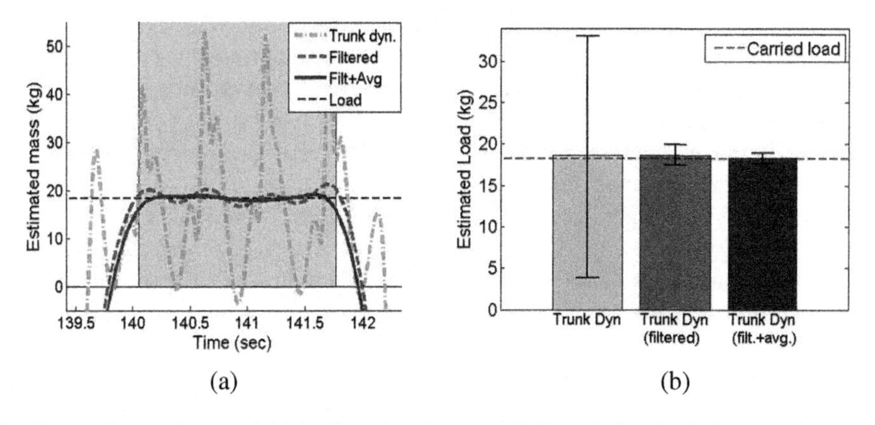

(a) (b)

Fig. 5. (a) Comparison of initial, filtered and averaged filtered signals during carrying event (shaded area). (b) Estimated mean and standard deviation of carried load estimation results for a representative lifting trial considering trunk only dynamics.

by 0.2% (18.2 ± 0.6 kg) for trunk only dynamics when compared to the actual carried load mass.

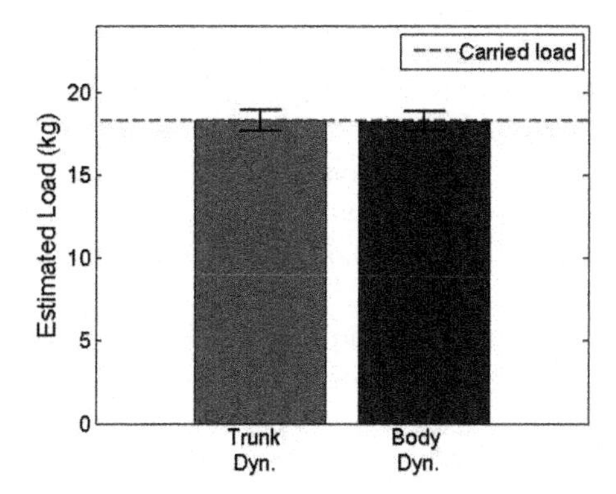

Fig. 6. Estimated average carried load considering whole body or trunk only dynamics.

4 Discussion

The purpose of this paper was to identify lifting events and extract parameters associated with injury risk, such as the weight of an object, lifting frequency and lift duration during MMH. The ability to identify lifting events primarily using GRF data represents an important step towards eliminating the need of motion capture systems used to detect these events. Signal processing of the normal component of the GRF vector, such as filtering and averaging, was necessary to suppress the dynamics of the upper and lower limbs. Processing the GRF signal using low frequency filters smoothens out the signal and allows for reliable extraction of the load information, however it introduces a limitation of not allowing detection of very short lifts or short load carrying events. The recommended duration of detecting lift and carry events when applying a filter with 1 Hz cutoff frequency should not be shorter than 1 s. The detection and load estimation of fast lifting and short carrying events (<1 s) is out of the scope of this paper as it would require the use of different data processing techniques and an additional validation study.

The signal processing parameters were selected based on our primary goal to detect all of the lifting events with 100% accuracy and to achieve the best accuracy of the lifting load estimation as the secondary goal. We tried to maintain the same parameters for lifting load estimations considering whole body and trunk only dynamics for comparison purposes. However, due to the large signal oscillations in the data considering trunk only dynamics, we had to choose a different averaging time interval window between the two to guarantee detection of all of the lifting events.

In our study we used the instantaneous frequencies ω_i as an additional validation of the computed continuous frequency. The results of the instantaneous frequency measurements show greater variations compared to the continuous frequency measurements for both methods. A similar observation was made when comparing to the reference frequency value. These observations show the importance of the continuous frequency measurements compared to the traditional approaches and suggest that such measurements can be used to improve monitoring of workers during MMH to obtain more complete exposure information.

An interesting observation from the study was also that the estimated lifting durations of the slower squat lifts at the beginning of the trial were largely overestimated compared to the faster, jerky and stoop lifts at the end of the trial. Potential reasons for the overall overestimations could lie in how the GRF signal was filtered and averaged. More accurate event detection and duration estimation of the faster lifting events could be due to greater instantaneous force generation, since we used a force threshold-based approach.

The limitations of this study include a small sample size and handling a single object during lifting and carrying. Further testing is required to identify the minimum weight of an object that the algorithm can reliably detect. We speculate, that for the average population, the performance of the algorithm's load estimation would improve, since the weight of the subject in our study was well above the average male. A lower subject's weight would lower the dynamics contributions from the limbs and would result in proportionally smaller GRF oscillations. Consequently, the algorithm would be more sensitive to the object weight and provide improvements in the load estimation.

The importance of the results in this paper lies in being able to sufficiently estimate risk parameters using a reduced number of measurements. Identifying the minimum number of sensors that still provide sufficient accuracy is particularly important and desirable for practical applications. Future work will use these results to develop a wearable system to track exposure to lifting related hazards and improve risk estimates during MMH.

Acknowledgement. Research reported in this publication was supported in part by the National Institute of Occupational Safety and Health of the National Institute of Health under award number T420H008414-11.

References

1. Parakkat J, Yang G, Chany AM, Burr D, Marras WS (2007) The influence of lift frequency, lift duration and work experience on discomfort reporting. Ergonomics 50(3):396–409
2. Office of Communications and Public Liaison (2014) Back pain fact sheet. https://www.ninds.nih.gov/Disorders/Patient-Caregiver-Education/Fact-Sheets/Low-Back-Pain-Fact-Sheet. Accessed 08 May 2017
3. Waters TR, Putz-Anderson V, Garg A, Fine L (1993) Revised NIOSH equation for the design and evaluation of manual lifting tasks. Ergonomics 36(7):749–776
4. Waters TR, Putz-Anderson V, Garg A (1994) Applications manual for the revised NIOSH lifting equation, Cincinnati, OH, US, pp 1–156. DHHS (NIOSH) Publication No. 94-110
5. Faber GS, Koopman AS, Kingma I, Chang CC, Dennerlein JT, van Dieën JH (2018) Continuous ambulatory hand force monitoring during manual materials handling using instrumented force shoes and an inertial motion capture suit. J Biomech 70(21):235–241
6. Winter DA (2009) Biomechanics and motor control of human movement, 4th edn. Willey, Hoboken

Assessment of Productivity and Ergonomic Conditions at the Production Floor: An Investigation into the Bangladesh Readymade Garments Industry

Abu Hamja[1,2(✉)], Miguel Malek Maalouf[1], and Peter Hasle[1]

[1] Sustainable Production, Department of Materials and Production, Aalborg University, Aalborg, Denmark
{abu,mml,hasle}@business.aau.dk
[2] Ahsanullah University of Science and Technology, Dhaka, Bangladesh

Abstract. The goal of this study is to investigate the change in productivity in the garments production floor initiated by lean and to assess the possible consequences for health and safety of workers. The study covers six garments factories where production lines with 250 sewing machine operators were included. Both quantitative and qualitative methods were applied. Three key lean tools (VSM, 5S, Time and Motion Study) have been applied and significant changes in productivity were found. Subsequently the actual changes initiated by lean have been assessed. The ergonomic assessment shows that the tangible lean changes tended to have either a neutral or a positive effect on OHS, whereas negative consequences were limited to attempts in pushing workers to work faster. The latter is not necessarily an inherent part of the lean methodology. The paper thus suggests that there may be possibilities for using lean to improve OHS and that the frequent critique of lean for intensifying work may not always be true. However, more research is necessary to study long term consequence for both productivity and OHS.

Keywords: Apparel · Lean · Ergonomics · Sewing

1 Introduction

The operational performance of a readymade garments (RMG industry) mainly depends on the management practices which generally focus on operational outcomes such as cost, quality, delivery, dependability, and flexibility [1]. OHS has traditionally not been high on the priority list of performance outcomes in the RMG industries. One of the main reason may be the difficulty in depicting the financial consequences and thereby showing a quantitative relationship with productivity. The consequence is that the operational departments in the manufacturing organizations may push for performance increases even at the cost of the health and safety of the workers [2].

Lean manufacturing methods have been gaining more popularity in the RMG industry during the last decades. The main reasons are the tough price competition, the increasing quality requirements and the fast changes in the fashion industry with still

S. Bagnara et al. (Eds.): IEA 2018, AISC 825, pp. 162–172, 2019.
https://doi.org/10.1007/978-3-319-96068-5_18

shorter seasons and thereby shorter lead time for the industry. Lean manufacturing is a methodology with a potential to meet these challenges [3, 4].

However, the effects on health and safety of workers are more ambiguous. Studies indicate that lean in repetitive assembly work tend to have a negative effect on workers' health [5]. The consequences for workers of applying lean in RMG are only studied to a limited extent without any convincing evidence [6]. The ergonomic literature suggests that interventions focusing on the redesign and appropriate modification of workstations, use of ergonomic seating position and training on low-risk methods and postures may considerably reduce the adverse effects [6]. Generally, a healthy and safe work environment helps in reducing costs associated with absenteeism and the loss of skilled staff, while increasing quality of work environment, workers' morale and job enrichment. This results to improved productivity and overall organizational performance [7]. However, it is still an open question whether lean will introduce such changes or mainly lead to an increase in work speed which may have negative consequences for workers' health and safety.

In this paper we study the practical consequences of lean implementation in order to assess the possible consequences for OHS, and thereby contribute to an discussion of the possibilities for simultaneously improvement of productivity and workers' health and safety.

2 Background

RMG is one of the most labor-intensive manufacturing industries, and as a results RMG industries are mainly found in countries where cheap labor is available in abundance. China is the largest supplier of RMG followed by other Asian countries such as Bangladesh, India, Sri-Lanka, Vietnam and Indonesia. All these countries can be characterized by cheap labor availability and a growing economy. However, as there is limited access to new low-income countries which can compete on low salaries, the international buyers are getting concerned about productivity and quality. For instance global retailers source from Bangladesh mainly due to price competitiveness whereas the industry is known to often miss the time-target on delivery [8]. According to the same study one of the major reasons for missing the time target is the poor labor productivity of the factories [8]. Some of the common causes pointed are lack of trained workers, inefficient mid-level management, poor operations management and poor working conditions in the production floor, no clear policy for employee benefits, poor housekeeping, improper storage system, disorganized production layout, lack of team–based work, substandard infrastructure, raw materials and political challenges [9–12]. In order to deal with these issues, the garment industries in several industrializing countries are implementing lean to improve performance [13, 14]. It is now well documented that lean has a positive influence on the productivity in RMG [15, 16] although there may be obstacles for extensive implementation [4, 17].

However, the work in the RMG industry is furthermore considered as potentially hazardous to workers health. They are exposed to repetitive work, poorly designed workstations and stairs as well as high work speed [18–20]. The consequence has a risk of repetitive strain injuries and other MSD disorders. The consequences of implementation of lean is only studied [4, 21] and it is still unclear how lean influences health and safety. This is even more that case as most studies of lean in assembly [5] as well as in garment do not specify how tangible elements in lean could influence OHS. Most studies simply measure a level of lean and a health outcome, without indicating the specific causes in lean for the adverse health effects. It is generally assumed, though, that lean would initiate a higher work speed and thereby increase the risk of repetitive strain injuries [22], but it is also a possibility that lean may remove obstacles in work which may both increase productivity and make ease the strain of the workers.

The importance of sorting out the consequences of lean is furthermore pertinent as international buyers are getting more serious regarding the compliance issues. After incidents like Rana-plaza and the Tazreen fire, which were all results of neglected control of OHS risk, there is an immense pressure from the buyer's side to improve compliance with basic working conditions and workers' rights issues in the RMG factories. International buyers are now very focused on compliance with codes of conduct before placing any import order [23]. Increases in MSD due to application of lean are therefore likely to create criticism from buyers and create an enhanced compliance pressure.

3 Methodology

This paper builds on an intervention project which the authors have initiated in 6 RMG factories in Bangladesh in order to study the possibilities for integration of productivity and ergonomics improvement by using lean methodologies. In this paper we present the initial lean intervention in detail and use the actual and tangible lean changes to assess the possible consequences for OHS of the worker in the production floor. The paper is part of a larger project with on-going data collection[1].

Snowball technique were used to select factories which have been in full operation for several years and expressed commitment to implement lean in pilot lines. The study covers a full sewing line in each of the six garment factories. The lines covered approximately 250 sewing machine operators, helpers and quality controllers.

The intervention in each factory included:

- Selecting a pilot line based on potential for benefits from lean.
- Lean implementation teams were formed in each factory comprising of people from multidisciplinary departments like industrial engineering, maintenance, planning, production, quality and compliance.
- Training of team members in four modules and one training designed for the workers.
- Application of three key lean tools: 5S, value stream mapping, bottleneck analysis by time and motion study [24–27].
- Systematic measurement of before situation building on production data, 5S audits, time measurements, photos, and videos.

[1] http://pohs-bd.org/.

- Suggestions for improvements were developed in collaboration between the researchers and the lean implementation teams.
- Suggested improvements were subsequently implemented by the lean improvement teams and after measurements carried out by the researchers.

According to the measurement of the productivity assessment was carried out on the possible consequences on OHS. The OHS assessment includes good postures and healthy workstation, repetitive hand and arm movement and prolonged work without break.

4 Results from the Lean Tool Application in the Pilot Lines

4.1 Supply Chain Analysis with VSM

VSM is used to identify waste, bottlenecks, flow of work and non-value-added activities. With the value stream mapping of the production process of the six garments factories, the reasons for lesser productivity was identified as lack of proper supply chain, proper planning and lack of management support. There were unsatisfactory results in almost all the factories due lack of coordination among Merchandising, IE and Production department are shown in the Table 1. During production this lack of coordination caused interrupted supply of fabric, trims and accessories. In one factory supply of gas was interrupted, which stopped steam flow in the line and the iron operation became shutdown. As that garment had about 60% of iron operation, there remained a huge waiting time which caused large amount of non-value activities in the process. However, consequence of application of VSM technique on the ergonomic or OHS conditions of the garments workers are insignificant.

Table 1. Most frequent changes initiated by VSM

Factory	VSM (Percentage of value of throughput time)		Example of changes/improvements
	Before	After	
F-1	3%	3%	Lack of proper supply chain, lack of management support
F-2	3%	3.50%	Bundle size, proper planning, effective maintenance works, less quality defects
F-3	3%	4%	Small bundle size, proper planning, effective maintenance works, less quality defects, Balancing the line
F-4	3%	2%	Lack of proper supply chain, proper planning, lack of management support, large bundle size
F-5	1%	2%	Bundle size, proper planning, effective maintenance works, less quality defects, Balancing the line
F-6	3%	3%	Lack of proper supply chain, proper planning, lack of management support

Another significant finding was that of the bundle size. In progressive bundle system bundles of garments parts are supplied to the sewing line from the cutting section. Larger bundle size was affecting the productivity negatively as it took relatively more time in picking and dropping the bundles.

4.2 5S

5S is a tool for securing efficient housekeeping and layout [24, 28]. An audit sheet developed for sewing lines was applied. The sheet was used to assess a 5S score as well as relevant improvements. The improvements were suggested to the lean teams who took responsibility for implementation. Afterwards new assessment of 5S score was carried out by the researchers. The results of the assessment as the most frequent changes are presented in Table 2.

Table 2. Most frequent changes initiated by 5S

Factory	5S score		Example of changes/improvements
	Before	After	
F-1	60%	65%	Labeling input rack, workstation redesign and making standardization of keeping the file in the shelves, emergency equipment was not maintained properly etc.
F-2	65%	70%	Labelling input rack and keeping labelled output rack, keeping standardized tool keeping system etc.
F-3	62%	69%	Labelling input rack, redesigning workstation and keeping labelled output rack and visual management tool, removing previous styles garment etc.
F-4	57%	60%	Removal of unnecessary thread cone, machine and obstacles from aisle, removing unnecessary box cartons from the line etc.
F-5	60%	70%	Keeping specific point for emergency equipment and keeping floor free from dust and other obstacles, keeping specific place for keeping the shoes etc.
F-6	65%	71%	Workstation redesigning and making the workstation free from dust, and machine free from oil leakage etc.

4.3 Time and Motion Studies and Bottleneck Analysis

Reduction of bottleneck is an important parameter for SMV. After identifying bottlenecks, method study was done which includes cellular layout, job sharing, reducing unnecessary activities and job merging. Wherever in the line, there remains bottleneck, a temporary cellular layout was formed in the line to improve the capacity. Job sharing also helped to improve the efficiency by involving workers having high capacity to another worker's activities having low capacity. In one factory, the worker of the side seam operation had high capacity. The side seam making operator was shifted for certain period of time at a regular interval to increase the capacity of marking operation. Besides these, workers' motivation was given to them who have low capacity.

The workers were motivated that they can make more amount if they work in more precise way. In this way, the capacity was increased, and it was observed that, those workers kept consistency in the style of their works. For example, the capacity of the workers of the operation named Dirt make and Belt Contrast make have been improved because of these motivations. Introduction of new tool guides, reducing unnecessary motion and keeping the marking of fabric just before stitching also helped to reduce bottleneck which helped in improving efficiency. In the Table 3 time saving and increased capacity are shown.

Table 3. Most frequent changes initiated by time and motion studies and bottleneck analysis

Change	Range of time saving	Increased capacity
In front part overlock operation, one helper was assigned to cut the outer thread	4 s in each cycle	9 pcs increased per hour
Motivating the worker about the method of doing side seam operation precisely	6 s in each cycle	13 pcs increased per hour
Keeping metal guide instead of paper guide and motivating about potential to increase capacity in lining joint operation in the front part of trouser	14 s in each cycle	5 pcs increased per hour
Changing habit of handling in Arm hole top stitch operation. Handling reduced from 6 times to 3 times	4 s in each cycle	6 pcs increased per hour

By reducing the bottleneck, capacity increased and finally increased the efficiency which is shown in the Table 4. Generally, the efficiency measurement is based on Standard Minute Value (SMV) and number of operators and further details are available on Quddus and Nazmul Ahsan, 2014. Lowering SMV and number of workers and

Table 4. Changes of efficiencies across the factories

Factory	Efficiency			Example of changes/improvements
	Before	After	Change	
F-1	49%	47%	–2%	Unprepared lean environment and lack of leadership
F-2	40%	49%	9%	By reducing three bottlenecks, job sharing and reducing unnecessary activities and 5S
F-3	54%	63%	9%	By reducing three bottlenecks and reducing manpower by forming temporary cellular layout and setting a guide tool in the sewing machine
F-4	57%	69%	11%	By reducing four bottlenecks, workstation redesign (space reducing), merging the work
F-5	59%	62%	3%	By reducing four bottlenecks and by applying 5S
F-6	54%	66%	12%	By reducing two bottlenecks by motivating by giving production bonus

increasing line capacity will tend to improve efficiency. In one factory (F-1) we found decreased efficiency due to lack of management commitment and no involvement of workers.

5 Assessment of OHS Consequence

5.1 VSM

Implementation of VSM in the garments production floor identified significant problems that have consequences regarding OHS of workers. One example is an observation which showed that worker's elbow or wrist bending is more than 30 and 45 degrees because of too much process inventory in their work. It results in more movement of hands and arms that increase the risk of work fatigue and musculoskeletal disorder. Another example, if you have excessive inventory in any factory, the shop floor where the inventory is stored becomes full, which reduces visibility and increases the likelihood of accidents. In this case, the waste has contributed to the increase of the number of accidents in the company.

If WIP increases proportionally, then the travelling distance also get increased, which creates a consequence on OHS. When workers pass garment bundle from one station to another, it becomes very difficult if it is a long distance. It has two demerits one is it is time consuming which is directly related to the efficiency and the other is they need more degree of bending and steeper movement of hand causes of pain in neck, arm and shoulder.

5.2 5S

Observations on the garments production floors in the surveyed factories show several OHS consequences of using 5S tools. Firstly, identification of both necessary and unnecessary tools in the workstation in a view to eliminate the unnecessary tools. For example, belt covers which are torn out, needle guards which are broken or useless, garments leftovers, thread cones, and other small equipment that are of no use in the workstation. Elimination of these unnecessary tools saves place, increases visibility, improves workstation environment; as a result, workers get more space and have less risk of injuries.

Secondly the necessary tools are set in order and everything has a defined place in the workstation so that the workers doesn't need extra time to look for things here and there. Thereby extra movements can be reduced. For example, some of the factories kept their garment outside of the marking aisle. For this reason, the workers often needed to bend at large angle to pick or drop the garments which can create pain in the different parts of the body. Besides this, keeping those boxes in the aisle is risky for those walking in the aisle.

In addition, cleaning up the workstation so that there is no dust, or fiber particles, insects in the workstation that again help to reduce exposure to the airways. Table 5 shows the OHS consequences of 5S changes.

Table 5. Assessment of consequences of 5S changes for workers

5S change	Likely consequence for OHS	Reason for assessment
Cleanliness	+	Reduces respiratory problem
Elimination of unnecessary things from workstation such as thread cone, previous style's thread and garments	+	Increases visibility, improves workstation environment
Keeping boxes for tags and labels and labeled boxes at output section	0	Reduces extra motion
Setting necessary items in the workstation in order such as scissor, tags, pattern, trims etc.	+	Reduces extra motion and fatigue
Removing extra box carton, pillow from the operators' chairs, operators shoes	+	Creating space having easy movement and reducing risk of MSD
Input area, output area, check points, line are labeled & maintained in proper way following the principles of X & Y axis	+	Reduces the hazardous conditions

–: Negative, 0: Neutral, +: Positive

5.3 Time and Motion Studies and Bottleneck Analysis

Efficiency with line balancing and bottlenecks have consequences regarding OHS of workers. For the bottlenecks analysis first attempt is to do method study, which helps to find out the actual movement or body posture of the worker. In several cases found that,

Table 6. Assessment of consequences of time and motion studies and bottleneck analysis for workers

Time, motion & bottleneck changes	Likely consequence for OHS	Reason for assessment
Reducing excess space between tables	+	Easier picking and dropping of bundle
More elbow angle to reach material	+	Reduces bad body posture and risk of MSD
Improve sitting position	+	Reduces pain in back and buttock
Changing handling style of the operator	0	Reduces extra motion
Reducing distance of chair and overlock machine table	+	Reduces back pain
Importance of accurate working method	+	Reduces bad body posture and risk of MSD

–: Negative, 0: Neutral, +: Positive

picking and dropping of job is not correct, more bending in the knee, sitting position that means chair and table height is not adjusted for this reason some of the cases, the height of the seats of workers are too low and they must use extra pillow to adjust the height. This is inappropriate and causes musculoskeletal disorders especially in the back pain, neck, shoulder etc. which is a serious health problem for workers. One of the researcher point out that seats that had high height created pain in the knee, lower legs and neck and seats having low height caused pain in the shoulders and neck [29]. Table 6 shows the consequences of time, motion studies and bottleneck change.

6 Conclusion

It is well documented that lean application has positive effects on productivity in garment [13, 14, 30]. In this study we can confirm that application of a few basic lean tools leads to productivity improvements in the short term.

The consequences for health and safety of the workers are less illuminated and lean is by some authors criticized for having a negative impact on workers [5, 31]. Lean is furthermore building on traditional rationalization methods such as time and motion studies which are also criticized for pushing workers to work fast.

However, in these studies the details on the changes initiated by lean are either not reported or very fragmented. We have indicated the actual changes which have been implemented, and we have as a first step in disentangling the relations between lean and OHS in garment, assessed what are the most likely consequences for OHS. The assessment indicate that most lean changes will have either a neutral or a positive effect on workers, whereas the negative effects are limited to attempts to push worker to work faster, which is not an inherent part of the basic lean methodology.

It is an exploratory study, and it is therefore needed with more in depth research where both the long term and more sustainable lean changes are followed, and the long-term effects are measured both in terms of ergonomic assessment and in terms of possible health consequences.

7 Limitations

As a preliminary study, the number of factory and the participants were relatively low and therefore not necessarily representative for the whole garment industry. Researchers also faced difficulties collecting data due to varied management commitment as well as worker who were reluctant to talk freely due to a fear that this might bring a threat to his or her job. To better understand the problems in Bangladesh garment industry, further research is needed for long term on the productivity and ergonomics conditions at the production floor.

Acknowledgement. This work was done under Aalborg University, Denmark and Ahsanullah University of Science & Technology POHS Research Project and financed by DANIDA.

References

1. Ward PT, Duray R, Leong GK, Sum C-C (1995) Business environment, operations strategy, and performance: an empirical study of Singapore manufacturers. J Oper Manag 13:99–115. https://doi.org/10.1016/0272-6963(95)00021-J
2. Von Thiele SU, Hasson H, Tafvelin S (2016) Leadership training as an occupational health intervention: improved safety and sustained productivity. Saf Sci 81:35–45. https://doi.org/10.1016/j.ssci.2015.07.020
3. Marudhamuthu R, krishnaswamy M, Pillai DM (2011) The development and implementation of lean manufacturing techniques in indian garment industry. Jordan J Mech Ind Eng 5:527. https://doi.org/10.1016/j.jom.2006.04.001
4. Maia LC, Alves AC, Leão CP, et al (2012) Design of a lean methodology for an ergonomic and sustainable work environment in textile and garment industry. In: Design, materials and manufacturing, Parts A, B, and C. American Society of Mechanical Engineers, Three Park Avenue, New York, NY 10016-5990, American Society of Mechanical Engineers, USA, p 1843
5. Hasle P, Bojesen A, Jensen PL, Bramming P (2012) Lean and the working environment: a review of the literature. Int J Oper Prod Manag 32:829–849
6. Kawakami T, Kogi K (2005) Ergonomics support for local initiative in improving safety and health at work: international labour organization experiences in industrially developing countries. Ergonomics 48:581–590. https://doi.org/10.1080/00140130400029290
7. Erdinc O, Vayvay O (2008) Ergonomics interventions improve quality in manufacturing: a case study. Int J Ind Syst Eng 3:727–745. https://doi.org/10.1504/IJISE.2008.020683
8. Kader S, Akter MKM (2014) Analysis of the factors affecting the lead time for export of readymade apparels from Bangladesh; proposals for strategic reduction of lead time. Eur Sci J 10:268–283
9. Berg A, Hedrich S, Kempf H, Tochtermann T (2011) Bangladesh's ready-made garments landscape: the challenge of growth
10. Hossain SD, Ferdous SR (2015) RMG's hot spot surrounded by challenges: a review landscape of Bangladesh ready- made garments (RMG) Segmentation of RMG's export market (2014–2015). J Sci Res Dev 2:29–38
11. Quelch JA, Rodriguez ML (2015) Rana plaza: workplace safety in Bangladesh (A) For the exclusive use of. J. Taylor. 44:1–14
12. Saha P, Mazumder S (2015) Impact of working environment on less productivity in RMG industries: a study on Bangladesh RMG Sector. Glob J Manag Bus Res G Interdiscip 15
13. Perera PST, Perera HSC (2013) Developing a performance measurement system for apparel sector lean manufacturing organizations in Sri Lanka. J Bus Perspect 17:293–301. https://doi.org/10.1177/0972262913505371
14. Kaur P, Marriya K, Kashyap R (2016) Assessment of lean initiatives: an investigation in the Indian apparel export industry. Indian J Manag 9
15. Wickramasinghe GLD, Wickramasinghe V (2016) Effects of continuous improvement on shop-floor employees' job performance in lean production the role of lean duration. Res J Text Appar 20:182–194. https://doi.org/10.1108/RJTA-07-2016-0014
16. Wickramasinghe G, Wickramasinghe V (2017) Implementation of lean production practices and manufacturing performance: the role of lean duration. J Manuf Technol Manag 28. https://doi.org/10.1108/jmtm-08-2016-0112
17. Hodge GL, Goforth Ross K, Joines JA, Thoney K (2011) Adapting lean manufacturing principles to the textile industry. Prod Plan Control 22:237–247. https://doi.org/10.1080/09537287.2010.498577

18. Pereira CCDA, Lpez RFA, Vilarta R (2013) Effects of physical activity programmes in the workplace (PAPW) on the perception and intensity of musculoskeletal pain experienced by garment workers. Work 44:415–421. https://doi.org/10.3233/WOR-131517
19. Habib MM (2015) Ergonomic risk factor identification for sewing machine operators through supervised occupational therapy fieldwork in Bangladesh: a case study. Work 50:357–362. https://doi.org/10.3233/WOR-151991
20. Pagnan AS, Câmara JJD (2012) Ergonomics analysis of the productive environment of fashion clothing firm in Belo Horizonte-MG. Work 1261–1267
21. Maia LC, Alves AC, Leão CP (2013) Sustainable work environment with lean production in textile and clothing industry. Int J Ind Eng Manag 4:183–190
22. Rampasso IS, Anholon R, Gonçalves Quelhas OL, Filho WL (2017) Primary problems associated with the health and welfare of employees observed when implementing lean manufacturing projects. Work 58:263–275. https://doi.org/10.3233/WOR-172632
23. Rahman A, Hossain S (2010) Compliance practices in garment industries in Dhaka city. J Bus Technol 5:71–87
24. Shamsi HS (2014) 5S conditions and improvement methodology in apparel industry in Pakistan. IOSR J Polym Text Eng 1:15–21
25. Lingam D, Ganesh S, Ganesh K (2015) Cycle time reduction for t-shirt manufacturing in a textile industry using lean tools. In: IEEE sponsored 2nd international conference on innovations in information, embedded and communication systems, pp 2–7
26. Quddus MA, Nazmul Ahsan AMM (2014) A shop-floor kaizen breakthrough approach to improve working environment and productivity of a sewing floor in RMG industry. J Text Apparel, Technol Manag, p 8
27. Yildiz EZ, Güner M (2013) Applying value stream mapping technique in apparel industry. Tekst ve Konfeksiyon 23:393–400
28. Haque KA, Chowdhury S, Shahw A (2014) Implementation of 5s and its effect in a selected garments factory: a case study. Bangladesh Res Publ J 10:291–297
29. Vandyck E, Fianu DA (2012) The work practices and ergonomic problems experienced by garment workers in Ghana. Int J Consum Stud 36:486–491. https://doi.org/10.1111/j.1470-6431.2011.01066.x
30. Wickramasinghe GLD, Wickramasinghe V (2017) Implementation of lean production practices and manufacturing performance. J Manuf Technol Manag 28:531–550
31. Bouville G, Alis D (2014) The effects of lean organizational practices on employees' attitudes and workers' health: evidence from France. Int J Hum Resour Manag 25:3016–3037. https://doi.org/10.1080/09585192.2014.951950

A Human Postures Inertial Tracking System for Ergonomic Assessments

Francesco Caputo[1], Alessandro Greco[1(✉)], Egidio D'Amato[1], Immacolata Notaro[1], Marco Lo Sardo[2], Stefania Spada[3], and Lidia Ghibaudo[3]

[1] Department of Engineering, University of Campania Luigi Vanvitelli, via Roma 29, 81031 Aversa, (CE), Italy
{francesco.caputo,alessandro.greco,egidio.damato, immacolata.notaro}@unicampania.it
[2] LinUp Srl, via ex Aeroporto c/o Consorzio il Sole, 80038 Pomigliano D'Arco, Italy
marco.losardo@linup.it
[3] FCA Italy – EMEA Manufacturing Planning and Control – Ergonomics, Gate 16, Corso Settembrini 53, 10135 Torino, Italy
{stefania.spada,lidia.ghibaudo}@fcagroup.com

Abstract. Since the early development for health purposes in 1950s, motion tracking systems have been strongly developed for several applications. Nowadays, using Micro Electro-Mechanics Systems (MEMS) technologies, these systems have become compact and light, being popular for several applications. Looking at the manufacturing industry, such as the automotive one, ergonomic postural analyses are a key step in the workplaces design and motion tracking systems represent fundamental tools to provide data about postures of workers while carrying out working tasks, in order to assess the critical issues according to ISO 11226 standard.

The aim of this work is to present an experimental wearable inertial motion tracking system, developed at the Dept. of Engineering of the University of Campania "Luigi Vanvitelli" in collaboration with Linup S.r.l., composed by several low-cost inertial measurement units (IMU).

The system allows to estimate the orientation of selected human body segments and to analyze the postures assumed during the working tasks. To increase the flexibility of use, the system is highly modular: it's composed by 4 independent modules in full-body configuration, each one made of 3 or 4 inertial units.

In this paper, the overall system is presented, supported by several test cases, carried out in Fiat Chrysler Automobile (FCA) assembly lines, to test the system reliability in industrial environments. Furthermore, an automatic posture analysis code is presented to evaluate the postural critical issue of the workplaces.

Keywords: Wearable devices · Inertial measurement unit
Industrial environment · Working postures

S. Bagnara et al. (Eds.): IEA 2018, AISC 825, pp. 173–184, 2019.
https://doi.org/10.1007/978-3-319-96068-5_19

1 Introduction

Ergonomic issues represent one of the main aspects characterizing manufacturing assembly lines. Workplaces' activities need continuous screening about the working conditions, especially when there are variations in the production that causes changes in cycle time and in the tasks to be performed by the workers.

Fundamental analyses concern the working postures assumed by the workers and, hence, posture angles analyses are necessary for assessing ergonomic indexes according to ISO 11226 [1].

In the last decades, human motion tracking related technologies have seen a great improvement. Vision-based systems, exoskeletons and inertial motion capture have become common tools, leaving the label of laboratory experimental equipment. Optical systems have been successfully used to quantify joint kinematics by tracking the trajectories and the positions of reflective markers during dynamic activities. Anyway, these systems need a wide equipped acquisition area, composed by several properly tuned cameras, making the system not portable and, for this reason, very difficult to use in an industrial environment, such as the automotive assembly lines. These limitations can be overcome by using motion measurement techniques based on low cost Inertial Measurement Units (IMU), composed by MEMS (Micro electro – Mechanical Systems) tri-axial accelerometer, magnetometer and gyroscope. These sensors can be worn over normal clothes, without interfering with working activities. This represents a great goal for industrial applications, due to the complexity of the working activity and the reduced space.

In [2] a survey of several human motion tracking systems with focus on possible applications is present, while in [3] the perspective is oriented towards technical aspects and [4] deals with sensor networks in human activity recognition.

About wearable inertial sensors, in literature, it is possible to find several examples of use: in [14] differences between static and dynamic activities have been investigated, by using only uniaxial accelerometers. In [15–17], the use of accelerometer and gyroscope has been exploited for analyze human kinematics.

These first approaches were extremely problem specific and they didn't deal with typical sensors issues as drift problem.

By using sensor fusion techniques as Kalman filtering [22–26] the whole data, acquired by IMUs, can be used to properly estimate the orientation, mitigating the effects of drift in gyroscope sensors. These methods are based on the idea of integrating inertial sensor data. For example, in [9], a pair of virtual sensors is located at two adjacent segments to measure the joint angle. The attitude of each segment is computed by using a combination of inertial accelerometers and gyroscopes. [10] is focused on optimizing the number of sensors to estimate upper-limb posture. In [11], an Extended Kalman Filter is used for the attitude estimation of each segment in the human upper limb, by using not only accelerometer and gyroscope, but also a magnetometer. In [12] the XSens® suite to estimate the total body posture is presented.

In robotic applications, a more active field of research is the fusion of heterogeneous sensors like IMUs, vision sensors, laser scanners, sonars, to strengthen attitude and position estimation [13, 18, 19, 20, 21].

This paper is aimed to present a body motion tracking system, already partially presented in [5, 6], composed by low cost IMUs and developed at the Department of Engineering of the University of Campania "Luigi Vanvitelli" in collaboration with LinUp Srl.

In the following section an introduction to inertial motion tracking is presented, with focus on previous literature and kinematic equations. In Sect. 3, an experimental session carried out in an automotive assembly line is shown, followed by a further data analysis in Sect. 4, concerning the evaluation of an ergonomic index related to working postures, to prove the reliability of the measurements in the chosen application.

2 Wearable Inertial Motion Tracking System: Theory and Description

The core of the proposed motion tracking system is an IMU composed by a tri-axial accelerometer, a tri-axial gyroscope and a tri-axial magnetometer. These inertial sensors represent the minimum set of measures to estimate the attitude of a rigid body system, inheriting techniques from typical Attitude and Heading Reference Systems (AHRS).

The human body can be considered as a system made of several rigid segments, connected by joints. Each IMU can be located on a single segment to estimate its pose. By concurrently monitoring every IMU it is possible to reconstruct the human postures over the time.

Neglecting hands and feet, schematically speaking the human body can be divided in 10 main segments: pelvis, trunk, two arms, two forearms, two upper legs and two lower legs.

Consider $\varepsilon = X_E Y_E Z_E$ as a fixed reference system, aligned at the initial time with the pelvis local frame and centered in the center of mass of the human body. Z_E-axis has the direction of gravity vector, X_E-axis points towards the front of the body and Y_E-axis is chosen according to a left-handed reference system (Fig. 1a).

On each segment equipped with one IMU, a local frame B_i can be defined. B_i is in agreement with the orientation of the relative segment and, at initial time, it is oriented like ε (Fig. 1b).

To avoid singularities during the estimation, a quaternion based orientation for each segment is used. Quaternion $\boldsymbol{q} = [q_1, q_2, q_3, q_4]^T$ can be defined as follows:

$$\begin{cases} \begin{pmatrix} q_1 \\ q_2 \\ q_3 \end{pmatrix} = \boldsymbol{r} \sin\frac{\phi}{2} \\ q_4 = \cos\frac{\phi}{2} \end{cases} \tag{1}$$

Where $\boldsymbol{r} \in \mathbb{R}^3$ is the unit vector and ϕ is the rotation of the reference system about $\boldsymbol{r}$.

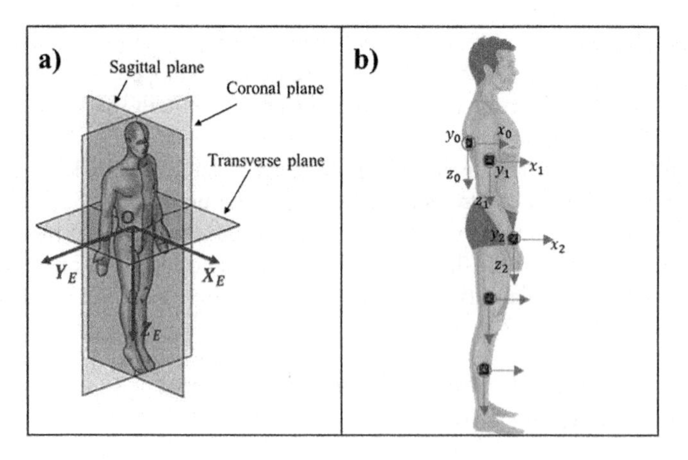

Fig. 1. (a) Human body reference system $\varepsilon = X_E Y_E Z_E$; (b) Local reference systems B_i.

The transformation of an arbitrary vector $\underline{x}$ between the fixed frame ($\mathcal{E}$) and the local frame ($\mathcal{B}$) can be written as follows:

$$\underline{x}^b = \underline{\underline{C}}_{BE}(q(t))\underline{x}^E \tag{2}$$

Where $\underline{q} = [q_1, q_2, q_3, q_4]$ is the quaternion vector and $\underline{\underline{C}}_{BE}$ is the rotation matrix defined as follows:

$$\underline{\underline{C}}_{BE}\left(\underline{q}\right) = \begin{bmatrix} q_1^2 - q_2^2 - q_3^2 + q_0^2 & 2(q_1q_2 - q_3q_0) & 2(q_1q_3 + q_2q_0) \\ 2(q_1q_2 + q_3q_0) & -q_1^2 + q_2^2 - q_3^2 + q_0^2 & 2(q_2q_3 - q_1q_0) \\ 2(q_1q_3 - q_2q_0) & 2(q_2q_3 + q_1q_0) & -q_1^2 - q_2^2 + q_3^2 + q_0^2 \end{bmatrix} \tag{3}$$

The attitude estimation relies on an Extended Kalman Filter (EKF) based on the following kinematic model:

$$\begin{cases} \dot{\boldsymbol{q}} = \frac{1}{2}\boldsymbol{M}\boldsymbol{q} \\ \dot{b} = -\frac{1}{\tau}\boldsymbol{b} \end{cases} \tag{4}$$

$$\boldsymbol{M} = \begin{bmatrix} 0 & -(p - b_p) & -(q - b_q) & -(r - b_r) \\ (p - b_p) & 0 & (r - b_r) & -(q - b_q) \\ (q - b_q) & -(r - b_r) & 0 & (p - b_p) \\ (r - b_r) & (q - b_q) & -(p - b_p) & 0 \end{bmatrix} \tag{5}$$

$$\begin{cases} \boldsymbol{a} = \boldsymbol{C}_{BE}\boldsymbol{g} \\ \boldsymbol{B} = \boldsymbol{C}_{BE}\boldsymbol{B}_E \end{cases} \tag{6}$$

where $\boldsymbol{b} = \left[b_p, b_q, b_r\right]^T$ is the gyroscope bias, $\boldsymbol{\omega} = [p, q, r]^T$ the angular velocity, $\boldsymbol{a}$ and $\boldsymbol{B}$ are respectively the acceleration and the magnetic field vector in the local reference system, and $\boldsymbol{g}$ and $\boldsymbol{B}_E$, are gravity vector and earth magnetic field in the fixed reference system. In the state space, the state vector is $= \left[\boldsymbol{q}^T, \boldsymbol{b}^T\right]^T$, the output vector is $\boldsymbol{y} = \left[\boldsymbol{a}^T, \boldsymbol{B}^T\right]^T$ and the input vector is $\boldsymbol{u} = \boldsymbol{\omega}$.

By realizing a simple kinematic model of the human body and by using attitude data it is possible to define posture angles, according to the bio-mechanical definition:

- Flection/extension forward, lateral bent and torsion of the trunk;
- Elevation, lateral rotation, rotation of the arms;
- Flection/extension of the elbows;
- Pronation/Supination of the wrists;
- Flection/extension, lateral rotation, rotation of the upper legs;
- Flection/extension of the lower legs;
- Subversion and inversion of the ankles;
- Rigid rotation of the pelvis.

An experimental wearable suite has been developed at the laboratory of Machine Design and Ergonomics of the Dept. of Engineering, based on several low cost IMUs with STM32F103 CPU, a MPU6050 (accelerometer and gyroscope) and a HMC5883L (magnetometer) as MEMS sensors.

A module is composed by three or four IMUs, linked to a Raspberry Pi based hub, representing the computing center and the responsible of synchronization with other hubs and the analysis application on the smartphone. In our application, we used four independent modules with four IMUs for the upper limb (pelvis, trunk, left/right arm, left/right forearm) and three IMUs for the lower limb (pelvis, left/right upper leg, left/right lower leg).

Fig. 2 shows the motion tracking system.

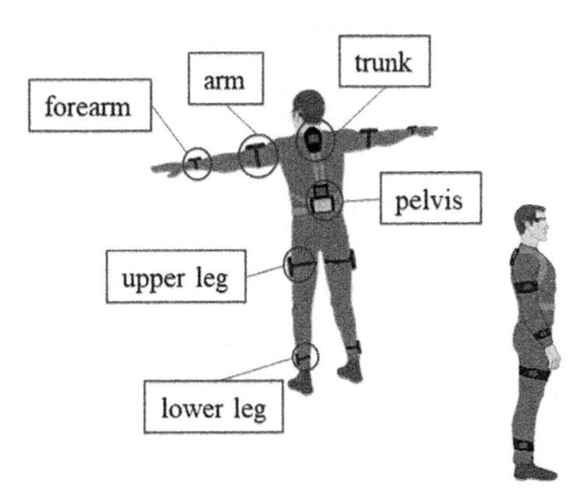

Fig. 2. Wearable motion tracking system in full-body configuration.

3 Industrial/Production Ergonomics Applications: An Experimental Session in Automotive Assembly Line

In order to test the reliability of the system in an industrial environment for ergonomics application, an experimental session has been carried-out at the assembly line of the FCA plant located in Melfi (Italy).

Acquired data have been used to assess indexes related to the working postures.

3.1 Working Activity and Procedure Description

The investigated workplace is related to the "central-cabinet assembly" in a mixed assembly line where Jeep Renegade and Fiat 500 X are produced. The assembly line is characterized by a cycle time of 60 s and a shift time of 8 h.

Figure 3 shows a description of the activity, that is the same for both car models, during each working cycle.

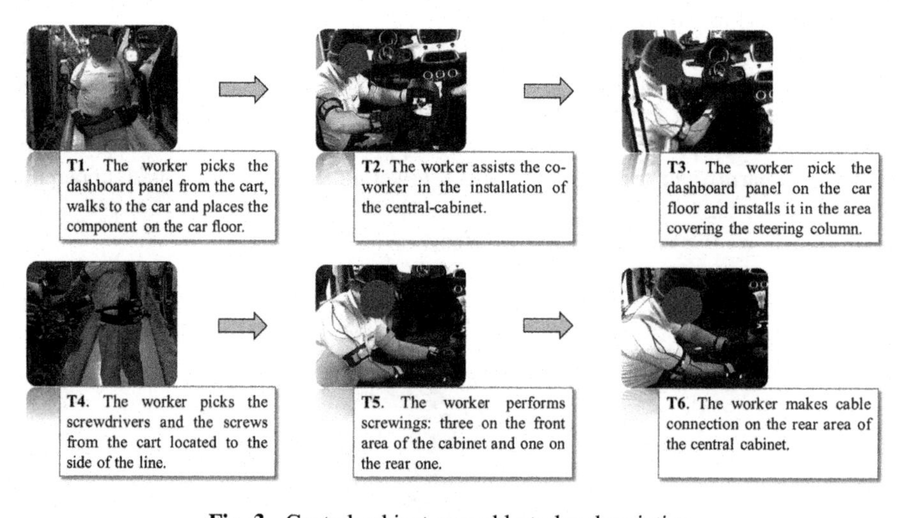

Fig. 3. Central cabinet assembly tasks: description.

Two workers have been chosen for the experimental session.

The first worker, named "worker A", is 180 cm high, with 95 kg weight, and he is very well trained and has a long experience in carrying out the investigated activity. He carries out the activity according to the task T1, T2, T3, T4, T5 and T6 described in Fig. 3.

The second worker, named "worker B", is 165 cm high, with 65 kg weight, and he has short experience in carrying out the activity that has been simplified for him. W.r.t. the description of Fig. 5, the worker B does not assemble the dashboard panel and the task are:

- T1*: The worker picks screwdrivers and places them on the car floor and he prepares for the central cabinet installation;
- T2;
- T5;
- T6.

Data have been acquired during the normal working activity every other cycle for 30 min per each worker. Out of a total of 30 working cycles acquired per both workers, 10 cycles per each of them have been selected for the analysis, excluding cycle with occurred delays or failures.

3.2 Ergonomic Method for Working Postures Study

As already mentioned, the study of working postures, assumed by workers, becomes fundamental for the design of the workplaces and for the continuous testing of them ergonomic goodness, especially in relation to production changes.

Purposely, during the design phase, FCA applies European Assembly Work Sheet - EAWS [7, 8] as 1st level ergonomic screening for the evaluation of the biomechanical overload for validating the design of the investigated workplace.

The EAWS is composed by four sections: static working postures, forces, manual material handling and repetitive actions with upper-limbs.

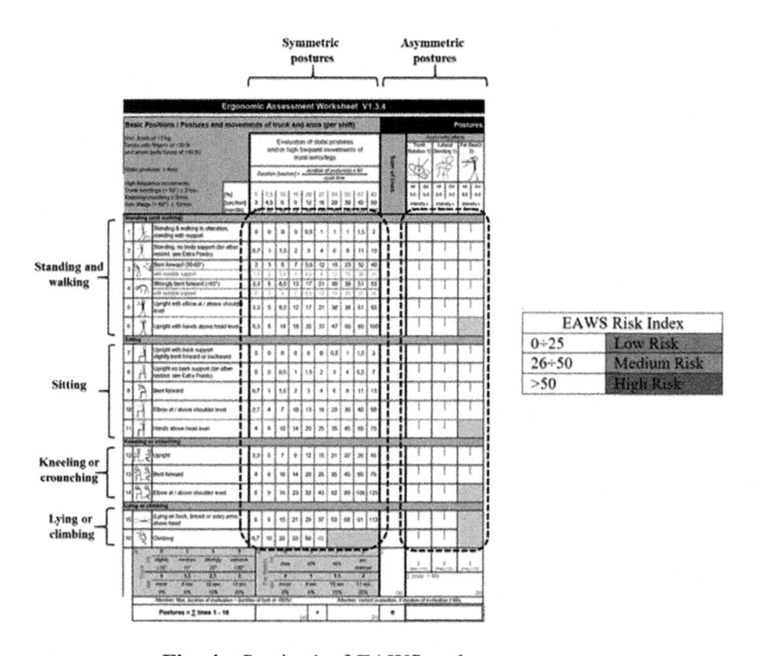

Fig. 4. Section1 of EAWS and score ranges.

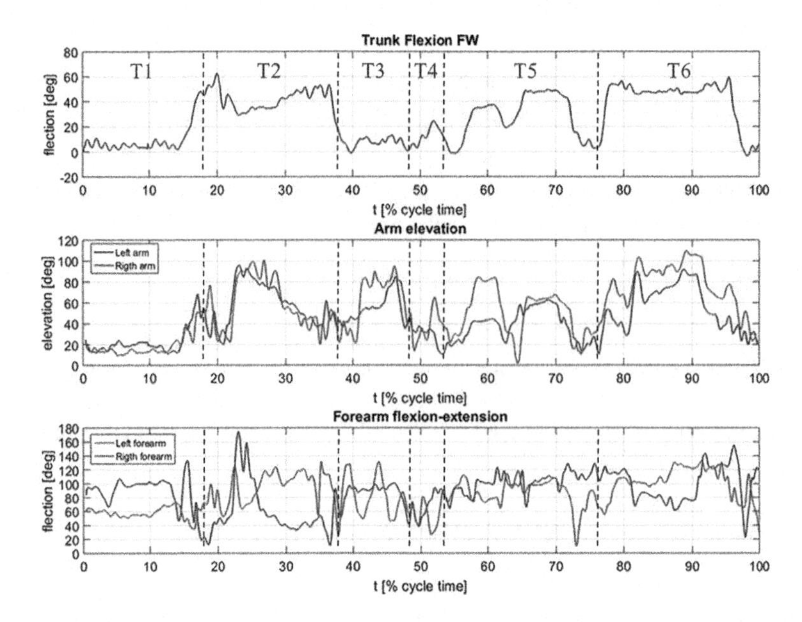

Fig. 5. Trunk flection fwd, arm elevation and elbow flection angles trends for worker A.

In particular, Sect. 1 – *static working postures* (Fig. 4) regards the evaluation of the working postures, assumed by the worker during his activity and movements with low additional physical effort, accordingly to the standard ISO 11226.

Postures, symmetric and asymmetric, are considered as static when they are at least 4 s consecutive long, without excessive and fast oscillations.

By combining data related to attitude quaternions and posture angles provided by the motion tracking system, the Sect. 1 of EAWS is evaluated by means of an algorithm coded by using MATLAB® programming language.

4 Data Analysis and Discussion

Experimental data, for this activity, have been acquired by using the tracking system in the upper-body configuration, being the activity repetitive and carried out in standing posture, where legs are used just for walking and moving around the workplace.

Figures 5 and 6 show the trends over the time of the most significant posture angles assumed, respectively by workers A and B, in carrying out the investigated activity in a working cycle: trunk flection forward, left and right arms elevation, left and right elbows flection. The angles are expressed in degree (*deg*) and the time is expressed in percentage of the cycle time and it is subdivided according to the tasks described in Sect. 4.1.

From the trends it is possible observing quickly that those ones related to worker B are noisier that those ones related to worker A, due to the less experience of the worker B.

Moreover, it is possible to observe that the different anthropometries influence as the workers reach different values of trunk flection and arm elevation angles in reaching

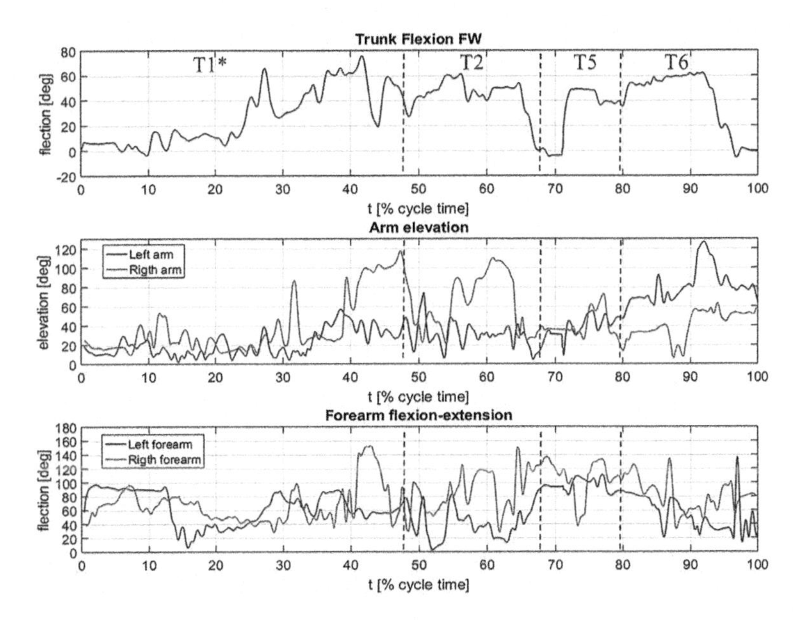

Fig. 6. Trunk flection fwd, arm elevation and elbow flection angles trends for worker B.

the working area, in the middle of the car. For this reason the values are higher for the worker B, that has a lower height than worker A.

Both workers are right-handed, but from the arm elevation angles trends it is easy to observe that the worker A uses both limbs almost in all tasks, most likely due to a higher experience in completing the working activity, while B tends to use only the right limb almost ever.

As already anticipated, posture angles and attitude data have been used to evaluate the Sect. 1 of EAWS that can be used to evaluate static working postures during a repetitive working activity, such as those ones typical of a manufacturing industry.

Figures 7 and 8 shows a statistical analysis of the EAWS – Sect. 1 scores for the two workers during 10 working cycles, evaluating the average and the standard deviation.

For both workers the scores are always below 25, the maximum value for low risk area (see Fig. 4), even if the postural contribution (Sect. 1) will be significant in the overall EAWS score, which is given by the sum of the scores of the Sects. 1, 2 and 3.

In detail, the average ($\bar{x}$) and standard deviation (σ) are:

- for worker A: $\bar{\boldsymbol{x}} = 21.2$, $\boldsymbol{\sigma} = 3.25$;
- for worker B: $\bar{\boldsymbol{x}} = 15.5$, $\boldsymbol{\sigma} = 4$.

Worker A is able to hold static postures, according to ISO 11226 definition, due to his higher experience in the investigated workplace, while worker B continuously changes his posture, in particular that one related to the trunk.

While for worker A, the EAWS - Sect. 1 score is mainly due to trunk flection forward angles, for worker B, in some working cycles, there is a contribution to the score due to posture with arms at/above shoulder level.

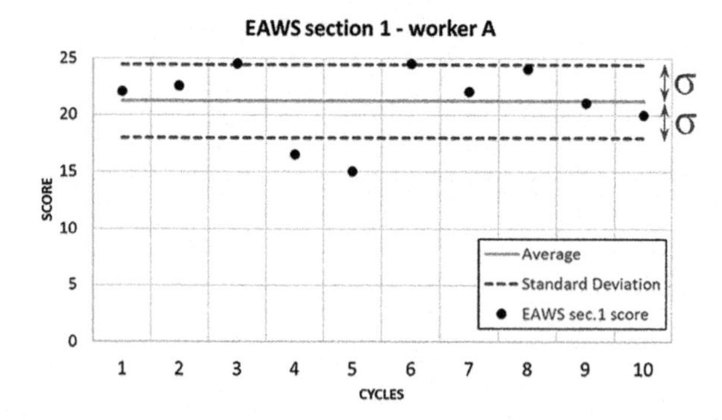

Fig. 7. Trend analysis of the EAWS Sect. 1 score in 10 working cycles for worker A: average and standard deviation.

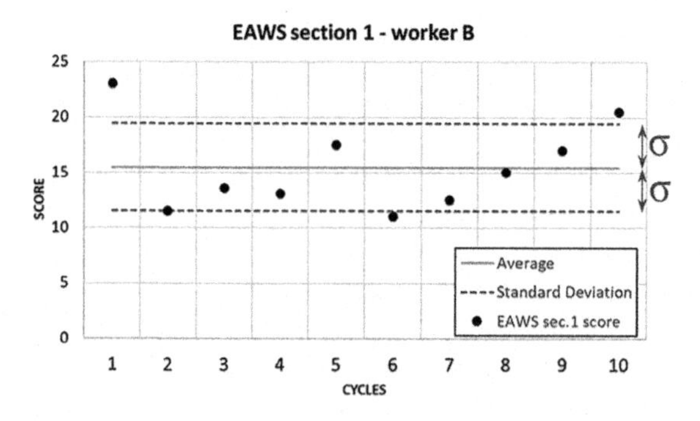

Fig. 8. Trend analysis of the EAWS Sect. 1 score in 10 working cycles for worker A: average and standard deviation.

From the data analysis, in particular the ergonomic scores, it is possible affirming that the motion tracking system, which measuring reliability has been already tested, ensures the repeatability of measurements.

5 Conclusions

The paper was aimed to propose an inertial motion tracking system for industrial application, in order to acquire data for human motion analysis and ergonomic indexes evaluation directly on the assembly line.

The system is composed by four independent modules in full-body configuration, with each module composed by 4 IMUs and a raspberry device for data registration.

To support the reliability of the system in an industrial environment, an experimental session, regarding a working activity in a FCA plant assembly line, has been carried out and described. The high number of acquired and processed data, by using the system in upper-body configuration, gave the chance to perform a detailed analysis of the human motion and a fast and automatic evaluation of the working postures according to the desired ergonomic method.

From data analysis it is possible to affirm that the system provides reliable and repeatable measures.

After the experimental session, an interview has been done to the workers, from which the system was found to be not invasive and not interfering with the normal working activity.

Acknowledgments. The authors would like to acknowledge the FCA – Fiat Chrysler Automobiles, EMEA Manufacturing Planning & Control – Ergonomics, and the LinUp S.r.l. for supporting the research work on which this paper is based.

References

1. I. 11226:2000 (2000) Ergonomics -Evaluation of Static Working Postures
2. Ahmad N, Ghazilla RAR, Khairi NM, Kasi V (2013) Reviews on various inertial measurement unit (IMU) sensor application. Int J Signal Proc Syst 1:256–262
3. Buke A, Gaoli F, Yongcai W, Lei S, Zhiqi Y (2015) Healthcare algorithms by wearable inertial sensors: a survey. China Commun 12:1–12
4. Gravina R, Alinia P, Ghasemzadeh H, Fortino G (2017) Multi-sensor fusion in body sensor networks: State-of-the-art and research challenges. Inf Fusion 35:68–80
5. Caputo F, Greco A, D'Amato E, Notaro I, Spada S (2017) A preventive ergonomic approach based on virtual and immersive reality. In: Advanced in Ergonomics in Design: Proceedings of the AHFE 2017, Los Angeles
6. Caputo F, Greco A, D'Amato E, Notaro I, Spada S (2018) Human posture tracking system for industrial process design and assessment. In: Advances in Intelligent Systems and Computing, Proocedings of the IHSI 2018 International Conference on Intelligent Human Systems Interaction, vol 722, Dubai
7. Schaub K, Caragnano G, Britzke B, Bruder R (2012) The European Assembly Worksheet. Theor Issues Ergon Sci 14(6):1–23
8. International MTM Directorate, EAWS - manuale applicatore
9. Dejnabadi H, Jolles BM, Aminian K (2005) A new approach to accurate measurement of uniaxial joint angles based on a combination of accelerometers and gyroscopes. IEEE Trans Biomed Eng 52(8):1478–1484
10. Hyde R, Ketteringham L, Neild S, Jones R (2008) Estimation of upperlimb orientation based on accelerometer and gyroscope measurements. IEEE Trans Biomed Eng 55(2):746–754
11. Yun X, Bachmann E (2006) Design, implementation, and experimental results of a quaternion-based Kalman filter for human body motion tracking. IEEE Trans Robot 22 (6):1216–1227
12. Roetenberg D, Luinge H, Slycke P (2009) Xsens MVN: Full 6DOF Human Motion Tracking Using Miniature Inertial Sensors, pp 1–7
13. Foxlin E, Harrington M (2002) Weartrack: a self-referenced head and hand tracker for wearable computers and portable VR. In: Proceedings of Fourth International Symposium on Wearable Computers, pp 155–162
14. Boonstra M, van der Slikke R, Keijsers N, van Lummel R, de Waal Malefijt M, Verdonschot N (2006) The accuracy of measuring the kinematics of rising from a chair with accelerometers and gyroscopes. J Biomech 39:354–358

15. Lyons G, Culhane K, Hilton D, Grace P, Lyons D (2005) A description of an accelerometer-based mobility monitoring technique. Med Eng Phys 27:497–504
16. Mayagoitia R, Nene A, Veltink P (2002) Accelerometer and rate gyroscope measurement of the kinematics: an inexpensive alternative to optical motion analysis systems. J Biomech 35:537–542
17. Najafi B, Aminian K, Paraschiv-Ionescu A, Loew F, Bula C, Robert P (2003) Ambulatory system for human motion analysis using a kinematic sensor: monitoring of daily physical activity in the elderly. IEEE Trans Biomed Eng 50:711–723
18. Honegger D, Meier L, Tanskanen P, Pollefeys M (2013) An open source and open hardware embedded metric optical flow cmos camera for indoor and outdoor applications. In: 2013 IEEE International Conference on Robotics and Automation, pp 1736–1741
19. Benini A, Mancini A, Longhi S (2008) An IMU/UWB/vision-based extended kalman filter for mini-UAV localization in indoor environment using 802.15.4a wireless sensor network. J Intell Robot Syst 24(5):1143–1156. https://doi.org/10.1007/s10846-012-9742-1
20. Mirzaei FM, Roumeliotis SI (2008) A kalman filter-based algorithm for imu-camera calibration: observability analysis and performance evaluation. IEEE Trans Robot 24 (5):1143–1156
21. Nützi G, Weiss S, Scaramuzza D, Siegwart R (2011) Fusion of IMU and vision for absolute scale estimation in monocular slam. J Intell Robot Syst 61(1):287–299. https://doi.org/10.1007/s10846-010-9490-z
22. D'Amato E, Mattei M, Mele A, Notaro I, Scordamaglia V (2017) Fault tolerant low cost IMUS for UAVS. In: 2017 IEEE International Workshop on Measurement and Networking (M N), pp 1–6
23. Yongliang W, Tianmiao W, JianHong L et al (2008) Attitude estimation for small helicopter using extended kalman filter. In: Proceedings of IEEE Conference on Robotics Automation and Mechatronics, pp 577–581
24. Kallapur AG, Anavatti SG (2006) UAV linear and nonlinear estimation using extended Kalman filter. In: Proceedings of International Conference on Computational Intelligence for Modelling Control and Automation and International Conference on Intelligent Agents Web Technologies and Internet Commerce, p 250
25. Marins JL, Yun X, Bachmann ER et al (2001) An extended Kalman filter for quaternion-based orientation estimation using MARG sensors. In: Proceedings of International Conference on Intelligent Robots and Systems, vol 4, pp 2003–2011
26. Choukroun D, Bar-ltzhack IY, Oshman Y (2006) Novel quaternion Kalman filter. IEEE Trans Aeros Electron Syst 42(1):174–190

Hutchinson Engaged for MSD'S Prevention Since 2006

D. Minard[1(✉)], P. Belin[2], and C. Desaindes[1]

[1] HSE Deputy Director Hutchinson, Paris, France
david.minart@paris.hutchinson.fr
[2] Director HSE Group Hutchinson, Paris, France

Abstract. Hutchinson make Ergonomics' analysis since 2007, to prevent professional diseases like Musculo-squelletics disease (MSD) but also to fight penibility at work.

This presentation illustrates the best practices of our plant dedicated to aerospace products, products that require many manual activities. The plant Hutchinson is located in Chemillé, France.

The manual tasks that are very present on the site are analyzed and rated to allow ergonomic evaluation of the activity. Both technical and organizational solutions are put in place to eliminate exposure to MSD and improve ergonomics at workstations.

The first step in the methodology used by the Chemillé site is the risk analysis and identification of the most demanding tasks in posture, handling and MSD (such as sewing activity). The second step is to set up a working group in collaboration with occupational medicine and employees. The third stage is deployment with the phase of understanding, diagnosis, awareness, action and valorization.

Concrete examples will be provided. The methodology in place since 2007 has made it possible to reduce the number of occupational diseases, even if it is still very difficult to estimate a figure. However, there are still some activities to improve or technical solutions are not easy to implement and sometimes need to revise the manufacture of our parts with the agreement of our customers.

Keywords: MSD · Ergonomics · Prevention

1 Context

The Hutchinson Group designs and produces custom materials and connected solutions in:

- anti-vibration systems, fluid management: specialist in rubber, plastic and TPE hoses for automotive and truck applications, including thermal management, noise reduction and pollution control
- sealing solutions: for precision, thermal management, engine components and under the hood, fuel line, transmissions, chassis and body parts
- transmission systems.

S. Bagnara et al. (Eds.): IEA 2018, AISC 825, pp. 185–188, 2019.
https://doi.org/10.1007/978-3-319-96068-5_20

The group distinguishes itself by a multi-market and multi-expertise intervention. Hutchinson has a turnover of 4.115 billion euros in 2017 and employs more than 43,000 people in 25 countries.

Chemillé: is a site of 18500 m^2 in the West of France. 350 employees produce thermal protection solutions for the aerospace market in particular. The processes are based on welding and sewing techniques with chemistry.

Jehier joined the Hutchinson Group in 2006. Prevention, a concern of the company's founders, was already well anchored in the organization. The hiring of an HSE referent and the deployment of group tools have given new impetus to a subject that has already been worked on.

We weren't starting from scratch. The HSE referent was able to meet the INRS expert Mr ZANA and benefit from training by our partner Ergorythme. The will of the management did the rest and site launched into the project in 2008.

2 Methodology

2.1 Overall Risk Assessment

As part of the risk assessment at the workstation, the site conducted a review of positions and tasks with a fairly standard criticality assessment. Nevertheless, this global and systematic approach made it possible to isolate critical positions/tasks, particularly for risk factors related to physical activity:

Mould handling
The postures
Repetitive gestures: pinning/cutting/welding/braiding

2.2 Physical Risk Assessment

To address these risks and work specifically on ergonomics, the Jehier site has formed a team.

The trinomial composed of at least 3 people: an HSE representative, maintenance and method has been trained and assigned to improve the identified positions.

The project was organized in different stages:

1- Explain the process to obtain everyone's support and collaboration
2- Diagnose more precisely the tasks previously identified
3- Improve positions
4- Sharing progress
5- Check the efficiency

The approach was explained to operators in order to obtain quality observations and information. Transparency was the key word to clearly define the reality of the tasks.

The diagnosis was carried out in the field, where the trained people worked with the operators. The occupational physician was also associated with this team.

The initial diagnosis was carried out on the basis of a classic OCRA method. Sometimes considered complex, the site gradually replaced it with a tool developed with the occupational physician. This pragmatic tool makes it possible to quote positions more quickly and especially with operators. It classifies the positions: green/orange/red: each colour corresponds to requests.

The method of observing 3/4 persons posts for about 2 h enabled the operators to improve their skills. Indeed, questions on postures, constraints allowed to share on the perception of the risk of muscle disorder and to share a common vocabulary. The last step is a question about the operator's perception. How does he feel on post?

The post improvement phase was organized pragmatically: what was simple was done immediately even if it concerned orange posts. More complex subjects were the subject of internal projects, sometimes with the development of new tools.

3 Improvement Projects Case Studies

Simple projects based on training or simply on information with observation elements. Operators complain of manual handling in difficult postures (Fig. 1a). During the action-training, the analysis of the activity allows them to become aware of their difficulties and to look for possible solutions (Fig. 1b).

Fig. 1. **a & b**: Improvement of manual handling constraints.

More complex projects requiring the development of tools for different repetitive tasks (Fig. 2).

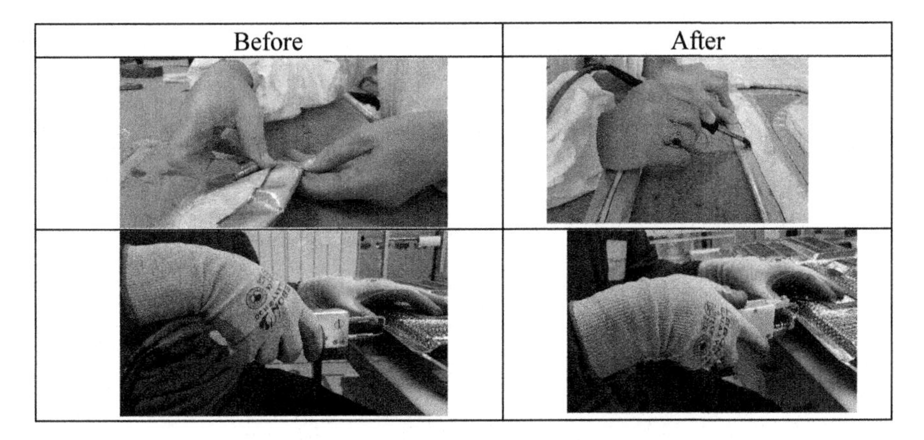

Fig. 2. Analysis and improvement of repetitive tasks.

4 Feedback from Experience

As a result of this effort and this approach of more than 3 years, the site is committed to a permanent practice that aims to treat all high-risk positions but above all to no longer introduce ergonomic risks into the process.

The time spent evaluating/correcting is now divided between continuous improvement and training: operators, technicians, methods and design.

There is no longer (or little) "spot ergo" but a systematic integration from methods to job validation.

Similar steps have been taken at the group's other sites since 2010. If there is no single tool, Hutchinson proposes a common methodology:

- Establishment of multidisciplinary groups
- Fieldwork
- Immediate treatment of critical situations
- Continuous improvement
- Integration of the subject from the development phase with training of method and industry technicians.

References

1. Health and Safety Executive (2000) Handle with care. Assessing musculoskeletal risks in the chemical industry - HSE Books, PO Box 1999, Sudbury, Suffolk CO10 2WA, Royaume-Uni, p 112
2. Zana J-P, Burgaud M (2009) Sustainable prevention approach to MSDs - commitment of three professional organizations actions of member companies, INRS 2009
3. NF EN 1005-5 - Safety of machinery—Human physical performance—Part 5: Risk assessment for repetitive handling at high frequency, AFNOR 2007

Assessment of Job Rotation Effects for Lifting Jobs Using Fatigue Failure Analysis

Sean Gallagher(✉), Mark C. Schall Jr., Richard F. Sesek, and Rong Huangfu

Auburn University, Auburn, AL 36849, USA
seangallagher@auburn.edu

Abstract. Job rotation is a common method employed by industry to reduce the risk of musculoskeletal disorders (MSDs). However, the efficacy of this technique has been open to question and methods of quantifying the effects of job rotation strategies have been scarce. However, recent evidence has suggested that MSDs may be the result of a fatigue failure process, and new risk assessment tools have been developed that have the capability to assess the effects of various job rotation strategies on MSD risk. The current analysis uses the Lifting Fatigue Failure Tool (LiFFT) to assess cumulative loading for a simulated job rotation scheme. Results of this analysis suggest that attempting to "balance" a high risk, medium risk, and low risk lifting job ends up creating three jobs that are all high risk. Rotation may somewhat reduce the risk associated with the worst job, but this will be accompanied by a steep increase in risk faced by all other workers in the rotation pool.

Keywords: Job rotation · Biomechanics · Fatigue failure · Risk assessment

1 Introduction

Many companies use job rotation as a method to control MSDs in the workplace. However, evidence regarding the effectiveness of this method in reducing MSDs is equivocal. A recent systematic review concluded that weak evidence existed regarding the efficacy of job rotation as a method for MSD reduction, with most showing no clear evidence [1].

A particular difficulty with respect to understanding the effect of job rotation on injury risk is quantification of MSD exposure associated with job rotation strategies. Fortunately, fatigue failure theory provides a method of calculating the effects associated with job rotation schemes. In the current analysis, we use the LiFFT low back risk assessment tool [2] to assess the efficacy of using a job rotation strategy to balance three jobs: one involving high low back risk, one medium, and one low back risk. The LiFFT risk assessment tool is based on fatigue failure theory and uses the peak load moment and the number of task repetitions associated with a lifting task to develop a "daily dose" of exposure. This tool has been validated against two epidemiological studies and has been shown to be significantly associated with low back outcomes [2].

S. Bagnara et al. (Eds.): IEA 2018, AISC 825, pp. 189–192, 2019.
https://doi.org/10.1007/978-3-319-96068-5_21

2 Method

In the present job rotation analysis, we developed examples of three different "mono-task" lifting jobs having 20%, 40%, and 60% probability of having a high risk job [3] using the LiFFT cumulative damage measure (each assuming an 8 h workday). Each mono-task job required 486 lifts per workday, with peak load moments of 27.3, 58.5, and 97.5 foot-pounds for low, medium, and high-risk jobs, respectively. These are summarized in Table 1.

Table 1. Estimates of cumulative damage derived from the LiFFT risk assessment tool, which is based on fatigue failure principles.

Task	Peak moment	Repetitions	Cum. damage	Low back risk
Low risk	27.3	486	0.0032	20%
Med. risk	58.5	486	0.0151	40%
High risk	97.5	486	0.1103	60%

The job rotation scheme to balance the risk of these jobs was designed such that each worker would rotate through each job (a third of the time would be spent at each workstation). Thus, workers would each perform 162 lifts at each job. The LiFFT tool was used to calculate the estimated risk associated the job rotation scheme.

3 Results

Table 2 provides the assessment of the risk associated with the job rotation method. As can be seen from this table, the rotation scheme results in an exposure that is high risk (>50% injury risk) for all three employees participating in the rotation. In terms of the cumulative damage, the high-risk job accounted for 86.8% of the total damage, while the medium job accounted for 11.6% of the damage, and the easy job accounted for 2.6% of the damage accrued in the job rotation scheme.

Table 2. Results of cumulative damage estimates from job rotation scheme. The total cumulative damage (0.0429) for the rotation is associated with Low Back Risk of 50.1%

Task	Peak moment	Repetitions	Cum. damage	Pct. total damage
Low risk	27.3	162	0.0011	2.6%
Med. risk	58.5	162	0.005	11.6%
High risk	97.5	162	0.0368	86.8%
Total		486	0.0429	100%

To assess the original situation (each employee assigned to their own mono-task lifting job) versus the job rotation method, simple probability methods were used to assess the efficacy of the redesigned method. If one assumes that the risk for each worker is independent, the probability that all three workers would actually be in a high-risk job (as defined by Marras et al. [3]) for the original circumstance would be (.2)(.4)(.6) or .048. However, in the job rotation scenario, the probability that all three workers would actually be in a high-risk job would be (.5)(.5)(.5) = .125. Thus, the job rotation scheme ends up increasing the risk that all three workers would be in a high-risk job by approximately two-and-a-half times.

4 Discussion

The results of the job rotation scheme are that the low-risk job becomes high risk, the medium-risk job becomes high risk, and the high-risk job stays high risk. This result stems from the fact that exposing the lower-risk workers to the high-risk task (even for a portion of the workday) hugely impacts their total daily load. It must be borne in mind that the exponential relationship between stress and damage incurred per cycle is exponential in nature, not linear. Thus, the damage per cycle incurred in the high-risk task is not tripled, but is more than 17 times that of the low-risk task. This is why the high-risk task accounts for 86% of the cumulative damage sustained during the job rotation arrangement.

Examining the probabilities of experiencing a low back outcome, it can be seen that attempting to "balance" the risk actually results in a higher probability that the three workers would experience back injuries according to the model. In fact, in probabilistic terms, balancing job risk will always result in the maximum risk of the three workers experiencing an injury, even if the risk relationship is linear. For example, assume that we had risks of .2 (low), .4 (moderate), and .6 (high), but we could manage to "balance" risks at .4 (moderate) for each. The probability that the three workers would each experience an injury would be .048 for the unbalanced scenario, but 0.064 for the balanced scenario. Thus, balancing the risk is worse than having variable risks, even in a linear situation. The exponential relationship associated with cumulative damage development simply adds an additional penalty to the attempt to balance risk, resulting in a much higher risk of injury. This finding suggests that companies wishing to reduce risk of injury should not attempt to balance the risk of multiple jobs in this fashion.

This analysis suggests that the first reaction of responsible individuals to reduce the risk associated with these three jobs should be to reduce the risk of the high-risk job using ergonomic redesign principles. This would appear to be the only viable approach to reducing risk of employee low back outcomes. If the high-risk job could be redesigned, say to reduce the probability of a high-risk job to 0.3, the probability that all three workers would experience a low back outcome would be (.2)(.3)(.4) or .024, half of the risk of the original scenario.

It should be noted that job rotation might have some benefits in terms of employee job satisfaction. Certainly, job rotation can help reduce the monotony associated with performing the same task for an 8- or 10-h shift. Job rotation can certainly be recommended for rotations where tasks all have light physical demands. However, most job rotation schemes appear to be based on balancing biomechanical demands, task complexity, or both [1]. The current analysis suggests that attempts to balance jobs based on biomechanical demands will instead lead to significantly increased injury risk for those rotating through higher demand jobs, increasing the overall risk of injury for participants of the job rotation pool.

5 Conclusions

On the basis of this assessment, the following conclusions are drawn:

1. Due to the exponential nature of cumulative damage development at different force levels, attempting to balance biomechanical demands using job rotation appears to be ill advised.
2. While the job rotation may lead to some stress by the worker in the highest stress job, workers in the medium and low stress jobs see large increases in risk due to exposure to the high stress job.
3. Job rotation between jobs of relatively low risk would be acceptable (and encouraged) to provide improved job satisfaction.

References

1. Padula RS, Comper MLC, Sparer EH, Dennerlein JT (2017) Job rotation designed to prevent musculoskeletal disorders and control risk in manufacturing industries: a systematic review. Appl Ergon 58:386–397
2. Gallagher S, Sesek RF, Schall MC Jr, Huangfu R (2017) Development and validation of an easy-to-use risk assessment tool for cumulative low back loading: the lifting fatigue failure tool (LiFFT). Appl Ergon 63:142–150
3. Marras WS, Lavender SA, Leurgans SE, Rajulu SL, Allread WG, Fathallah FA, Ferguson SA (1993) The role of dynamic three-dimensional trunk motion in occupationally-related low back disorders: the effects of workplace factors, trunk position, and trunk motion characteristics on risk of injury. Spine 18:617–628

From Prescription to Regulation: What Workers' Behavior Analyses Tell Us About Work Models

Lisa Jeanson[1,2(✉)], J. M. Christian Bastien[1], Alexandre Morais[2], and Javier Barcenilla[1]

[1] Université de Lorraine, PErSEUs, EA 7312, Metz 57045, France
lisa.jeanson@mpsa.com
[2] PSA Group, Paris, France

Abstract. In 2004, PSA developed *PSA Excellent System*, an organizational system based on the *Lean Manufacturing* principles to optimize vehicle production. One of the pillars of this system is the follow up of a "*work standard*" designed by the methods engineers. In theory, work standards allow for the balancing of shifts, i.e. the organization of tasks that operators can perform within a given period. The objective here is to maintain the operators' performance and health.

Despite this approach, errors and complaints on the assembly lines have emerged. In order to understand these phenomena, we carried out a detailed analysis of the operators' activity on workstations. For the data collection, we adopted a bottom-up approach by combining several methodologies: hierarchical task analysis of the prescribed and actual operators' tasks, filmed observations, behavioral coding and interviews with the operators and methods engineers.

Data analyses revealed discrepancies between work standards and actual tasks. "*Anticipated*" and "*individual*" regulations could explain these differences. Although these regulations are essential to the production, they involve additional actions and are not taken into account in the design of workstations. Similarly, they escape the methods engineers and are symptomatic of a dichotomy between the Lean Manufacturing rules shaping workstation design and real production constraints. We attempt to identify these constraints to improve the PSA Excellent System's predictive performance models. In this paper, we present the results of an analysis on a workstation in the assembly plant in Sochaux (France).

Keywords: Mental workload · Regulations at work · Behavior analysis
Work model

1 Introduction

In the 2000s, the PSA Group, a French car manufacturer, acquired the PSA Excellent System (PES), a work organization system based on the principles of Lean Manufacturing (Morais and Aubineau 2012). One of the pillars of the PES is the

S. Bagnara et al. (Eds.): IEA 2018, AISC 825, pp. 193–202, 2019.
https://doi.org/10.1007/978-3-319-96068-5_22

standardization of workstations. This results in the design of working standards composed of manufacturing sequences, themselves composed of phases; i.e. elementary operations to be carried out on a workstation in a given time.

Everything starts from the piece. Indeed, engineers design a part for one or more types of vehicle(s) or component(s) and then list the phases and time required to assemble it, either by timing it or by using the Measurement Time Method 2 or MTM2 (Antis et al. 1973; Maynard et al. 1952; Maynard 1956). The phases are then grouped into manufacturing sequences. Finally, the manufacturing sequences and the complementary operations (these include moving, taking information, taking and depositing pieces) are dispatched in the working standards. Each station has a working standard and a prescribed time or cycle time, per type of vehicle or component on which the operator must work. To this cycle time, a few extra seconds are added, an availability including a provision of time to manage hazards and a recovery coefficient calculated according to the difficulty to operate the workstation. The objective is that the average of the cycle times per types of vehicle or component, the theoretical times for carrying out the prescribed tasks, should be less than the running time of the vehicle or component; this is the physical time available to the operator to intervene on the vehicle or component.

Standardization of workstations has several advantages. On the one hand, it allows for the balancing and sizing of workstations and ensures that they are equivalent to each other in terms of workload. On the other hand, it contributes to the elimination of waste by limiting process variability and providing the operator with the right amount of time to perform his tasks to reduce unnecessary resource consumption (Morais and Aubineau 2012).

Ergonomists are present throughout the workstation design process and monitor workstations once they are set up on production lines. They intervene on problems related to physical solicitations (posture, load bearing, assembly feasibility, etc.) but also related to non-physical solicitations (information retrieval, memorization, feedback, etc.). From the prescription point of view, everything is therefore planned and controlled to maintain the operators' performance and health.

However, non-physical complaints and recurrent defects have recently emerged on assembly lines (Jeanson et al. 2017), suggesting that some workstation characteristics are not taken into account in the design models. The origin of these complaints and the appearance of malfunctions could be explained in particular by variations in mental workload. In order to test this hypothesis, we have undertaken detailed analyses of operators' activity. We present this approach here using an example that of the "air connection" workstation situated in an area where operators intervene on motors.

2 Study of the Motor Assembly "Air Connection" Workstation

2.1 Data Collection

Automobile assembly is a complex activity to analyze because of the production speed and the repetitiveness of operations. Operators repeat the same operations up to 400 times a day until these tasks become automatic and thus executed at a subconscious level (Lindblom and Thorvald 2014). Therefore, it is difficult for them to talk about their activity and to debrief about its constraints. Moreover, the operators on the workstations form a community with their own codes and well-kept secrets, especially when it comes to addressing possible discrepancies between prescribed and actual work. In this context, we initially adopted an internal point of view of the activity by doing a participating observation of the "air connection" workstation in the motor area. Participant observation allows the researcher to become an actor in his field and to study phenomena that are difficult to access or cannot be directly observed (Lapassade 2016), such as cognitive processes underlying operators' activity (Clot and Fernandez 2005; Theureau 1992). We therefore followed the training modules, as did all new entrants, namely the "safety", "ergonomics" and "basic" modules (learning the basic types of operations: connecting, screwing, shaping of cable bundles, and assembling parts). We were also trained directly by an operator at the air connection workstation for a week, by taking daily records on the significant events, feelings and difficulties encountered. This internship allowed us to explore the conditions of knowledge transmission, to "feel" and identify constraints on the workstation and to establish a relationship of trust with the operators.

Then we have analyzed the work standards in order to collect information on the prescribed tasks. We completed this analysis with interviews with the engineers in charge of designing the manufacturing sequences and workstations.

Finally, we made video recordings of operators in a working situation, at the rate of two video sequences per operator, making sure beforehand that the rhythm of the assembly line was not subject to unusual variations. During these recordings, we manipulated two working conditions: a 10-min sequence "without instructions" where we filmed the operators in their normal working conditions, and a 10-min sequence "with the standard" where we asked the operators to work strictly following the working standards.

We filmed two operators: a 44-year-old novice, on an interim contract in the company for 6 months and having spent about 4 months at the "air connection" station; this one is qualified at the workstation and can therefore keep it alone in a nominal situation; and a 26-year-old expert operator, on an interim contract in the company for 15 months and on the workstation for 12 months, who, in addition to being qualified at the workstation, knows it well enough to face alone all the hazards encountered and to train other operators. We then conducted individual self-confrontation interviews (verbalizations on the work from the filmed tracks) with the observed operators. These interviews enabled us to complete the behavioral data collected, by providing details on the strategies implemented, the training followed and the operators' feelings.

2.2 Data Analysis

The recorded videos were analyzed using The Observer XT software. This software is used to organize, analyse and present behavioral data (Morrison et al. 2003; Noldus 1991; Noldus et al. 2000). Thus, we were able to collect and analyse data on the operations carried out by the two operators (differentiating the operations planned by the standard from those that were not), the order and time of execution of the tasks, the behavior of the installations (assembly line and servo screwdriver), the errors on the workstation (forgetting to take a piece, falling piece, or others) and the direction of the operator's glances.

Furthermore, in order to compare the mental models of the job designers with those of the operators, we conducted hierarchical task analyses, on both work standards and actual activity. This method consists of describing tasks in terms of goals and sub-goals in order to identify gaps that could be indicators of the workload on workstations (Stanton 2006). The lowest level of description chosen for analysis is that of the phase (for example "clipping the connector" or "screwing the left screw onto the turbo"). The comparison between the working standard and the actual activity makes it possible to identify the activity regulations implemented by operators on workstations to maintain a given level of performance or to compensate for variations in activity (Hart and Staveland 1988; Loft et al. 2007; Kostenko et al. 2016). Moreover, thanks to the analysis of glances and errors, we were able to quantify the cost of these regulations in terms of information gathering and performance.

3 Results

3.1 Presentation of the "Air Connection" Workstation and the Temporal Regulation of Work Among the Observed Operators

The "air connection" workstation is located in the mechanical area reserved for motor assembly. The running time at the workstation (the time available to the operator to perform all operations on an engine) is 56.12 s. This workstation mainly consists of screwing pieces (metering connection, air connection, turbo, heat shield supplement, probe and attenuator), connecting connectors, and attaching pipes and harnesses to the motors using rubber bands.

There are 4 types of motors (DV6, EP6, EB2 and DW10) according to which the cycle times and operations foreseen in the working standards differ (see Fig. 1 below). The DV6 and EP6 engines are the richest ones (those for which there are the most operations). They have the longest cycle times and they have the particularity of including controlled screwing operations, i.e. the operator must wait until the motor (which advances at constant speed on the line) is at a precise location to receive a screwing authorization from the installed system. We have found that, in general, the theoretical and actual times differ by more than 10% for the DV6, EP6 and EB2 engines (see Fig. 1 below). This first level of analysis therefore demonstrates the regulations put in place by operators.

3.2 Other Forms of Regulation Identified on the Air Connection Workstation

Data analyses reveal several types of operators' regulations.

Early collective regulations common to all operators on the workstation: the chronological analysis of the activity shows that although the workstation standard provides for a sequential sequence of motor activities (one motor after the other), the

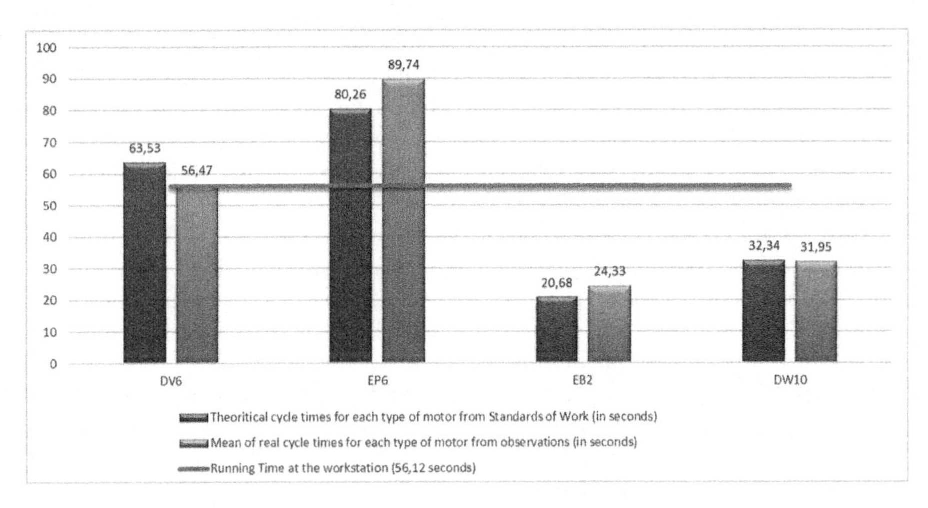

Fig. 1. Average theoretical (blue) and actual (red) cycle times in seconds, depending on engine type. A *green line represents the standard running time of the «air connection» workstation (time between the start of production of one unit and the start of production of the next unit during which operators can act on motor).* (Color figure online)

operators work on several motors at the same time and reorganize the operations to be carried out on the DV6 and EP6 motors into several "sections" (see Fig. 2 below).

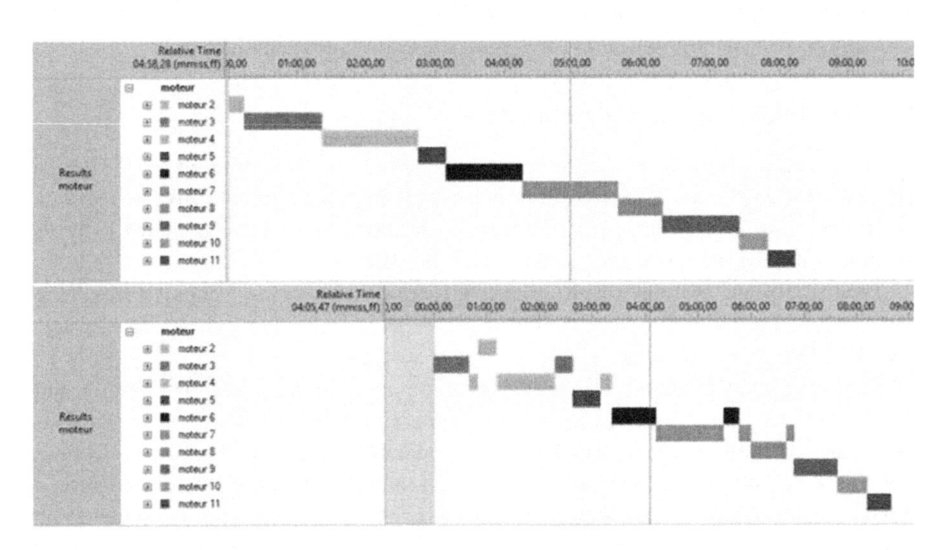

Fig. 2. Engine sequence on the "air connection" workstation (The Observer XT). *The color indicates the type of motor: green = DV6, red = EP6, yellow = EB2 and blue = DW10. The upper diagram describes the order of the motors from the point of view of the prescribed task and the lower diagram describes the chronology of the actual activity on the motors present at the workstation.* (Color figure online)

For example, while operators must work sequentially on engines 5, 6, 7, 8, from the point of view of prescribed work, we see that they alternate their activity from one engine to another (5, 6, 7, 5, 7, 8, 7, etc.). In a context where operators work on an average of 55 vehicles per hour, operators fill the waiting times generated by the servo-controlled screwdriver (not taken into account in the dimensioning of the station) with operations carried out on the following motors.

Anticipatory collective regulations common to all operators on the station and which may affect adjacent stations: the comparison of the hierarchical analyses of the prescribed and actual tasks reveals that the operators occupying the previous workstation in fact carry out certain operations on the DV6 and EP6 engines initially planned for the air connection workstation.

Anticipatory collective regulations common to the whole sector and characterised by the modification of the physical environment of the workstation: for example, the production teams had desoldered the stops surrounding the servo-controlled screwdrivers fixed on rails above the «air connection» workstation.

Individual regulations specific to each operator appearing in the event of hazards or when the operator wants to save time in the execution of certain tasks:

- When the novice operator wishes to perform a servo screwing and the installation has not yet given him the authorisation to screw, he pulls the motor (DV6 or EP6) in order to obtain the screwing authorisation more quickly.
- If two DV6 engines follow each other, and the two previous engines do not have servo screwing, the expert operator will perform the operations on these engines by grouping the operations similar to several engines (screwing the bypasses of the two engines, then conditioning the beams of the two engines, and so on).

3.3 The Impact of Regulations on Physical and Cognitive Load

We then tried to quantify the impact of the regulations on the job in terms of physical and cognitive workload. Indeed, operators perform tasks that are not taken into account in the design of their workstation. However, any operation not accounted for by the organisation of work represents an additional burden.

From the point of view of physical load, working on several engines simultaneously requires additional operator movement, especially for the richest engines, EP6 and DV6 (see Fig. 3 below).

From the point of view of cognitive load, this also implies that the operator constantly monitors the lamp that gives the operator the authorization to screw. Thus during an observation sequence of 10 engines, operators look at the lamp up to 18 times (see Fig. 4 below). This lamp monitoring is each time accompanied by an evaluation of the remaining distance (and therefore time) between the motor and the screwdriver. This information gathering and these treatments are not taken into account by the work standards.

Moreover, the comparison of the hierarchical analyses of the prescribed tasks and the actual tasks shows us that the operators reorganize their activities according to the types of tasks (autonomous screwing, servo screwing, beam storage and connections) and not simply according to the types of engines as recommended by the work

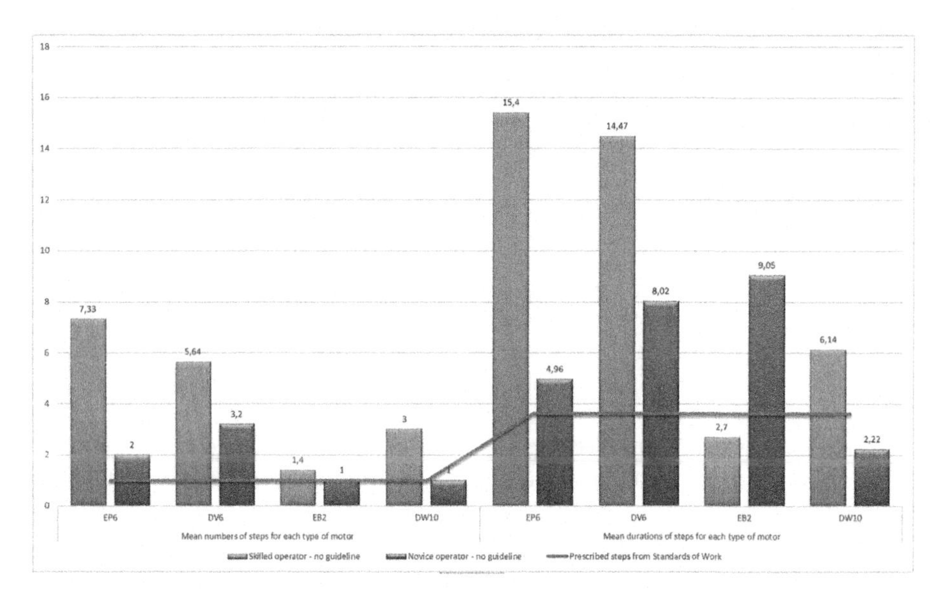

Fig. 3. The number and duration of movements (in seconds) of the operators according to their expertise and the type of engine

	Skilled operator	Novice operator
Number of motors during each period of observation	10	10
Number of glances at the light of the automatic screwer	18	17
Mean number of glances at the light of the automatic screwer per motor	1,8	1,7

Fig. 4. The number of glances on the lamp of the servo-screwdriver compared to the number of engines on the observation sequence at the expert and novice operator

standards. It is a form of regulation that allows operators to save their energy by grouping tasks sharing common operations and to maintain a continuous activity.

Operators may also misjudge the distance between the motor and the servo-screwdriver. In this case, it causes a break in the activity, and this leads to a risk of error and/or load increase. For example, during our observations, we observed that the novice operator arrived too early under the lamp. To avoid wasting a few seconds waiting without doing anything, the latter pulled the engine (weighing several tens of kilograms) to force the authorization to screw. As a result, he screwed the engine incorrectly and made a mistake.

Finally, these regulations are totally beyond the control of the designers of work-stations, pieces and production tools and are in conflict with product and operator safety. Thus the PSA Q282100 standard guaranteeing ‘the conformity of assembly, screwing and tightening operations of the elements, participating in Safety and Security assemblies' cannot be respected because the operators have unbolted the stops of the servo-screwdriver for the area covered by the latter.

4 Discussion

Individual regulations are a phenomenon well known to ergonomists, and considered as the expression of operators' creativity (Faverge 1977) as part of a deliberate activity (Clot 2010). In mental load models, **they correspond to strategic choices aimed at minimizing the impact of peak loads on performance**. Thus, Hart and Staveland (1988) describe a closed loop evaluation that operators perform between task constraints, perceived actual performance and strategies implemented.

This theory will be replicated and supplemented in the dynamic and multidimensional model of mental load, according to which operators compare elements of the activity before adopting the best possible strategy (i.e. the least costly in terms of cognitive load, the most effective and the most relevant) (Kostenko et al. 2016). Similarly, for Leplat (2006) and Sperandio (1971), operators regulate their activity according to mental load using two feedback and control loops in which they evaluate the gap between expected results and actual performance. In these models, deviations from the prescribed task are therefore the result of adaptation and control of the activity by the operators, which aims to regulate the workload and/or its impact on performance.

In Lean Manufacturing companies, on the other hand, regulations and the resulting deviations from standards are a source of variability in manufacturing processes and constitute a risk for product quality (Ohno 1988). Indeed, as their name indicates, these companies are characterized by the constant search for an ideal of production that would be optimized in terms of costs (Ross et al. 1990). This optimization is largely based on the standardization of workstations. This reduces variability in the manufacturing process and non-production times. These are the main sources of waste according to Taiichi Ohno, former engineer at Toyota and creator of this new form of work organization (Ohno 1988). In addition, if non-quality problems or risks arise, the standards are reviewed. Thus, at PSA, **the monitoring of work standards and continuous improvement projects should guarantee, in theory at least, optimal vehicle production while ensuring that operators' health is maintained**.

However, for the air connection workstation presented above, **this theoretical ideal comes up against real production constraints**. Operators could theoretically handle the arrival of a rich engine at 90 s of cycle time by getting ahead or by catching up with a poor engine requiring only 24 s of intervention while respecting work standards. However, **enslaving rich engines to a screwdriver** forces operators to be at a given location at a given time, making it **impossible to manage cycle time variability**.

Thus, even though the air connector workstation meets the Group's design criteria, time constraints are such that regulations become the sine qua non condition for engine production. In a context where deviations from standards are not allowed and where it is the production system itself that imposes these deviations, the continuous improvement process cannot be engaged. Thus, far from reducing the impact of workload on performance, as is the case for individual regulations, anticipated collective regulations become the norm. These unofficial practices are costly for operators who see their physical and cognitive load increased by micro-tasks not taken into account in the design of workstations and which introduce ruptures in the automatisms set up in the activity (over-surveillance of the lamp, evaluation of the distance between the engine and the servo-screwdriver, movements to pass from one engine to another, memorization of the operations remaining to be done on the interrupted engines). They are also costly for the production system because they are completely beyond its control, jeopardizing engine quality monitoring.

5 Conclusion

Thanks to the analysis of the operators' behaviour, we were able to describe in detail the real activity of the operators in order to compare it with work standards. For example, we identified individual regulations at the motor engine air connection workstation that enabled operators to cope with hazards, but also anticipated collective regulations. The latter were necessary for the performance of the tasks although they represented an additional cost in terms of workload and could compromise the monitoring of product quality.

The results of our analyses reveal a dichotomy between the Lean Manufacturing rules governing workstation design and actual production constraints. As this study is part of a global study of assemblers' activity within the PSA Group, we hope, in the long run, to improve performance prediction models by ensuring that the constraints identified are taken into account from the design stage.

References

Antis W, Honeycutt JM, Koch EN (1973) The basic motions of MTM, 4th edn. Maynard Foundation, Mogadore

Clot Y (2010) Au-delà de l'hygiénisme: l'activité délibérée 'Beyond hygienism: free activity'. Nouvelle revue de psychosociologie 2:41–50

Clot Y, Fernandez F (2005) Analyse psychologique du mouvement: apport à la compréhension des TMS 'Psychological Analysis of movement to understand musculoskeletal deseases'. Activites 02(2)

Faverge J-M (1977) Analyse de la sécurité du travail en termes de facteurs de risque 'Security at work in terms of risk factors'. Revue Épidém. et Santé publique 25:229–241

Hart SG, Staveland LE (1988) Development of NASA-TLX (Task Load Index): results of empirical and theoretical research. In: Peter AH, Najmedin M (eds) Advances in Psychology, pp 139–183. North-Holland

Jeanson L, Morais A, Barcenilla J, Bastien C (2017) N'oubliez pas de brancher les connecteurs: automatismes et traitement de l'information 'Don't forget to plug in the connectors: automatims and information processing'. In: 53ème Proceedings Congrès de la SELF - Présent et Futur de l'Ergonomie, pp 229–234, Toulouse

Kostenko A, Rauffet P, Chauvin C, Coppin G (2016) A dynamic closed-looped and multidimensional model for mental workload evaluation. IFAC-PapersOnLine 49(19):549–554

Lapassade G (2016) Observation participante 'Participating observation'. In: Vocabulaire de psychosociologie, pp 392–407. ERES
Leplat J (2006) La notion de régulation dans l'analyse de l'activité 'The concept of regulation in work analysis'. Perspectives interdisciplinaires sur le travail et la santé (8–1)
Lindblom J, Thorvald P (2014) Towards a framework for reducing cognitive load in manufacturing personnel. Adv Cogn Eng Neuroergonomics 11:233–244
Loft S, Sanderson P, Neal A, Mooij M (2007) Modeling and predicting mental workload in en route air traffic control: critical review and broader implications. Hum Factors 49(3):376–399
Maynard HB, Stegemerten GJ, Schwab J (1952) MTM [Method-Time-Measurement]: méthodes de travail et tables de temps 'MTM [Method-Time-Measurement]: work method and time tables'. Gauthier-Villars, Paris
Maynard HB (2001) Industrial engineering handbook, 5th edn. McGraw-Hill, New York
Morais A, Aubineau R (2012) Articulation entre l'ergonomie et lean manufacturing chez PSA 'Articulation of ergonomics and lean manufacturing in PSA Group'. Activites 9(2)
Morrison RS, Hemsworth PH, Cronin GM, Campbell RG (2003) The social and feeding behaviour of growing pigs in deep-litter, large group housing systems. Appl Anim Behav Sci 82(3):173–188
Noldus LP (1991) The observer: a software system for collection and analysis of observational data. Behav Res Meth Instrum Comput 23(3):415–429
Noldus LP, Trienes RJ, Hendriksen AH, Jansen H, Jansen RG (2000) The observer video-pro: new software for the collection, management, and presentation of time-structured data from videotapes and digital media files. Behav Res Meth Instrum Comput 32(1):197–206
Ohno T (1988) Toyota production system: beyond large-scale production. CRC Press, Boca Raton
Ross D, Womack JP, Jones DT (1990) The machine that changed the world, 1st edn. Free Press, New York
Sperandio JC (1971) Variation of operator's strategies and regulating effects on workload. Ergonomics 14(5):571–577
Stanton NA (2006) Hierarchical task analysis: Developments, applications, and extensions. Appl Ergon 37(1):55–79
Theureau J (1992) Le cours d'action: analyse sémio-logique: essai d'une anthropologie cognitive située. P. Lang, Bern

Assistive Robots in Highly Flexible Automotive Manufacturing Processes

A Usability Study on Human-Robot-Interfaces

Tim Schleicher[1(✉)] and Angelika C. Bullinger[2]

[1] BMW AG, BMW Allee 1, 04349 Leipzig, Germany
tim.schleicher@bmw.de
[2] Chemnitz University of Technology, 09111 Chemnitz, Germany

Abstract. Increasing demand for more product variants combined with shorter life cycles lead to higher complexity in production processes. One opportunity to overcome these challenges is seen in the combination of manual and automated work within usable instructive Human-Robot-Collaboration (iHRC). An assistive robot receives instructions from a worker, where and how to execute the next production steps. This allows a flexible adaptation to various production conditions without altering the robot programming. In order to realize such production systems, usable Human-Robot-Interfaces (HRIs) are necessary. Frameworks for the conceptual and methodological design as well as evaluation of such systems cannot be found in scientific literature or practice. This paper reports the development of an iHRC system for the automotive sector with the use of a mixed-method user-centered design framework. We show multi-dimensional usability aspects of HRIs as well as reflect on the suitability of the chosen development methods and their use in practice. The resulting prototype, the executed analysis and usability studies provide a basis for usable iHRC. First design principles for further developments of HRIs in the industrial context can be derived.

Keywords: instructive Human-Robot-Collaboration · Human-Robot-Interface Usability

1 Introduction

The increasing complexity in industry requires more flexible production processes [1]. In that context, fully automated, highly productive production lines push against their limits due to their high configuration and setup effort [2]. With the ongoing technical developments, a new kind of collaborative robots enables the tradeoff between flexible, manual and highly productive, automated work in so-called hybrid production systems [3, 4]. In order to use such systems in flexible production environments, the configuration effort of robots has to be reduced significantly [5]. In this sense, Human-Robot-Interaction (HRI) can be pointed out as a key technology for future production systems [6–8]. Usable interfaces between human and robot are crucial for an easy reconfiguration and with that adaption to new conditions. Given the background of decreasing

S. Bagnara et al. (Eds.): IEA 2018, AISC 825, pp. 203–215, 2019.
https://doi.org/10.1007/978-3-319-96068-5_23

lot sizes and its economic challenges in production, besides the configuration for a new production batch, an interaction during production gets more and more interesting [9]. The intuitive programming of collaborative robots changes to an instruction during the ongoing production process. The robot receives instructions containing necessary information for the execution of the next production steps. Conventionally rigid robot skills are used in a flexible manner without altering the robot programming. Within the frame of this research work, such systems are defined as instructive Human-Robot-Collaboration (iHRC).

With this in mind, the development and testing of usable HRI in the flexible industrial field is important for mastering future challenges in production [9–11]. In this context, user-centered research is missing [12–14] and with that the extensive consideration from a human factors perspective [15]. In order to close that gap, we present the application of a mixed-method user-centered design (UCD) framework [16] on the development, iterative evaluation and improvement of a surface finishing robot in the automotive sector.

2 Related Work and Conducted Research Design

The need for usable human-robot systems and their use in flexible production processes brings short-term interactions in a transitional way into focus. They are characterized by the transfer of information and/or material for the handover of a task. Human and robot work independently after a short interaction. Monotonous and strenuous tasks can be transferred to an assistive robot. Documented research in this category with an industrial context [5, 13, 17–26] focusses on the demonstration of application-specific feasibility. The application of UCD frameworks for the development of such systems in the industrial field as well as their methodological embodiment is currently missing. In order to ensure external validity with their development, literature explicitly claims the need of valid subject groups as well as the detailed description of empirical studies to support replication [14]. Further requirements of the industry are seen in the use of standardizable and with that transferrable concepts, easy-to-use and resource-efficient development methods as well as description of characteristics and design principles for future systems [16].

The explicit lack of documented research in this context requires a UCD framework for the design of instructive human-robot collaboration. The development of this framework follows the research paradigm of design science research (DSR) according to Hevner et al. [27]. Relevant practical challenges are tackled with the rigorous design and evaluation of new and innovative artifacts in an iterative manner, strictly using scientific methods. The goal is to extend the scientific knowledge base as well as deliver solutions to the practical field of use [27]. In this paper we present the application of such a UCD framework [16] on the development of an assistive surface-finishing robot in the automotive sector and analyze its utility. On the one hand, we discuss the practical use of analysis, design and evaluation methods. On the other hand, we report the results of their application.

3 Basics of an Instructive Human-Robot Collaboration

The concept of an iHRC is derived from the interaction paradigm of robot programming, especially skill-based programming in the industrial context. It enables fast and intuitive programming with the use of a library of robot skills. Skills can easily be composed and adapted to an application-specific robot task with the use of graphical programming methods [1, 28–30]. Transferred to the idea of iHRC, the parametrization of predefined robot skills via usable human-robot interfaces enables the flexible use of an assistive robot during ongoing production processes. The instruction of the robot itself is carried out in two phases. During the first interaction phase, position-relevant data of a task is delivered. It tells the robot where to do the work. This can be a desired position of a robot-carried tool, a reference position on a workpiece or a reference object. During the second phase, further adjustments of a robot task can be carried out. This tells the robot how to do the work, e.g. the parametrization of robot speed in absolute values in mm/s or in predefined categories such as slow/medium/fast.

4 User-Centered Development of an iHRC

The user-centered design of iHRC systems consists of an analysis phase followed by the iterative design and evaluation. Each of the two phases is split into three sequential steps, which allows the preparation of user studies in a first step, the execution of user studies in a second step and the interpreting expert-based analysis in a third step. The analysis phase iteratively gains knowledge about the context of use, user requirements and possible solutions. The design and evaluation phase iteratively gains knowledge about the experiences of the use of an iHRC. A first design and evaluation loop focuses on positioning for the fulfillment of a robot task at a specified position, a second on the parametrization of a robot task [16].

The focus of this paper lies on the application of such a UCD framework on the development of a surface-finishing robot, reporting the overall analysis as well as the design and evaluation of the robot positioning.

4.1 Analysis of the Context of Use and User Requirements for the Use of a Surface-Finishing Robot

The characteristics of flexible production processes, not completely standardized due to constantly changing input, are seen in the manual rework, e.g. the surface-finish after the paint shop.

Analysis of the Context of Use. The analysis of the context of use is based on a preparative document analysis of the process descriptions and standards as well as an additional participating observation. In the finishing process, workers control the painted parts and adjust their rework according to the detected surface defects. After detection, workers start sensitively sanding the defect, e.g. inclusions. With a rough surface being left, the polishing process brings back the shine of the clear coat. Process factors of these processes constantly change, depending on the position, shape and size of the defect. With that in mind, finish workers may use an assistive robot to handover the time-consuming and monotonous process of polishing. Based on the observations, such an iHRC has the potential of improving ergonomic, economic as well as quality aspects. These first insights assist in preparing an interview guideline for a semi-structured interview with potential users of the iHRC in order to gather more detailed information about user requirements and preferred technical concepts.

User Requirements Analysis. The semi-structured interview is based on an interview guideline presented in [31] and adapted to the already gained process knowledge. It is extended with requirement-specific questions regarding the information transfer from the human to the robot and the preferred interaction interface with a surface-finishing robot. Audio-recorded interviews with male finish workers (N_{FW} = 15; 0,75 h average on the context of use; 0,75 h average on the user requirements; aged from 22 to 54; M_{age} = 34.94; SD_{age} = 5.33; having finishing experience of M_{exp} = 4.83, SD_{exp} = 2.77 years) were transliterated and analyzed using a grounded theory approach according to [32]. In total, the following aggregated specific requirements could be extracted with respect to the iHRC use in industrial production:

- human control over the robot (e.g. start signals)
- ease of use, high efficiency and failure robustness (high usability)
- avoid speech input due to production noise and dialects
- avoid movable devices (probability of losing them)
- avoid battery-driven devices (probability of running out of battery)
- avoid major changes and additional devices to current process setup

The interview results showed rather non-specific preferences, and with that no precise wishes for human-robot interfaces. This underlines the fact, that potential users can hardly imagine future solutions without experiencing them [33].

Expert Analysis. The last analysis step within a focus group brings together a group of experts. It consists of one production improvement engineer from the OEM (M_{age} = 30) and three scientific usability experts (UE) in the industrial engineering field (N_{UE} = 3; M_{age} = 36, SD_{age} = 5.72). The common analysis of the already gained context and requirement knowledge leads to the prototypical design and empirical use case for evaluation. These results are extensively described in the following chapters.

Critique of Methodology. Besides these first results, the method selection for the analysis phase could be revised. The process-specific document analysis and the participating observation are seen as easy-to-use and assisted with the preparation of the interview guidelines. The context of use interview section may be excluded in future developments. Although it gained relevant insights to possible process improvements, there was no additional information compared to the gathered data from the document analysis and participating observation dealing with the development of an iHRC. Further, the analysis of the audio-recorded data needed little academic expertise in transliteration and qualitative analysis. Due to that, this time-consuming process may be reduced to the analysis of handwritten interview protocols in the future, only concentrating on the user requirements. The final analysis within a focus group was seen as a great benefit to finally harmonize the future development steps, from usable technical as well as process-oriented organizational perspectives.

4.2 Prototypical Design of the Positioning of a Surface-Finishing Robot

The prototypical design focuses on the positioning of a surface-finishing robot to a sanded spot on the surface of a workpiece. With regard to the previously gained process knowledge, gathered user requirements as well as preferences, the following HRIs are selected for empirical testing:

Hand-Guiding (HG). The user manually guides the robot to the specified work positions in a zero-gravity mode. The positions are logged in with a touch-on procedure, which also serves as haptic feedback. Lifting the robot above a certain height will serve as a start signal.
Mouse Positioning (MP). Clicking on the specified positions on a camera view will log in the positions within a custom graphical user interface (GUI). Feedback is provided with markers on the clicked points on the screen. A click on a start symbol will start the robot.
Touch Positioning (TP). Touching the GUI surface will create a point on the camera view. Moving it to the desired position while still touching the surface will enable precise positioning. Lifting up the finger will log in the point. Feedback is provided with markers on the screen. Touching the start symbol will start the robot.
Marker Detection (MD). Applied markers on the workpiece will be recognized by the camera system. A click on an accept button will log in the camera-recognized points. A further click on a start symbol will start the robot.
Pointing Gesture (PG). Marker-based pointing gestures log in points with staying still for one second. The marker is placed on the finger tip on the work glove. An acoustic signal serves as log in feedback. The start signal is given to the robot with pulling the hand away from the workpiece and not exposing the marker to the camera for one second.

4.3 Usability Evaluation of the Positioning of a Surface-Finishing Robot

Design. Based on the need for resource-efficient development, the iterative design and evaluation is split up into a preparative laboratory with students (novice users [NU]) and an application-specific study with potential users (finish workers [FW]). The focus of the laboratory study lies on functionality and usability tendencies. Major usability and functionality issues lead to first system improvements. The follow-up user study tries to get deeper insights into usability aspects within the context of use. As it could be the case, that there is one or more surface defect to be reworked, the experiments gather data for the positioning to either one or five predefined spots on the workpiece. The interaction starts after marking either one or all five spots. Marking is done with the placement of aims in the laboratory or the application of polishing paste in the application-specific use-case. With the use of a point-positioning gauge (symmetrical zigzag five-point pattern with overall 50 cm × 25 cm), the pre-specified points stay constant in their position. Using this design for the laboratory as well as application-specific use-case makes the gathered data comparable. Each participant is shortly introduced to the experiment. A latin square design is used to randomize the use of the interfaces in a within-subject design in order to eliminate sequence effects. After a short introduction to the current interface, the participants have the chance to shortly make themselves familiar with the robot control. Immediately afterwards, the experiment starts. The participants are asked to mark one spot in a first round or all five spots in a second round for each interface.

Equipment. The mentioned HRIs are all implemented with the use of a Kuka LBR iiwa 14 R820 equipped with either a pen or a polishing tool developed by Visomax Coating GmbH. A Microsoft Kinect v1 serves as camera together with a custom software (GUI) running on a Siemens Simatic HMI 19″ IPC. The goal of the study is to let the robot either touch on a specified point on the plate surface (laboratory) or to polish a specified spot on the workpiece (application). Both setups imitate the real workplace, a flat table to finish painted parts and a close-by monitor for handling process specific IT-systems. The application-specific experiment is executed right next to the original finish workplace. The two systems are shown in Fig. 1.

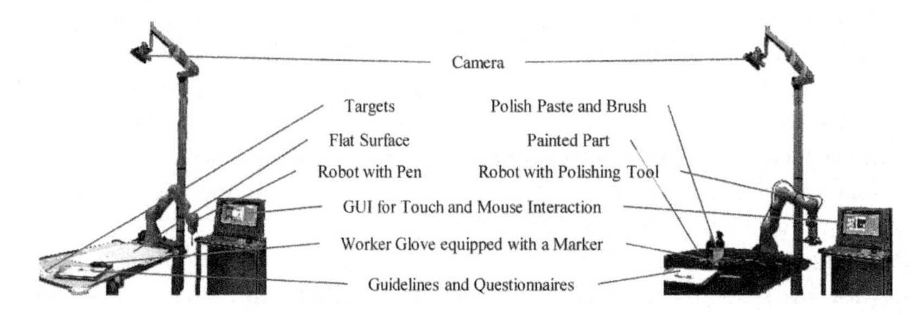

Fig. 1. Used systems for the laboratory and application-specific Study on iHRC positioning

Methodological Approach. The interaction time, interpreted as the time span from marking the last spot (either the first or the fifth spot) and the start of the robot, as well as handling errors are measured. Afterwards, an additional run through the procedure gives the chance to gather qualitative data with the use of the thinking aloud method. Moreover, participants are asked to fill out questionnaires after each interface experiment, realizing a mixed-method approach. The System Usability Scale (SUS) [34] is used in both scenarios. It delivers an easy and resource-efficient way to evaluate the pragmatic quality of an interactive system. The NASA-TLX [35] is additionally used in the application-specific study with potential users. It enables a more detailed insight in subjective task loads during the interaction.

Participants. We invited 20 students (Novice Users [NU], 13 male, 7 female) for the laboratory study. They were aged from 19 to 37 (M_{age} = 25.1, SD_{age} = 2.81). We asked each participant to rate their experiences in working together with a robot using a seven item Likert-Scale. High experience is scored with 6 points, no experience with 0 points. The participants stated to have low to medium experience in working together with a robot (M_{exp} = 1.8, SD_{exp} = 1.64).

The application-specific case study was executed with a prototypical surface-finishing robot. We invited 35 male finish workers (FW). They were aged from 22 to 55 ($M_{age} = 36.14$, $SD_{age} = 6.47$) and have finishing experience of $M_{exp} = 4.3$ years on average ($SD_{exp} = 2.19$). The participants stated to have low to medium experience on working together with a robot ($M_{exp} = 2.17$, $SD_{exp} = 1.97$).

All participants were able to understand, read and write German language, since the instructions and questionnaires were in German.

Quantitative Results. Quantitative data was statistically analyzed using the software IBM SPSS Statistics 23. The data, each measurement for each group, mainly shows a non-normal distribution based on a Shapiro-Wilk test ($p < 0.05$). Due to that fact and with respect to some outliers, a Friedman test is used to analyze potential differences. In each case, a post-hoc Dunn-Bonferroni test reveals detailed insights.

Positioning Time (t). We statistically compared the positioning times for each group, interface (see Sect. 4.2) and task (one or five spots), shown in Fig. 2.

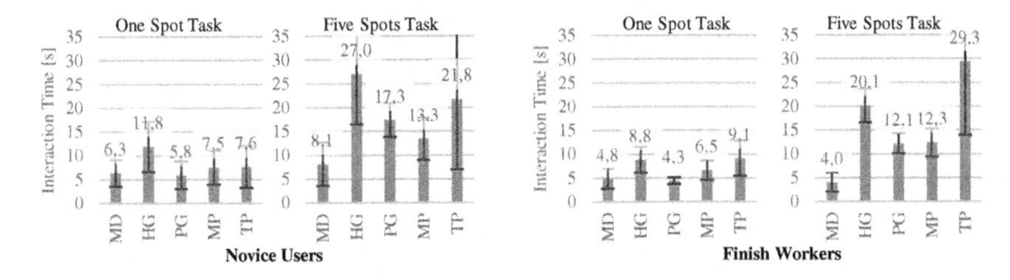

Fig. 2. Positioning times from the laboratory (left) and application-specific study (right)

The positioning times analysis shows significant differences within all participant groups for each task ($p = .000$). In the laboratory, the use of HG for the one spot task resulted in significantly higher positioning times compared to all other interfaces ($p < .05$). Similar effects are seen in the laboratory five spots task. MD and MP significantly performed better than HG (both with $p = .000$). MD shows outstanding performance, having significantly lower interaction times than PG, HG and TP (all with $p = .000$). These first insights indicate superior positioning time performances of MD, with its low additional positioning effort after the markers being placed, and MP as a very familiar interface for the PC-experienced group of students.

Analogical results can be found in the comparison of the finish worker tasks. For the one spot task, PG and MD performed with a similar performance (no significant differences, $p = 1.000$), turning out pointing in the real world with a finger or on a camera view with a mouse to be similar processes. TP and HG performed similarly bad (no significant differences, $p = 1.000$). This again points out MD to be a very efficient interface (significant differences to all other interfaces, $p < .05$). For the five spot finish worker task, pointing in the real world (PG) and on a camera view (MP) also perform with similar results (no significant differences, $p = 1.000$). HG and TP again perform worst and show no significant difference between each other ($p = 1.000$).

The application-specific use-case underlines the inferiority of HG and TP with respect to the positioning times. As the markers are placed with the application of aims or polish paste at the spots of interest in any case, MD does not need any further positioning information. The interaction just needs a log in of the recognized points from the camera system and the start signal. The system takes the positioning information from recognizable environment changes. MD is followed by PG and MP, having similar characteristics with a pointing gesture on the part in the real world or a click on the camera view.

Interaction Errors (ie). An overall interaction error count for each task in each group only reveals significant differences in the finish worker group between the interfaces ($p = .000$, $n = 35$). The TP interface, with a mean interaction error of $M_{ie,FW,TP} = 1.6$ per finish worker and $SD_{ie,FW,TP} = 1.786$, shows the highest numbers. TP interface has significantly more interaction errors (e.g. repositioning of badly or unintendedly placed points) than MD, MP and PG ($p < .05$). This underlines the bad accuracy and with that usability of TP interface for their use in an iHRC system and gives a reason for the high positioning times compared to other interfaces (see Fig. 2).

System Usability Scale (SUS). Figure 3 shows the SUS scores for each group.

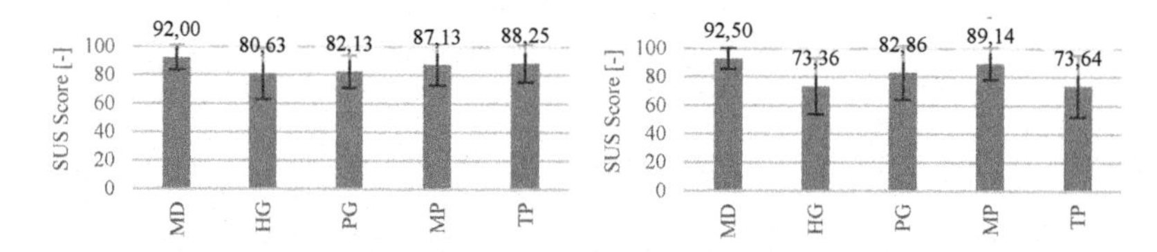

Fig. 3. SUS Scores for Novice Users (NU, left) and Finish Workers (FW, right)

The only significant differences in the novice user group are found between MD and PG ($p = .005$) leading to some technical improvements of the marker detection robustness in the used algorithm. The similarity of the other SUS scores can be traced back to the novelty effect of interacting with a robot and the non-use-case-specific experiment. Interfaces are mainly rated according to their task fulfillment. As there are no major differences in interaction errors, all interfaces are rated with a high usability. In the application-specific use case, TP and HG are rated significantly worse than MD, MP and PG ($p < .05$), mainly based on the bad accuracy of TP and comparably high effort of HG.

NASA-TLX. The NASA-TLX questionnaire was only filled out by the finish workers. All interfaces show significant differences between the six evaluated categories, mental, physical and temporal demand, performance, effort and frustration (all with $p = .000$, $N = 35$). Although the data scatters a lot, the most meaningful peaks are found in following data. Temporal demand of TP ($M = 35.00$, $SD = 22.687$) is significantly higher compared to all the other interfaces ($p < .05$). This can be traced back to the bad accuracy and with that high correction effort. The highest physical demand can be seen in the HG data ($M = 29.86$, $SD = 24.055$). Because of this significantly higher physical effort compared to all other interfaces ($p < 0.5$), HG cannot be recommended for the use over a full work shift duration.

Qualitative Results. Qualitative data is taken from the protocolled thinking aloud method. Results are clustered in categories, shown in Table 1.

Table 1. Clustered qualitative results of the usability studies.

HRIs	Strengths	Weaknesses	Improvements
Hand Guiding	operation without any additional devices (e.g. cameras), high precision positioning abilities	missing visual feedback compared to GUI interfaces, robot´s inertia and high physical demand, positioning limitation due to arm length, unusual direct contact to a robot	handle for a better grip, visual system feedback in order to recognize the logged points and the status of the robot
Mouse Positioning	ease of use and accuracy in combination with a cross line cursor and customizable sensitivity	dilemma between concentration effort and accuracy, a small rack to place the mouse is needed	elimination of unnecessary mouse movements, e.g. use of context menus
Touch Positioning	ease of use, possible integration into production terminals	Inaccuracy and long positioning resulting in stress	use of bigger displays and/or zoom functions
Marker Detection	ease of use, time savings due to no additional positioning effort	missing trust in automated recognition systems under any given conditions (size of the marker and exposure to light or reflections)	use of alternative and faster start signals (e.g. buttons), direct visual feedback on the workpiece in order to get rid of GUI interaction
Pointing Gesture	direct interaction with the part, no need for extra displays, pointing is easy to learn	audio feedback was often not recognized due to loud machine noises in production, missing visual feedback, unintended interaction errors	direct visual feedback, more direct control of the robot (e.g. explicit start signal)

Final Expert Analysis. A final analysis and discussion is done within a focus group of usability experts ($N_{UE} = 4$, $M_{age} = 34$, $SD_{age} = 6.04$) and the supervising production improvement engineer. Evaluations and results are interpreted in order to identify design principles and give recommendations for future developments:

Use Process-Specific Markers. As the MD interface stands out with respect to positioning time, SUS score and user feedback, we recommend to use markers from the original process in order to position a robot on the surface of a workpiece. Further examples are drilled holes as a position marker for a thread cutting robot and the application of a fluid for positioning an ultrasonic inspection robot and a lot more.

Interact with the Workpiece. The direct interaction with the workpiece is evaluated superior to the interaction with a display. The dependability to a fixed display was criticized by many participants. In contrast to that PG and HG can be operated from any side. This is seen as a main advantage, enables flexibility and reduces travel paths. Due to missing visual feedback of MD, PG and HG, we recommend the use of projection systems in order to transfer the GUI onto the part. This also enables MP interfaces for direct interaction on the workpiece and may additionally inform the user about the actual system state and already logged points. With the absence of visual feedback, the participants often did not know, if they actually logged in a point or when to robot is going to start. Context bars, visualization of user inputs and robot movements or next process steps may assist the user with the interaction.

Design simple, Effortless and Robust Interfaces. We highly recommend the use of understandable and consistent interfaces. Mouse movements and clicks should be reduced to a minimum, so we recommend the use of context menus. Besides, full body movements are involved in the use of HG and PG. As the participants` arm reach vary, it is strenuous for smaller participants to reach points being further away from their standpoint. Reaching out the arms to their maximum over an eight hour shift is seen as a non-convenient action. Additionally, the guidance of a robot, even in zero-gravity mode, was strenuous to many participants. Further, as unintended start signals frighten the user, especially the start of the robot needs a clear signal, e.g. pressing a button. The users should be able to undo unintended commands before the robot starts.

Critique of Methodology. The split into non-specific and application-specific studies shows advantages with respect to functional improvements and method testing. It allows the better preparation of studies with potential users. For example, the invariant SUS evaluation in the laboratory study indicates the necessity of further evaluation tools. User experience questionnaires such as UEQ [36] may be used in future iHRC developments. The final focus group was again seen as a great benefit with the common analysis of the gathered data.

5 Conclusion and Future Work

In this paper, we empirically developed and evaluated an iHRC system with the use of a mixed-method UCD framework. Anchored in the design science research paradigm, this allowed the iterative revisal of the methodological embodiment for their practical use on the one hand, and the iterative development of an iHRC system on the other hand. We described the analysis phase and user studies on the positioning with the development of an assistive surface-finishing robot in the automotive sector. The latter were executed in two steps, a preparative laboratory study and an application-specific use case. The iterative development with its mixed-method usability measures assists in improving the iHRC system as well as deriving design principles for further system developments. With the help of the gathered data, automated marker detection, mouse positioning and pointing gestures could be identified as superior interfaces for positioning a robot in order to fulfill a task at a given spot on a workpiece in an industrial context. This will serve as the basis for further studies.

With the outstanding development of the parametrization within an iHRC, the research results in an empirically tested design concept of iHRC systems, containing a UCD framework for their iterative development as well as derived design principles. Within our future research, we additionally want to enable the parametrization of the system, not only instructing the robot where, but also how to do its work. With these first insights into usability studies, similar industrial production processes can more easily be assisted with iHRC systems.

References

1. Bänziger T, Kunz A, Wegener K (2017) A library of skills and behaviors for smart mobile assistant robots in automotive assembly lines. In: Mutlu B, Tscheligi M (eds) Proceedings of the companion of the 2017 ACM/IEEE international conference on human-robot interaction. ACM, pp 77–78
2. Kahl B, Bodenmüller T, Kuss A (2016) Technologien für flexible Robotersysteme: Wirtschaftliche Automatisierungslösungen (nicht nur) für kleine und mittlere Produktionsgrößen. Ind 4.0 Manag 32(2):11–14
3. IFR Statistical Department (2016) Industrial robots 2016
4. Steegmüller D, Zürn M (2017) Wandlungsfähige Produktionssysteme für den Automobilbau der Zukunft. In: Vogel-Heuser B, Bauernhansl T, Hompel M ten (eds) Handbuch Industrie 4.0 Bd.1: Produktion, 2., erweiterte und bearbeitete Auflage. Springer, Heidelberg, pp 27–44
5. Akan B, Cürüklü B, Spampinato G et al (2010) Towards robust human robot collaboration in industrial environments. In: Proceedings of the 5th ACM/IEEE international conference on human-robot interaction: HRI 2010. ACM, IEEE, Piscataway, pp 71–72
6. Forge S, Blackman C (2010) A helping hand for Europe: the competitive outlook for the EU robotics industry
7. Guerin KR, Riedel SD, Bohren J et al (2014) Adjutant: a framework for flexible human-machine collaborative systems. In: Proceedings of the 2014 international conference on intelligent robots and systems. IEEE, RSJ, pp 1392–1399
8. Michalos G, Makris S, Spiliotopoulos J et al (2014) ROBO-PARTNER: seamless human-robot cooperation for intelligent, flexible and safe operations in the assembly factories of the future. Proc CIRP 23:71–76. https://doi.org/10.1016/j.procir.2014.10.079
9. Naumann M, Dietz T, Kuss A (2017) Mensch-Maschine-Interaktion. In: Vogel-Heuser B, Bauernhansl T, Hompel M ten (eds) Handbuch Industrie 4.0 Bd.4: Allgemeine Grundlagen. Springer, Heidelberg, pp 201–216
10. Huber W (2016) Industrie 4.0 in der Automobilproduktion: Ein Praxisbuch. Springer Fachmedien Wiesbaden, Wiesbaden
11. Tsarouchi P, Makris S, Chryssolouris G (2016) Human-robot interaction: review and challenges on task planning and programming. Int J Comput Integr Manuf 29(8):916–931. https://doi.org/10.1080/0951192x.2015.1130251
12. Buchner R, Mirnig N, Weiss A et al (2012) Evaluating in real life robotic environment: bringing together research and practice. In: Proceedings of the 2012 21st IEEE international symposium on robot and human interactive communication: IEEE RO-MAN 2012. IEEE, Piscataway, pp 602–607
13. Weiss A, Huber A, Minichberger J et al (2016) First application of robot teaching in an existing industry 4.0 environment: does it really work? Societies 6(3):20. https://doi.org/10.3390/soc6030020

14. Baxter P, Kennedy J, Senft E et al (2016) From characterising three years of HRI to methodology and reporting recommendations. In: Proceedings of the 2016 11th ACM/IEEE international conference on human robot interation: HRI 2016. ACM, IEEE, pp 391–398
15. Sheridan TB (2016) Human-robot interaction: status and challenges. Hum Factors 58 (4):525–532. https://doi.org/10.1177/0018720816644364
16. Schleicher T (2018) Leitfaden zur Gestaltung einer gebrauchstauglichen instruktiven Mensch-Roboter-Kollaboration am Produktionsarbeitsplatz (in Veröffentlichung). In: Bullinger-Hoffmann AC (ed) innteract 2018: innovation der Innovation - neu gedacht, neu gemacht. aw&I Conference. aw&I Wissenschaft und Praxis
17. Akan B, Cürüklü B, Asplund L (2008) Interacting with industrial robots through a multi-model language and sensor system. In: Proceedings of the 39th international symposium on robotics. ACM, pp 66–69
18. Vogel C, Walter C, Elkmann N (2012) Exploring the possibilities of supporting robot-assisted work places using a projection-based sensor system. In: Zug S (ed) Proceedings of the 2012 IEEE international symposium on robotic and sensors environments (ROSE). IEEE, Piscataway, pp 67–72
19. Rohde M, Pallasch A, Kunaschk S (2012) Effiziente Automatisierung mittels Industrierobotern: Potenziale für gering standardisierte Aufgabenstellungen in der Logistik. Prod Manag 17(2):21–24
20. Švaco M, Šekoranja B, Jerbić B (2012) Industrial robotic system with adaptive control. Proc Comput Sci 12:164–169. https://doi.org/10.1016/j.procs.2012.09.048
21. Sekoranja B, Jerbic B, Suligoj F (2015) Virtual surface for human-robot interaction. Trans Famena 39(1):53–64
22. Sobaszek Ł, Gola A (2015) Perspective and methods of human-industrial robots cooperation. AMM 791:178–183. https://doi.org/10.4028/www.scientific.net/AMM.791.178
23. Barbagallo R, Cantelli L, Mirabella O et al (2016) Human-robot interaction through kinect and graphics tablet sensing devices. In: Proceedings of the 24th mediterranean conference on control & automation. IEEE, pp 551–556
24. Maurtua I, Pedrocchi N, Orlandini A et al (2016) FourByThree: imagine humans and robots working hand in hand. In: Proceedings of the 2016 IEEE 21st international conference on emerging technologies and factory automation (ETFA). IEEE, pp 1–8
25. Bdiwi M, Rashid A, Pfeifer M et al (2017) Disassembly of unknown models of electrical vehicle motors using innovative human robot cooperation. In: Mutlu B, Tscheligi M (eds) Proceedings of the companion of the 2017 ACM/IEEE international conference on human-robot interaction. ACM, pp 85–86
26. Sadik AR, Urban B, Adel O (2017) Using hand gestures to interact with an industrial robot in a cooperative flexible manufacturing scenario. In: Proceedings of the 3rd international conference on mechatronics and robotics engineering: ICMRE 2017. ACM, pp 11–16
27. Hevner AR, March S, Park J et al (2004) Design science in information systems research. MIS Q 28(1):75–106
28. Schou C, Damgaard JS, Bogh S et al. (2013) Human-robot interface for instructing industrial tasks using kinesthetic teaching. In: Proceedings of the 2013 44th international symposium on robotics (ISR). IEEE, pp 1–6
29. Alexandrova S, Tatlock Z, Cakmak M (2015) RoboFlow: a flow-based visual programming language for mobile manipulation tasks. In: IEEE Proceedings of the 2015 IEEE international conference on robotics and automation (ICRA). IEEE, Piscataway, pp 5537–5544
30. Pedersen MR, Nalpantidis L, Andersen RS et al (2016) Robot skills for manufacturing: from concept to industrial deployment. Robot Comput Integr Manuf 37:282–291. https://doi.org/10.1016/j.rcim.2015.04.002

31. Deutsche Akkreditierungsstelle GmbH (2010) Leitfaden Usability: Gestaltungsrahmen für den Usability-Engineering-Prozess
32. Glaser BG, Strauss AL (1967) The discovery of grounded theory: strategies for qualitative research. Observations. Aldine Pub. Co., Chicago
33. Roser T, Samson A, Valdivieso EC (2009) Co-creation: new pathways to value. An overview. Promise, London
34. Brooke J (1996) SUS: a quick and dirty Usability Scale. In: Jordan PW (ed) Usability evaluation in industry. Taylor & Francis, London, pp 189–194
35. Hart SG, Staveland LE (1988) Development of NASA-TLX (Task Load Index): results of empirical and theoretical research. Adv Psychol 52:139–183. https://doi.org/10.1016/s0166-4115(08)62386-9
36. Laugwitz B, Held T, Schrepp M (2008) Construction and evaluation of a user experience questionnaire. In: Holzinger A (ed) USAB 2008, vol 5298. LNCS. Springer, Heidelberg, pp 63–76. https://doi.org/10.1007/978-3-540-89350-9_6

Validation of the Lifting Fatigue Failure Tool (LiFFT)

Sean Gallagher(✉), Richard F. Sesek, Mark C. Schall Jr., and Rong Huangfu

Auburn University, Auburn, AL 36830, USA
seangallagher@auburn.edu

Abstract. Manual material handling is common in industry and has demonstrated a strong association with the development of low back disorders (LBDs). Several risk assessment tools exist in the literature to assess acceptable lifting limits, and/or the development of improved design of manual lifting tasks. However, recent evidence has strongly suggested that LBDs (and other MSDs) may the result of a process of mechanical fatigue failure. Prior tools have not used fatigue failure methods to assess risk, which may be beneficial is these disorders are indeed the result of such a process. The purpose of this paper is to describe a new risk assessment tool for manual lifting (LiFFT) and to provide validation of this tool using two existing epidemiological databases. Results demonstrate that the LiFFT cumulative damage measure is significantly associated with low back outcomes in both epidemiological studies.

Keywords: Low back disorders · Risk assessment · Fatigue failure Epidemiology

1 Introduction

Low back pain (LBP) affects roughly 40–80% of the population at some point during their lifetime [1–4]. Results from the Global Burden of Disease 2010 Study indicate that LBP is the greatest contributor to global disability when measured in number of years lived with disability and the sixth highest contributor to "overall burden" when measured in disability-adjusted life years out of all 291 conditions studies [3].

Recent evidence strongly suggests that MSDs (including LBP) are the result of a fatigue failure process [5]. If this is indeed the case, risk assessment should be performed based on fatigue failure principles. However, previous tools have not used these principles in their risk assessment techniques. This paper describes a new practitioner-friendly risk assessment tool for manual lifting (LiFFT), describes validation of the tool against two separate epidemiological studies, and describes recommended use of the tool for the assessment of low back pain risk.

S. Bagnara et al. (Eds.): IEA 2018, AISC 825, pp. 216–223, 2019.
https://doi.org/10.1007/978-3-319-96068-5_24

2 Methodology

2.1 Model Rationale

In basic concept, assessment of risk using fatigue failure theory is quite straightforward. All one needs is an estimate of the magnitude(s) of stress experienced by the tissue of interest (relative to maximum stress), and the number of repetitions performed at each stress magnitude. Results of assessments can be summed to estimate the cumulative damage resulting from the stress/repetition combinations experienced.

The LiFFT tool uses a fatigue failure curve derived from previous work on cadaver spinal motion segments [6, 7]. Peak load moment is used to estimate the compression load on the spine and these values are compared to the average compressive strength of a mixed male/female sample of cadaver spines [8]. Given a peak load moment, an estimate of compression is derived and compared to the average spine load to develop an estimate of cumulative damage per cycle for that task. This value is multiplied by the number of repetitions to obtain an overall estimate of cumulative damage for a given task. If multiple tasks are performed during the workday, cumulative damage estimates for each task are summed to get an overall "daily dose" of cumulative damage for the workday.

Figure 1 provides an example of the use of LiFFT for a job involving multiple lifting tasks. As can be seen from this figure, 5 lifting jobs are performed during the worker's day. The maximum lever arm and load are determined as well as the number of repetitions for each task. The total cumulative damage for this task is 0.0086 corresponding to a 43.32% probability that this job is a high-risk job for low back disorders. The last column provides the percentage of the total cumulative damage associated with each task, which can be used to prioritize job redesign decisions.

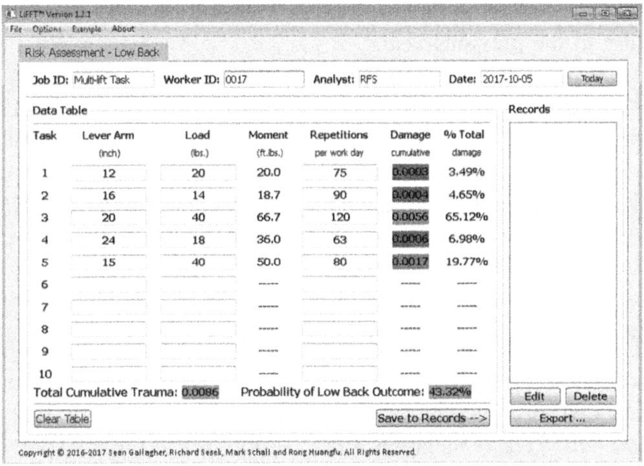

Fig. 1. Example of LiFFT for a multi-task job.

2.2 LiFFT Validation

We used two databases to validate LiFFT. Both contained the requisite information regarding PLM and the number of lifts performed for each task by individuals as well as the associated LBD or LBP health outcomes. The first was the LMM database described by Marras et al. [9] as presented in Zurada et al. [10]. This database consists of two levels of LBD risk. Low-risk jobs were categorized as those having no injuries and no turnover for the preceding three-year period. High-risk jobs were those having at least 12 injuries per 200,000 h of exposure. The former category had a total of 124 jobs and the latter consisted of 111 jobs. It should be noted that risk was assessed at the job level, not at the individual level for this study. As a result, it was not possible to control for individual characteristics such as age, gender, or body mass index (BMI) for this study.

The second database was from an epidemiological study involving a large automotive manufacturer. Data were analyzed from a database consisting of 667 manufacturing jobs. The database included historical injury data for the analyzed jobs as well as symptom interviews for 1,022 participants [11]. A total of 304 jobs involving manual material handling tasks were analyzed in the LiFFT tool validation. Of these, six subsets (each having different LBP case/control definitions) were analyzed. These subsets ranged from 179–304 subjects. Symptoms were categorized into one of five categories attributed to the job or to other jobs or factors: Category (1) "job related symptoms," symptoms originated on the subject's job; Category (2) "job aggravated," symptoms aggravated by the subject's current job, but not originating on job; Category (3) "no change or improvement in subject symptoms" while on the job, but symptoms not originating on the job; Category (4) "symptom improvement on the job," symptoms not originating on present job and improving; and Category (5) "no symptoms present." Symptoms were self-reported both on the day of the subject interview and retrospectively for the previous year. Case definitions included various combinations of symptoms and LBP-related injuries on the job in the previous year. Controls were subjects lacking job-related symptoms and/or working on jobs with no LBP-related reports of injury in the previous year.

While the LMM database consisted of jobs where workers performed the same job throughout the day, the automotive database contained numerous instances of jobs involving several different tasks (up to six), for which CD was calculated by summing the exposure calculated for each task. The PLM for each task was calculated by multiplying the horizontal distance to the load (measured according to RNLE method) and the known load mass. Repetitions for each task were available and were multiplied by the DPC to get the CD on a task basis.

A total of six separate low back case and control definitions were used in the automotive database. These included:

(1) LBP *Today* (Categories 1 & 2) + Job Reported Injury (previous year) vs. No or Improving LBP (Categories 4 & 5) and no Job Related Injury (previous year)
(2) LBP *in the Last Year* (Categories 1 & 2) + Job Related Injury (previous year) vs. No or Improving LBP (Categories 4 & 5) and no Job Related Injury (previous year)
(3) LBP *Today* (Categories 1 & 2) vs. No LBP *Today* (Categories 4 & 5)
(4) LBP *Today* (Category 1 only) vs. No LBP *Today* (Category 5 only)

(5) LBP *in the Last Year* (Categories 1 & 2) vs. No LBP *in the Last Year* (Categories 4 & 5)
(6) LBP *in the Last Year* (Category 1 only) vs. No LBP *in the Last Year* (Category 5 only).

Logistic regression models were developed to analyze the relationship between the log of the LiFFT Cumulative Damage score (LogCD) and outcomes of both epidemiology studies. In addition, 2 × 2 contingency tables were analyzed using a Cumulative Damage cut point of 0.03 for all outcomes in both studies.

3 Results

Results of the logistic regression for the log of the continuous LiFFT CD measure were significantly associated with the probability of a high-risk job in the LMM database (OR = 2.78, 95% CI = [2.02, 3.83]). Figure 2 illustrates the relationship between the LiFFT LogCD measure and the probability of a job being considered high risk (i.e., 12 + low back disorders per 200,000 h).

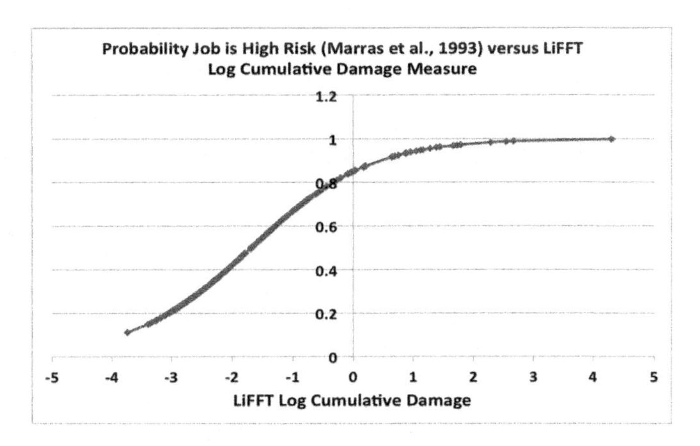

Fig. 2. Probability of a High Risk Job as a function of LiFFT log Cumulative Damage Measure based on logistic regression model.

Table 1 provides logistic regression results for four outcomes from the automotive database validation. These include outcomes #3–#6 above. All of these outcomes were significantly associated with the LiFFT Log Cumulative Damage in both crude and adjusted analyses. Outcomes #1–#2 above were not significant in the logistic regression analyses.

Table 1. Crude and adjusted logistic regression analyses for four outcomes from the automotive epidemiology study.

Outcome	Analysis	Cases	N	Variable	df	Chi-sq	*P*	OR	95% CI
Low Back Pain Last Year (1,2 v. 4,5)	Crude	84	296	**Log CD**	**1**	**4.80**	**0.028**	**1.364**	**(1.013, 1.813)**
	Adjusted	84	296	**Log CD**	**1**	**4.79**	**0.029**	**1.427**	**(1.030, 1.978)**
				Site	**5**	**14.00**	**0.016**	**Var**	**Var**
				Gender	1	0.23	0.633	0.855	(0.451, 1.621)
				Age	1	0.27	0.607	1.167	(0.649, 2.099)
				BMI	1	2.12	0.146	0.649	(0.359, 1.172)
Low Back Pain Last Year (1 v. 5)	Crude	67	270	**Log CD**	**1**	**6.22**	**0.013**	**1.479**	**(1.074, 2.036)**
	Adjusted	67	269	**Log CD**	**1**	**6.26**	**0.012**	**1.219**	**(1.038, 1.431)**
				Site	**5**	**11.98**	**0.035**	**Var**	**Var**
				Gender	1	0.02	0.874	0.944	(0.461, 1.934)
				Age	1	1.17	0.280	1.428	(0.748, 2.726)
				BMI	1	3.28	0.070	0.550	(0.284, 1.067)
Low Back Pain Today (1,2 v. 4,5)	Crude	40	242	**Log CD**	**1**	**4.94**	**0.026**	**1.469**	**(1.030, 2.095)**
	Adjusted	40	242	**Log CD**	**1**	**6.03**	**0.014**	**1.768**	**(1.094, 2.858)**
				Site	**5**	**133.33**	**<0.001**	**Var**	**Var**
				Gender	1	0.003	0.962	0.979	(0.396, 2.418)
				Age	1	3.63	0.057	2.176	(0.970, 4.880)
				BMI	1	1.14	0.286	0.684	(0.296, 1.446)
Low Back Pain Today (1 v. 5)	Crude	53	259	**Log CD**	**1**	**5.82**	**0.016**	**1.621**	**(1.068, 2.460)**
	Adjusted	53	259	**Log CD**	**1**	**4.29**	**0.038**	**1.513**	**(1.008, 2.272)**
				Site	5	9.44	0.093	Var	Var
				Gender	1	0.20	0.656	0.839	(0.369, 1.809)
				Age	1	1.66	0.198	1.584	(0.785, 3.195)
				BMI	1	0.84	0.360	0.726	(0.363, 1.453)

Table 2 provides the results of an analysis of the odds ratios, sensitivity, specificity, positive and negative predictive values a As can be seen from this table, all LBP outcomes demonstrate significant odds ratios. Odds ratios were particularly strong for the case definitions of High Risk versus Low Risk group from the LMM database (OR = 9.50, 95% CI: 4.78, 19.18), and Injury Case plus Pain Today (OR = 9.65, 95% CI: 2.38, 39.15) and Injury Case plus Pain Last year (OR = 8.28, 95% CI: 2.60, 26.31). Each of the six case definitions from the automotive database demonstrated significant odds ratios with the CD measure, ranging between 2.80–9.65. Overall, the cut point of 0.03 resulted in high agreement (71.5–86.8%), high specificity (0.90–0.96), high negative predictive values (0.67–0.90), but more modest positive predictive values (0.33–0.82).

Table 2. Results of the analysis for a LiFFT CD cut point of 0.03 for low back outcomes in both the LMM and Automotive Databases.

	Marras (1993), Zurada (1997)	(Sesek 1999)					
	High Risk vs. Low Risk	Injury Case + Pain *Today* (4/5 vs. 1/2)	Injury Case + Pain in *Last Year* (4/5 vs. 1/2)	Pain *Today* (4/5 vs. 1/2)	Pain *Today* (5 vs. 1)	Pain *Last Year* (4/5 vs. 1/2)	Pain *Last Year* (5 vs. 1)
OR	9.50	9.65	8.28	2.80	2.91	2.81	2.94
95% CI	(4.71, 19.18)	(2.38, 39.15)	(2.60, 26.31)	(1.14, 6.87)	(1.09, 7.75)	(1.28, 6.17)	(1.29, 6.72)
Agreement	71.5	90.7	86.2	77.7	80.9	71.5	74.5
Prevalence	0.47	0.08	0.14	0.20	0.16	0.28	0.25
Sensitivity	0.50	0.31	0.28	0.17	0.18	0.16	0.18
Specificity	0.90	0.96	0.96	0.93	0.93	0.94	0.93
PPV	0.82	0.36	0.50	0.39	0.33	0.50	0.46
NPV	0.67	0.94	0.89	0.81	0.85	0.74	0.77
N	235	172	181	264	246	302	274

4 Discussion

We have presented a new risk assessment tool, LiFFT, for LBP/LBDs based upon fatigue failure principles. The tool has been designed to be easy to use and should be accessible to all safety and ergonomics practitioners. The only measurements required to use the tool are the load(s) being lifted, the maximum horizontal distance from the spine to the load, and a count (or estimate) of the number of daily repetitions. Extension of the tool to more complicated jobs that involve multiple tasks performed in a workday is quite easy and the "daily dose" of CD associated with LBP/LBD risk may be estimated by summing the individual subtask CD estimates. This provides an easy way to evaluate and compare risks associated with various tasks, and to determine priorities for job redesign.

The CD metric used by the tool demonstrated significant relationships for all LBP/LBD outcome measures in both epidemiological databases used for validation. Logistic regression of this measure against the LMM database outcome left only 8% of the deviance unexplained, and four outcome measures of LBP (both current and over the last year) in the automotive database explained between 72.3–94.6% of deviance. Two outcome measures in the automotive database (LBP case plus pain today or during the last year) did not demonstrate statistical significance in logistic regression; however, when examining a CD cut point of 0.03 both outcomes demonstrated substantial ORs (9.65 and 8.28, respectively).

Closer scrutiny of the data reveals the importance of the prevalence for specific outcomes and the strength of the relationship observed in logistic regression analyses. The outcomes involving LBP case + pain in the automotive database (current pain or pain in the last year) had very low prevalence rates (<14%), reflective of a small number of cases (22 and 26, respectively). The relative paucity of cases may explain why logistic regression failed to demonstrate significance for these outcomes, and why significant ORs for the CD measure were only observed in the cut point analysis. Outcomes involving pain had higher numbers of cases (49–86) and prevalence rates (ranging from 19–28%). A dose-response relationship between the LiFFT CD measure and outcomes emerged strongly and consistently for these cases. The LMM database outcome had by far the highest prevalence (47.2%) and demonstrated a very strong relationship to our CD measure, as mentioned above. The strength of relationship for the LBD outcome from this database may be due in part to the fact that low risk and high risk outcomes were separated by eliminating the medium risk category, making the relationship more pronounced. Additionally, the range of predicted CD was much wider for the LMM database.

5 Conclusion

This paper describes an easy-to-use manual lifting risk assessment tool, "LiFFT", that requires only the PLM (horizontal distance from spine to load times the weight of load) for a lifting task and number of repetitions performed for that task. A measure of CD is calculated based upon these values and can be summed across numerous tasks to derive a "daily dose" of CD. This measure was validated against two databases (Sesek 1999; Zurada et al. 1997) and the following results were obtained:

1. Logistic regression analysis for the LMM database (composed of mono-task jobs) demonstrated a strong dose-response relationship between jobs categorized as high risk and the log of the LiFFT cumulative damage metric, with a continuous odds ratio of 2.78.
2. Logistic regression models demonstrated significant relationships between our CD measure and outcomes involving both current LBP and LBP in the past year in an automotive manufacturer epidemiology study involving jobs where the risk associated with multiple tasks were summed.
3. A cut point of 0.03 for the CD measure was found to result in significant odds ratios for all outcomes (i.e., both databases). Odds ratios ranged from 2.43 to 9.97.
4. Results strongly support the use of a CD measure (derived using fatigue failure techniques) to assess the risk of a wide variety of LBD/LBP outcomes.

References

1. Balagué F, Mannion AF, Pellisé F, Cedraschi C (2012) Non-specific low back pain. Lancet 379:482–491
2. Calvo-Muñoz I, Gómez-Conesa A, Sánchez-Meca J (2013) Prevalence of low back pain in children and adolescents: a meta-analysis. BMC Pediatr 13:1
3. Hoy D, Bain C, Williams G, March L, Brooks P, Blyth F, Woolf A, Vos T, Buchbinder R (2012) A systematic review of the global prevalence of low back pain. Arthritis Rheum 64:2028–2037
4. Woolf AD, Pfleger B (2003) Burden of major musculoskeletal conditions. Bull World Health Organ 81:646–656

5. Gallagher S, Schall MC Jr (2016) Musculoskeletal disorders as a fatigue failure process: evidence, implications and research needs. Ergonomics 60:255–269
6. Brinckmann P, Biggemann M, Hilweg D (1988) Fatigue fracture of human lumbar vertebrae. Clin Biomech 3:S1–S23
7. Gallagher S, Marras WS, Litsky AS, Burr D, Landoll J, Matkovic V (2007) A comparison of fatigue failure responses of old versus middle-aged lumbar motion segments in simulated flexed lifting. Spine 32:1832–1839
8. Jager M, Luttmann A (1991) Compressive strength of lumbar spine elements related to age, gender, and other influencing factors. Electromyographical Kinesiol 291–294
9. Marras WS, Lavender SA, Leurgans SE, Rajulu SL, Allread WG, Fathallah FA, Ferguson SA (1993) The role of dynamic three-dimensional trunk motion in occupationally-related low back disorders: the effects of workplace factors, trunk position, and trunk motion characteristics on risk of injury. Spine 18:617–628
10. Zurada J, Karwowski W, Marras WS (1997) A neural network-based system for classification of industrial jobs with respect to risk of low back disorders due to workplace design. Appl Ergon 28:49–58
11. Sesek RF (1999) Evaluation and refinement of ergonomic survey tools to evaluate worker risk of cumulative trauma disorders (Dissertation)

Between Ergonomics and Anthropometry

Massimo Grandi(✉)

Martini Alimentare S.r.l, Gatteo, Italy
m.grandi@martinigruppo.com

Keywords: Anthropometry · Jobs · Reachable positions · Tools

1 Between Production and Health of Workers

1.1 Production Needs Regarding Food Sectors

The food industry sector production of Martini Alimentare Srl lives actually a period of big transformation influenced by the constant changing of the market conditions.

In order to maintain the competitiveness now the company must be flexible in production sector; this flexibility consists first in giving fast answers to the customers, then in guaranteeing elasticity and finally in investing in new equipments and machinery.

Despite the constant investments, the production domain needs always a large use of employees. This manpower works in several phases of production cycle.

1.2 Problems of the Company to Stay on the Market

Having the company a mass production, from the point of view of workers' health, the most diffused problems are relative to the manual cargo-handling and to the repetitive handling in factory; the risk is essentially the deterioration of staff people. If not managed in a good way the manpower can be exhausted by this job.

The main aim of this project is to search and guarantee a better staff management but also to make department heads and Company Management conscious of this situation.

More awareness is required.

1.3 The Company Restrictions

The main difficulties regarding manpower management are:

- Workers belong often to different nationalities (they are from heterogeneous countries): this aspect is very important because human sizes can change according to the country of origin (for example staff coming from Africa has very different body sizes if compared to asian one).

M. Grandi—First Author.

S. Bagnara et al. (Eds.): IEA 2018, AISC 825, pp. 224–229, 2019.
https://doi.org/10.1007/978-3-319-96068-5_25

- The working population is very heterogeneous in terms of age: the middle age of the company staff is 47 years old; this fact demonstrates that there are two categories of workers: the first one with people who is growing old having worked enough years behind and another category of young people having a short employment history.
- Another important aspect is represented by the percentage of the restrictions on work given by the doctor during the periodic visits. This factor is very relevant because sometimes doctor's restrictions avoid to the employer to carry out some tasks. In this case, in order to satisfy production exigences the rest of employment staff is overloaded from other tasks. The most problematic and frequent restrictions concern the shoulder because the limitation of shoulder movement influence the different positions that the hand can reach.
- The equipments and machineries do not respect the ergonomic requirements: a bad posture on workstations, especially during activities of repetitive handling and movements, is one of the main factors contributing to accelerate disorders at upper limbs' charge.

All these aspects bring staff management to face up a series of difficulties concerning manpower, for example employees' placement on some activities to carry out but also job rotation with the aim to preserve health's employees.

2 Some Problems Found and Their Solution: "The Right Person in the Right Place"

One of the missions of Prevention and Protection Service of the company is to help production area to put the right person in the right place.

A worker who doesn't feel good on its workplace is a person who is not able to give good performances.

To place the right person on the right place means giving the worker the possibility to stay well on the workplace guaranteeing him the less discomfort possible.

How is it possible to put an employee in the right place?

In order to answer to this question we start from restrictions already mentioned in the paragraph above and from ISO 11228_3 Norm concerning the position of the arm as against the shoulder.

2.1 The Project's Subject

This project is inspired from the following question: if doctor's judgement limits the height of the arm as against the shoulder, in this case where could the hand arrive?

The project "the right person in the right place" has the aim to allow to the department head to use a tool able to manage the placement of the staff on the different workstations.

The main purpose is interpolate the restrictions resulting from the workstation as against the sizes of each person and to check and verify if that position is included into the activity area of the worker's arm.

The idea is that if the surface reached by the hands, taken into consideration the following limits like for example the corner of the arm as against the body less than 80°, include the layout and geometry of the workplace, consequently it is reasonable to limit situations causing possible damages at upper limbs' and shoulder's charge, in particular if we refer to the restriction corner we have mentioned above (see Image 1).

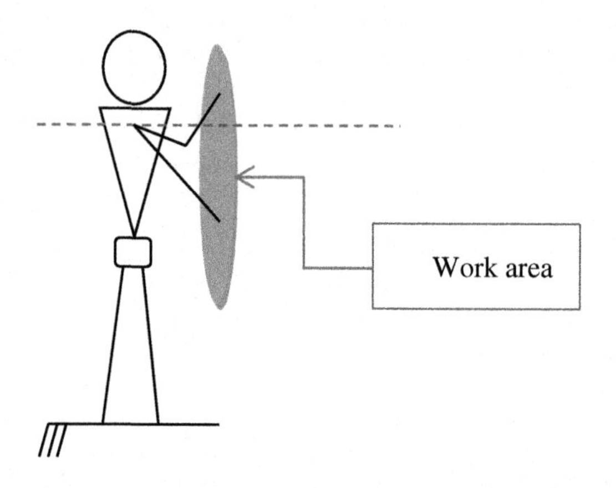

Image 1.

2.2 The Mathematical Model

In order to identify the mathematical model we start taking into consideration the system of the arm, on the basis of the shoulder, like a robotic system based on two restrictions.

With this model it appears that a point in the space P with respect to a reference system with centre area on the shoulder can be represented with a trigonometric system (Image 2):

$$X = L_G * \sin(\alpha) + L_P * (\sin(\beta) * \cos(\alpha) - \cos(\beta) * \sin(\alpha))$$

$$Y = H_S - L_G * \cos(\alpha) + L_P * (\sin(\beta) * \sin(\alpha) + \cos(\beta) * \cos(\alpha))$$

With:

- X = coordinate with regard to axis x;
- Y = coordinate with regard to axis y;
- H_S = shoulder's height from the floor;

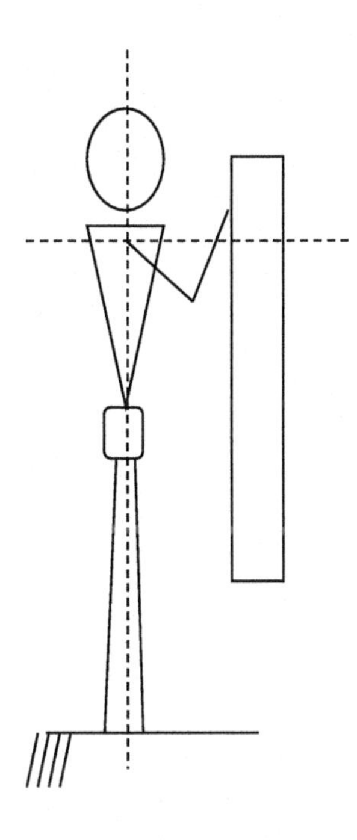

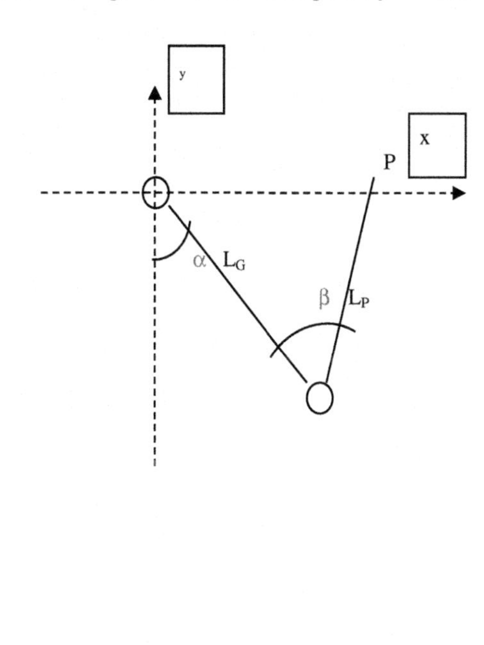

Image 2.

- L_G = Length of elbow in accordance with ISO 547_2 norm;
- L_P = Length of the arm from the elbow to the point of taking calculated like as the difference between the point of taking and the floor and the elbow distance from the floor in accordance with ISO 547_2 norm;
- α = corner between axis of the body and the arm; α is included $-15° \leq \alpha \leq 75°$ in order do not have aggravating with the score OCRA on the shoulder;
- β = corner between forearm and arm; the interval of β è: $20° \leq \beta \leq 180°$.

It is clear that the limit of the model results from the fact that the work area is described only in two dimensions.

2.3 Doctor's Contribution

The company doctor, Dott. Vincenzo Musumeci, during visits identifies anthropometric data of the staff employed in accordance with ISO 7250 norm.

The measures taken are: height of the person, height of elbow as against to the floor and height of the catch of an object having a diameter of 20 mm with regard to the floor. With these data it has been calculated the measure of the forearm. See information in the following Excel paper (Image 3):

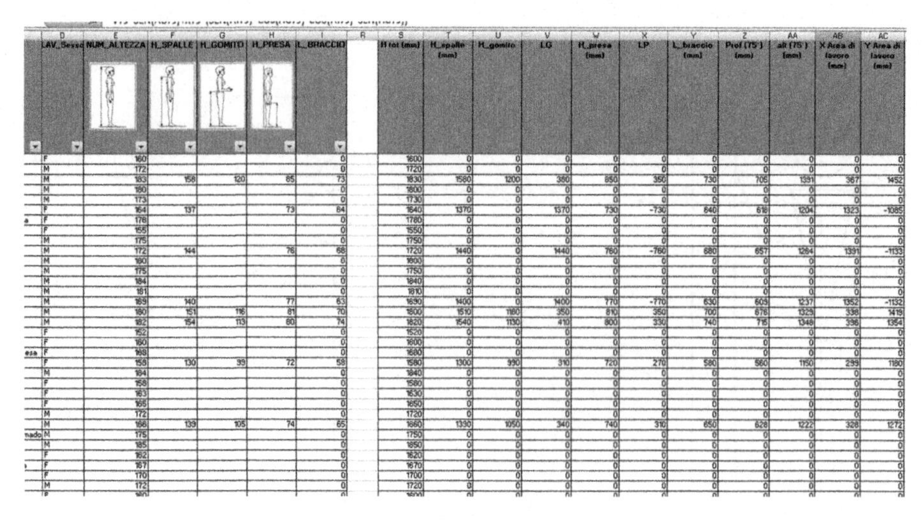

LAV_Sesso	NUM_ALTEZZA	H_SPALLE	H_GOMITO	H_PRESA	L_BRACCIO	H tot (mm)	H_spalle (mm)	H_gomito	LG	H_presa (mm)	LP	L_braccio (mm)	Prof (75°) (mm)	alt (75°) (mm)	X Area di lavoro (mm)	Y Area di lavoro (mm)
F	160				0	1600	0	0	0	0	0	0	0	0	0	0
M	172				0	1720	0	0	0	0	0	0	0	0	0	0
M	183	158	120	85	73	1830	1580	1200	380	850	350	730	705	1391	367	1452
M	180				0	1800	0	0	0	0	0	0	0	0	0	0
M	173				0	1730	0	0	0	0	0	0	0	0	0	0
F	164	137		73	84	1640	1370	0	1370	730	-730	840	618	1204	1323	-1085
F	178				0	1780	0	0	0	0	0	0	0	0	0	0
F	155				0	1550	0	0	0	0	0	0	0	0	0	0
M	175				0	1750	0	0	0	0	0	0	0	0	0	0
M	172	144		76	68	1720	1440	0	1440	760	-760	680	657	1264	1391	-1133
M	180				0	1800	0	0	0	0	0	0	0	0	0	0
M	175				0	1750	0	0	0	0	0	0	0	0	0	0
M	184				0	1840	0	0	0	0	0	0	0	0	0	0
M	181				0	1810	0	0	0	0	0	0	0	0	0	0
M	169	140		77	63	1690	1400	0	1400	770	-770	630	609	1237	1352	-1132
M	180	151	116	81	70	1800	1510	1180	350	810	350	700	676	1329	338	1419
M	182	154	113	80	74	1820	1540	1130	410	800	330	740	716	1348	398	1354
F	152				0	1520	0	0	0	0	0	0	0	0	0	0
F	160				0	1600	0	0	0	0	0	0	0	0	0	0
F	168				0	1680	0	0	0	0	0	0	0	0	0	0
F	158	130	99	72	58	1580	1300	990	310	720	270	580	560	1150	299	1180
M	184				0	1840	0	0	0	0	0	0	0	0	0	0
F	158				0	1580	0	0	0	0	0	0	0	0	0	0
F	163				0	1630	0	0	0	0	0	0	0	0	0	0
F	165				0	1650	0	0	0	0	0	0	0	0	0	0
M	172				0	1720	0	0	0	0	0	0	0	0	0	0
M	166	139	105	74	65	1660	1330	1050	340	740	310	650	628	1222	328	1272
M	175				0	1750	0	0	0	0	0	0	0	0	0	0
M	185				0	1850	0	0	0	0	0	0	0	0	0	0
F	162				0	1620	0	0	0	0	0	0	0	0	0	0
F	167				0	1670	0	0	0	0	0	0	0	0	0	0
F	170				0	1700	0	0	0	0	0	0	0	0	0	0
M	172				0	1720	0	0	0	0	0	0	0	0	0	0

Image 3.

2.4 The Company Evaluations

The Prevention end Protection Service has registered every workstation and has displayed the data on the electronic paper (until today two factories have been examinated for a total of 629 workstations). The data extracted have been: the height and the depth of catch and release points.

2.5 Data Interpolation

With the geometric limits which have been described in the trigonometric functions in the paragraphs above, leaving from the configuration of the body the most far possible, you can obtain a series of values in millimeters (it has been chosen a value from corner $\alpha = 70$ e $\beta = 75°$). If the value obtained with the geometric configuration is higher than the point to reach it means that the person has not a lot of difficulties and the risk caused from repetitive movements, at least from posture's point of view, is reduced (Image 4).

2.6 Limits and Experimental Results

The project is still object of study but the heads of departments have really appreciated the idea and they are beginning to use the tool. The doctor in charge, the Prevention and Protection Service, supported by Marco Cerbai and his collaborators of the company Safety Work S.r.l., are observing and monitoring the issues obtained in order to check if better results can be got on a medium and long term.

A	B	C	D	E	F	BO	BP
id	nome post	H_Presa	H_terra	L_spazio	D_Lato		
3_2_2_5	Scarico nastro	850	740	300		OK	OK
3_2_2_5	Scarico nastro	850	890	300		OK	OK
3_2_2_5	Scarico nastro	850	1040	300		OK	980
3_2_2_5	Scarico nastro	850	1190	300		1130	980
3_2_2_5	Scarico nastro	850	1340	300		1130	980
3_2_2_5	Scarico nastro	850	1490	300		1130	980
3_2_2_5	Scarico nastro	850	1640	300		1130	980
3_2_4_1	Carico nastro Gran sapore	140	900	300		OK	OK
3_2_4_1	Carico nastro Gran sapore	290	900	300		OK	OK
3_2_4_1	Carico nastro Gran sapore	440	900	300		OK	OK
3_2_4_1	Carico nastro Gran sapore	590	900	300		OK	OK
3_2_4_1	Carico nastro Gran sapore	740	900	300		OK	OK
3_2_4_1	Carico nastro Gran sapore	890	900	300		OK	OK
3_2_4_1	Carico nastro Gran sapore	1040	900	300		OK	980
3_2_4_1	Carico nastro Gran sapore	1190	900	300		1130	980
3_2_4_1	Carico nastro Gran sapore	1340	900	300		1130	980
3_2_4_1	Carico nastro Gran sapore	1490	900	300		1130	980
3_2_4_1	Carico nastro Gran sapore	1640	900	300		1130	980
3_2_4_2	Legatura Gran sapore	800	880	350		OK	OK
3_2_4_3	Formazione casse Gran sapore	1800		400		1130	980
3_2_4_4	Incassettamento Gran sapore	900	1020	400		OK	980
3_2_4_5	Bilancia e formazione pallet Gran sapore	900	140	450		OK	OK
3_2_4_5	Bilancia e formazione pallet Gran sapore	900	290	450		OK	OK
3_2_4_5	Bilancia e formazione pallet Gran sapore	900	440	450		OK	OK
3_2_4_5	Bilancia e formazione pallet Gran sapore	900	590	450		OK	OK
3_2_4_5	Bilancia e formazione pallet Gran sapore	900	740	450		OK	OK
3_2_4_5	Bilancia e formazione pallet Gran sapore	900	890	450		OK	OK
3_2_4_5	Bilancia e formazione pallet Gran sapore	900	1040	450		OK	980

Image 4.

Actually limits are in the fact that the description of the hand position with only two corners is not yet exhaustive but if the project will obtain good issues, the program will be extended to the third degree of freedom (the shoulder in its lateral raising).

References

1. ISO 7250: Basic human body measurements for technological design
2. Tesi di laurea di Massimo Grandi: Progetto di un interfaccia robotico per il rilevamento di superifici virtuali (anno accademico 2000/2001)

Passive Upper Limb Exoskeletons: An Experimental Campaign with Workers

Stefania Spada[1], Lidia Ghibaudo[1], Chiara Carnazzo[1], Laura Gastaldi[2], and Maria Pia Cavatorta[2](✉)

[1] Fiat Chrysler Automobiles - EMEA Region – Manufacturing Planning & Control – Direct Manpower Analysis & Ergonomics, 10135 Turin, Italy
{stefania.spada,lidia.ghibaudo,chiara.carnazzo}@fcagroup.com
[2] Politecnico di Torino, 10129 Turin, Italy
{laura.gastaldi,maria.cavatorta}@polito.it

Abstract. Wearable exoskeletons are currently evaluated as technological aids for workers on the factory floor, as suggested by the philosophy of Industry 4.0. The paper presents the results of experimental tests carried out on a first prototype of a passive upper limbs exoskeleton developed by IUVO. Eighteen FCA workers participated to the study. Experimental tests were designed to evaluate the influence of the exoskeleton while accomplishing different tasks, both in static and dynamic conditions.

Quantitative and qualitative parameters were analyzed to evaluate usability, potential benefits and acceptability of the device. Results show, on average, that wearing the exoskeleton has a positive effect in increasing: (i) endurance time while holding demanding postures with raised arms and/or having to lift and hold small work tools, (ii) endurance time and accuracy execution in precision tasks. The users also declared a lower perceived effort, while performing tasks with the exoskeleton.

Keywords: Upper limb exoskeleton · Human-robot cooperation
Usability

1 Introduction

An exoskeleton is a wearable robotic device, powered with passive or active systems, that allows limbs or trunk movement with increased strength and/or endurance [1, 2]. The type of actuation, the number of degrees of freedom and the body regions involved are extremely variable; hence, for the design of these devices it is necessary to target both the type of application and the users.

The robotic exoskeleton technology has acquired a rapid development starting from late 20th century with the advances in the technology in mechanical engineering, biomedical engineering, electronic engineering and artificial intelligence. The different existing exoskeletons, both commercial and laboratory prototypes can be classified considering the human body parts that are supported: lower body, upper body and full body exoskeletons. Furthermore, they can be divided into active and passive: the first

S. Bagnara et al. (Eds.): IEA 2018, AISC 825, pp. 230–239, 2019.
https://doi.org/10.1007/978-3-319-96068-5_26

ones have actuators to augment the user's power, while the latter use elements that store energy harvest by the wearer (e.g. mechanical springs, dampers, flexible materials) and return it during movements to assist in posture and to perform physical movements.

Quite recently the employment of exoskeletons has been extended from the rehabilitation [3, 4] and military field [5, 6] to the industrial setting [7–10]. This is in line with the philosophy of Industry 4.0, in which humans can be assisted by technological devices in difficult or unsafe tasks [11].

The research in industrial exoskeleton is still at an early stage; however, predictably exoskeletons can be useful when other preventive measures are not feasible or effective to lower workers' fatigue. Potential benefits of the introduction of exoskeleton as supporting devices in the industrial manufacturing environment are expected to increase worker's alertness, productivity and work quality; to support experienced personnel in the work force for longer; and to reduce work related musculoskeletal disorders.

Various passive upper limb exoskeletons have been developed in the last years for industrial applications. In particular, occupational tasks that require postures with elevated arms or overhead works, and hence represent a high risk factor for musculoskeletal disorders, are considered. Passive exoskeleton devices give a fixed contribution, independently from the external applied load. Usually they are designed to compensate, partially or totally, the gravity forces acting on the limb or on the trunk. However only few studies investigated effectiveness, usability, comfort, drawbacks and biomechanical strains associated to the use of upper limb exoskeleton in manufacturing tasks. Effectiveness, usability, comfort and drawbacks of the Levitate exoskeleton [7, 8] and of EksoVest [10, 12] are assessed in laboratory simulated operational tasks; an evaluation of the biomechanical strains, using electromyography, is presented in [10, 12, 13]. In [7, 8] the use of the exoskeleton shows a positive effect and on average there is an increase of the 30% both in duration and in quality of the performed tasks. However also some drawbacks associated to specific weight handling are disclosed. In [12] a drilling task was simulated: completion time decreased by nearly 20% with the exoskeleton, while in contrast precision decreased.

Both in [12] and [13] a significant muscle EMG activity reduction is recorded, when executing a task or handling a load. Also, no significant negative effects on the lower body are reported. However in [13] tasks were very short and the current prototype is not suited for industrial application.

Effectiveness, usability and comfort strictly depend on the design of the exoskeleton, but also on the tasks workers are required to perform. In this paper a first prototype of a new passive upper limb exoskeleton developed by IUVO [14], is tested. The new device was specifically designed for manufacturing uses, and it presents a reduced weight, a wearable, robust and compact design. Test campaign was intended to evaluate the effectiveness of the exoskeleton to enhance the duration of arm elevation in static posture and to assist the user in repeated arm elevation, with the same test designed proposed in [7, 8]. Tests were performed in the laboratory by automotive workers and they were designed to mimic manufacturing tasks.

2 Materials and Methods

2.1 Participants and Ethical Approval

Eighteen healthy male team leaders from Fiat Chrysler Automobiles Industries (FCA) volunteered to participate to the experimental campaign with the upper limb exoskeleton (Means ± SD: age 43.0 y ± 11.1 y; height 176.9 cm ± 5.5 cm; mass 77.3 kg ± 9.1 kg). All the workers had no limitation in strength or musculoskeletal disorders at the upper limbs.

Participants were completely informed about the nature of the study. All of them signed an informed consent and were free to interrupt the tests at any moment. The research methods and the protocols were standard and the measurements were performed in accordance with the declaration of Helsinki.

2.2 Exoskeleton Description

The exoskeleton, developed by IUVO [14], is passive and provides mechanical support to shoulders and arms, while forearms are not involved. The exoskeleton is worn similarly to a backpack (padded pelvis belt and two straps that go over the shoulders), as presented in Fig. 1. A metallic frame, paired with the trunk and the arms is present.

Fig. 1. Image of the exoskeleton worn by an operator during assembly tasks.

This first prototype of the exoskeleton, used in the test campaign, is consisting of:

- Human Interface: structure and materials that are in direct contact with the user's body. The main metallic structure comprises a T-shaped metallic frame for the core, with the transversal bar positioned on the back of the trunk in correspondence of the shoulders. On this bar are articulated two padded armbands that support the arms. In this way the metallic frame allows the transfer of the arm's weight to the trunk and

pelvis, partially relieving arm, shoulder, neck and upper trunk muscles. Soft materials that are in direct contact with the body are removable and washable;
- Core technology: a mechanism composed of elastic elements that harvests energy and provides a variable assistive torque for the flexo-extension of the arm. The assistive torque has a peak at a flexo-extension angle of about 90°. The return of the arm to a neutral position is assured by the weight of the arm itself.

The exoskeleton assistance given to the users is adjustable and it was intended to support up to 70% of the user's arm weight. Furthermore, the adjustability is useful to suit different working tasks. Depending on the frequency of the upper-limb flexo-extension movements during the working cycle, the user may prefer a higher or lower level of assistance.

In order to fit the entire population, the exoskeleton has two available sizes: S/M, L/XL and each size can be further adjusted. Perfect fitting is mandatory to assure comfort, control and force handling.

2.3 Test Description and Procedure

All tests were led in Ergolab, Ergonomics Laboratory of FCA Manufacturing Engineering and they were performed similarly to what is extensively described in [7]. Tests were video recorded using a frontal and a lateral camera.

The experimental activities consisted of three different types of test to evaluate the contribution of the exoskeleton to assist in: (a) holding a static posture with extended arms, (b) repeated manual handling task and (c) performing precision task.

Tests were first performed without the exoskeleton and then with the exoskeleton.

Type of task and without/with exoskeleton conditions followed systematically the order reported above.

Moreover, at the end of the trials, semi-structured interviews were administered to the workers and rating scales were used to assess usability and user's acceptance of the device. All the tests and interviews were conducted in the same day for each participant. A short description of the tests is here reported, more details are stated in [7]:

(a) holding a static posture with extended arms

The test was designed to evaluate the potential benefit introduced by the exoskeleton on the onset of muscular fatigue during a demanding prolonged static action.

The worker was required to maintain a static posture: standing upright with extended arms (90° with respect the trunk) while holding a load (Fig. 2a), having a mass of 3.5 kg. The weight was placed on the forearm, so that the wrist was not involved.

The worker was requested to stop when feeling fatigue or discomfort.

(b) repeated manual handling task

The test was designed to evaluate the potential benefit introduced by the exoskeleton during a manual material handling activity vs. possible restriction to movements (e.g. frequent muscle contraction, shoulder abduction-adduction) [15].

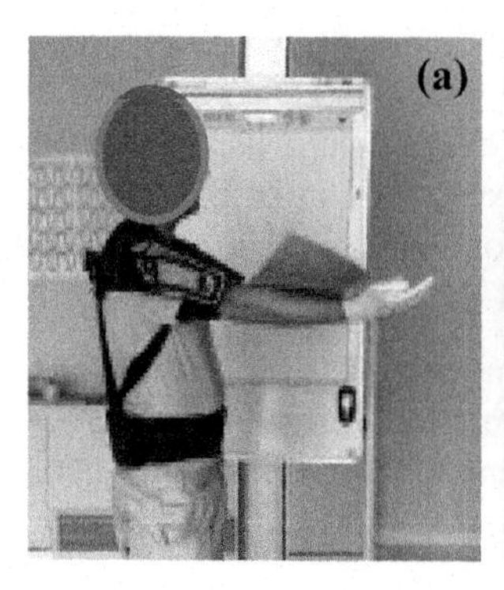

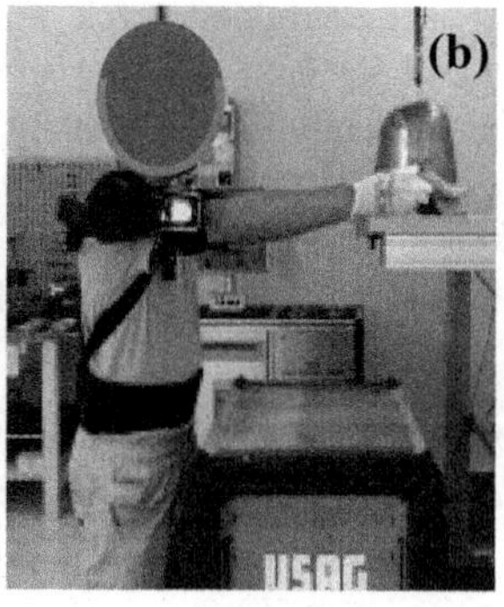

Fig. 2. Experimental activities wearing the upper-limb exoskeleton: (a) holding a static posture; (b) repeated manual material handling task; (c) precision task.

To simulate a real working task, the participant was requested to move an object with mass of 3.4 kg between two positions having different heights (Fig. 2b). Movements were paced at 30 action/min using a metronome.

(c) perform a precision task

The test was designed to evaluate the potential benefit introduced by the exoskeleton (lessen muscle strain, higher comfort rating and dexterity) during a precision task with significant static load on shoulder joint.

A sealing operation was simulated by asking to the participants to trace a continuous wavy line between two premarket traces, on a paper fixed on a billboard. The subject was standing, with his arm almost extended (Fig. 2c) and was not allowed to lower its arm, except at the end of the task. On the billboard, 5 wavy rows, having 27 arches each, were present and they were custom-placed at a different height with respect to the participant's shoulder. The end of the task was at subject's will or at the end of the premarket guides.

In Table 1 the main features of the tasks are summarized, in particular time duration and data collected are reported.

Table 1. Main features of the tasks

Task	Aim	Task duration	Data collected
(a) Holding a static posture	Evaluate the potential benefit on the muscular fatigue during prolonged static action	At subject's will	• Maintenance time of the static posture • Perceived exertion
(b) Manual material handling	Evaluate the potential benefit during a material handling activity vs. possible restriction to movements	600 s or at subject's will	• Number of lifting • Assessment of the pace • Perceived exertion
(c) Precision task	Evaluate the potential benefit during a precision task with a significant static load on shoulder joint	End of the pre-marked guides or at subject's will	• Number of arches traced • Precision score • Execution time • Perceived exertion

2.4 Procedure

At the worker's arrival into the lab, the functioning of the exoskeleton was explained. He was also informed about the test protocol and that he could stop whenever he wanted, without completing the entire task or the whole protocol. Personal data and anthropometric measurements were also collected.

The participant was then asked to perform the static, repeated manual material handling and precision tasks without the exoskeleton. Between the tests, adequate time was left to the worker to rest.

The exoskeleton was adjusted to fit the user, it was worn and regulated.

The participant was then allowed to familiarize with the exoskeleton, before asking him to repeat the tests with the exoskeleton (Fig. 2).

The Borg Rating of Perceived Exertion Scale [16] was administrated to the participants to assess usability and acceptability of the exoskeleton. The subject was requested to quantify the intensity level of the activity at the end of each task, both without and with the exoskeleton.

3 Results and Discussion

Due to organisational glitches two workers did not finished the entire test campaign and hence they were discarded.

According to the main aim of the test, quantitative and qualitative parameters were analysed.

(a) Holding a static posture

Posture maintenance was controlled by visually inspecting the recorded video images and possibly, if this changed during the trial, the corresponding time was assumed as the end of the trial itself. In addition, a comparison of the postures without and with the exoskeleton was made, with particular attention to arms and the spine postures. No substantial differences were detected.

In Table 2 mean values and SD of the task duration with and without exoskeleton were reported. Also the time variation interval ($\Delta T = T_{EXO} - T_{NO\_EXO}$) and the relative variation ($\Delta T\% = (T_{EXO} - T_{NO\_EXO})/T_{NO\_EXO}$) were evaluated.

Table 2. Results of the static task

	T_{NO_EXO} [s]	T_{EXO} [s]	ΔT [s]	$\Delta T\%$
Mean	108.6	157.8	49.2	56%
SD	±59.4	±76.1		

The operators maintained the static posture for a mean time of 108.6 s and 157.8 s without and with the exoskeleton respectively, with a 56% relative longer time length in the second case. All participants, except one, increased their endurance time.

The average score of the Borg scale was 3 and 1.6 without and with the exoskeleton respectively.

(b) Repeated manual material handling task

In this trial, no operator went out of pace. However, in both conditions, without and with the exoskeleton, none of the participants accomplished the entire test duration (600 s), they all stopped before.

12 operators increased the endurance time when wearing the exoskeleton, while 4 decreased the time. The average score of the Borg scale was 3 and 2 without and with the exoskeleton, respectively.

These results suggest that no or minimal restriction to movements was introduced by the exoskeleton.

(c) Precision task

Also in this case posture maintenance, without and with the exoskeleton, was controlled by visually inspecting the recorded videos and no substantial differences were detected.

To analyse and quantify the performance obtained during this task, the wavy line was segmented into arches. The maximum number of arches to be filled was 135, distributed on 5 lines. In Table 3 the mean number of arches traced by each operator is reported without and with the exoskeleton.

To overcome the ceiling effect on the number of arches that the current task may present, also the average time necessary to trace the single arch was calculated. A significant increase of the number of traced arches and a decrease of the time execution for each arch can be observed when the workers worn the device.

Table 3. Results of the precision task

	N° arches NO EXO	N° arches EXO	Time/arch NO EXO [s]	Time/arch EXO [s]	Δ N° arches	Δ% N° arches
Mean	100.6	127.3	2.2	2.0	15.6	26.5%
SD	±41.2	±18.1	±0.53	±0.51		±30%

Arches were then examined in order to assign a precision score. A full score of 10 was assigned to each arch, with an overall possible score of 1350. Four types of errors were identified based on how much the line traced by the worker was out with respect to the pre-marked traces. Weight factors were then assigned for type of error and row position. A Precision Index (PI) was calculated according to (1):

$$PI = \frac{1350 - \Sigma weight \cdot error}{1350} \% \tag{1}$$

The Precision Index increased when the task was executed with the exoskeleton with respect to when executed without it, being PI = 600 and PI = 475 respectively.

The average score of the Borg scale in this task was 2.7 and 1.6 without and with the exoskeleton respectively.

In general, we can observe that in this task the presence of the exoskeleton was beneficial for the perceived fatigue, for the time execution and for the precision with which the assignment was performed.

4 Conclusions and Future Perspectives

Passive exoskeletons are taken into account for possible introduction as supporting devices in the industrial manufacturing environment. FCA has planned a testing campaign to define applicability, usability and implementation of exoskeletons in working tasks.

In this paper, the experimental activity on the first prototype of the passive exoskeleton developed by IUVO is presented. Sixteen workers from a FCA automotive plant participated in the tests. Qualitative and quantitative results show a positive effect of the exoskeleton for those activities that involve a posture with raised arms. Workers increased their endurance time when wearing the exoskeleton and also declared a lower perceived fatigue. Moreover, also execution precision increased when using the device.

In general feedbacks from the use of exoskeletons are positive [7, 8, 10, 13], but before a systematic employment in manufacturing environment more aspects have to be enquired. Comparing the results obtained by experimental activities carried out by various authors is not straightforward, since the test campaign and the exoskeleton solutions are different. However, some general remarks can be pointed out.

In all the experimental tests [7, 8, 10, 12, 13], users were allowed to familiarize with the exoskeleton only for short periods of time. This involves a not optimal use of the device, but it may also hold back discomfort that only longer periods of use might reveal. With a longer training and daily usage the workers can develop specific strategies that might results in better performance, but also unexpected biomechanical load in anatomical regions other than the one directly interested by the exoskeleton (i.e. upper limbs or trunk).

It is important that the participants are experienced workers [7, 8]. Tests conducted with non-experienced workers could be misleading, since they might not be skilled at manual works and/or have not developed personal strategies to minimise the effort to accomplish the task. In case of experienced workers, a lower gap of the results obtained with and without the exoskeleton can be expected. Moreover, the sample participants seldom reflect the average range of the working population for age and body weight and build.

From a biomechanical point of view, posture and kinematics changes with and without exoskeleton have to be monitored as well and in this case long-term observation is very important. In the present study, only qualitative visual inspection was considered, comparing postures without and with exoskeleton, and no evidence of changes in posture was found. Considering the industrial environment, a mo-cap system based on inertial sensors can be effectively used to quantitatively assess postures and kinematics. Related to the biomechanical strains, electromyography can be helpful, but in general, only the activities of superficial muscles can be analysed. Considering that there can be dynamic changes in human body regions not directly interested by the exoskeleton (e.g. thighs, pelvis), electromyography evaluation needs to consider a large number of muscles. Biomechanical models can be useful to estimate the overall biomechanical changes [17] and to suggest which are the body regions that have to be further examined.

Another important aspect is users' acceptance, which usually can be considered only if the test campaign is conducted with experienced workers. In fact, not only comfort has to be assured, but also psychological aspects, compatibility with the work environment and potential benefits in every day routine activity have to be considered.

Finally, laboratory tests are a necessary step to have a first evaluation of the potential pros and cons of exoskeleton-assisted work, but the following necessary step is the evaluation of the devices directly into the manufacturing environment. For this reason, FCA is planning a test campaign directly in the plant. Volunteer workers will wear an improved prototype of the exoskeleton while executing their standard tasks. Daily time of usage will progressively increase up to the full shift duration in the 5th day. The campaign seeks for a holistic evaluation of the device considering long-term trials. It also investigate how the presence of other workers, equipment, limited spaces may interfere with the exoskeleton.

References

1. Gopura RARC, Kiguchi K, Bandara DSV (2011) A brief review on upper extremity robotic exoskeleton systems. In: 2011 6th international conference on industrial and information systems. IEEE, pp 346–351
2. Borzelli D, Pastorelli S, Gastaldi L (2017) Elbow musculoskeletal model for industrial exoskeleton with modulated impedance based on operator's arm stiffness. Int J Autom Technol 11:442–449
3. Yang C-J, Zhang J-F, Chen Y, Dong Y-M, Zhang Y (2008) A review of exoskeleton-type systems and their key technologies. Proc Inst Mech Eng Part C J Mech Eng Sci 222:1599–1612
4. Belforte G, Sorli M, Gastaldi L (1997) Active orthosis for rehabilitation and passive exercise. In: International conference on simulations in biomedicine, Proceedings, BIOMED. Computational Mechanics Publ, pp 199–208
5. Zoss AB, Kazerooni H, Chu A (2006) Biomechanical design of the Berkeley lower extremity exoskeleton (BLEEX). IEEE/ASME Trans Mechatron 11:128–138
6. Liang P, Yang C, Wang N, Li Z, Li R, Burdet E (2014) Implementation and test of human-operated and human-like adaptive impedance controls on Baxter robot. Presented at the conference towards autonomous robotic systems
7. Spada S, Ghibaudo L, Gilotta S, Gastaldi L, Cavatorta MP (2017) Investigation into the applicability of a passive upper-limb exoskeleton in automotive industry. Proc Manuf 11:1255–1262
8. Spada S, Ghibaudo L, Gilotta S, Gastaldi L, Cavatorta MP (2018) Analysis of exoskeleton introduction in industrial reality: main issues and EAWS risk assessment

9. de Looze MP, Bosch T, Krause F, Stadler KS, O'Sullivan LW (2016) Exoskeletons for industrial application and their potential effects on physical work load. Ergonomics 59:671–681
10. Kim S, Nussbaum MA, Mokhlespour Esfahani MI, Alemi MM, Alabdulkarim S, Rashedi E (2018) Assessing the influence of a passive, upper extremity exoskeletal vest for tasks requiring arm elevation: Part I – "Expected" effects on discomfort, shoulder muscle activity, and work task performance. Appl Ergon 70:315–322
11. Romero D, Stahre J, Wuest T, Noran O, Bernus P, Fast-Berglund Å, Gorecky D (2016) Towards an operator 4.0 typology: a human-centric perspective on the fourth industrial revolution technologies. In: CIE 2016: 46th international conferences on computers and industrial engineering
12. Kim S, Nussbaum MA, Mokhlespour Esfahani MI, Alemi MM, Jia B, Rashedi E (2018) Assessing the influence of a passive, upper extremity exoskeletal vest for tasks requiring arm elevation: Part II – "Unexpected" effects on shoulder motion, balance, and spine loading. Appl Ergon 70:323–330
13. Huysamen K, Bosch T, de Looze M, Stadler KS, Graf E, O'Sullivan LW (2018) Evaluation of a passive exoskeleton for static upper limb activities. Appl Ergon 70:148–155
14. IUVO – Wearable Technologies. https://www.iuvo.company/
15. Macdermid JC, Ghobrial M, Badra Quirion K, St-Amour M, Tsui T, Humphreys D, Mccluskie J, Shewayhat E, Galea V (2007) Validation of a new test that assesses functional performance of the upper extremity and neck (FIT-HaNSA) in patients with shoulder pathology. BMC Musculoskelet Disord 8:42
16. Borg G (1998) Borg's Perceived exertion and pain scales. Human Kinetics
17. Spada S, Ghibaudo L, Carnazzo C, Di Pardo M, Chander DS, Gastaldi L, Cavatorta MP (2018) Physical and virtual assessment of a passive exoskeleton. In: 20th congress international ergonomics association, Florence

Ergonomics Management Program: Model and Results

C. M. C. Varella(✉) and M. A. L. Trindade

Samsung Eletronics Ltda., Thomas Nielsen Street, 150, Pq. Imperator, Campinas, São Paulo, Brazil
carcoe@gmail.com, manoela_a@hotmail.com

Abstract. The objective of this work is to present the Ergonomics management program of a multinational company based in the state of São Paulo (Brazil), and its positive results, which has generated an improvement in the working conditions and the quality of life of the operators who work there. The program involves several actions (planned, controlled and documented) based on the Ergonomics of the Activity (whose main objective is to understand the work to transform it) and in the national legislation, covering the ergonomic workplace analyzes (which contemplate the three dimensions of ergonomics - physical, cognitive and organizational), execution and validation of projects of ergonomic improvements (conception and correction), investigation of work-related out-patient complaints, follow-up of return to work processes and inclusion of people with disabilities in workstations, trainings (both work-related and non-work related issues) and actions aimed at the well-being and quality of life of the employees, with relaxation, strengthening and postural alignment activities in a place inside the company equipped and with professionals specialized in Physical Education and Physiotherapy. Therefore, with the combination of the Ergonomics concepts of the activity and the management model of the PDCA cycle, it is possible to propose steps for an Ergonomics program that seeks the continuous improvement of the work processes and the constant validation of the actions performed by it.

Keywords: Ergonomics · Management · Program

1 Introduction

The industrial sector is an important segment for the world economy, and can be considered one of the pillars of economic development, manufacturing goods and generating jobs. The evolution of this sector has occurred since the beginning of industrial revolutions and the introduction of new systems models, resulting in the advancement of production and management systems within companies [1].

However, if on the one hand there is an increasingly competitive industry, generate production and profit, on the other, there is human labor, fundamental for any system productive. Regardless of the type: consumer goods industry, basic, intermediate, to the service sectors, all these organizations need man to ensure that their processes are fully implemented [1].

S. Bagnara et al. (Eds.): IEA 2018, AISC 825, pp. 240–246, 2019.
https://doi.org/10.1007/978-3-319-96068-5_27

In this scenario, we highlight the discipline Ergonomics, which has as main objective the understanding of the interactions between human beings and other elements of a system, being the profession that applies theory, principles, data and methods to design in order to optimize human well-being and overall system performance [2].

Currently, the effective application of Ergonomics in the worksystem, is seen as a facilitator within the reach of a balance between the characteristics of the workers and job requirements, improving worker productivity, providing better security (physical and mental), and job satisfaction. Second [3], several studies have demonstrated positive effects of application Ergonomics in the workplace, including machines, structural projects.

However, in a study on management guidelines in ergonomics, [4] states that it was possible to identify, through theoretical review and interviews with Ergonomics professionals, that there is no uniformity in the performance of Ergonomics programs in companies, each company does Ergonomics in a different way, and often, not following a management model, such as the PDCA cycle (Plan, Do, Check, Action). In addition, this same study carried out the investigation of norms related to ergonomics management, and identified that there are many international norms that mention Ergonomics, however, none of them addresses the issue of management.

1.1 Objective

The objective of this work is to present a management model in ergonomics that has been certified in Brazil by the representative body of ISO (International Organization for Standardization), ABNT (Brazilian Association of Technical Norms) and has been showing good results, through documented, registered and controlled actions, respecting the PDCA cycle management system.

2 Ergonomics Management Program

The ergonomics activities in this company have existed since 2009, however, from 2015 onwards that the management program was set up and the controls and records began to be made in a systematic way. With this, it is possible to monitor all program results, create indicators and check the PDCA (Plan, Do, Check, Act) management cycle, after all, management models need to facilitate the achievement of high levels of efficiency, effectiveness and effectiveness.

The ergonomics program in question contains several actions, in order to meet the requirements of the Ergonomics of the Activity and also to comply with local legislation. Therefore, the main actions will be described below.

- Registered Program Procedures: all the actions mentioned above are registered and documented according to the procedures of Ergonomics registered in company's internal system ("Ergonomics Management Procedure" and "Procedure Ergonomic Guidelines for Workplaces). Through this control, it is possible to identify the PDCA cycle of development of the management system, being possible to make comparisons with previous years and to identify the evolution of the program. These procedures contain all workflows that involve ergonomics and other areas of the company;

- Annual census of all the existing workstations in the company, for later elaboration of the annual planning for the accomplishment of the Ergonomic Work Analyzes;
- Elaboration and control of visits and technical opinions requested by the medical department or the company's areas: according to the need of the clinic or the area, ergonomics follows up complaints through employees, returns to work or professional rehabilitation, of pregnant and nursing mothers, medical restrictions of musculoskeletal origin and even the support of skills that may involve the ergonomic question;
- Request, follow-up and validation of ergonomic improvements: all the ergonomic improvements made are validated by the operators who are involved in this improvement;
- Participation in the projects of new models and new workstations of the company: this task was the biggest challenge for the area of Ergonomics, because in most cases, ergonomics was triggered for corrective situations. However, now the Ergonomics started to be included in the company's projects involving new products, new workstations or new production lines;
- Training: Ergonomics-related training is carried out from the first moment the employee enters the company, that is, in the Integration Lectures there is the "Ergonomics" module, up to the day-to-day work, through security dialogues, and specific training for some groups within the company, such as training for leaders, process engineering, maintenance and innovation teams (Fig. 1);

Fig. 1. Trainings carried out in the last 6 months for different audiences on various topics related to Ergonomics.

- Support to other programs: the company has a program called "Insert", which is responsible for all recruitment, selection, reception, insertion and monitoring of people with disabilities in the workplace. There is also the "Maternal Program", dedicated to the reception and follow-up of the pregnant and nursing mothers of the factory;
- Management of the Labor Gymnastics Program and the "Ergonomics Center": space for quality of life at work, where physical activities of different modalities (Pilates classes, functional training, postural orientations, relaxation, among others)

are carried out for all interested employees, who even receive gifts for their participation (Fig. 2).

Fig. 2. Photos of the activities of Labor Gymnastics and Center of Ergonomics of Samsung.

3 Results

All actions carried out by this company's ergonomics program seek to comply with Brazilian legislation and Ergonomics of Activity, which has as a characteristic to understand the work to transform it, through the Ergonomic Work Analysis, whose essential characteristic is to examine the complexity of the work activity, taking care that it agrees with the vast majority of those who will occupy it [5].

The fact that all the actions of the program are described in internal procedures, allows the program to be conducted in a standard way and is not compromised in cases of exchange of professionals, for example.

All actions are documented and recorded in control worksheets, which allows generating graphs and reports of results and comparisons of all actions of the program, besides helping in the annual planning of the actions of the management system. At the end of the year it is possible to demonstrate comparisons of the number of analyzes carried out, visits and opinions elaborated, improvements implemented and validated, actions requested by other areas monitored. It is also possible to draw comparisons of the actions of the management system between the years, as shown in the figure below (Fig. 3).

Figure 4 is an example of the results that can be generated by the management system, referring to the results of the ergonomic analyzes in terms of criticality, in the years 2015 to 2017, i.e. in 2015, for example, there were two ergonomic analyzes with high ergonomic risk. As early as 2017, there was no analysis with this risk.

The Fig. 5 shows an example of how the results are presented to the Ergonomics area management. Graphs with ergonomic risk indicators are presented in the areas analyzed, indicators of the number of ergonomic complaints by company area, number of training and ergonomics trained workers during the year, among others.

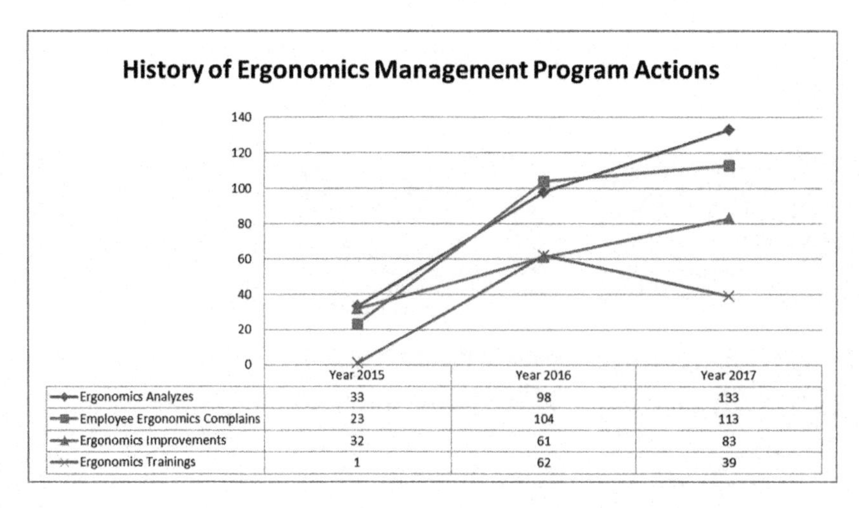

Fig. 3. Comparative history of the actions of the management program in ergonomics in the years 2015 to 2017.

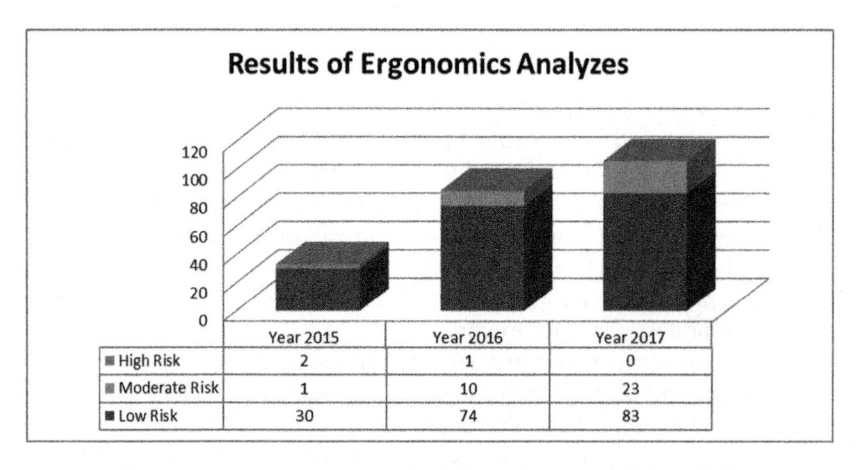

Fig. 4. Results of ergonomic analyzes in the years 2015 to 2017.

This presentation with the annual results is usually performed every beginning of the year, with the ergonomics team and its management and board present. In this way, it is possible to present to top management the effectiveness of the ergonomics program.

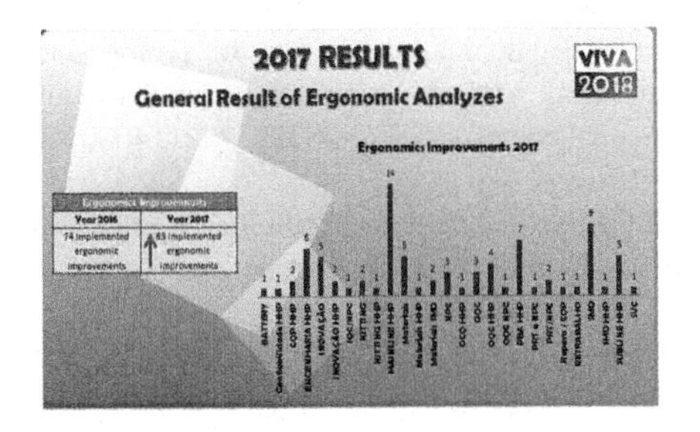

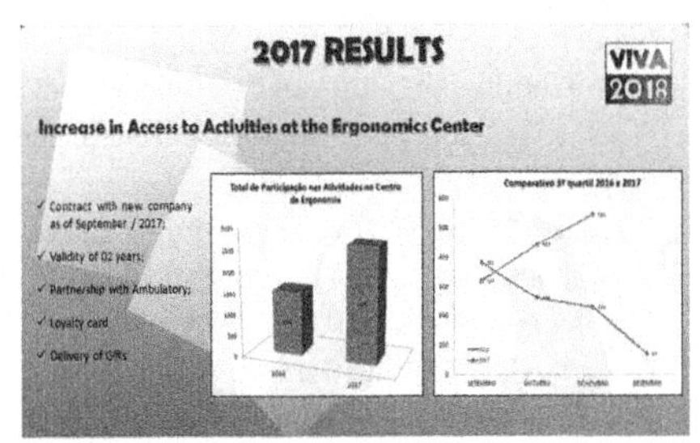

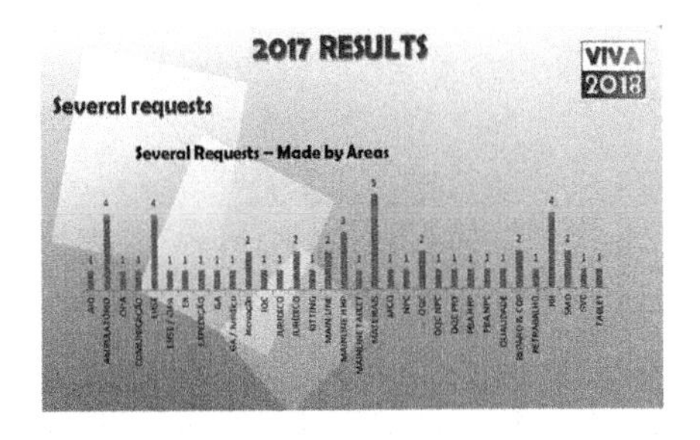

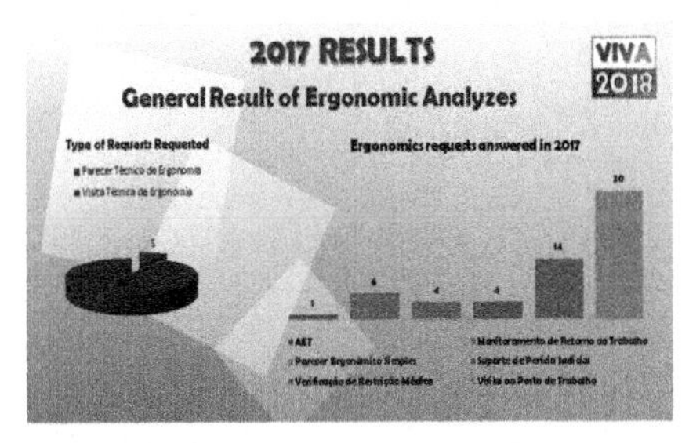

Fig. 5. Examples of graphs of the annual results of the Ergonomics Management System

4 Conclusion

Through this work, it was intended to present a management model in ergonomics that covers the concepts of Ergonomics of the Activity and meets the legal and normative requirements of Brazil, and can be presented as a suggestion of a program for other companies.

The fact that all the actions of the program were documented also helped to present the results of the program to the top management, showing the importance of the program for the entire organization and its search for better working conditions.

In this way, Ergonomics can be treated within companies as the other management systems (quality, environment, health and safety), given their importance and strategic role within the organization.

References

1. Mattos DL (2015) Evaluation of an ergonomics management model based on lean production practices: focus on the absenteeism index in a cardboard packaging company in Santa Catarina. 2015. 196 f. Dissertation (Masters in Production Engineering). Graduate Program in Production Engineering, Federal University of Santa Catarina, Florianópolis
2. International Ergonomics Association (IEA) (2016) Definition. http://www.iea.cc/whats/index.html. Accessed 10 May 2016

3. Azadeh A, Sheikhalishahi M (2014) An efficient Taguchi approach for the performance optimization of health, safety, environment and ergonomics in generation companies. Saf Health Work 6(2):77–84. http://www.sciencedirect.com/science/article/pii/S2093791114000948. Accessed May 2016
4. Trindade MAL (2017) Ergonomics management guidelines: standardization and practice in companies. 2017. 214 f. Thesis (PhD in Production Engineering) – Department of Production Engineering, Federal University of São Carlos, São Paulo
5. Wisner A (2004) Epistemological issues in ergonomics and labor analysis. In: Daniellou et al (ed) Ergonomics in search of its principles: epistemological debates. Edgard Blucher, São Paulo, pp 199–216

Physical and Virtual Assessment of a Passive Exoskeleton

Stefania Spada[1], Lidia Ghibaudo[1], Chiara Carnazzo[1], Massimo Di Pardo[2], Divyaksh Subhash Chander[3], Laura Gastaldi[3], and Maria Pia Cavatorta[3(✉)]

[1] Fiat Chrysler Automobiles - EMEA Region – Manufacturing Planning & Control – Direct Manpower Analysis & Ergonomics, 10135 Turin, Italy
{stefania.spada, lidia.ghibaudo, chiara.carnazzo}@fcagroup.com

[2] Centro Ricerche Fiat – WCM Research & Innovation, Orbassano, Italy
massimo.dipardo@crf.it

[3] Politecnico di Torino, 10129 Turin, Italy
{chander.divyaksh, laura.gastaldi, maria.cavatorta}@polito.it

Abstract. The paper describes the testing activity carried out on a commercial passive lower limb exoskeleton: the Chairless Chair, a wearable sitting support that allows workers to switch between a standing and a sitting posture. Tests were carried out with FCA workers who volunteered for the study. Laboratory trials served to familiarize the users and to obtain an initial feedback on the usability of the device in the assembly line. At a second step, virtual modelling of a few static postures was carried out, reproducing the anthropometry and the postural angles of the worker while using the exoskeleton. A main output of the model is the estimate of what forces are exchanged between the subject and the exoskeleton. In the case of the lower limb exoskeleton, an important parameter to consider is the percentage of the subject's weight that is sustained by the exoskeleton frame. The higher is this percentage, the lower will be the strain on the subject's lower limbs. First comparison between experimental and simulated results showed good agreement and auspicious advantages of exoskeletons in relieving the strain on workers.

Keywords: Exoskeleton · Simulation · Postural comfort

1 Introduction

Musculoskeletal discomfort and risk factors have been studied in detail in recent years [1–3] and continue to motivate innovative methods to identify [4–6] and solve the problem [7, 8]. In the current era of the fourth industrial revolution, thrust has been on assisting operators through collaborative robots or exoskeletons in carrying out tasks that are difficult to automate [9–11].

An exoskeleton can be defined as a wearable, external mechanical structure that enhances the strength and endurance of the user. Exoskeletons can be classified into

S. Bagnara et al. (Eds.): IEA 2018, AISC 825, pp. 247–257, 2019.
https://doi.org/10.1007/978-3-319-96068-5_28

two types: active and passive. Active exoskeletons consists of powered (electric, hydraulic, pneumatic or other) actuators to augment the human capabilities. Instead, passive exoskeletons consists in unpowered systems (using materials, springs or dampers) to store energy from human motion and release energy when required to support a posture or a motion [8].

Exoskeletons could also be classified based on the field of application. They could be used for military, rehabilitation or industrial purposes. Most commercially available exoskeletons have been developed for rehabilitation purposes. In terms of industrial exoskeletons, instead, active exoskeletons are still in a research and development stage, while passive exoskeletons are already available in the market [12]. Yet another method of classification of exoskeletons could be on the basis of the supported body area: upper limbs, lower limbs, spine, or a combination of body parts.

In this work, a commercially available passive industrial exoskeleton, Chairless Chair, has been analyzed. The Chairless Chair, as the name suggests, is a wearable sitting support for standing workplaces allowing the user to conveniently switch between a standing and a seated posture. As the user sits, the device locks in place at a user-selected height and provides sitting support.

Industrial exoskeletons have been generally analyzed by comparing the without-exoskeleton and the with-exoskeleton case. Biomechanical assessments are made using electromyography (EMG) of the primary muscle group involved in performing the task (or receiving assistance) [8]. Additionally, regions of the body that are in contact with the exoskeleton and exchange forces with the exoskeleton or are stressed to counteract the action of the exoskeleton might be investigated for potential adverse effects [8]. For most exoskeletons, it is a straightforward comparison as the assistance is provided maintaining the same working strategy. Thus, the active muscle groups remain the same in both the cases and improvements have been documented. However, the analysis of the Chairless Chair presents an interesting challenge in the way it changes the strategy of work. The Chairless Chair is intended to replace trunk bending with a seated posture. Therefore, the active muscle groups shift from the back in the without-exoskeleton case to the legs in the with-exoskeleton case. This change in working strategy could affect the comparison of without-exoskeleton and with-exoskeleton case in unknown ways. Therefore, a detailed analysis of the Chairless Chair is required.

In this work, the Chairless Chair is tested using industrial operators of Fiat Chrysler Automobiles (FCA), who volunteered for the testing campaign. The laboratory trials served to familiarize the users and to obtain an initial feedback on the usability of the Chairless Chair in the assembly line for different working tasks. Additional laboratory trials were intended to support biomechanical assessments through musculoskeletal modeling. It is believed that virtual assessment of biomechanical parameters could be especially useful to obtain an initial comparison of the without-exoskeleton and with-exoskeleton due to the resulting change of working strategy. These virtual assessments could be used to plan subsequent trials of the Chairless Chair. While biomechanical simulations have been used extensively for analyzing rehabilitative exoskeletons [13–15], their use in the analysis of industrial exoskeletons has been scarce. Thus, this work is intended to serve as a first step towards developing and validating the musculoskeletal model of the with-exoskeleton case.

2 Materials and Methods

2.1 Exoskeleton

Chairless Chair is a flexible wearable ergonomic sitting support, developed by the Swiss company Noonee, in partnership with automotive companies. Chairless Chair is indeed designed for manufacturing companies, particularly those where workers have to stand for long periods and traditional sitting supports are not suitable, as they would interfere with the work area [16]. While wearing the sitting support, workers can walk along the line for short distances, stop and sit at various heights when needed.

Designed to allow freedom of movement with a fast switch between sitting, standing and walking, the use of Chairless Chair should prevent strenuous postures such as squatting, bending or crouching. The seat is adjustable at different heights that can be locked in the desired position. The chair frame can be regulated to suit people of different stature and body build, and it is worn through braces and belt. The device weights 3.5 kg and can support a user's weight up to 115 kg.

2.2 Experimental Testing

The testing campaign was organized in different steps. An initial set of tests was run at FCA Ergolab and was carried out by a FCA team leader during a two-day campaign, with repetitions of tasks with and without the exoskeleton on both days. Instead of statistical significance, an initial survey of intended users was sought. As said, the change in posture may reflect also on the work strategy, so that the comparison of without-exoskeleton and with-exoskeleton tasks is not always straightforward for the subject. Nor it was always feasible to find quantitative parameters to assess the exoskeleton effectiveness and efficacy. Thus, laboratory trials primarily served to obtain an initial feedback on its usability in the assembly line as well as to guide the selection of suitable workstations in the plant.

Figure 1 contains a collection of snapshots taken during some of the tests that were carried out. In particular, in the top row, Fig. 1a depicts an assembly task that does not imply any reachability issue and a task that the operator can perform with the arms at elbow height, while Fig. 1b pictures a precision task, which simulates a sealing operation and that involves both lateral and forward reaching. In the bottom row, two different moments of a seal assembly operation on door frames are reproduced. The seal operation is performed with the Chairless Chair set at two different heights: in particular, the mounting of cables and seals in the door frame in the low seat configuration (c) and the mounting of clips and window seal in the high seat configuration (d). Both mounting operations on the door frame involve lateral reaching with both hands and, in the case of the window seal, working with raised arms.

Some tests were set up specifically to highlight potential issues such as limitations in reachability and force requirements, as well as the operator's perception on precarious stability or excessive steadiness during different operations. Feedbacks were indeed very important for the selection of suitable workstations on which to carry out the second part of the testing activity.

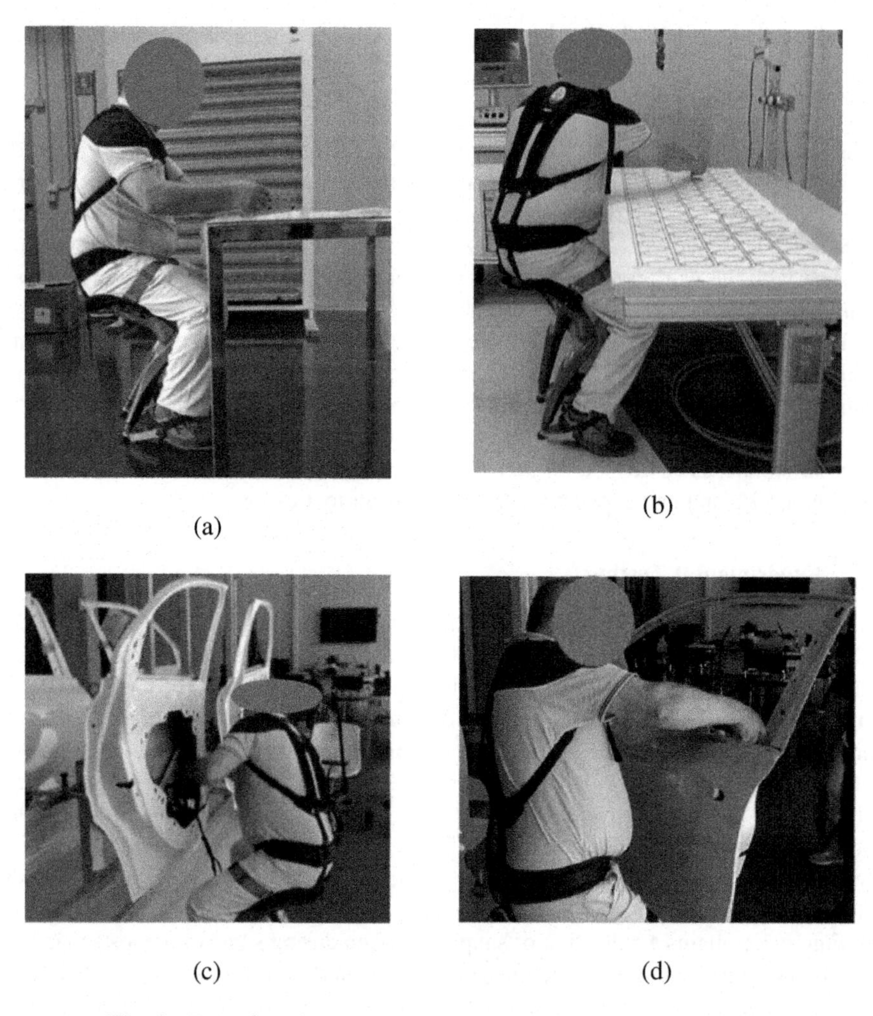

(a) (b) (c) (d)

Fig. 1. Examples of the testing activities carried out at FCA Ergolab.

2.3 Musculoskeletal Modeling

The musculoskeletal modeling was performed using AnyBody Modeling Systems (AMS) software [17]. AMS version 7.0.1 was used to develop the Human Standing template from the AnyBody Managed Model Repository (AMMR) version 1.6.6 to simulate the use of Chairless Chair. Static simulations were developed replicating the posture of the subject while using the exoskeleton. In line with previous work [18], the mannequin was scaled to the body height and weight of the subject using the length-mass-fat scaling law [19] and inverse dynamics analysis was performed using the min/max criterion [20] and the 3-element (3E) muscle model.

A 3D CAD model of Chairless Chair was developed in SolidWorks 2012. The model of Chairless Chair was subsequently translated into a script file that could be read by AMS using the AnyExp4SOLIDWORKS plugin (version 1.1.0) for SolidWorks (see Fig. 2a). This plugin conserves the geometry, mass, inertia, and joint properties as defined in the SolidWorks assembly file during the translation into the script file.

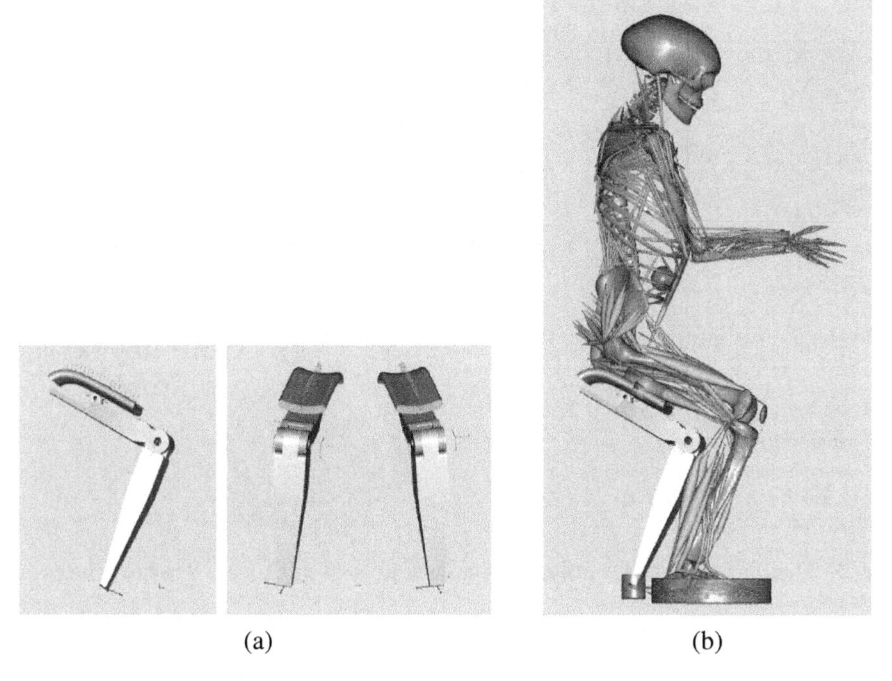

Fig. 2. AnyBody Modeling Systems (a) CAD model of Chairless Chair imported in AMS. (b) Musculoskeletal model of using the Chairless Chair. The red sphere represents the center of mass of the mannequin

The Human Standing template was modified by adding the model of Chairless Chair and establishing contact between the various interfaces (see Fig. 2b). Existing contact between the mannequin's feet and the ground was replaced by a modified version of the method reported in [21] to predict ground reaction forces and moments (GRF&Ms), as also adopted in other works [22–24]. 25 contact points were created at each foot. This method provided a better picture of the force distribution at the feet as compared to the existing contact between the feet and the ground that treated the foot as a single rigid body and calculated the overall reaction forces and moments.

Additionally, the method of Fluit et al. [21] was also adopted to simulate the contact between the exoskeleton and the mannequin's thigh and the exoskeleton and the ground. In the case of the contact between the exoskeleton and the thigh of the mannequin, the seat of the exoskeleton was modeled as the ground by creating a virtual force plate and 21 contact points were created on the thigh in the zone of contact. Instead, in the case of the contact between the exoskeleton and the ground, 11 contact points were created on each leg of the exoskeleton along with its corresponding virtual force plate on the ground. Thus, in total, six contact interfaces were modeled: two each (right and left side) to model contact between the mannequin and the ground, mannequin and the exoskeleton, and the exoskeleton and the ground (see Fig. 3). Each interface had its own virtual force plate to simulate contact with its corresponding element. A friction coefficient of 0.5 was used at each of the three interfaces.

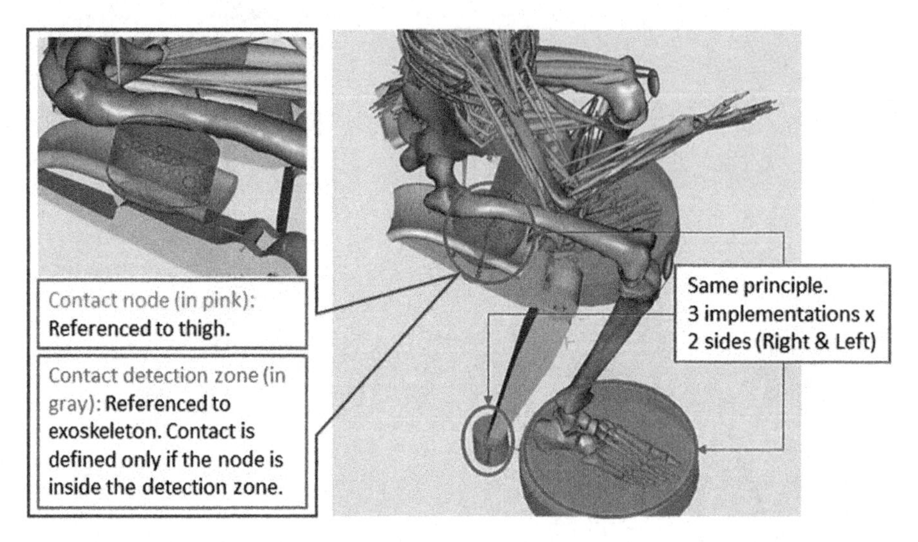

Fig. 3. Modeling of contacts using the method to predict Ground Reaction Forces and Moments.

The model was validated by comparing the predicted weight distribution between the exoskeleton and the mannequin's feet with experimental weight distribution. Experimental measurements were taken at the ergonomics lab in CRF (FCA Research Center) and involved 14 technical employees who volunteered for the testing. The experimental set-up was very simple and constituted of a weighing scale that was positioned below the subject's feet while the exoskeleton supports lay on a rigid structure at the same level of the scale. The main purpose of this set of tests was to verify if the distribution of the subject's weight between the exoskeleton and the subject's feet would remain within a given range, regardless of the subject's stature and body build and of the height selected for the chair frame. For this latter aspect, three configurations were considered: the highest seat selection, the lowest seat selection, and a seat height selected by the subject as the most comfortable. A minimum of three measurements were taken for each subject at any given seat height. For one subject, who had had significant training in the use of the Chairless Chair, measurements were also taken with different hand postures to simulate the reaching forward, the reaching with raised arms and the lateral reaching with both hands.

3 Results and Discussion

Results from the weighing tests pointed out that the body weight distribution during static sitting was neither significantly influenced by the subject's stature and body build, nor by the height chosen for the chair frame. On the contrary, training and user confidence appeared a requisite. After the subjects had acquired the necessary confidence in the use of the device, the weight distribution was not far from the 70 : 30 ratio declared by the producer, where 70 is the percentage of body weight supported by the exoskeleton and 30 is the percentage of body weight through the subject's feet (as measured by the scale). This optimal weight distribution was recorded during static sitting. As expected, the percentage of body weight through the subject's feet increased as the subject's arms stretched or moved laterally. Weight distribution with the same sitting postures was also tested using the AnyBody musculoskeletal model.

Table 1 shows the empirical and simulated results of the percentage of body weight supported by the legs of the subject while using Chairless Chair.

Table 1. Percentage of body weight supported by the legs of the subject.

	working at elbow height	working with raised arms	lateral reaching (to the right)
Percentage of body weight through subject feet.			
Empirical	33 %	53 %	66 %
Simulated	34 %	55 %	61 %

The results show a good match between the empirical conditions and their simulations using musculoskeletal modeling. They confirm the validity of the musculoskeletal model in various conditions of use of the Chairless Chair. Further, these results show that the percentage of body weight supported by the legs of the subject increases as the subject attempts to reach a distant point. Reaching for distant points results in bending forward, thereby, displacing the center of mass of the body towards the legs of the subject and away from the legs of the exoskeleton.

Table 2 shows the reaction forces exchanged between the exoskeleton and the thigh of the subject. The reaction forces are expressed in the local reference system XYZ (right hand) that is depicted in the table. The Y-axis is parallel to the length of the seat of the Chairless Chair, while the X-axis is perpendicular to the seat surface. The forces in the YZ plane represent the frictional forces at the interface between the subject and the exoskeleton. Their values depend on the value of the coefficient of friction at the interface. In the model, the coefficient of friction was 0.5 and, as can be seen from Table 2, the frictional forces in Y-axis are approximately half the normal forces (in X-axis). Forces in Z-axis are negligible.

Table 2. Predicted reaction forces (in percentage of body weight) at the subject-exoskeleton interface on the right (R) and left (L) side.

	working at elbow height		working with raised arms		lateral reaching (to the right)	
Predicted reaction force (in percentage of body weight). 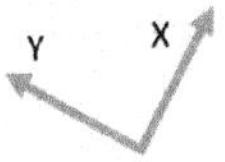 Y X Local Reference System	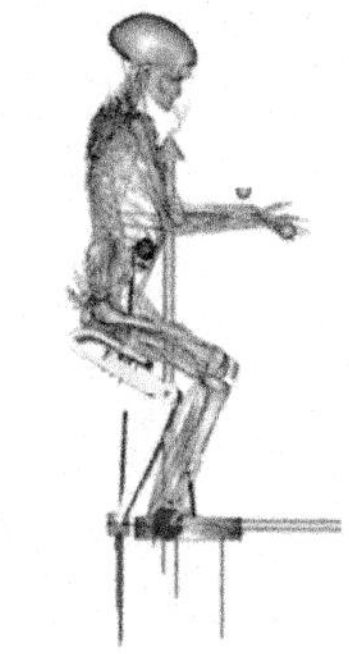		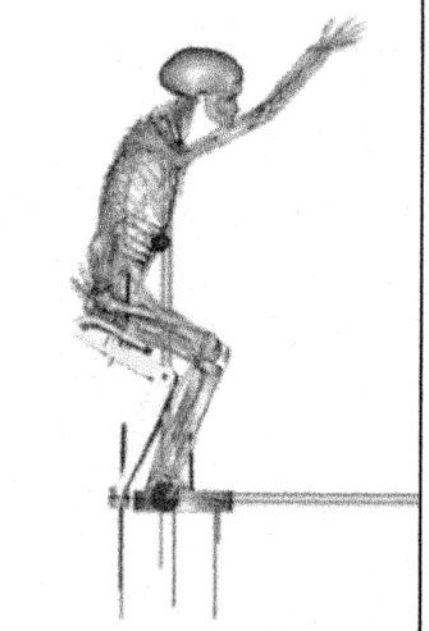			
X	R: 30 %	L: 30 %	R: 20 %	L: 20 %	R: 34 %	L: 2 %
Y	R: 14 %	L: 14 %	R: 10 %	L: 10 %	R: 17 %	L: 1 %

The sensitivity of the model to the coefficient of friction was studied separately in a parametric study by varying the coefficient of friction at each of the three interfaces (Subject-Ground, Subject-Exoskeleton, and Exoskeleton-Ground) from 0.2 to 0.8 in steps of 0.1. We found that the effectiveness of the exoskeleton, which was measured by the percentage of body weight supported by the exoskeleton, declined sharply with a coefficient of friction at the subject-exoskeleton interface lower than 0.4. The effectiveness of the exoskeleton was relatively independent of the coefficients of friction between mannequin and ground, and exoskeleton and ground.

Equilibrium stability was indeed a main issue also in the feedback given by workers. As the exoskeleton requires the users to balance their body weight between their legs and the legs of the exoskeleton, it is critical that the users are given adequate time to gain confidence with the exoskeleton. The confidence level of the user would directly affect the distribution of the body weight between the legs of the user and the legs of the exoskeleton. Thus, the body weight supported by the legs of the user serves as an important measure in these initial trials and it is directly affected by the location of the center of mass of the body with respect to the legs of the exoskeleton.

Proper workplace and work task selection is mandatory for testing exoskeletons. Replacing trunk bending with a seated posture reflects also on the work strategy and workplace layout, as it affects the reachability plane of the worker as well as the preferable direction for force application. Thus testing of the Chairless Chair in the assembly line was not always straightforward. Figure 4 depicts a worker at the fly-wheel assembly workstation, where the Chairless Chair was successfully tested. The compatibility of the device with the workstation and the work task ensured the right comfort for the worker as well as an improved visibility of the work area.

Fig. 4. Flywheel assembly workstation.

4 Conclusions

The field of exoskeleton research has rapidly enlarged in the last few years. There have been many studies carried out and several are still in progress. The majority of exoskeletons that target the industrial use case are under development at a prototype phase and yet not ready to go into production. Development requires active support from potential customers.

The paper focused on a commercial passive lower limb exoskeleton: the Chairless Chair, a wearable sitting support that allows workers to switch between a standing and a sitting posture. The seated posture may require a change in work strategy and workplace layout, as it affects the reachability plane of the worker and the preferable direction for force application. Thus testing of the Chairless Chair in the assembly line was not always straightforward. Nor was the comparison of without exoskeleton and with-exoskeleton case for the users.

For this reason, a musculoskeletal model was also developed to analyze the human-exoskeleton interface. Biomechanical simulations are a promising tool for better understanding the biomechanical load of the different body segments in exoskeleton-assisted work tasks. In particular, musculoskeletal modelling has the potential to provide an estimate of both the torque and reaction at the joints, and the muscle activity level. First results obtained with the Anybody Modeling Systems demonstrated a good agreement between simulated and empirical data for the distribution of the body weight between the legs of the user and the legs of the exoskeleton in a few static postures.

References

1. Aziz FA, Ghazalli Z, Mohamed NMZ, Isfar A (2017) Investigation on musculoskeletal discomfort and ergonomics risk factors among production team members at an automotive component assembly plant. In: IOP conference series: materials science and engineering, vol 257
2. Stock SR, Nicolakakis N, Vézina N, Vézina M, Gilbert L, Turcot A, Sultan-Taieb H, Sinden K, Denis M-A, Delga C, Beaucage C (2018) Are work organization interventions effective in preventing or reducing work-related musculoskeletal disorders? A systematic review of the literature. Scand J Work Environ Health 44:113–133
3. dos Reis DC, Tirloni AS, Ramos E, Moro ARP (2018) Risk of developing musculoskeletal disorders in a meat processing plant. In: Goonetilleke RS, Karwowski W (eds) Advances in physical ergonomics and human factors. Springer, Cham, pp 271–278
4. Chander DS, Cavatorta MP (2017) An observational method for Postural Ergonomic Risk Assessment (PERA). Int J Ind Ergon 57:32–41
5. Pavlovic-Veselinovic S, Hedge A, Veselinovic M (2016) An ergonomic expert system for risk assessment of work-related musculo-skeletal disorders. Int J Ind Ergon 53:130–139
6. Savino M, Mazza A, Battini D (2016) New easy to use postural assessment method through visual management. Int J Ind. Ergon 53:48–58
7. Maurice P, Padois V, Measson Y, Bidaud P (2017) Human-oriented design of collaborative robots. Int J Ind Ergon 57:88–102
8. de Looze MP, Bosch T, Krause F, Stadler KS, O'Sullivan LW (2016) Exoskeletons for industrial application and their potential effects on physical work load. Ergonomics 59:671–681
9. Huysamen K, Bosch T, de Looze M, Stadler KS, Graf E, O'Sullivan LW (2018) Evaluation of a passive exoskeleton for static upper limb activities. Appl Ergon 70:148–155
10. Spada S, Ghibaudo L, Gilotta S, Gastaldi L, Cavatorta MP (2017) Investigation into the applicability of a passive upper-limb exoskeleton in automotive industry. Proc Manuf 11:1255–1262
11. Spada S, Ghibaudo L, Gilotta S, Gastaldi L, Cavatorta MP (2018) Analysis of exoskeleton introduction in industrial reality: main issues and EAWS risk assessment. In: Goonetilleke RS, Karwowski W (eds) Advances in physical ergonomics and human factors. Springer, Cham, pp 236–244
12. Huysamen K, de Looze M, Bosch T, Ortiz J, Toxiri S, O'Sullivan LW (2018) Assessment of an active industrial exoskeleton to aid dynamic lifting and lowering manual handling tasks. Appl Ergon 68:125–131
13. Guan X, Ji L, Wang R, Huang, W (2016) Optimization of an unpowered energy-stored exoskeleton for patients with spinal cord injury, pp 5030–5033

14. Zhou L, Li Y, Bai S (2017) A human-centered design optimization approach for robotic exoskeletons through biomechanical simulation. Rob Auton Syst 91:337–347
15. Jensen EF, Raunsbæk J, Lund JN, Rahman T, Rasmussen J, Castro MN (2018) Development and simulation of a passive upper extremity orthosis for amyoplasia. J Rehabil Assist Technol Eng 5:205566831876152
16. NOONEE. http://www.noonee.com. Accessed 17 Apr 2018
17. Damsgaard M, Rasmussen J, Christensen ST, Surma E, de Zee M (2006) Analysis of musculoskeletal systems in the AnyBody Modeling System. Simul Model Pract Theory 14:1100–1111
18. Chander DS, Cavatorta MP (2018) Multi-directional one-handed strength assessments using AnyBody Modeling Systems. Appl. Ergon 67:225–236
19. Rasmussen J, De Zee M, Damsgaard M, Christensen ST, Marek C, Siebertz K (2005) A general method for scaling musculo-skeletal models. In: 2005 international symposium on computer simulation in biomechanics, vol 3
20. Rasmussen J, Damsgaard M, Voigt M (2001) Muscle recruitment by the min/max criterion - a comparative numerical study. J Biomech 34:409–415
21. Fluit R, Andersen MS, Kolk S, Verdonschot N, Koopman HFJM (2014) Prediction of ground reaction forces and moments during various activities of daily living. J Biomech 47:2321–2329
22. Skals S, Jung MK, Damsgaard M, Andersen MS (2017) Prediction of ground reaction forces and moments during sports-related movements. Multibody Syst Dyn 39:175–195
23. Eltoukhy M, Kuenze C, Andersen MS, Oh J, Signorile J (2017) Prediction of ground reaction forces for Parkinson's disease patients using a kinect-driven musculoskeletal gait analysis model. Med Eng Phys 50:75–82
24. Peng Y, Zhang Z, Gao Y, Chen Z, Xin H, Zhang Q, Fan X, Jin Z (2018) Concurrent prediction of ground reaction forces and moments and tibiofemoral contact forces during walking using musculoskeletal modelling. Med Eng Phys 52:31–40

Holistic Planning of Material Provision for Assembly

Leif Goldhahn and Katharina Müller-Eppendorfer(✉)

InnArbeit – Zentrum für innovative Arbeitsplanung und Arbeitswissenschaft, Hochschule Mittweida, Technikumplatz 17, 09648 Mittweida, Germany
{goldhahn,mueller7}@hs-mittweida.de

Abstract. Due to the rising variety and increasing flexibility of products and processes for assembly, the planning processes and the work systems need to be designed in a goal oriented and adaptive way. Especially the manual assembly and the provision of materials associated are taken into consideration. Scientists of the Institute "InnArbeit" prepared the "concept for the systematic planning of the material provision". This approach guarantees the organized and structured planning for the material provision. For holism the approach considers up- and downstream processes.

Potentials of the Virtual Reality (VR) – Technology are utilized for defining appropriate steps of the approach and include ergonomic evaluations. Tools for the continuous planning (e.g. "Stratergie-Matrix", "Förderkette" and "Digitaler Erweiterbarer Katalog für Bereitstellequipment"), which are included in the approach to assist the process planners within the material provision were also created by the institute.

Keywords: Material provision · Planning · Ergonomics · Manual assembly Virtual Reality

1 Introduction

To meet the challenges of high variance, rapid pace of change and variation of products and main processes (division: assembly, Goldhahn and Müller-Eppendorfer 2012), related processes like logistic processes also have to be focused on. With an improvement of logistic processes, the assembly could be enhanced and the costs reduced (see Goldhahn et al. 2014; Adolph and Metternich 2016).

However, herein lies the difficulty for the process planner to deal with the increasing requirements of the planning process: there are no suitable methods and procedures for material provision planning (mpp) (Blöchl et al. 2017). Additionally, further publications like Pause et al. (2016), which addresses the "Wahl der richtigen Beschaffungsstrategie" (choosing the right procurement strategy) or Dudczig et al. (2016) which raises the question of the intra logistics of the future "Wie sieht die Intralogistik der Zukunft aus?", focus on the logistic processes and approaches to improve these essential secondary processes.

The first concepts for material provision have been established by Bullinger and Lung (1994, p. 225 f.) and were expanded and improved by Lotter and Wiendahl

S. Bagnara et al. (Eds.): IEA 2018, AISC 825, pp. 258–266, 2019.
https://doi.org/10.1007/978-3-319-96068-5_29

(2012, p. 289 f.). The current techniques of the processing of information and visualisation are not considered yet. That is the reason why the VR-technique is integrated in the established holistic "concept for the systematic planning of the material provision" to use the potentials of processing of information and visualisation (Goldhahn and Thümer 2014). In this context Wirth et al. (2014) describe VR as an interactive environment generated by computers, which depicts a representation of physical reality in real time. Technical aids help to activate the natural sense of the observer (immersion). Imagination, the third attribute of VR, will be achieved if the virtually designed work system is assessed as an exact and perceptible system (Burdea and Coiffet 2003).

2 Approach for Material Provision Planning

The relevance and the crucial factor for the systematic procedure within the planning of material provision have already been described. The establishment of the holistic concept for the systematic planning of the material provision is described below Fig. 1.

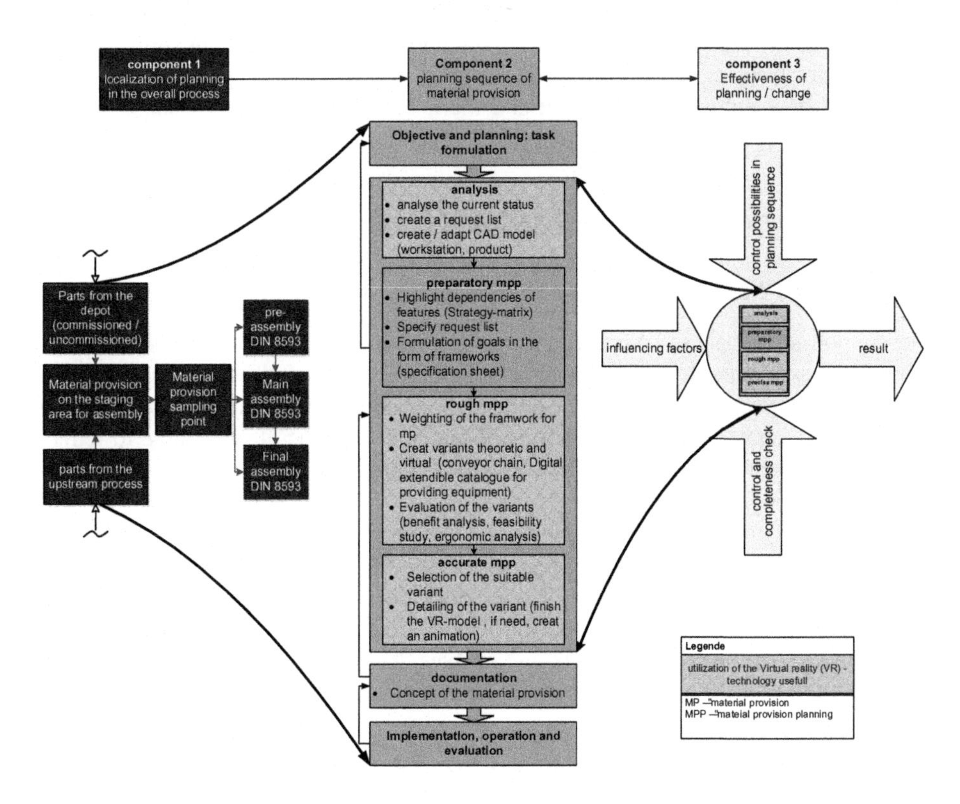

Fig. 1. Concept for the systematic planning of the material provision (copyright HS Mittweida)

The concept consists of three components and meets the fields of holistic nature (component 1), consistency and integration (component 2) as well as effectiveness and completeness (component 3). Within component 1, the localization of i.e. the divisions where the material provision is carried out as well as the identification of related processes which might influence the material provision (e.g. container systems, work piece carrier, shelf systems, transportation). This leads to the consideration of processes, equipment and staff necessary and can therefore help to improve the planning process in component 2.

The essential planning process for material provision is assigned to component 2 and split into the following seven parts: problem definition, analysis, preparatory material provision, rough material provision, accurate material provision, documentation and implementation/operation. This procedure occurs iteratively for the approximation to a optimal planning state as far as possible. The separate parts are supported by proven as well as newly established approaches (vgl. Goldhahn et al. 2016). Special attention must be set on to the VR-technique which is applied here and integrated within this component (Fig. 1 – light grey).

Component 3 is responsible for controlling the effectiveness and completeness if the planning changes respectively. Within that, the extent of acceptance for the result of planning can be verified or the necessity for implementing more improvements with other control features can be identified. Thereby, the general acceptance of the alternative and new outcomes is avoided and the need of consequential changes can be considered.

3 Tools for the Continuous Planning

3.1 Strategy-Matrix

For supporting the phase "preparatory mpp" of component 2, the "strategy matrix" is considered as a suitable tool for planning. In this phase numerous features are structured with their characteristic values and their dependence is shown. Based on the requirements created within the analysis of the current status, it is possible to retrace these requirements respectively to their characteristic values (dependence to other features), see Fig. 2.

The strategy matrix textually builds on various relevant sources as well as own considerations (Nyhuis 2009; Wirth et al. 2014; Pfohl 2010; Lotter and Wiendahl 2006). To apply the strategy matrix, it is necessary to choose an initial feature, which is determined based on the phase "analysis". This initial feature (e.g. demand-oriented) can be chosen in the strategy matrix and enables deriving all dependencies. Based on those features and the characteristics, the further planning can be designed.

3.2 Digital Extendible Catalogue for Providing Equipment

The digital extendible catalogue for providing equipment (DEKaB) represents a tool of the "rough mpp" phase of component 2, which is subordinated in the concept for systematic planning of the material provision. The catalogue functions as interface

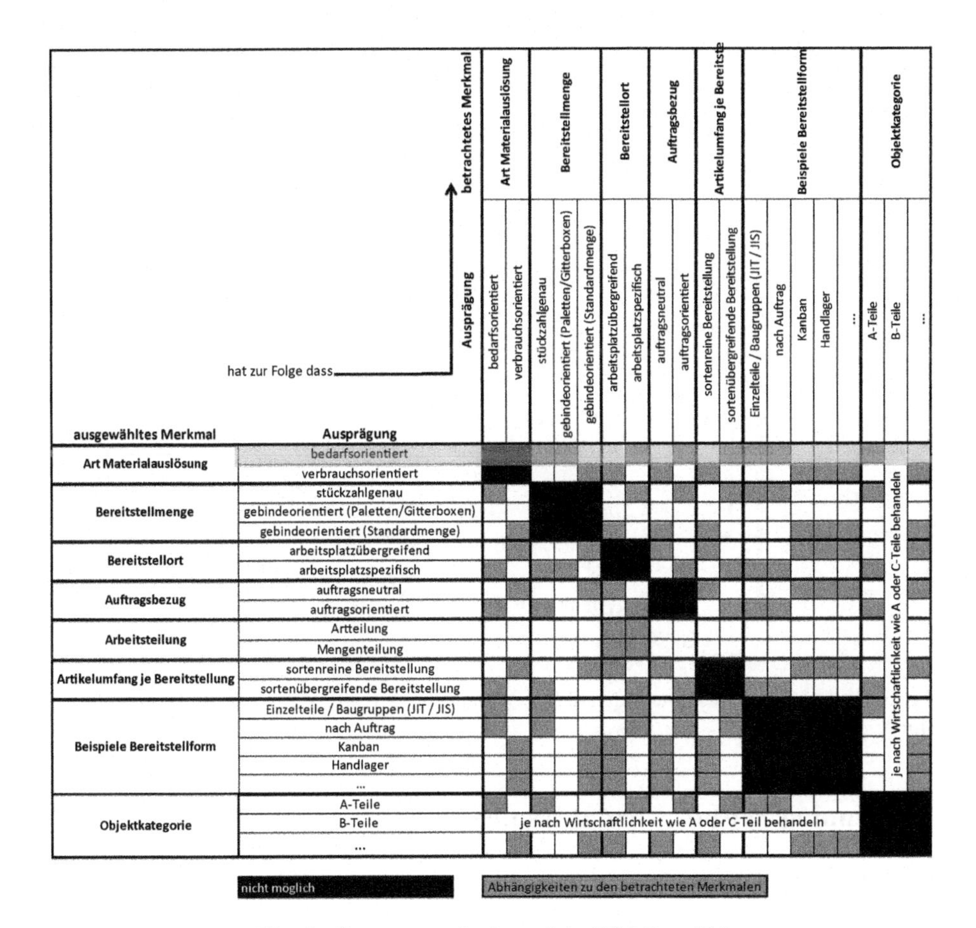

Fig. 2. Strategy matrix (copyright HS Mittweida)

between the theoretical planning and the visualisation through VR and especially through the software IC.IDO (ESI GmbH). Within this software, there is no actual library for the provision of basic elements. Therefore, the DEKaB serves as a useful tool to design different variants and as a means for bundled transmission to the existing VR-software.

Preparing the catalogue, "PHP5" was defined as programming language and an Access-Database was set as data base. This database is extendible individually. The library-like design (structured and sorted) and the fact that it is web- and data-based are further characteristics of the digital extendible catalogue for providing equipment. The DEKaB is structured in the following five main groups: handling aids (Transporthilfsmittel), provision mediums (Bereitstellmittel), provision facility (Bereitstelleinrichtung), conveyors (Fördermittel) and handling means (Handhabungsmittel), Fig. 3. Subgroups and characteristics are assigned to these main groups.

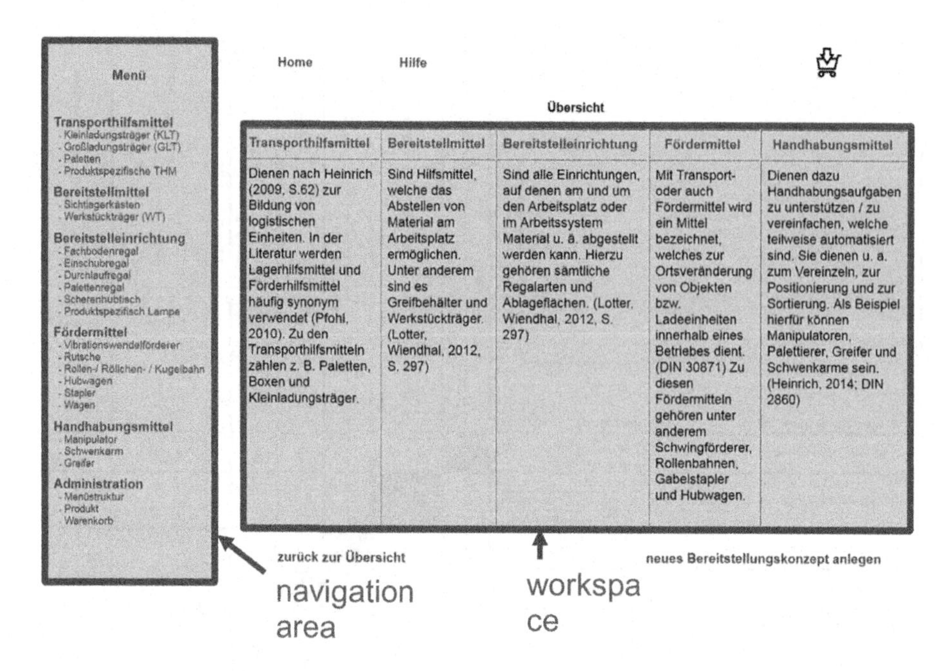

Fig. 3. User interface "digital extendible catalogue for providing equipment" (copyright HS Mittweida)

4 Planning Within Virtual Reality (VR)

4.1 Planning Support

Virtual design and the presentation of work systems and objects are the potentials of VR-technique. The advantages are the large-scale presentation, the interaction with the models and the participation of several project team members during the planning sessions.

Especially in the planning process of the material provision, the high quality, accurate visualization and design are special features of this technique. This results in an improvement of quality as well as quantity regarding the primary activities within the division assembly. Furthermore, the virtual comparison of the variants with the different options of material provision can be realized (Fig. 4).

Based on this procedure, two variants can be combined into one preferred option within the virtually designed work system. This option can then be discussed with the project team members within the planning sessions. In addition, the human-oriented planning will be supported by the VR-manikin. This will be explained in the next chapter.

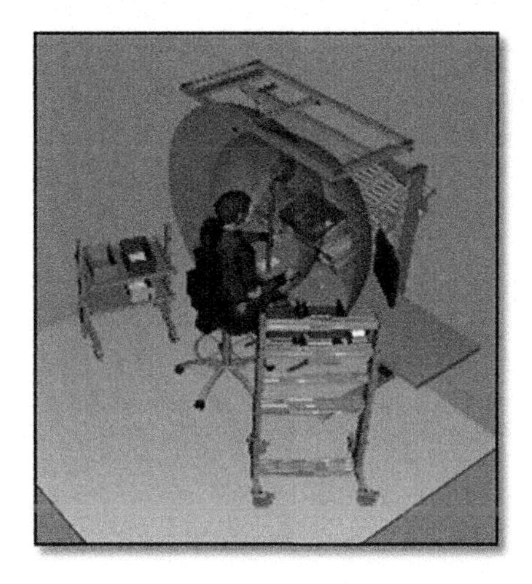

Fig. 4. Variants to display the material provision of a single workstation with VR-technique (copyright HS Mittweida)

4.2 Ergonomic Evaluation

An evaluation of the work system regarding the arrangement of provision mediums like pallets, containers or racks, can be conducted using the VR-technique and the IC. IDO software, Fig. 5. Manikins are displayed for the evaluation of ergonomic aspects such as the display of grabbing or vision ranges in the IC.IDO module "Ergonomics" (Figs. 6 and 7).

Fig. 5. Arrangement of containers and swivel arm (copyright HS Mittweida)

In addition to that, the manikins can be adapted with regard to percentile and types of physiques. This means that the suitability of the work system can be verified depending on different groups of people. Different variants of planning are assessed and evaluated with these functions of the software.

The DEKaB can be used for rapid changes of the variants of material provision as well as for new variants which can be designed.

Fig. 6. Display of reach on the single assembly workstation (copyright HS Mittweida)

Fig. 7. Display of vision ranges on the single assembly workstation (copyright HS Mittweida)

5 Example Applications

The concept for the systematic planning of the material provision was applied within pilot projects for two different workstations (single assembly workstation and assembly system with belt conveyor and workpiece carrier (Fig. 8)) and four different products (3-fold socket, round light, vice and engine component group (Fig. 9)) and it has been verified positively.

The development of further case studies on these workstations and products was carried out and demonstrates the practicability and goal orientation of the planning concept and the application of VR-technique for assembly processes.

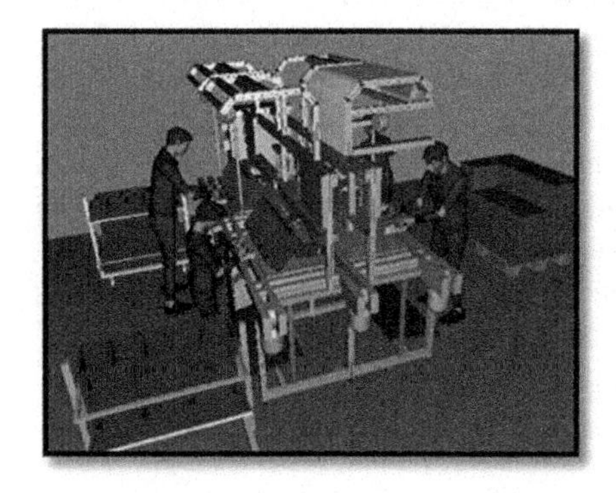

Fig. 8. Representation of the assembly system (with belt conveyor and workpiece carrier) in the IC.IDO software (copyright HS Mittweida)

Fig. 9. Representation of the material provision for the example "engine component group" (copyright HS Mittweida)

6 Conclusion

Solution-oriented strategies and tools for material provision (Strategy Matrix and Digital extendible catalogue for providing equipment) have been demonstrated. Furthermore, the VR-technique has been integrated into the planning procedure in a more goal-oriented way. It has also been used more particularly for the ergonomic evaluation through the module "Ergonomics".

A consistent procedure for the determination of material provision, especially for the division of manual assembly with its high variance and large variety of parts, is enabled by the structured and precise planning. Based on the application of the procedure and the established tools within pilot projects, the practicability and goal-orientation for assembly processes have been verified and the improvements of effectiveness and quality of the assembly processes have been achieved.

The goal-oriented use of the VR-technique is a considerable factor that has contributed to this result.

References

Adolph S, Metternich J (2016) Materialbereitstellung in der Montage – Eine empirische Analyse zur Identifikation der Anforderungen an künftige Planungsvorgehen. Zeitschrift für wirtschaftlichen Fabrikbetrieb 111(1–2):15–18

Blöchl St-J, Schneider M, Stolz K (2017) Industrie 4.0 in der Materialbereitstellung – Eine empirische Analyse der Herausforderungen von Prozessplanern. Zeitschrift für wirtschaftlichen Fabrikbetrieb 112(1–2):91–93

Bullinger H-J, Lung M (1994) Planung der Materialbereitstellung in der Montage. Teubner, Stuttgart

Burdea G, Coiffet P (2003) Virtual reality - technology, 2nd edn. Wiley, New Jersey

Dudczig M, Schumann M, Klimant P, Lornez M (2016) Wie sieht die Intralogistik der Zukunft aus? – Kommissionierung unterstützt durch sensorbasierte Fahrerlose Transportfahrzeuge. Zeitschrift für wirtschaftlichen Fabrikbetrieb 111(7–8):449–452

Goldhahn L, Müller E, Müller-Eppendorfer K (2016) Strategie und Werkzeuge zur Planung der Materialbereitstellung für manuelle Montageprozesse. In: Müller E (Hrsg) Smarte Fabrik & Smarte Arbeit - Industrie 4.0 gewinnt Kontur. Tagungsband VPP 2016 TU Chemnitz: Wissenschaftliche Schriftenreihe des IBF Sonderheft 22, pp 239–248

Goldhahn L, Müller E, Müller-Eppendorfer K (2014) Planung der Materialbereitstellung in der Montage. In: Müller E (Hrsg) Produktion und Arbeitswelt 4.0 - Aktuelle Konzepte für die Praxis. Tagungsband TBI 2014 TU Chemnitz: Wissenschaftliche Schriftenreihe des IBF Sonderheft 20, pp 205–214

Goldhahn L, Müller-Eppendorfer K (2012) Planning-modules for manual assembly using Virtual Reality techniques. In: Conference proceedings 4th international conference on applied human factors and ergonomics (AHFE) 2012. USA Publishing, Louisville, pp 2740–2749. ISBN 978-0-9796435-5-2

Goldhahn L, Thümer C (2014) Design and technical construction of virtual-reality-supported learning elements for manual assembly. In: Ahram T, Karwowski W, Marek T (eds) Proceedings of the 5th international conference on applied human factors and ergonomics AHFE 2014, Kraków, Poland, 19–23 July 2014, pp 5748–5760. ISBN 978-1-4951-1572-1

Lotter B, Wiendahl H-P (2012) Montage in der industriellen Produktion, 2nd edn. Springer, Heidelberg

Lotter B, Wiendahl H-P (2006) Montage in der industriellen Produktion. Springer, Heidelberg, New York

Nyhuis P (2009) Konzepte und Möglichkeiten einer bestandsarmen, zeitnahen Materialbereitstellung. Präsentation zur 5. Sitzung AWF AG moderne Produktionslogistik, Wiesloch, 20/21 January 2009

Pause D, Starick Ch, Adema J, Kraut A, Blum M, Hinrichs N (2016) Die Wahl der richtigen Beschaffungsstrategie – Modellierung einer Supply Chain zur Bewertung unterschiedlicher Beschaffungsstrategien. Zeitschrift für wirtschaftlichen Fabrikbetrieb 111(12):806–808

Pfohl H-Chr (2010) Logistiksysteme, Betriebswirtschaftliche Grundlagen (8. Überarbeitete und erweiterte Auflage). Springer, Heidelberg, Dordrecht, London, New York

Wirth S, Schenk M, Müller E (2014) Fabrikplanung und Fabrikbetrieb, Methoden für die wandlungsfähige, vernetzte und ressourceneffiziente Fabrik (2. Überarbeitete und, erweiterte edn. Springer, Heidelberg

Digitalization in Manufacturing – Employees, Do You Want to Work There?

Caroline Adam(✉), Carmen Aringer-Walch, and Klaus Bengler

Technische Universität München,
Garching Boltzmannstraße 15, 85748 Garching, Germany
caroline.adam@tum.de

Abstract. IT-driven transformation processes in manufacturing result in major changes for employees' daily work and they are ever-present in media and research, called *inter alia* digitalization, computerization or Industry 4.0. A variety of research focuses on these new hardware and software technologies. Research dealing with motivational processes and attitudes of employees in manufacturing, however, is lacking. How do they face these changes?

Data were collected using a standardized questionnaire (KFZA) that was returned by n = 109 apprentices working in manufacturing. The questionnaire queried the actual state of the work situation, as well as the target state. Additionally, fears, opportunities, and general aspects participants associate with a highly digitalized work environment, were also investigated.

Results are divided. Firstly, they show that young employees are negatively opposed to the changes of Industry 4.0 and digitalization in manufacturing as they fear massive job losses. Then again, they favour the idea of learning something new, having a greater degree of self-determination, and versatility, which is also connected to future work tasks.

Thus, early clarification of realistic risks and chances connected to digitalization in manufacturing is necessary in order to prevent young employees from being resigned and disillusioned before they even start their professional career.

Keywords: Manufacturing · Digitalization · Employees · Industry 4.0

1 Background

These days, manufacturing companies are confronted with a variety of challenges, e.g. globalization, shortening of product life cycles, new technologies, and digitalization [1]. In Germany, the term Industry 4.0 is the epitome for these changes and is linked to the use of cyberphysical systems and new technologies that facilitate communication and the use of information concerning production processes and machines in real time. As these technologies become more widespread in manufacturing environments, the role of humans will change and work complexity will increase – even on the shop floor [2]. It is assumed that simple, manual tasks in production will decrease [3]. The literature describes different prognoses for the development of the workplace situation in Germany. Some authors predict a massive job loss, while others assume that new jobs will gain [4, 5]. However, reliable forecasts are difficult.

S. Bagnara et al. (Eds.): IEA 2018, AISC 825, pp. 267–275, 2019.
https://doi.org/10.1007/978-3-319-96068-5_30

So far there is a variety of research that focuses on technological feasibility [6–8] or concepts to support workers [9, 10] in increasing digitalized manufacturing environments (e.g. assistance systems). Additionally, there is particular emphasis on which skills are needed to be able to cope with these changes [11, 12]. A future work environment in smart factories is generally associated with lifelong learning, codetermination, self-determination, or versatility [13]. Required competencies therefore are e.g. problem solving and supervision, personal responsibility and holistic thinking [2]. There is, however, a lack of research dealing with motivational processes and attitudes of employees in manufacturing. How do they face these changes? Do young employees like the idea of an increasingly digitized job? Do they see chances for their work future like increasing self-organization, more qualitative work or responsibility [3]? To answer these questions for the group of apprentices, and thereby future employees in manufacturing, the following study was designed.

2 Research Questions

To ascertain how future employees face digitalization developments in manufacturing, the following research questions were formulated:

1. Is there a difference in the perception of current work situation between apprentices in small and medium enterprises and those in large corporations?
2. Do apprentices like a work environment that is characterized by Industry 4.0-related future remit?
3. What do apprentices associate with Industry 4.0, what opportunities do they expect and what risks do they fear?

3 Study Design

Below is a description of the study design to examine the research questions.

3.1 Sample

To account particularly for employees who are very likely to work in highly digitalized manufacturing within the coming years, this analysis focused on apprentices in the metalworking industry. The questionnaire was distributed to seven classes of apprentices and was filled out by n = 135 participants. After preclusion of incomplete questionnaires, n = 109 participants were included in the analysis. Of these, n = 57 participants had been working in small and medium enterprises (SMEs) with less than 499 employees, and n = 49 had been working in large corporations with more than 500 employees (n = 3 n.a.). In large corporations, the degree of digitalization was higher (according to participants, in 50% of the large corporations Industry 4.0 is implemented mostly or fully) than in SMEs (according to participants, in 15.6% of the SMEs Industry 4.0 is implemented mostly).

The sample size consisted of n = 12 women and n = 97 men, representing a common ratio in manufacturing, aged 16 – 31 years (m = 19.6 years). Participants were training to become industrial mechanics (n = 47), precision mechanics (n = 31), mechatronics technicians (n = 28) and others (n = 3). Data were collected in July 2017.

3.2 Measuring Instrument and Approach

Since the aim of this study was to gain information about the actual state of the work situation, as well as the target state, the standardized and validated questionnaire KFZA (Kurz-Fragebogen zur Arbeitsanalyse) was used, which is a German short questionnaire for work analysis, developed by Prümper [14]. The questionnaire consists of 26 items to be answered on a 5-point Likert scale (1 = strongly disagree; 5 = strongly agree). These 26 items measure the following scales: *scope of action*, *versatility*, *holistic nature*, *social support*, *cooperation*, *qualitative workload*, *quantitative workload*, *work interruption*, *environmental stress*, *information and codetermination*, and *operational benefits*.

For each question, participants were asked to give an estimation of the actual state of their work situation and, additionally, on their desired target state of work situation. The scales obtained are connected to future work scenarios in highly digitalized manufacturing and therefore results allow first conclusions on the attitudes of apprentices concerning work scenarios characterized by e.g. high degrees of codetermination, versatility and qualitative workload.

Three additional open qualitative questions were asked on the fears, opportunities, and general aspects the apprentices associate with a highly digitalized Industry 4.0 work environment. Furthermore, the degree of digitalization in their companies was queried based on five categories (Industry 4.0 is: 1 = not implemented; 5 = fully implemented). Finally, demographic data were collected.

Although the questionnaire has previously been used for data collection from apprentices [15], a pre-test was conducted in order to ensure the questionnaire was suitable for the target group. Afterwards, the questionnaire was distributed in three different schools and no introduction to Industry 4.0 and digitalization in manufacturing was given to prevent falsification of the answers. Instructions concerning completion of the questionnaire were standardized on the first page. Participants were assured that data would be treated anonymously and that participation was voluntary. The difference between statements on their actual and target work situation state was explained.

3.3 Analysis

Concerning the statistical analysis, total values were determined for all scales by calculating the mean value for each scale and participant. This procedure resulted in 11 values for the actual state and 11 values for the target state per participant.

In order to test the first research question, an independent samples t-test was conducted. In doing so, potential differences between apprentices in large corporations and SMEs, and thus in corporations with higher and lower degrees of digitalization, concerning their actual work situation, were calculated.

To answer the second research question, a comparison based on descriptive statistics revealed differences between actual situation and target state.

Further open qualitative questions on fears, opportunities, and general aspects associated with digitalization in manufacturing and Industry 4.0 (3rd research question) were evaluated qualitatively. For each question, the content was coded using inductive categorization. In doing so, frequencies of different codes could be illustrated in charts (Figs. 1, 2, and 3).

4 Results

Differences between participants working in large corporations and SMEs in terms of their actual work situation should be revealed by means of an independent samples t-test. When testing variables for normal distribution, according to Kolmogorov-Smirnov-Test variables *operational benefits*, *scope of action* and *cooperation* are normally distributed, while the others are not. Nevertheless, a t-test was conducted as, for example [16] recommended using a t-test for large samples. Heteroscedasticity is given in all variables (Table 1 Levene's test). The data showed significant effects between participants in large corporations and SMEs for the variables *quantitative workload* (sig. = .002), *cooperation* (sig. = .035) and *operational benefits* (sig. = .000), with high effects for *quantitative workload* (d = .63, Power = .90) and *operational benefits* (d = .97, Power = .99). For the complete list of values see Table 1.

Table 1. Levene's Test, independent samples t-test, effect size and power to identify differences between participants working in large corporations and SMEs in terms of their actual work situation (n = 109).

	Levene's test		Independent samples t-test				Effect size (Cohen's d)	Power
	F	Sig.	T	df	Sig. (2-seitig)	SF		
Versatility	.309	.579	1.354	104	.179	.14698	.26	.27
Holistic nature	1.640	.203	.521	104	.603	.14359	.10	.08
Qualitative workload	.930	.337	−.293	104	.770	.15757	−.06	.06
Quantitative workload	2.071	.153	−3.256	104	.002*	.17516	−.63	.90
Work interruption	.074	.787	−1.023	104	.309	.15203	−.20	.17
Environmental stress	.677	.412	−.800	104	.425	.16200	−.16	.12
Scope of action	.044	.835	.499	104	.619	.17066	.10	.08
Social support	.285	.594	−1.232	104	.221	.15107	−.24	.23
Cooperation	.875	.352	2.136	104	.035*	.13061	.42	.56
Information and codetermination	.771	.382	.425	104	.672	.16513	.08	.07
Operational benefits	.343	.560	4.985	104	.000*	.18104	.97	.99

The questionnaire also queried differences in the actual situation and target state of the participants' work situation. Results can be seen in Table 2. The biggest differences between actual situation and target state showed up in the *work interruption*, *environmental stress*, *versatility, information and codetermination*, and *operational benefits* aspects. The size of the difference is illustrated by the darkness of the colour in the boxes.

Table 2. Mean values and standard deviations (SD) of current work situation and target state, and differences in participants' actual work situation and target state (n = 109).

	Mean current situation	SD current situation	Mean target state	SD target state	Difference current/target state
Versatility	3,85	0,75	4,47	0,44	0,62
Holistic Nature	3,87	0,74	4,38	0,60	0,51
Qualitative Workload	2,32	0,80	1,92	0,86	0,40
Quantitative Workload	2,37	0,93	2,04	0,84	0,33
Work Interruption	2,28	0,77	1,52	0,64	0,76
Environmental Stress	2,40	0,83	1,62	0,69	0,78
Scope of Action	2,80	0,86	3,69	0,67	0,89
Social Support	4,11	0,77	4,71	0,47	0,60
Cooperation	3,44	0,69	3,87	0,62	0,43
Information and Codetermination	3,44	0,85	4,33	0,63	0,89
Operational Benefits	3,38	1,02	4,49	0,62	1,11

Next, the results of the open qualitative questions are described. The five most common codes per question are displayed. All aspects reported by participants were coded, therefore more codes per participant per question are possible.

General aspects associated with Industry 4.0 by young employees are presented in Fig. 1. Most frequently assigned codes were automation, digital networking and digitalization.

When participants were asked what opportunities they see in Industry 4.0 (Fig. 2), most answered aspects that were coded as easier or less work. The second most frequent answer was no opportunities at all.

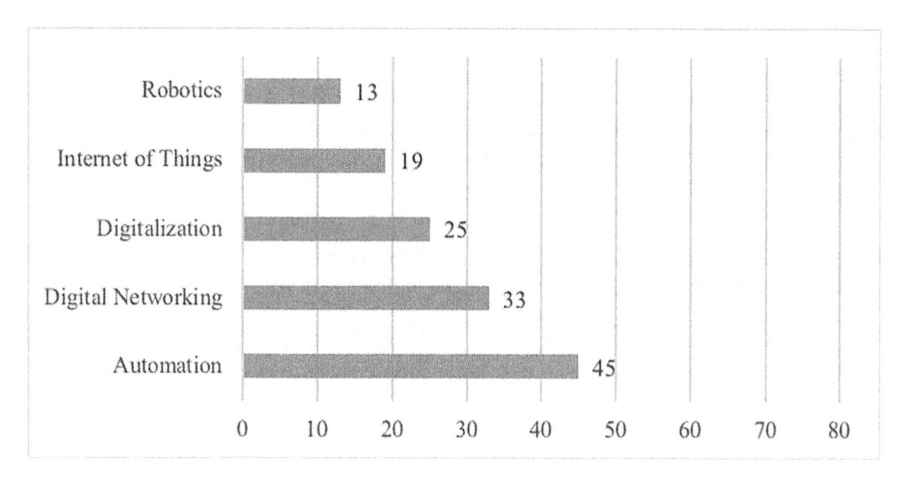

Fig. 1. Most frequent codes for the question concerning general aspects participants connected to Industry 4.0 (n = 98).

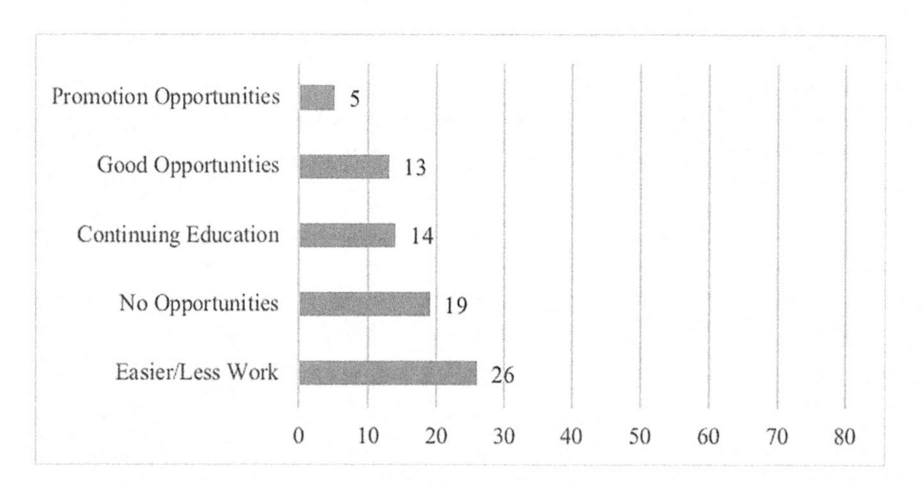

Fig. 2. Most frequent codes concerning opportunities due to Industry 4.0 (n = 79).

Finally, participants were asked what risks they expect due to increasing digitalization and Industry 4.0 (Fig. 3). Almost all participants who answered this question referred to job loss (n = 72). Sporadically, aspects such as monotonous work, hacking attacks and task complexity were mentioned.

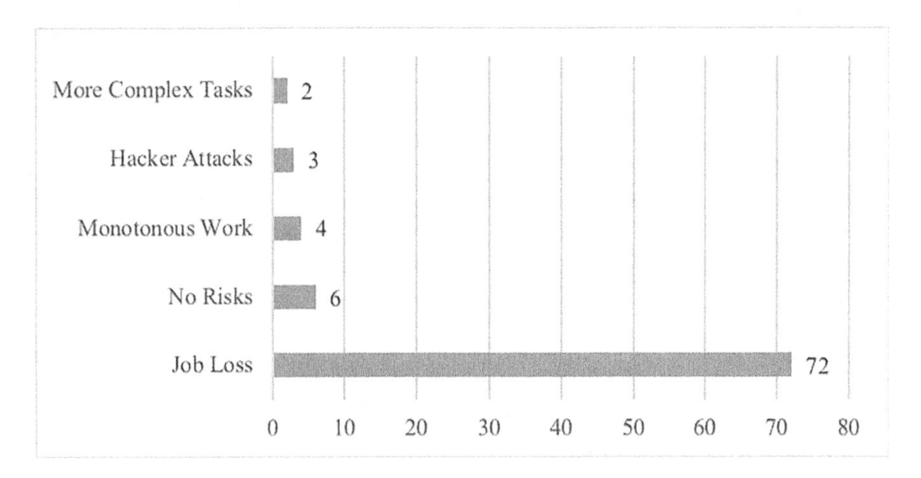

Fig. 3. Most frequent codes concerning expected risks due to Industry 4.0 (n = 81).

5 Discussion

Apprentices working in large corporations reported a higher degree of digitalization in their company and they have significantly less quantitative workload and better operational benefits according to their self-assessment. These results indicate better work conditions in larger corporations and, in the case of this study, higher digitalized corporations. Effects in aspects that are connected to Industry 4.0, such as versatility, holistic nature or scope of action, were not significant. Nevertheless, the differences between current situation and target state exemplify that participants would enjoy a future workplace which offers this. It emerges that young employees would prefer higher degrees of versatility and holistic nature for their work tasks. Additionally, increased self-determination with respect to the order of steps, type of work or independent planning of one's work are important to young employees and also in Industry 4.0. Social skills are becoming increasingly significant, particularly for digitalization in manufacturing and Industry 4.0, which was assessed positively by the participants. An increase in teamwork, close relationships with colleagues and cohesion in their department were also positively connoted.

What is more, Fig. 2 shows that participants are not reluctant to continue learning. Some participants see opportunities in continuing education and associated career opportunities.

Considering the results as a whole, it can be seen that future employees strive for a work environment characterized by greater versatility, codetermination and information. This is not, however, what they associate with a future work environment in highly digitalized manufacturing. The main aspect they connect to digitalization at work is job loss and fewer opportunities for their future. These negative expectations may lead to frustration and demotivation at the very beginning of their work life. As Industry 4.0 or digitalization in manufacturing leads to a fast-moving work environment, eagerness to learn and a high degree of motivation are necessary. Therefore, it is important to communicate not only the risks due to digitalization but also the opportunities, in order to motivate young employees to learn how to work in future manufacturing. Based on the learning motivation, education opportunities should be offered in combination with possibilities to promote oneself.

It is likely that in future especially simple tasks will be replaced by automation. This aspect is also frequently reported in media. Hence, there should be an increased focus on educating employees, especially apprentices, to enable and to motivate them to work in jobs that are more holistic, demanding and ambitious.

6 Limitations

One major limitation of this study is that participants of only three different schools were included in the analysis. All of these schools were in Bavaria and therefore results are not transferable one-to-one to other federal states or even other countries. The results do, however, provide an insight into reactions of future employees in manufacturing in the matter of digitalization in manufacturing and Industry 4.0. Furthermore, aspects of social desirability cannot be completely ruled out when using questionnaires. Results can only be transferred cautiously to other situations.

7 Conclusion

This study highlighted the great fear of young employees towards terms such as Industry 4.0 or digitalization. Few positive aspects were stated by future employees associated with a workplace strived for by smart factories. The opposite results, obtained concerning future work tasks characterized by Industry 4.0, underline the strong need for clarification of what a future workplace in digitalized environments stands for. In order to create a work environment that motivates rather than frightens young employees, these latter need to be enlightened concerning the opportunities of future work. By acquiring a deeper understanding of future work scenarios, employees would have the chance to get ready for highly digitalized manufacturing.

References

1. Adolph S, Tisch M, Metternich J (2014) Challenges and approaches to competency development for future production. J Int Sci Publ Educ Altern 12:1001–1010
2. Prinz C, Morlock F, Freith S, Kreggenfeld N, Kreimeier D, Kuhlenkötter B (2016) Learning Factory modules for smart factories in Industrie 4.0. Proc CIRP 54:113–118
3. Kagermann H, Helbig J, Hellinger A, Wahlster W (2013) Recommendations for implementing the strategic initiative INDUSTRIE 4.0: securing the future of German manufacturing industry; final report of the Industrie 4.0 Working Group. Forschungsunion
4. Spöttl G, Gorldt C, Windelband L, Grantz T, Richter T (2016) Industrie 4.0–Auswirkungen auf Aus-und Weiterbildung in der M + E Industrie. bayme vbm Studie

5. Wolter MI, Mönnig A, Hummel M, Schneemann C, Weber E, Zika G, Helmrich R, Maier T, Neuber-Pohl C (2015) Industrie 4.0 und die Folgen für Arbeitsmarkt und Wirtschaft: Szenario-Rechnungen im Rahmen der BIBB-IAB-Qualifikations-und Berufsfeldprojektionen
6. Wang S, Wan J, Li D, Zhang C (2016) Implementing smart factory of Industrie 4.0: an outlook. Int J Distrib Sens Netw 12(1):3159805
7. Gubbi J, Buyya R, Marusic S, Palaniswami M (2013) Internet of Things (IoT): a vision, architectural elements, and future directions. Future Gener Comput Syst 29(7):1645–1660
8. Zuehlke D (2008) Smartfactory–from vision to reality in factory technologies. IFAC Proc Vol 41(2):14101–14108
9. Valdez AC, Brauner P, Schaar AK, Holzinger A, Ziefle M (2015) Reducing complexity with simplicity-usability methods for Industry 4.0
10. Paelke V (2014) Augmented reality in the smart factory: supporting workers in an Industry 4.0. environment. In: 2014 IEEE Emerging technology and factory automation (ETFA). IEEE, pp 1–4
11. Gorecky D, Schmitt M, Loskyll M, Zühlke D (2014) Human-machine-interaction in the Industry 4.0 era. In: 2014 12th IEEE international conference on industrial informatics (INDIN). IEEE, pp 289–294
12. Guo Q (2015) Learning in a mixed reality system in the context of 'Industrie 4.0'. J Tech Educ (JOTED) 3(2)
13. Reinhart G, Bengler K, Dollinger C, Intra C, Lock C, Popova-Dlugosch S, Rimpau C, Schmidtler J, Teubner S, Vernim S (2017) Der Mensch in der Produktion von Morgen. In: Handbuch Industrie 4.0: Geschäftsmodelle, Prozesse, Technik. Carl Hanser Verlag GmbH Co KG, pp 51–88
14. Prümper J, Hartmannsgruber K, Frese M (1995) KFZA. Kurz-Fragebogen zur Arbeitsanalyse. Zeitschrift für Arbeits-und Organisations psychologie A&O 39(3):125–131
15. Mir E, Kada O, Brunkel H, Kohlmann E, Kohlmann CW (2016) Wie nehmen Auszubildende der Altenpflege die Arbeits-und Organisationsstrukturen in der Praxis wahr? HeilberufeScience 7(2):83–87
16. Lumley T, Diehr P, Emerson S, Chen L (2002) The importance of the normality assumption in large public health data sets. Ann Rev Pub Health 23(1):151–169

Systematic Approach to Develop a Flexible Adaptive Human-Machine Interface in Socio-Technological Systems

Julia N. Czerniak[1(✉)], Valeria Villani[2], Lorenzo Sabattini[2], Frieder Loch[3], Birgit Vogel-Heuser[3], Cesare Fantuzzi[2], Christopher Brandl[1], and Alexander Mertens[1]

[1] Institute of Industrial Engineering and Ergonomics, RWTH Aachen University, Bergdriesch 27, 52062 Aachen, Germany
{j.czerniak,c.brandl,a.mertens}@iaw.rwth-aachen.de

[2] Department of Sciences and Methods for Engineering, University of Modena and Reggio Emilia, Reggio Emilia, Italy
{valeria.villani,lorenzo.sabattini,cesare.fantuzzi}@unimore.it

[3] Institute of Automation and Information Systems, Technical University of Munich, Boltz-mannstraße 15, 85748 Garching, Munich, Germany
{frieder.loch,birgit.vogel-heuser}@tum.de

Abstract. Modern automatic machines in production have been becoming more and more complex within the recent years. Thus, human-machine interfaces (HMI) reflect multiple different functions. An approach to improve human-machine interaction can be realised by adjusting the HMI to the operators' requirements and complementing their individual skills and capabilities, supporting them in self-reliant machine operation. Based on ergonomic concepts of information processing, we present a systematic approach for developing an adaptive HMI after the MATE concept (Measure, Adapt & Teach). In a first step, we develop a taxonomy of human capabilities that have an impact on individual performance during informational work tasks with machine HMI. We further evaluate three representative use cases by pairwise comparison regarding the classified attributes. Results show that cognitive information processes, such as different forms of attention and factual knowledge (crystalline intelligence) are most relevant on average. Moreover, perceptive capabilities that are restricted by task environment, e.g. several auditory attributes; as well as problem solving demand further support, according to the experts' estimation.

Keywords: Adaptive HMI · Human abilities · Human-machine interaction Taxonomy · MATE concept · Performance

1 Introduction

When interacting with machines the operators' task is to capture the system status based on directly or indirectly processed information, to predict potential developments of the essential variables, and, to initiate actions, if necessary [1]. These tasks can be

S. Bagnara et al. (Eds.): IEA 2018, AISC 825, pp. 276–288, 2019.
https://doi.org/10.1007/978-3-319-96068-5_31

overstraining for the operator due to an information overload or complex activities to handle critical situations. Machine users have to learn machine-operating by adapting to the given framework and parameters of the machine. However, due to the complexity of automated machines, this can be a difficult and time-consuming process. For a user-centred system design, it is fundamental to gain knowledge about the users' needs regarding their skills and capabilities to support the user properly. Regarding to individual requirements, the success in learning processes and efficiency while operating can be very different. To support the learning process of machine-operating and reduce the strain level while operating with machines, an innovative interaction concept is required that relieves the users.

We discuss a first attempt towards this approach [2] with the human-controller approach, by seizing the fundamental controller-human theory of Marienfeld [3]. In contrast to Marienfeld, we model the control loop between two interacting system mechanisms reversed, by controlling the human's strain level by the machine.

In this paper we present a concept for the development of an adaptive HMI that addresses this approach, by adapting to the user's requirements. Individual human reactions during machine-operating represent the key element for the concept. The definition of a classification for user attributes that implies relevant, for machine adaptable parameters hereby becomes mandatory. Further, a human-machine interaction classification is introduced, by using a common classification scheme and has been established for more complex system descriptions: a taxonomy, which summarizes available data hierarchically. Further, we evaluate and cluster included attributes according to their relevance, before design recommendation for adaptive features are formulated.

1.1 Adaptive HMI Approach

The aim of the new developed concept for an adaptive HMI is to ergonomically support the operator in self-reliant machine operation by adapting towards the user's needs and requirements. The HMI, before adapting, needs to process specific data about possible influencing factor, which cause different human behaviours. We therefore introduce the MATE concept (**Me**asure, **A**dapt & **Te**ach) [4]:

Measure Approach

In a first step the user's individual characteristics and strain level whilst operating will be assessed. The user data has to be stored in the system via user profiles.

The following three user profiles will be implemented in the system:

A-priori user profile: Contains the user's innate and evolved skills and capabilities, which can be evaluated beforehand and are applied continuously for the individual. This static user profile includes constitutional and dispositional performance characteristics, as well as given information processing abilities, as far as assessable.

Real-time user profile: Measures the actual strain of the operator whilst working and proceeds the data directly to the adaption module of the system. The system will directly adapt towards the user's behaviour, either when a critical situation arises or the user is overwhelmed by the current operation. The real-time user profile contains dynamic strain and emotional information of the situational characteristics, when currently conducting the operating task.

Longitudinal user profile: This user profile can be understood as a continuous monitoring module, which contains data about the user performance and his capabilities development. The tracked evolution can be used to give long time support to the operator, depending on gained education or experience. The longitudinal profile is in charge of the qualification and competence characteristics and their development in working life. In this regard data protection regulations have to be considered.

Adaption Approach

Interpreted user information is directly translated into respective adaption rules. Specifically, adaptation can be provided with respect to measured perceptive and cognitive abilities of the operator and, in addition, with respect to the most appropriate interaction means. The aim of adaptation is to allow the system to be able to automatically adjust the optimal HMI configuration based on the user's needs.

Perception adaptation refers to the fact that sensorial (e.g., visual, auditory, haptic, motoric) capabilities of the user are accommodated and information is presented accordingly. Also, environmental influences, such as noise, illumination, or dust are considered. Generally speaking, this level of adaptation refers to how information is presented.

As regards cognition adaptation, user's ability to understand information is considered. This is influenced by user's skills and current emotional status and the kind of interaction task. This level of adaptation refers to what information is presented to the user. Adaptation rules at this level consider information selection (i.e., what information is relevant to be shown to the user, given her/his current status), organization of alarms (i.e., establishing a priority among alarms and deciding the quantity and the kind of alarms to present), guidance procedures (to guide less skilled operators or those needing assistance), and functionality enabling (when the user is experiencing cognitive difficulties, the system needs to reduce the number of available functionalities, in order to avoid confusion, to simplify the task for the user, and to reduce the possibility of mistakes.

Finally, with regard to interaction adaptation, depending on user's sensorial and physical capabilities, the best interaction means are selected to allow a smooth interaction. This level of adaptation refers to how interaction is enabled.

Teaching Approach

The third part of the MATE concept is a teaching system that adaptively trains the user according to their qualification and competence skills, which are stored in the longitudinal user profile. Teaching is provided online, whilst operating, and offline before starting a working session. Both parts can be adapted to the needs of the user, for instance by providing non-verbal symbols or adapting the presentation or the interaction mechanisms of the training system.

The offline training is in charge of familiarizing the user, e.g. unskilled personnel, with the system and the task to perform. It therefore applies a virtual environment that provides training in a secure environment with low distraction to meet the requirements of vulnerable user groups [5]. Different representation techniques allow providing training systems that meet the requirements of operators and the trained tasks.

Additionally, the user is provided with continuous online training that provides real-time guidance. Different training mechanisms, for instance augmented reality or speech assistance, e.g. by means of augmented reality are provided to support different user or task requirements.

Moreover, this module provides a social network system networking between different users. Problems can directly be discussed between users and expertise expert and knowledge can be obtained [6].

1.2 Literature Review on Human Ability Classification

Several fundamental taxonomies exist to describe human abilities or limitations in operating work tasks. For cognitive abilities Fleischman et al. [7] developed a comprehensive framework of 52 human abilities, separated into two main categories, perceptual-motor abilities and physical proficiency abilities. These depict cognitive resources available to an individual in performing any task. Fleishman's taxonomy approach distinguished between capabilities related to general performance capacity and skills for specific tasks [8]. Rasmussen [9] developed a taxonomy for human performance and error, which is based on the input-output response model for human-machine interaction. According to this model, error mechanisms and failure models depend on mental functions, such as perception, cognition and action, and knowledge, triggered by subjective factors. These have to be interfered from the task characteristics or measured by means of suitable techniques. Another taxonomy addressing variables influencing human performance in complex systems was created by Gawron et al. [10] through combining, adjusting and simplifying multiple existing taxonomies on the model of human information processing. Hogan et al. [11] also referred to these requirements for system development and the variability of human abilities regarding performance as dependent variables of information processing depend on the description of stimulus-process-result (S-P-R) performance in a task descriptive language. [11]. The research of Rantanen et al. [12] aimed at summarizing efforts to map intervention technologies onto error categories and to develop a conceptual framework for the dimensions human operator, task and environment. Their approach for the dimension human operator relies on two major models of information processing, such as the human information processing framework of Wickens [13] and the error taxonomy of Berliner [14].

In summary two major principles for human abilities during machine interaction become obvious: human performance and information processing, which are strongly correlated according to literature [15]. Moreover, different dimensions for influencing factors can be identified, referring to internal, external, and situational variables. However, no comprehensive taxonomy for human-machine interaction currently exists, which combine performance and information processing parameters for the use in production machines.

2 Theoretical Background

Human-machine interaction can be described as informational work, in which the users at first receives information about the current status of the machine and work progress through the HMI. Second, an action has to be prepared after making a decision. Finally, by giving information input into the HMI, machine movements are released [1]. These processes basically can be referred to three stages of information processing: perception, cognition, and action [16]. Several models for human information processing are known in literature, e.g. sequential models, capacity models, mathematic functional models and control measure models, which are mainly limited by one-dimensional perspectives [1]. In contrast to these models Wickens and Hollands [17] introduce a multiple resources model that provides different correlating information capacity sources, with different interacting effects on the allocation of available resources [18]. For example regarding different sensory modalities, it is easier to simultaneously attend towards a visual and auditory stimulus compared to two visual or two auditory stimuli at the same time, due to the fact that auditory perception uses different resources than visual perception (Isreal [19] cited after Schlick et al. [1]). As abovementioned, informational processes correlate with performance, which is strongly depended on individual's personal characteristics. A classification scheme for superior performance categories is provided by Luczak [20]. The classification distinguishes between four different categories, *constitutional characteristics* that are parameters for performance that cannot be changed during a life cycle; *dispositional characteristics* that are defined as changeable parameters regarding the lifecycle, but not directly influenceable; *qualification & competence* that describe earned capabilities that can be influenced willingly, and *adaptable characteristics* which are the most variable characteristics, dependent on the current situation. A central characteristic of performance is it's inter- and intraindividual variability, with variances of up to 1000% between different working persons for the same task. However, through compensating strategies these deviations sometimes do not become noticeable [20]. Figure 1 visualises the discussed parameters of informational work.

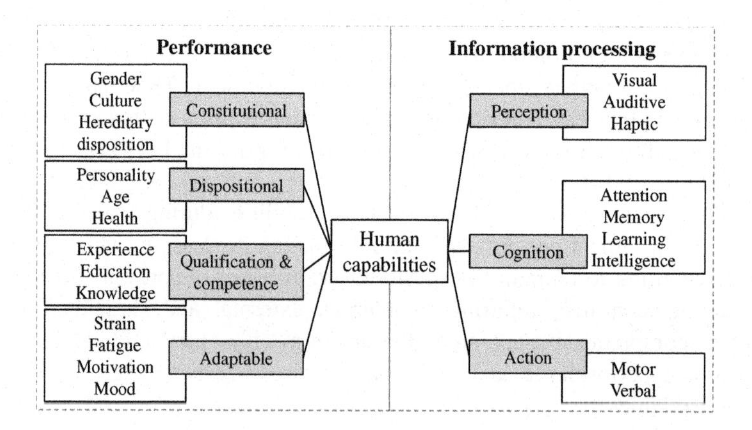

Fig. 1. Parameters influencing human-machine interaction

3 Method

By means of an extensive literature research on human capabilities, performance, and information processing we further complete the influencing parameters (Fig. 1) with relevant attributes, developing a comprehensive taxonomy with up to four depth levels. The concepts considered for the current approach combine different ergonomic methodologies of human centred system design [21], for example guidelines for information management on HMIs (e.g. see Standard DIN EN ISO 9241), task and workplace requirements [22], and ergonomic research on human capabilities, such as age-related changes in cognitive processes. The complete taxonomy can be found in Table 2 of the Annex.

To evaluate this taxonomy, we investigated three representative use cases for human-machine interaction with production machines. Following this aim, we conducted pairwise comparisons [23]. For the 37 taxonomy attributes of the information processing dimensions perception, cognition and action. The method of pairwise comparison allows evaluating a multitude of N parameter correlations by reducing system complexity through single-pair matching. These pairs are directly compared (1 = more important, 0 = similar, −1 = less important) by spanning an NxN matrix. With the resulting sum, the parameters can be ranged according to their relative importance. We did not consider performance characteristics for the assessment, because with regard to the adaptive interface, these attributes can only indirectly be influenced by the HMI through adapting informational variables, e.g. by increasing font size or contrast. Each pairwise comparison was conducted by an expert team of the use case owning company. Participants were instructed to assess the attributes according to their influence regarding stressors while conducting the use case tasks. The following use cases were assessed [24]:

Use case A Woodworking machine for a small artisans' companies that are typically run by elderly employees

Use case B Automated robotic cell mostly operated manually for panel bending and to be applied in companies in developing countries, and

Use case C Industrial plant for bottling and labelling.

4 Results

Figure 2 shows results for pairwise comparisons of the three use cases and the mean value. Total values of attributes were translated into positive values by adding the minimum sum of all attributes incremented by one. In the following we will distinguish the assessed attributes into three clusters by mean value, with regard to adaptable feature of an HMI.

Several suggestions can be derived from the mean value. The first thirds of the attributes with the highest mean (mobility of joints (sum 42) to precision (sum 34.33)) mainly include cognitive attributes, besides attributes referring to information response processes (action). Cognitive attributes in this third include active vocabulary, focused,

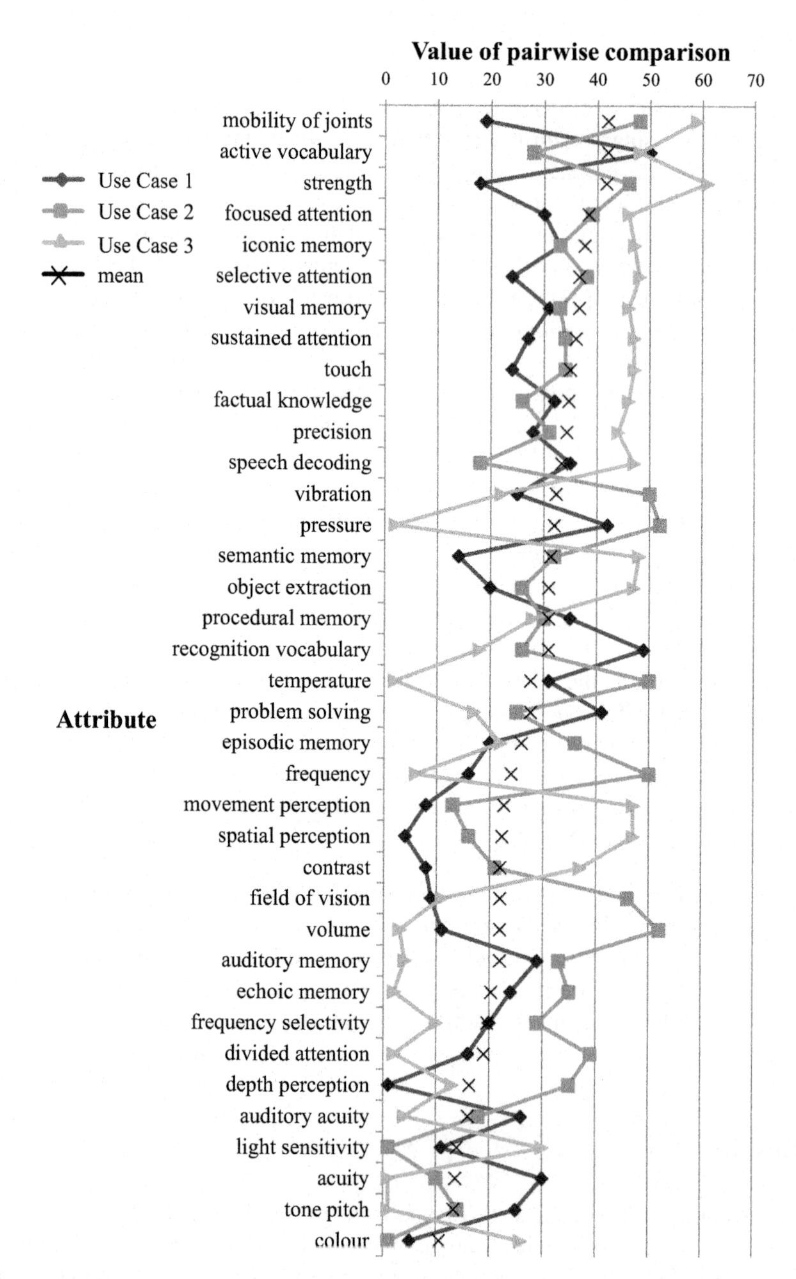

Fig. 2. Results of pairwise comparison: Value of information processing attributes in three different use cases sorted by mean. The value for each attribute is calculated by the attribute's sum, added with the minimum attribute sum + 1 of the use case to generate positive values.

selective, and sustained attention, iconic and visual memory, as well as factual knowledge. Since, physical condition cannot be compensated by adapting the HMI, we will not consider them further for the clustering.

The middle third of attributes (speech decoding (sum 33.33) to spatial perception (sum 22.33)) mainly includes perceptive attributes like auditory perception, such as the ability to decode speech input, distinguish tone frequencies, and spatial perception. Moreover, haptic perception, such as vibration, pressure, and temperature, and visual perception (movement) is included in this cluster. Cognitive attributes are the ability to solve problems and episodic memory.

The last thirds (contrast (sum 22) to colour (sum 10.67)) focuses on visual perception attributes like contrast, field of vision, depth perception, light sensitivity and colour perception, as well as auditory attributes, for instance volume, frequency selectivity, auditory acuity, and tone pitch. Cognitive attributes in this cluster consider auditory memory, echoic memory, and divided attention.

5 Discussion

Findings show that the greatest need for action regarding an adaptive HMI lies in cognitive support for machine operators' capabilities. According to the findings, the use cases require high concentration over a certain period of time (focused, selective and sustained attention), experience (factual knowledge), the ability to visually perceive, remember (iconic and visual memory), and understand (active vocabulary) information. This means that implementing the adaptive HMI should, on the one hand, focus on lack of concentration of the operators, e.g. by measuring strain or fatigue. On the other hand, it should adapt to visual information overload, by reducing the information presented, if possible, and providing perceptive support, such as clear presentation of important information, if necessary. Missing factual knowledge can be compensated by providing instructions and training by the HMI. The operators' vocabulary can be addressed by a speech interface that is able to recognise individually used language. The high mean of the action attributes "mobility of joints", "strength" and "precision" show that experts still estimate motoric capabilities with high importance for machines tasks.

Further, perceptive user capabilities matter with regard to the second cluster. Results imply that especially auditory perception seems to be restricted, but of importance for machine operating. One explanation is the constant noise level in machine halls, in addition to hearing protection that make it hard to receive auditory information properly. An adaptive HMI concept should consider this fact and support the operators concerning this matter, e.g. by using single sound sources and distinguishable sounds. In addition, problem solving can be improved by the HMI by means of the teaching module, whereas helping features, such as tool tips can compensate a lack of individual experiences (episodic memory).

Visual perceptive attributes were assessed relatively low. These parameters are easily adaptable, and since they are detailed considered by design recommendations given by Standards (e.g. EN ISO 9241), they are usually already implemented in state of the art interfaces. These for example provide features to adjust the contrast, colour, illumination, or sound volume of the HMI manually. Further, the impact of the relevance to remember specific sounds (auditory memory, echoic memory) is rated relatively unimportant. Since the investigated use cases consider single-machine operation, divided attention in these cases is not relevant for the adaptive HMI. Discussed design recommendations are summarized in Table 1.

Table 1. Design recommendations for an adaptive HMI with regard to the clustered information processing attributes, as results of the investigated pairwise comparisons.

Cluster	Dimension	Attribute	Design recommendations
1	Cognition	Attention (focused, selective and sustained attention)	Measure strain or fatigue Reduce information density or provide clear information presentation
	Cognition	Factual knowledge	Provide instructions and training
	Action	Active vocabulary	Adapt to individual vocabulary
2	Perception	Decode speech input, distinguish tone frequencies, and spatial perception	Single sound sources and distinguishable sounds
	Cognition	Problem solving (episodic memory)	Provide helping features, such as tool tips
3	Perception	Visual (contrast, colour, illumination, volume)	Consider design recommendations (e.g. DIN EN ISO 9241)

However, results show a high variance of several attributes between the three use-cases. This indicates that generalization of their importance is difficult. Thus, we suggest that the selection of concrete attributes remain a compromise between relevance for the use-case and practicability for implementation, and requires a case-by-case examination. Further, we found that several relevant informational stressors, influencing the operators' performance, refer to environmental circumstances or user abilities that cannot be influenced by HMI design itself.

6 Conclusion

In this paper we provided a systematic approach to develop a flexible adaptive human-machine interface. We introduce the MATE system as an implementation approach, consisting of three main pillars, which are to measure relevant skills and capabilities, to adapt the behavior of the HMI towards assessed limitations and to teach the operators in handling the system according to the operators' qualification and competence characteristics. The aim of the concept is to support users in complex machine operations by the HMI. The interface's main advantage is its ability to offer assistance to a user in self-reliant machine operation, instead of solely requiring users to learn unintuitive complex human-machine interaction, as presently demanded in modern automated production machines. In addition, the system is, on the one hand, able to detect exhausting situations and to provide solutions for long-term strain; and, on the other hand, recognizes suddenly occurring unexpected user problems, by means of exceptional increasing physiological indicators. The implementation of such an HMI requires a classification of user attributes relevant to human-machine interaction. Therefore, we developed a human-machine interaction taxonomy for user attributes that was evaluated regarding the importance of the attributes.

The developed clusters give first recommendations for the implementation of an adaptive HMI, and suggest focusing on cognitive and perceptive capabilities, which are not yet considered in current HMI design. We conclude that the introduced MATE approach contains necessary features to cover the requirements for an adaptive HMI. However, results show that an overall industry standard cannot be provided easily, due to high variance in use-case requirements. A case specific consideration should be conducted for practical implementation. Further, for the implementation of an adaptive HMI prototype, it is not feasible to consider the whole set of user attributes at once. A brief selection of case relevant attributes according to task requirements has to be conducted. It is worth mentioning that the feasibility for real working environments decreases with the degree of customization of the system. Moreover, it is not necessary to individualize all different emphasizes of user attributes. A more application-driven approach is the realization of clusters for the adaption of similar attribute characteristics, e.g. clustering age spans after age-related perceptive or cognitive changes, a practice well-known in literature. However, utilizing the cluster approach entails the risk of a lack of outliers to consider or individual cases. A clustering solution still takes into account most user percentiles.

Another important factor for the implementation of an adaptive HMI is to include ethical, legal, and social implications. With regard to assessing and storing personal and health data in the user profiles, legal requirements have to be met and an abuse of the data to the detriment of the workers has to be prevented. In many cases the assessment of performance data can also cause low user acceptance. To tackle these issues, different scenarios for data profiling are conceivable, ranging from mostly real-time data, with no or minimum a-priori assessment ("The machine does not know the operator") to a complete a-priori, real-time, and longitudinal evaluation of a maximum of user capabilities ("The machine knows the operator").

Acknowledgements. The research is carried out within the "Smart and Adaptive Interfaces for INCLUSIVE Work Environment" project, funded by the European Union's Horizon 2020 Research and Innovation Program under Grant Agreement N723373. The authors would like to express their gratitude for the support given.

Annex

Table 2. Taxonomy for human capabilities

Dimension	Parameter	Category	Attribute
Perception	Visual		Acuity
			Colour
			Contrast
			Depth perception
			Movement perception
			Object extraction
			Light sensitivity
			Field of vision
	Auditory		Auditory acuity
			Tone pitch
			Volume
			Frequency
			Spatial perception
			Speech decoding
			Frequency selectivity
	Haptic		Pressure
			Vibration
			Touch
			Temperature
Cognition	Attention		Selective attention
			Focused attention
			Sustained attention
	Memory	Sensory memory	Iconic memory
			Echoic memory
		Working memory	Visual memory
			Auditory memory
		Long-term memory	Episodic memory
			Semantic memory
			Procedural memory
	Intelligence		Fluid (problem solving)
			Crystalline (factual knowledge)

(*continued*)

Table 2. (*continued*)

Dimension	Parameter	Category	Attribute
Action	Motor		Mobility of joints
			Strength
			Precision
			Active vocabulary
			Recognition vocabulary
Constitutional			Culture
			Ethical background
			Gender
			Hereditary disposition Bio rhythm
Dispositional			Age
			Personality
			Health
Qualification & Competence			Experience
			Knowledge
			Training
			Competence
			Education
			Talent
Adaptable			Strain
			Motivation
			Emotion
			Satisfaction
			Daily condition Fatigue

References

1. Schlick C, Bruder R, Luczak H (2018) Arbeitswissenschaft, 4th edn. Springer, Heidelberg.
2. Czerniak JN, Brandl C, Mertens A (2017) Designing human-machine interaction concepts for machine tool controls regarding ergonomic requirements. IFAC-PapersOnLine. https://doi.org/10.1016/j.ifacol.2017.08.236
3. Marienfeld H (1970) Modelle für den Regler Mensch - Ein Praktikumsversuch (Engl.: Models for the human controller - a practical trial). In: Oppelt W, Vossius G (eds) Der Mensch als Regler (Engl.: The human controller). VEB Verlag Technik, Berlin, pp 19–41
4. Villani V, Sabattini L, Czerniak JN, Mertens A, Fantuzzi, C (2018) MATE robots simplifying my work. The benefits and socioethical implications. IEEE Robot Automat Mag https://doi.org/10.1109/mra.2017.2781308
5. Loch F, Vogel-Heuser B (2017) A virtual training system for aging employees in machine operation. In: 2017 IEEE 15th international conference on industrial informatics (INDIN), University of Applied Science Emden/Leer, Emden, Germany, 24–26 July 2017: proceedings. 2017 IEEE 15th international conference on industrial informatics (INDIN), Emden. IEEE, Piscataway, pp 279–284. https://doi.org/10.1109/indin.2017.8104785

6. Villani V, Sabattini L, Czerniak JN, Mertens A, Vogel-Heuser B, Fantuzzi C (2017) Towards modern inclusive factories. A methodology for the development of smart adaptive human-machine interfaces. http://arxiv.org/pdf/1706.08467
7. Fleishman EA, Quaintance MK, Broedling LA (1984) Taxonomies of human performance: the description of human tasks, vol 51. Academic Press Inc, Orlando
8. Furnham A, Stumm S, Makendrayogam A, Chamorro-Premuzic T (2009) A taxonomy of self-estimated human performance. J Individ Differ 4(30):188–193
9. Rasmussen J (1982) Human errors. A taxonomy for describing human malfunction in industrial installations. J Occup Accidents. https://doi.org/10.1016/0376-6349(82)90041-4
10. Gawron VJ, Drury CG, Czaja SJ, Wilkins DM (1989) A taxonomy of independent variables affecting human performance. Int J Man Mach Stud. https://doi.org/10.1016/0020-7373(89)90020-5
11. Hogan J, Broach D, Salas E (2009) Development of a task information taxonomy for human performance systems. Mil Psychol. https://doi.org/10.1207/s15327876mp0201_1
12. Rantanen EM, Palmer BO, Wiegmann DA, Musiorski KM (2006) Five-dimensional taxonomy to relate human errors and technological interventions in a human factors literature database. J Am Soc Inf Sci. https://doi.org/10.1002/asi.20412
13. Wickens C (1984) Processing resources in attention. In: Parasuraman R, Davies DR (eds) Varieties of attention. Academic Press, New York, pp 63–102
14. Berliner DC, Angell D, Shearer JW (1964) Behaviors, measures and instruments for performance evaluation in simulated environments. In: Subcommittee on Human Factors (ed.) Proceedings of the symposium and workshop on the quantification of human performance, Alberquerque
15. Wickens CD, Hollands JG, Banbury S (2016) Engineering psychology and human performance. Prentice-Hall, Upper Saddle River
16. Luczak H (1975) Untersuchungen informatorischer Belastung und Beanspruchung des Menschen. VDI-Zeitschrift H 02(10)
17. Wickens CD, Hollands JG (1999) Engineering psychology and human performance. Prentice-Hall, Upper Saddle River
18. Wickens CD (2002) Multiple resources and performance prediction. Theor Issues Ergon Sci 3(2):159–177
19. Isreal JB (1980) Structural interference in dual task performance: behavioral and electrophysiological data. Dissertation, University of Illinois
20. Luczak H (1989) Wesen menschlicher Leistung. In: Institut für angewandte Arbeitswissenschaft e.V. (ed.) Arbeitsgestaltung in Produktion und Verwaltung. Taschenbuch für den Praktiker. Wirtschaftsverlag Bachem, Köln, pp 54–65
21. Salvendy G (2012) Handbook of human factors and ergonomics, 4th edn. Wiley, Hoboken
22. Landau K (1978) Das arbeitswissenschaftliche Erhebungsverfahren zur Tätigkeitsanalyse - AET (Engl.: The ergonomic task analysis). Dissertation, Technische Hochschule Darmstadt
23. Eversheim W, Kuster J, Liestmann V (2003) Anwendungspotenziale ingenieurwissenschaftlicher Methoden für das service engineering. In: Bullinger HJ, Scheer AW (eds) Service engineering. Springer, Heidelberg, pp 417–441. https://doi.org/10.1007/978-3-662-09871-4_17
24. Sabattini L, Villani V, Czerniak JN, Mertens A, Fantuzzi C (2017) Methodological approach for the design of a complex inclusive human-machine system. In: CASE 2017: Proceedings of the IEEE conference on automation science and engineering

Proposal of an Intuitive Interface Structure for Ergonomics Evaluation Software

Aitor Iriondo Pascual[1(✉)], Dan Högberg[1], Ari Kolbeinsson[1], Pamela Ruiz Castro[1], Nafise Mahdavian[1], and Lars Hanson[1,2]

[1] School of Engineering Science, University of Skövde, Skövde, Sweden
aitor.iriondo.pascual@his.se
[2] Scania CV, Södertälje, Sweden

Abstract. Nowadays, different technologies and software for ergonomics evaluations are gaining greater relevance in the field of ergonomics and production development. The tools allow users such as ergonomists and engineers to perform assessments of ergonomic conditions of work, both related to work simulated in digital human modelling (DHM) tools or based on recordings of work performed by real operators. Regardless of approach, there are many dimensions of data that needs to be processed and presented to the users.

The users may have a range of different expectations and purposes from reading the data. Examples of situations are to: judge and compare different design solutions; analyse data in relation to anthropometric differences among subjects; investigate different body regions; assess data based on different time perspectives; and to perform assessments according to different types of ergonomics evaluation methods. The range of different expectations and purposes from reading the data increases the complexity of creating an interface that considers all the necessary tools and functions that the users require, while at the same time offer high usability.

This paper focuses on the structural design of a flexible and intuitive interface for an ergonomics evaluation software that possesses the required tools and functions to analyse work situations from different perspectives, where the data input can be either from DHM tools or from real operators while performing work.

Keywords: Ergonomics · Interface · Data management

1 Introduction

This paper presents initial results from an ongoing research and development project. The project is established to develop a system that facilitates ergonomists to analyze physical work conditions on the shop floor in an efficient way, e.g. in order to identify needs for improvements of work stations from a physical ergonomics point of view. The system can also be used to give operators feedback on their work techniques. The system captures data on exposure variables and compares those exposures to exposure limits. The outcome is then presented through a software interface, in a manner that is

S. Bagnara et al. (Eds.): IEA 2018, AISC 825, pp. 289–300, 2019.
https://doi.org/10.1007/978-3-319-96068-5_32

useful for the ergonomist, or to the operator if the system is used to give recommendations for how to improve work techniques.

The area of ergonomics has grown in importance in the industry over the decades due to an increase of public health problems and economic costs to employers and workers [1], with a cost in EU of €476 billion per year, resulting in a loss of 3.9% of GDP (Gross domestic product) [2].

Ergonomics evaluation methods are commonly used in industry, where many of the methods are based on the assessment of postures of workers. An example is Rapid Upper Limb Assessment (RULA) [3], which is used to analyze risks for work-related disorders in the upper extremities by assessing work postures. RULA facilities a quick screening of current conditions, and the outcome can be used to identify needs for improvements of work stations or work techniques.

RULA is an example of an observation-based posture evaluation method. Such methods require an ergonomist to visually observe and judge the work, directly or by viewing video recordings. Data on the work characteristics, e.g. joint angles at extreme or common work tasks, are then entered into the tool, and the tool uses algorithms to calculate risks for work-related disorders. Performing such observations is time consuming and requires special training in ergonomics in order to be performed correctly. Basing assessments on observations is also subjective, and research has shown that observation-based ergonomics assessments often leads to different results also when the assessments are performed by experts [4].

Also, as making such visual assessments is a time-consuming work, it can be automated using different technologies. Instead of visual assessments, cameras and image pattern recognition techniques can be used [5]. Another option to obtain the data to make the assessments is by using position sensors. In the case of sensors, the measurement of angles becomes more accurate and faster than evaluating images. In the last few years technologies like sensors have become more important in the area of ergonomics since they have improved their accuracy and duration, making it easier to create virtual environments and real-time analyses [6].

Direct measurement methods using sensors, apart from preventing the ergonomist from having to collect all the information manually, also allow a greater collection of data and therefore a continuous evaluation using quantitative data collection and analysis [7]. And it is also possible to collect data for several operators in parallel. The data collection and analysis can be done after the recording session has been closed, or it can be done in more of less real-time, offering the possibility to perform live evaluations. The latter solution facilitates giving feedback to operators on work techniques, e.g. when training unexperienced operators in performing work tasks. Obtaining data through sensors helps obtain more information than with visual feedback, however the amount of information generated by these systems is too large [8] to manage unless it is properly structured.

Also, nowadays different ergonomics software, a.k.a. digital human modelling (DHM) tools, are used to generate simulations of human work in virtual environments in order to ensure good ergonomics for future designs [9]. In such cases posture data comes from the computer manikin in the simulation software rather than from data collected from real persons, but a tool is still necessary to organize the information generated virtually, and for the simulation tool to facilitate the tool user to perform the ergonomics evaluations.

This article describes the creation of a software tool solution that manages collected posture data. A software tool interface that presents the data and the ergonomics evaluation outcomes to the ergonomist is also described. The tool is designed to be able to handle data being collected from real users as well as data collected from a computer manikin in a DHM tool. Hence, the tool facilitates evaluations of work in a real industrial environment as well as work performed in a virtual environment within a digital human modelling tool. However, this paper focuses on evaluations of work in real industrial environments.

2 Methods

In order to design the software tool, including its intuitive interface, a first step was to collect requirements from representative users of the tool, i.e. ergonomists (Fig. 1). Based on these requirements, use cases for the interface were built. Then, through these use cases it was possible to define the necessary data to process the information required by the ergonomists. The next step was to generate design proposals for how to present the data in a way so that it would be intuitive for the user to find the desired data and interpret the data for the desired purpose. Design proposals were then tested on representative users in order to evaluate and iteratively improve the design (indicated by arrow 1, 2, 3, and 4 in Fig. 1).

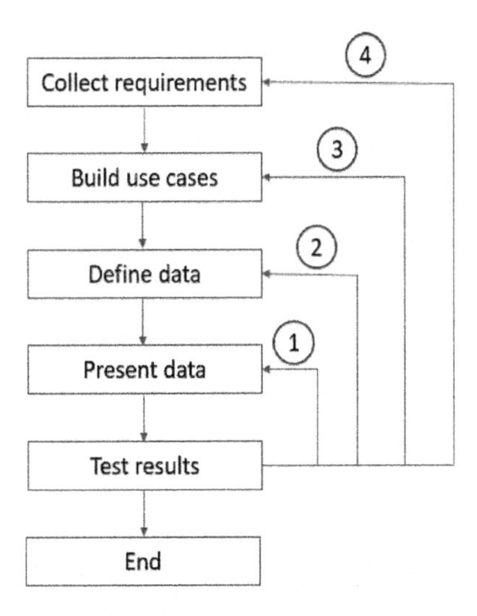

Fig. 1. Method flowchart

2.1 Collect Requirements and Build Use Cases

Ergonomists will be the users of the information provided by the software tool. Hence, ergonomists were consulted about their needs and requirements on the tool (3 and 4 in Fig. 1).

2.2 Define Data

After defining the use cases for the interface, it was necessary to define the type of data that would be needed to be able to perform them (2 in Fig. 1). For this purpose, a study was made based on evaluation methods such as RAMP [10] and RULA [3], to define the type of input that was necessary to be able to evaluate work. Some aspects in these evaluation methods are subjective, and usually cannot be done with information obtained only through sensors. Therefore, manual input by the user will be necessary in some cases. For the rest of the evaluation aspects, for which manual input was not needed, in general, the necessary information was the joint angles of the parts of the body to be evaluated and the forces exerted by the operator.

2.3 Present Data

Afterwards, the way of representing this information in each case was studied, since depending on the purpose, the need of the ergonomist can vary. Sometimes quick and simple information was needed. These valuations are called first level evaluations. Other times it was needed to analyze the data in depth and in greater detail, which corresponds to a second level evaluation. For this, an iterative process of widgets design and testing with ergonomists was started until arriving at a tool that presented the information in a clear and simple way despite the complexity of it.

3 Results

3.1 Definition of Use Cases

The different cases of use that may occur during the project have been recorded and have been summarized in the following needs:

Individual cases:

- Visualization of the raw posture data generated (real or virtual). This includes visualization of every body part of every user during time and several statistics such as mean, standard deviation and percentiles.
- Visualization of evaluations. There is a need of using different methods such as REBA [11], RULA [3] or EAWS [12] visualize their score (the value defined by the methodology, i.e. in RULA from 1 to 7) during time and get statistics from them.
- Comparison between users. Comparing both raw data and scores from different methods. It has to be possible to compare body parts, groups of body parts (i.e. upper limb, lower limb etc.) and total scores of the users.

Group cases:

- Visualization of the scores of the users in a group. Statistics are also necessary in order to have a general vision of the group without the need to analyze each individual belonging to the group.
- Visualization of the group score at a general level to measure the average behavior of several people in the same environment.
- Comparison between groups using different methodologies to evaluate them and with the possibility of focusing on a specific part of the body, or evaluations in order to compare ergonomic problems within groups (i.e. critical anthropometrics such as the height in cases of lower back pain in specific workstations).

Work station cases:

- Comparison of different designs of work stations used by one or several groups of operators, to optimize the design of them and reduce the risks of damage taking into account diversity aspects.

Real-time cases:

- Visualization of the information live so that the ergonomist can assess the ergonomics and work techniques of the operators and give them feedback to improve their working postures, taking the role of a "human coach".
- An automatic system that sends audio feedback to operators without the need of an ergonomist to assess the work. In this case the system takes the role of a "virtual coach".

3.2 Generation of a Tree Model for the Data

Sensors are used to capture the characteristics of the work and to be able to process the sensor data through a software. These sensors can be position, velocity and force sensors. By means of these sensors, joint angle and force data is obtained. The anthropometric information is added manually since the sensors in the current system are not able to measure different values such as height, weight or length of the arms. Age data is also entered manually. This information is especially important for the ergonomist at the time of performing individual evaluations, or for forming groups with similar characteristics.

The amount of information to be stored, organized and analyzed is very large. With usual use, an information vector is available for each simulation frame or for every second of real time. This means that for a simulation of one hour, 3600 vectors are created.

Apart from that information, the ergonomist wants to add information about users, create groups and organize the data according to calendar (time, day etc.). This causes the need to structure the information so that it has a logical order and is easily accessible for analysis.

Figure 2 shows the developed data structure.

There are four main levels in the data structure:

1. *Session level*: Each session includes one or several groups. In the industry this is used to name different tests, workplace tests or factories. All of them tested in

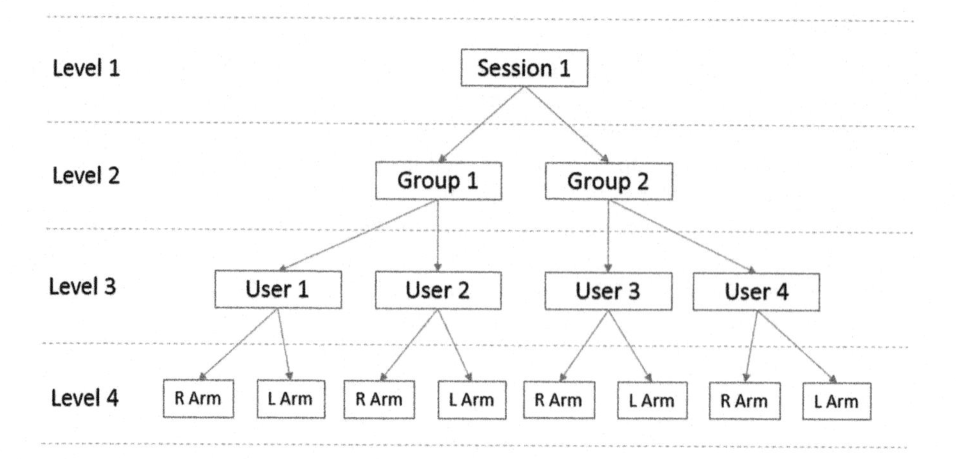

Fig. 2. Data structure

different groups of operators with different anthropometrics, methods and/or types of feedback.

2. *Group level*: Each group is formed by one or several operators. Commonly it is desired to be able to group operators with similar anthropometrics, methods, types of feedback and/or work stations in order to be able to compare them against other groups with different characteristics.
3. *User level*: Each user contains one or several exposure evaluations and individual characteristics such as medical difficulties/recommendations or anthropometrics.
4. *Exposures level*: This level is the one that contains the data obtained from the real-world sensors or from the virtual simulations. Raw data such as angles of body parts, forces and speeds are available to analyze in this level. Also, if an evaluation with a specific method (such as RULA) is done, the results will be available in this level. Higher levels such as user level, group level and session level will include the final scores evaluated in this level. Also, extra information can be input manually to add more information, for example in RULA some information has to be a manual input such as "muscle use score", since there is no sensor measuring that in the current system.

Time is a dimension added to all data levels, and the most complicated to deal with at the levels of this structure. To this end, differentiation has been added using work cycles (or *takt time* in industry) and work status (working time and break time, or different routines), through which previously defined time intervals (by manual input) can be studied.

Given the variety of evaluation methods, the information needed to perform the evaluations may vary. Therefore, it is difficult to define a structure that supports all evaluation methods. In order to solve this problem, it has been proposed to add each evaluation as one more exposure, even though in some cases they do not represent a specific body part.

3.3 Proposed Interface

Two completely different types of use for this interface have been identified:

- The data being generated is to be used to provide feedback, either through a human coach or through a virtual coach.
- The data being generated is stored to provide for later analyzes.

For these two use cases it has been decided to build two independent modules, the first module is called "Live" and the second one is called "Lab". Each of these two modules have different characteristics, since the Live module seeks to display the information of a single group in a clear and simple way, while the Lab module seeks to show information in detail and at all levels of data. Apart from these two modules, it has been decided to add a third module to facilitate the user the construction of the aforementioned data structure. This last module is named "Set". Therefore, the interface contains a main menu like the one shown in Fig. 3.

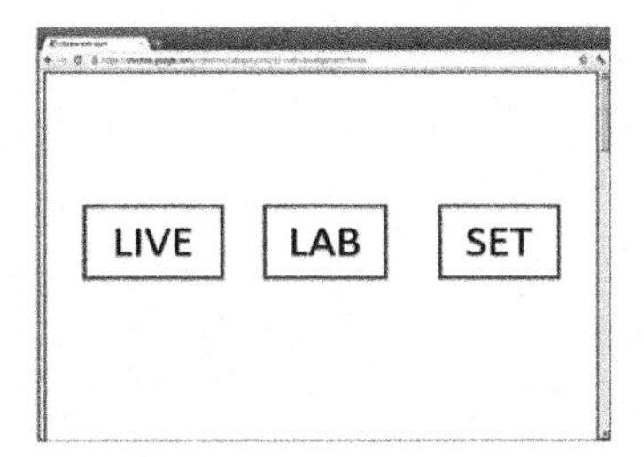

Fig. 3. Main menu

Explanations of each module:

Live. In the Live mode the information can analyzed by a human coach in more of less real-time, in order to provide feedback to operators of how to improve work techniques.

To read the status of operators' work techniques more quickly, the information is presented clearly and concisely. Depending on the number of sensors that are being used and the evaluation method, the information can vary from 10 to 30 values, multiplied by the number of people being analyzed. The feedback has to be as visual as possible, avoiding the use of numbers and text and reducing the results to colors, preferably related to the defined evaluation method and to the part of the body to which they correspond.

After iterating the design with the comments obtained by the ergonomists, it was decided that a panel of buttons is the solution for how to select the operator that it is wanted to be seen. These buttons change color depending on the evaluation of the operator. These colors follow a green/yellow/red scale (Fig. 4) based on the threshold of the selected method.

For the viewing of each operator it has been decided to use a dummy in which each part of the body is colored depending on the evaluation (Fig. 5). Given the variety of evaluation methods, some of which contain abstract evaluations that do not represent a part of the body (repetitions, physical exhaustion, lung capacity etc.), an extra panel with icons of variable color representing the evaluations are used.

Fig. 4. Buttons to select operator to analyze (Color figure online)

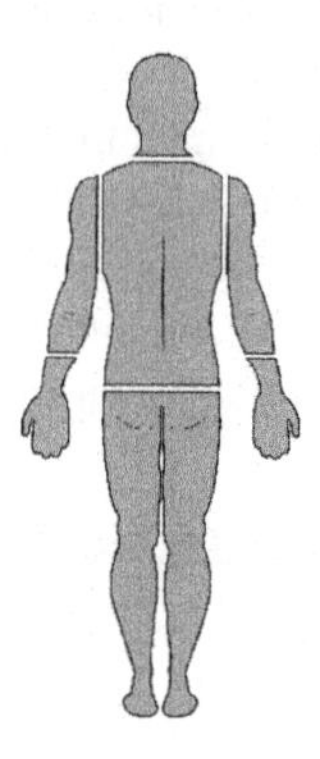

Fig. 5. Dummy/mannequin for color evaluations

It has also been decided to add the possibility of exporting operators' information as well as their evaluations for further studies. The preferred format for ergonomists to access data from different software is C.S.V. (comma separated values).

Lab. In order for the Lab module to be intuitive, it must follow the same flow as the information that is being treated. Therefore, the page starts at the session level and as it goes down, the group, user and exposures levels are loaded (Fig. 6). As the time window, work cycle and work status are options that must be available at all levels the menu for the selection of these options must accompany the user to throughout the entire interface. This will allow the user to select the information that is desired and re-evaluate with different conditions in every level.

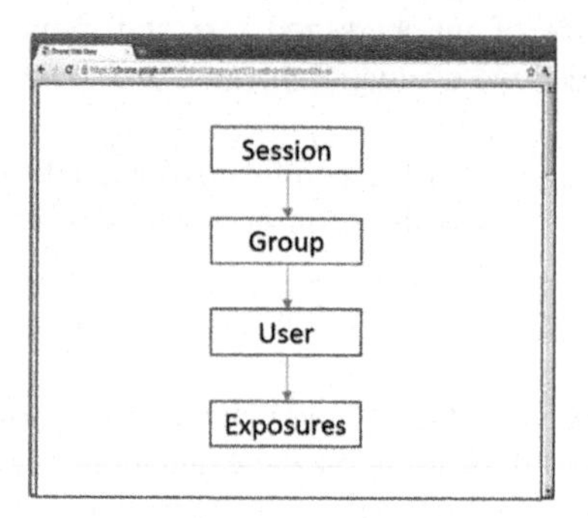

Fig. 6. Interface Lab workflow

At the session level, the groups belonging to the session and their evaluations are shown. At this level of comparison of groups, it has been seen that the use of statistics such as percentiles, mean, median and percentages can indicate the work behavior of a group in a general way. In addition to this it is also necessary to indicate the operator with the most extreme evaluation within the group. It is also important to indicate the average anthropometric characteristics of the group, as this can give information about why certain groups have worse evaluations.

At the group level, it has been seen that a table of colors is an intuitive way to show all the evaluations of each operator in the group, so that the colors guide the ergonomist when deciding which operator to study in depth (Fig. 7). It also shows the anthropometric characteristics of each member of the group as well as some brief statistics about their evaluations. In this way different members can be compared quickly without having to go down to the user level.

USER	RIGHT ARM	LEFT ARM	RIGHT ARM SPEED
SMA4548			
SMB1788			
SMC1032			

Fig. 7. Table of colors for group level evaluation (Color figure online)

At the user level the operator can be studied in more detail, being able to see the statistical results of each of the evaluations as well as selecting each of them to generate graphs at the level of exposures. The maximum and minimum levels of the evaluations, the raw data and percentages of time of the state of each part of the time are accessible, being able to analyze how long the postures have been extreme or how many times the same movement has been repeated.

At the exposure level the angles, velocities and forces of each of the body parts are shown, as long as the information is available. In turn, the graphs show with colors if the evaluation in that point is good or bad, so that it allows to analyze intervals in which the operator has carried out tasks in a risky position. In order to be able to select these intervals, a selectable double graphic tool is offered, so that a small line chart shows all the time recorded and selecting a section, this is represented in the large line chart (Fig. 8). This large line chart is also scrollable so the navigation is more intuitive.

When structuring the information in so many levels the user of the interface can be confused. To avoid this, a superior navigation bar is included in which the user can easily navigate between the different levels, at the same time that he/she can select the time and the work cycle to analyze, which affects all levels.

To make comparisons, several studies can be carried out in parallel, moving to the desired levels and adding several values to the same graphs, so that they can be compared by superimposing them. All comparisons can be made through the structure.

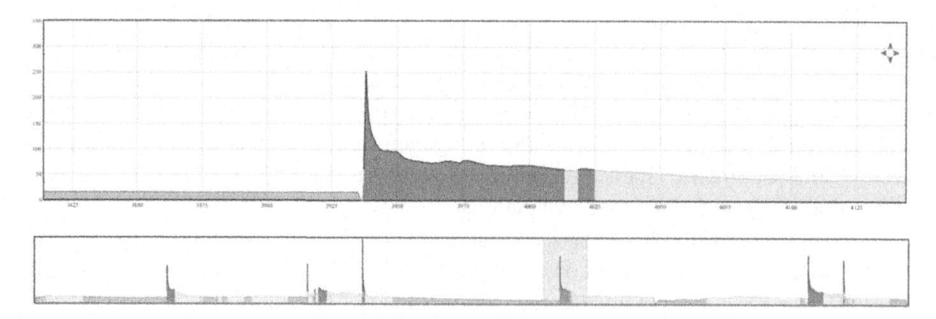

Fig. 8. Line charts for exposure level

For example, to compare groups of operators they can be grouped by anthropometric characteristics and then compared with the created groups.

The Lab module also accepts the import of data in xml or csv formats, so that data can be imported from simulations made in other software. The selected data can also be exported in csv format so that more specific analyses can be done in other statistical software. In addition to this, PDF with the evaluations previously made can be created so that the studies made on the page can be saved.

Set. In this section the different groups of operators and the sessions in which they are included is built. The information is saved so that saved groups can be reused for further analysis. In this way, statistical comparisons for different work stations or groups of people with different anthropometries are easier to perform and easily repeatable.

To structure this information, it will be done in the way that has been explained above: a group is formed by grouping one or several operators and a session is formed by joining one or several groups.

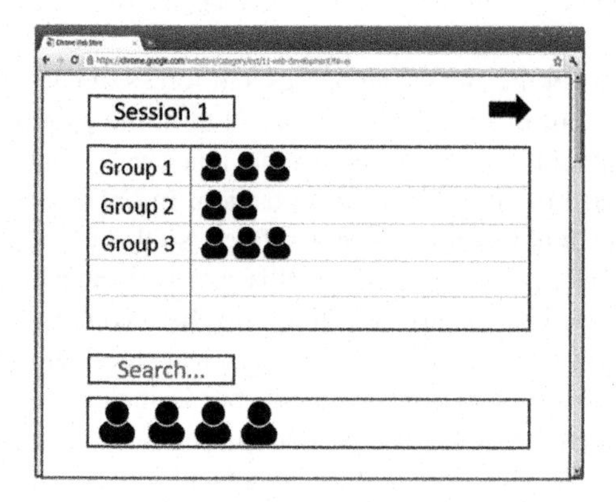

Fig. 9. Group and session build interface

After asking to the possible users of the interface it was seen that a drag and drop interface is the best option. In this case, a search of operators is available to be able to add them to a grid, from which they can dragged to different lines of another grid to form groups. Everything included in the second grid forms a session (Fig. 9).

4 Discussion

There are DHM tools to make virtual simulations, hardware to capture real human data and tools to evaluate this information. However, these tools have been mostly designed for the evaluation of a single user, and if it is needed to study different groups of people the ergonomists often end up going to spreadsheets. Therefore, it is beneficial to create a tool that allows to carry out the most common studies that ergonomists usually do, without resorting to statistical calculations in other software.

Ergonomic data generated by sensors or by simulations is large and it can be considered a big data problem [8]. That is why a structure is required to organize this data so that it can be accessible when required, without duplicating it and so that it can be reused without losing information. The structure offered in this paper has managed to organize the information in an appropriate way, complying with the use cases defined by the ergonomists.

However, once this information is organized, it must be presented to the ergonomist, which is complicated due to all the levels of information that exist, with the added variable of time. Due to the use cases defined, it was seen that it is necessary to divide the interface into two levels. First level evaluations are considered *Live* and second level evaluations are considered *Lab*, since one requires fast and simple evaluations and the other requires deeper and more accurate studies.

For the Live section, buttons have been designed that change color according to the evaluation of the operator so that it is not necessary to select it to obtain general information about it. Once an operator has been selected a mannequin shows the evaluations by colors on the body parts.

In the Lab section the difficulty is given by the amount of information that must be presented, since in this section more accurate and deep information is required. To solve this problem, a page is presented that follows the flow of information from top to bottom, presenting it in the same way. An upper navigation bar allows adjusting the time variable to what is required by the ergonomist. This bar also presents the information selected so far, so that it is always known what information is being viewed.

The information must be presented in each level in different ways, being more general information at higher levels and more specific information at lower levels. For the session level, it was decided to use only statistical values that define the groups the session contains, so that general information can be obtained in a simple way. For the group level, it was preferred to use a table of colors, in which each operator and the evaluations of their exposures appear. For the user level statistics of more accurate user evaluations and in greater depth are presented. Finally, at the exposure level, a double line graph has been created to be able to select the required information with greater accuracy.

For future lines, it should be studied in greater depth how to create standards between different evaluation methods so that the integration with the interface is more natural and intuitive.

Acknowledgments. This work has been made possible with the support from Vinnova/UDI in Sweden, in the project Smart Textiles for a Sustainable Work Life, and by the participating organizations. This support is gratefully acknowledged.

References

1. Sultan-Taïeb H, Parent-Lamarche A, Gaillard A, Nicolakakis N, Hong QN, Vezina M, Coulibaly Y, Vézina N, Berthelette D (2017) Economic evaluations of ergonomic interventions preventing work-related musculoskeletal disorders: a systematic review of organizational-level interventions. BMC Public Health 17(1):935
2. EU-OSHA. https://osha.europa.eu/en/about-eu-osha/press-room/eu-osha-presents-new-figures-costs-poor-workplace-safety-and-health-world. Accessed 27 May 2018
3. McAtamney L, Nigel Corlett E (1993) RULA: a survey method for the investigation of work-related upper limb disorders. Appl Ergon 24(2):91–99
4. Eliasson K, Palm P, Nyman T, Forsman M (2017) Inter- and intra- observer reliability of risk assessment of repetitive work without an explicit method. Appl Ergon 62:1–8
5. Tang A, Lu K, Wang Y, Huang J, Li H (2015) A real-time hand posture recognition system using deep neural networks. ACM Trans Intell Syst Technol 9(4): 23
6. Lin JH, Kirlik A, Xu X (2018) New technologies in human factors and ergonomics research and practice. Appl Ergon 66:179–181
7. Jayaram U, Jayaram S, Shaikh I, Kim Y, Palmer C (2006) Introducing quantitative analysis methods into virtual environments for real-time and continuous ergonomic evaluations. Comput Ind 57(3):283–296
8. Drury CG (2015) Human factors/ergonomics implications of big data analytics: chartered institute of ergonomics and human factors annual lecture. Ergonomics 58(5):659–673
9. Demirel HO, Duffy VG (2017) Applications of digital human modeling in industry. In: Digital human modeling, pp 824–832
10. Lind C, Forsman M, Rose L (2017) Development and evaluation of RAMP I : a practitioner's tool for screening of musculoskeletal disorder risk factors in manual handling. Int J Occup Safety Ergon: 1–16
11. Hignett S, McAtamney L (2000) Rapid entire body assessment (REBA). Appl Ergon 31 (2):201–205
12. Schaub KG, Mühlstedt J, Illman B, Bauer S, Fritzsche L, Wagner T, Bullinger-Hoffman A (2012) Ergonomic assessment of automotive assembly tasks with digital human modelling and the "ergonomics assessment worksheet" (EAWS). Int J Hum Factors Modell Simul 3(3–4):398–426

Proposal of a Guide to Select Methods of Ergonomic Assessment in the Manufacturing Industry in México

López Millán Francisco Octavio(✉),
De la Vega Bustillos Enrique Javier, Arellano Tanori Oscar Vidal,
and Meza Partida Gerardo

TecNM/Instituto Tecnológico de Hermosillo, 83170 Hermosillo, Sonora, Mexico
lopezoctavio@yahoo.com.mx

Abstract. This paper presents the particular situation on the manufacturing industry in Mexico, a sector of the economy that provides important figures for the economy and that generates an important number of jobs, and the relevance of ergonomics in the health care of workers, in particular the Work Muscle Skeletal Disorders (WMSD). In recent years there has been an increase in the number these cases. In addition, the new legal regulations that require companies to identify and evaluate Ergonomics Risk Factors (ERF) have taken effect and it is needed take actions to minimize the risk of exposure to a WMSD. The paper presents a review of the ERF and the WMSD described in the literature as well as the proposal of a methodological guide to select the Ergonomic Assessment Methods (EAM) with greater application for the manufacturing industry, distinguishing two main activities; repetitive work and manual handling of loads. For each of these, the EAMs with the simplest application in two levels are suggested; the recommended EAM and at a second level the alternative EAMs to complement the evaluation. The guide is presented graphically and a brief description of each method and its scientific reference is provided.

Keywords: Work Muscle Skeletal Disorders · Ergonomics Risk Factors
Ergonomic Assessment Methods

1 Introduction

Work has been a means for society to access certain levels of well-being. The industry and in particular the manufacturing industry in Mexico are of great importance for the economy of the country, it is estimated that 1.5 of the gross domestic product (GDP) is generated due to this industry [1]. In this sector of the economy, the automotive industry (AI) is of great importance, according to the Mexican Association of the Automotive Industry [2], the country is the seventh producer of automobiles worldwide and generates more than seven hundred thousand jobs; At an international level, according to the International Organization of Automobile Manufacturers [3]. In addition to the AI, there are other types of industries that make an important contribution to the economy. In the Sector Diagnosis [4], the country is highlighted as the

S. Bagnara et al. (Eds.): IEA 2018, AISC 825, pp. 301–309, 2019.
https://doi.org/10.1007/978-3-319-96068-5_33

main exporter of flat screen televisions and the fourth exporter of computers and microphones and loudspeakers, just to name a few.

Continuous improvement and make more profitable production systems has increasing production rates associated with the needs of optimize the process, even human resources, could lead to the intensification of work, in several surveys it is perceived and reported by workers, in the United States, a study on working conditions [5], 40% of respondents answered that at least half of the work schedule, they performed activities with demand for an accelerated work rhythm and with the pressure of time to complete their activity. In Spain, in a similar study [6], it was found that 33% of workers must work always or almost always at high speed and 35% must meet tight deadlines with the same frequency. In Europe, the percentage in general reported by the workers coincides [7], and they mention as the index of labor intensification the combination, among others, of the following indicators; working at speed and to tight deadlines and not having enough time to do the job. The demand for production of goods and the demands of the speed up of work and the deadlines to comply with the work may be related to the presence of Work-Related Musculoskeletal Disorders (WMDs).

It is important to mention that in Mexico, the federal authority on occupational health makes it mandatory to identify the worker's exposure to WMSD risks and to carry out control actions to mitigate that risk.

2 Work-Related Musculoskeletal Disorders (WMSD)

According to the information of the International Labor Organization [8], work-related musculoskeletal disorders (WMSD) continue being a global concern, mentioning that the cost of work-related diseases in the European Union it is estimated at least on 150 billion euros per year, in France, the government estimates a cost between 1.3 and 1.9 billion euros per year, in Korea, the WMSD represent .7% of GDP. In England, the Health and Safety Executive [9], estimates 8.9 million lost days at work due to WMSD. In Mexico, according to data from the Mexican Institute of Social Security [10], an institution that registers and attends to health at work, reports an annual increase in WMSD greater than 20%, in 2016, 4,607 WMSD were reported. As a consequence of industrial economic activity, it can be assumed that the number of cases diagnosed as WMSD has gradually increased. It is estimated that the loss of work days caused by the WMSD is approximately 120 days, considering the reported cases can be estimated than an amount greater than 550 thousand days lost per year. The results show in general the women most susceptible to suffer some MSDS, being more evident the difference in Carpal Tunnel Syndrome, De Quervain's Tenosynovitis, while in men the Dorsopathies and Arthrosis are more recurrent. In the affectation by gender, WMSD on distal upper extremities affects more to women while in men, the affection is in the lower back. WMSD are considered occupational diseases, and they occur when the worker is exposed to certain occupational risk factors.

3 The Risk Factors Associated with the MSDs

Risk factors associated with MSDs, are also known as ergonomic risk factors (ERF), although some inaccuracy in the term can be highlighted, it is common that this expression is used, they are varied and the studies mention the combination of these as the most likely cause of a MSDs.

Simoneau et al. [11], describe as modulators ERF; the intensity effort, frequency and repetition of awkward postures and, time of the sustained efforts when a work activity is developed, as risk enhancers to develop a MSDs. The following are recognized as some of the ERF at work related MSDs:

- The forced postures;
- Effort and strength;
- The repetition and invariability of the work;
- Static muscle work;
- Exposure to certain physical aggressors such as vibration;
- Organizational factors such as the pace of work and recovery times.

Additionally, they can be considered as ERF unrelated to work;

- Age;
- Unpaid work;
- Extra work regularly in informality;
- The propensity to degenerative diseases;
- Recreational activities such as sports.

The repetitive work is still present as an activity of workers today, it could be assumed that in XXI Century, working conditions should have improved by attenuating the ERF in the design of work, however data show another reality; in countries with more advanced economies, repetitive work prevails and occupies a preponderant place in the perception of workers; Maestas et al. [5], presents the results of the survey on working conditions in America 2015, and highlights the opinion of 74.8% of respondents where repetitive work is mentioned, Pinillas et al. [6], with data from 2015, found that 73% of workers in industries mentioned doing repetitive work. 63% of the workers who answered the European survey on working conditions, on Parent-Thirion [7], is mentioned that doing repetitive work is at least 25% of the working day.

Static muscle work is also considered a risk factor, it is described as a dangerous posture held for a certain time, usually in seconds, and it limits the flow of blood to the shoulder muscle group accelerating the presence of local muscle fatigue. In Kodak [12], it is recommended to perform ergonomic research when a sustained effort in the shoulder exceeds 5 s in duration.

Simoneau et al. [11] mention vibration and contacts with surfaces or objects as potential ERF, Bernard [13], also refers to vibration as a risk factor to WMSD in the lower back.

Organizational factors such as the relationship between workload and recovery time or factors such as the hurry to comply with a work objective or the work schedule or rhythm, recently have appeared as potential ERF.

Few evidence was found on the WMSD relationship with the permanence of a work activity, in the document of the Health Council of the Netherlands [14], it is mentioned that Carpal Tunnel Syndrome is present in workers with more than 20 years of seniority in a job, it also refers to the effects of lateral epicondylitis in women when they perform the same activity more than 75% of the working day.

Repetitive work is a basic activity of work and a risk factor in combination with other ERF, the other activity of equal importance in the analysis of work is the manual handling of loads (MMH) and the ERF associated with greater recurrence are the awkward postures and the efforts or application of force to mobilize some type of load causing potential affectations in the lower back.

In Colombini et al. [15], there are distinguish three types of sub-activities in the MMH:

- Lifting or lowering objects weighing more than 3 kg.
- Carrying a load and
- Push or pull a load.

In the survey on working conditions in America 2015, Maestas et al. [5], 45.1% of the respondents mention that they carry heavy loads as part of their work. In Pinillas et al. [6], they found that 44% of workers in industries mentioned carrying or moving heavy loads. Meanwhile, 34% of workers who responded to the European survey on working conditions, Parent-Thirion [7], mentioned moving or moving loads at least 25% of the working day. Even when mechanized technology is available for cargo handling, it is still a relevant percentage of workers who perform this activity.

4 The Process of Identification and Evaluation of the ERF

The intervention process of occupational ergonomics includes identification and assessment of the risk of a TME due to work, usually the evaluation is done through some ergonomic assessment method (EAM). A wide variety of ergonomic assessment methods are available, however, there is no method that includes all the risk factors according to the main work activities.

In the previous paragraphs, the definitions were mentioned and the importance of the main work activities in the industry was highlighted; repetitive work and manual handling of loads. The prevalence of MSD related to these work activities is highlighted.

Identification of the ERF, and the measures, control and mitigation of the ergonomic risk must be part of the process of occupational ergonomics, that is to say, it does not have an end in function of time, it must be constant in the workplace.

Some considerations should be made prior to the ergonomic process; It is advisable to start from the identification of the ERF in the workstations, this entails the knowledge of the ERF. It is useful to develop a checklist that includes all the potential ERF, it is also pertinent to consider the information coming from the medical services regarding the consultations related to the symptoms of the MSDs and to follow up the cases that have been diagnosed with. It is advisable to make notes and record the entire initial state of the ergonomic process on an electronic sheet, highlighting the jobs that are considered for the evaluation process with ergonomic methods.

It is necessary to ensure, at the time of performing the ergonomic assessment, that the incorrect work practices should not be evaluated, as well as making sure that the approved work method is followed. It is just as important to make sure to evaluate the properly trained worker to perform the work activity.

It is pertinent to have the most information available, for example;

- The time and motion studies;
- Production standards;
- The process description sheets;
- The weight of materials, tools, and containers;
- Dimensions of the workplace, height of work surfaces, location and dimensions of containers and tools;
- Distances traveled, when carrying loads;
- It is good to have available video recordings of the workstations being evaluated, preferably several work cycles;
- Length of the working day and scheduled breaks;
- If there is a job rotation plan, it is necessary to know it;
- It is important to make observations about other aspects, for example, if there is supports on the wrists, use of bands, ribbons on the fingers, gestures after doing the work activity;
- It is important to give notice to those involved in the management of the production process and, where appropriate, to the representatives of workers union.

4.1 Ergonomics Assessment Methods (EAM)

Refer to the risk without making an evaluation is speculation. The practice of ergonomics changed favorably when the EAM was used. A widely variety of ergonomic assessment methods are available: In Stanton et al. [16], Marras and Karwowski [17], Malchaire [18], Stack et al. [19], among others, information can be obtained about the different EAM. David [20] and Garg and Kapellush [21], present comparisons among some EAM. The EAM of greatest use is distinguished between ergonomists and depending on various working conditions. The use of the EAM is recommended to be carried out by properly trained personnel.

5 Proposal of a Guide to Select of EAM

For the focus of this work, the proposal EAM depends on it use in general, first at all is the assessment of repetitive tasks, then the assessments of manual materials handling.

The offer of methods is extended and can be considered as tools, according to each situation an EAM is used. The guide proposal is considering the two major activities; repetitive work and manual handling of loads. It is considered repetitive work when the work activity cycles are short, from 4 to 30 s per cycle, the pattern of movements in the muscle groups is repeated and more than 4 h are worked per day in the same activity. If the shift exceeds 8 h of work, a special consideration must be made. For the MMH, the main type of activity to select the EAM would be considered.

For repetitive work, it is proposed to use the ART Tool [22] as a basic tool, including in the evaluation almost all the risk factors and muscle groups of the upper extremities, including the back. It does not evaluate the lower extremities however, in the data on the WMSD, the number of failed cases in that segment of the body is not significant. The evaluation time is shorter than for the other methods and is relatively simple to understand and apply it works in the field and allows to observe more details. It offers the possibility of including the extended shift greater than 8 h, a condition that is increasingly present in the manufacturing industry. If it is necessary to evaluate the duration of the sustained efforts and if it is required only for the elbow segment at hand the Strain Index [21] can be considered a second method, but if it is necessary to assess the sustained effort on the shoulder or back, a second Method can be the Muscular Fatigue Analysis [12, 16], it also has the advantage of driving in a more concrete way the search for solutions for high risk valuations of WMSD. Strain Index and HAL do not consider the type of grip on the fingers, a situation that is included in ART Tool and MFA. If the need is the postural assessment, RULA [23] and REBA [24] are viable alternatives.

For the MMH the proposal of EAM is more concrete; to evaluate lifting, lowering or transporting loads the MAC Tool [25], has some advantages, it is not limited to a symmetrical grip on the hands, it is easier to use, involves fewer calculations and includes more variables. It is of greater scope when considering two types of activity and with more options. The NIOSH lifting equation [26], is the most used tool on lifting or lowering tasks, but it is restricted by the symmetrical position on the hands. The Liberty Mutual Tables [27], also could be used to assess lifting or lowering a load, but just show the percent of population capable to safety do the job and it is restricted to some data on the basic variables of the activity.

For the activities of pushing the suggestion is to use the RAPP Tool [28], it does not have a reference for the level of intervention, however, it consider the weight of the equipment used to be pushed or pulled and the weight of the load, and depending on it, there are the limits on weight for a safety movement, it also include the other factors influencing in pushing or pulling tasks. The RAPP tool also considers push or pull loads without the use of equipment. The Liberty Mutual Tables [27], also could be used to assess pushing or pulling a load, but just show the percent of population capable to safety do the job and it is also restricted to some data on the basic variables of the activity.

The following figure shows the guide to use the EAM (Fig. 1).

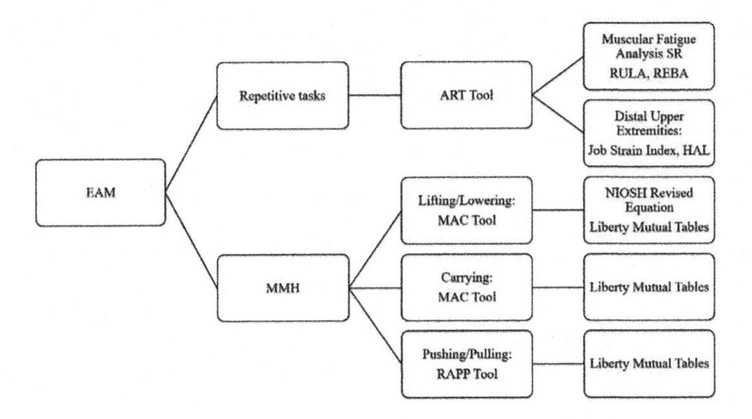

Fig. 1. Proposal Guide to EAM. Source; Own elaboration, 2018

6 Conclusion

The WMSD is still a health problem at work, the risk factors are present in the work activity and it is necessary to measure the risk in order to attend to it, assign a value or show the need to act to reduce or avoid the risk, fundamental for the advancement of occupational ergonomics. The AEM are an excellent tool to quantify the risk and a large number of options are already available, which is why this proposed selection guide for EAM according to the basic work activity can be very helpful.

Companies continue move forward and governments have the obligation to establish the relevant legal frameworks, but they can also generate guidelines for EAM that make it easy the inspection and prevention of occupational diseases, this guide can be an auxiliary in the process. For the ergonomist, it can shorten the time for the assessment and thus have more time to search for improvements and solutions in working conditions.

References

1. Instituto Nacional de Geografía, Informática y Estadística (INEGI) (2017) Estadísticas a propósito de la Industria Automotriz. http://www.amia.com.mx. Accessed 22 Oct 2017
2. Asociación Mexicana de la Industria Automotriz (AMIA). http://www.amia.com.mx. Accessed 26 Nov 2017
3. The International Organization of Motor Vehicle Manufacturers (OICA) (2017). http://www.oica.net. Accessed 26 Nov 2017
4. Promexico (2017) Diagnostico sectorial, industria electrónica. http://www.promexico.gob.mx/documentos/diagnosticos-ectoriales/electronico.pdf. Accessed 18 Jan 2018
5. Maestas N, Powell D, Wenger J (2017) Working conditions in the United States, Results of the 2015 American Working Conditions Survey. RAND Corporation. http://www.rand.org/t/RR2014. Accessed 08 Feb 2018

6. Pinila GJ, Almodovar MA, Galiana BM, Hervas RP, Zimmerman VM (2015) Encuesta Nacional de Condiciones de Trabajo. 2015 6ª EWCS – España. Instituto Nacional de Seguridad e Higiene en el Trabajo (INSHT). http://www.insht.es. Accessed 12 Feb 2018
7. Parent-Thirion A, Biletta I, Cabrita J, Vargas O, Vermeylen G, Wilczynska A, Wilkwns M (2017) Eurofound, sixth european working conditions survey – overview report (2017 update). Publications Office of the European Union, Luxembourg
8. International Labor Organization, The prevention of occupational diseases. http://www.ilo.org/publns. Accessed 13 Nov 2017
9. Health and Safety Executive, Health and safety statistics, http://www.hse.gov.uk/statistics/index.htm. Accessed 19 Nov 2017
10. Instituto Mexicano del Seguro Social (IMSS) (2016) Memoria estadística. http://www.imss.gob.mx/conoce-al-imss/memoria-estadistica-2016. Accessed 05 Nov 2017
11. Simoneau S, ST-Vincent M, Chicoine D. Les LATR- Mieux les comprendre pour mieux les prévenir.Institut de recherche Robert-Sauvé en santé et en sécurité du travail. http://www.irsst.qc.ca/media/documents/PubIRSST/RG-779.pdf. Accessed 11 Dec 2017
12. The Eastman Kodak Company (2004) Ergonomic design for people at work, 2nd edn. Wiley, Hoboken
13. Bernard B (1997) Musculoskeletal disorders and workplace factors; a critical review of epidemiologic evidence for work-related musculoskeletal disorders of the neck, upper extremity, and low back. Centers for Disease Control and Prevention, National Institute for Occupational Safety and Health (NIOSH)
14. Health Council of the Netherlands. Repetitive movements at work: Risk to health. The Hague: Health Council of the Netherlands. Publication no. 2013/05E. Health Council of the Netherlands. http://www.healthcouncil.nl. Accessed 11 Dec 2017
15. Colombini D, Occhipinti E, Alvarez-Casado E, Waters T (2013) MANUAL LIFTING: a guide to study of simple and complex lifting tasks. CRC Press, Boca Raton
16. Stanton NA, Hedge A, Brookhuis K, Salas E, Hendrick HW (2006) Handbook of human factors and ergonomics methods. CRC Press, Boca Raton (2006)
17. Marras WS, Karwowski W (2006) Fundamentals and assessment tools. CRC Press, Boca Raton
18. Malchaire J, Cock N, Vergratch S (2011) Review of factors associated with musculoskeletal problems in epidemiologic studies. Int Arch Occup Environ Health 74:79–90
19. Stack T, Ostrom LT, Wilhelmsen CA (2016) Occupational ergonomics: a practical approach. Wiley, Hoboken
20. David GC (2005) Ergonomic methods for assessing exposure to risk factors for work-related musculoskeletal disorders. Occup Med 55:190–199
21. Garg A, Kapellush JM. Job analysis techniques for distal upper extremity disorders. Rev Hum Factors Ergon. http://rev.sagepub.com/content/7/1/149. Accessed 14 Dec 2017
22. Health and Safety Executive. Assessment of repetitive tasks (ART Tool). http://www.hse.gov.uk/pubns/indg438.htm. Accessed 05 Oct 2017
23. McAtamney L, Corlet EN (1973) RULA: a survey method for the investigation of work-related upper limb disorders. Appl Ergon 24(2):91–99
24. Hignett S, McAtamney I (2000) Rapid entire body assessment (REBA). Appl Ergon 31:201–205
25. Health and Safety Executive. Manual handling assessment charts (MAC Tool). http://www.hse.gov.uk/pubns/indg383.htm. Accessed 08 Oct 2017

26. Waters TR, Putz-Anderson V, Garg A. NIOSH Applications Manual for the Revised NIOSH Lifting Equation. https://www.cdc.gov/niosh/docs/94-110/pdfs/94-110.pdf. Accessed 10 Oct 2017
27. Liberty Mutual Tables. Manual material handling tables. https://libertymmhtables.libertymutual.com/CM_LMTablesWeb. Accessed 10 Dec 2017
28. Health and Safety Executive. Rapid assessment for pushing and pulling (RAPP Tool). http://www.hse.gov.uk/pubns/indg478.htm. Accessed 11 Dec 2017

Analysis of Physical Workloads and Muscular Strain in Lower Extremities During Walking "Sideways" and "Mixed" Walking in Different Directions in Simulated U-Shape in the Lab

Jurij Wakula(✉), Stefan Bauer, Sören Spindler, and Ralph Bruder

Institute of Ergonomics and Human Factors, Technical University Darmstadt, Darmstadt, Germany
wakula@iad.tu-darmstadt.de

Abstract. The muscular strain at the lower extremities was analysed in the IAD-lab using the simulated U-shape with short-cycle tasks (approx. 80 s.) with walking "sideways" and "mixed" walking (sidesteps and normal steps). Also focus was on analysis of the effects of "walking sideways counter clockwise" vs. "turn clockwise sideways" on the muscular strain in the three selected muscles in the right and the left leg. Four different scenarios were tested. The U-shape consisted of five work stations, was 2 m long and 1.4 m wide in scenarios walking with "sidesteps" (A, B) only. In scenarios with "mixed" walking the assembly U-shape was about 3,2 m long and 1.4 m wide. The EA-activities in selected three leg muscles in the left and right legs were analysed using surface EMG-method. Six test subjects, between 19 and 30 years old, without experience in assembly work took part in the study. The results complement the study Wakula et al. (2017a,b) and show that walking "sideways" counter-clockwise (CC) cause the selected right leg muscles more strain compared to the left leg muscle by some test persons. When walking clockwise (C) two muscles in the left leg were more stressed compared to the right leg muscles. Changing the direction of moving at the U-line: CC → C → CC → C is positive for the muscular strains - it brings some balance of the EA values in analyzed right and left leg muscles. Walking with "mixed" (lateral and two-three normal) steps in the analyzed U-shape did not reduce muscular strains in the legs compared to walking with "sidesteps" only.

Keywords: U-shape · Walking "sideways" · "Mixed" walking
Muscular strain

1 Introduction

For customer-specific reasons, more and more products with many parts and product variants are being assembled today, which requires a consumption-related provision of numerous different individual parts and tools. Here, the use of classic single assembly spaces reaches its limits. Many individual parts and tools make the workplace confusing and a division into individual workplaces increases the transport, the waiting time and the material buffers. Nevertheless, to assemble the product with the highest

S. Bagnara et al. (Eds.): IEA 2018, AISC 825, pp. 310–322, 2019.
https://doi.org/10.1007/978-3-319-96068-5_34

possible quality, at low cost and in the shortest possible time, the flow principle of "One-Piece-Flow" is used.

This principle has its origins in the "Toyota-production system" and is based on the just-in-time principle of the "lean production" approach. The advantage is that the multi-stage machining of a part takes place without intermediate storage at different workstations. In this way, despite a high variety of variants, the flexibility of the output quantity can be ensured. To avoid long distances between the start and end stations, the workstations are typically arranged in a U-layout (see Fig. 1). The entirety of the workstations in this arrangement is also called "U-shape". Two types of U-shapes are distinguished: the "caravan" system and the "hand-transfer" system. The advantage of the "caravan" system is that it allows several workers to work without disruption at higher volumes. The employees go through the entire assembly process one after the other in each cycle. In contrast, the manual transfer system provides a clear delimitation of the individual work areas, whereby the transfer points result according to the required quantity.

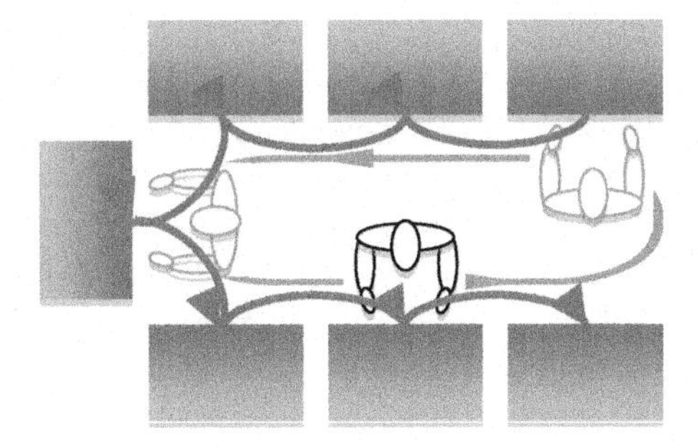

Fig. 1. U-shape with seven workstations and operating mode – "caravan" system.

The prerequisite for the use of such systems is, on the one hand, the comprehensive qualification of the employees, who are able to carry out all tasks at the work stations, but on the other hand quickly convertable workstations or tools in order to achieve the flexibility (Lotter and Wiendahl 2006). Finally, there are opportunities to expand the job (job enlargement) or to a rotating employment (job rotation) of the employees, as well as to the formation of autonomous groups, which leads to an improvement of the general sense of working, in contrast to monotonous tasks at individual stations (Günther et al. 2005).

During U-shape activities, both upper limb and shoulder area stresses and strains occur through short movement cycles with highly dynamic hand-arm movements, as well as in the lower extremities by standing and walking constantly between work-stations. In addition, "lateral walking" (walking with sidesteps) and "mixed walking" are common. Based on the findings of previous studies on load and stress analysis of selected leg muscles in "normal" and "lateral" walking (Wakula et al. 2016) and "lateral" walking with different directions of movement - going counter-clockwise (CC) and going in a clockwise direction (C) (Wakula et al. 2017a,b), a laboratory study was designed and conducted at the Institute of Ergonomics of the TU Darmstadt

(IAD) within the framework of the "U-Line Assembly Systems" project funded by BGHM (Berufsgenossenschaft Holz und Metall) and BGETEM (Berufsgenossenschaft Energie Textil Elektro Medienerzeugnisse).

1.1 Objectives of the Laboratory Study

As far as we are aware, in industrial practice during the design of the U-Shape a product value creation is usually carried out with a circulation direction in the counterclockwise (CC) direction. This could possibly lead to unilateral, asymmetric stresses on the upper and lower extremities at work. In this context, one focus of this study was to analyze the effect of "lateral walking" (with "sidesteps") in different directions of rotation in the U-shape - counterclockwise (CC) and clockwise (C) - on the muscular strain on the left and right legs. The direction of rotation for the movement of each subject in the U-shape was changed after approx. 10 min. The workstations of the U-shape with cycle times of approx. 80 s were initially run counterclockwise (CC), then clockwise (C) and then again CC. As a result, on the one hand, conclusions can be drawn about the stress caused by the respective running direction and, on the other hand, the change of direction.

In tight U-shape in practice, where several people work at the same time, it is often possible to change the workstations only by lateral (side-) steps. For the lines that are sufficiently well-spaced, the workers can move between the workspaces/stations with both normal and lateral/side steps. Therefore, in scenarios C and D in our study, we focused on the effect of the type of walking at the station change in U-shape - pure "lateral walking" with sidesteps versus "mixed walking" (sidesteps and 2–3 normal steps) - on the muscular strain of the legs.

2 Methodology and Experimental Setup

2.1 Setup of the Simulated U-Shape, Scenarios, Measurement Procedure

The simulated U-Shape consisted of five workstations. For scenarios A and B the distances between the stations were arranged so small (from 0.15 m to 0.4 m) that test person could switch between the tables with the approx. 10 sidesteps ("lateral walking") (Fig. 2a).

The five workstations are equipped for assembling with nuts and bolts and small metal details. The subject has a working time of 15 s per station. Meanwhile, subject should perform simple assembly operations with the provided material. The following four scenarios were analyzed in the study:

- A. "Walking sideways" in simulated U-shape with several steps counterclockwise-CC (Fig. 2a)
- B. " Walking sideways " with several steps in a clockwise (C) direction (Fig. 2a);
- C. "Mixed walking" counterclockwise (Fig. 2b)
- D. "Mixed walking" clockwise (Fig. 2b)

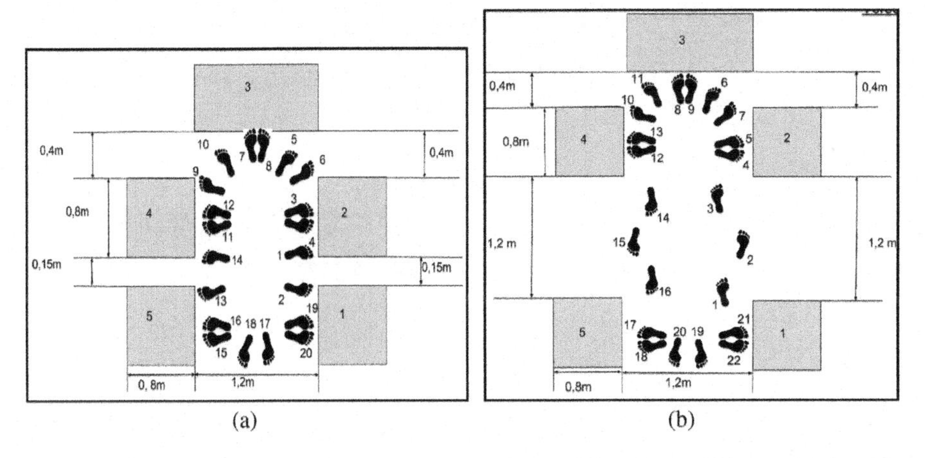

Fig. 2. a: Simulated U-shape with five workstations for "walking sideways" (scenarios A and B), b: Simulated U-shape with five workstations for "mixed walking" (scenarios C and D)

In scenarios C and D stations 1 and 5 were about 1.2 m apart (Fig. 2b). The subject must walk from station 1 to 2 and from 4 to 5 by two to three "normal" steps and the remaining stations by 7–8 "side steps" (mixed walking). Standing and assembly work at each workstation was about 15 s. Meanwhile, the test person should use the provided material at each station to perform simple assembly work with both hands. The total cycle time was about 80 s.

The trial included six series of measurements (SM) with whole duration more than 2 h including recovery breaks. Each subject first had the opportunity to familiarize themselves with building the line and getting to know the motion strategy in a few trial cycles. Afterwards, each subject has eight working cycles (about 12 min). Counter-clockwise (CC, scenario A, MS1), then 8 cycles clockwise (C; scenario B, SM2) and then another 8 cycles (about 12 min) is moved counterclockwise (CC, scenario A, SM3) with "lateral" steps. After each series of measurements, there was a five-minute break in sitting position for muscle recovery. The sequence of walking in scenarios C and D was maintained and the rest duration between each measurement series remained unchanged.

3 Methods

3.1 Surface Electromyography (OEMG)

Objective physiological strain data were measured and analyzed on selected leg muscles (left and right side of the body) using surface electromyography (OEMG, AWMF 2013). A portable TeleMyo 2400 G2 device from Noraxon company was used for the EMG recording (Wakula et al. 2016). The myoelectric signals were measured

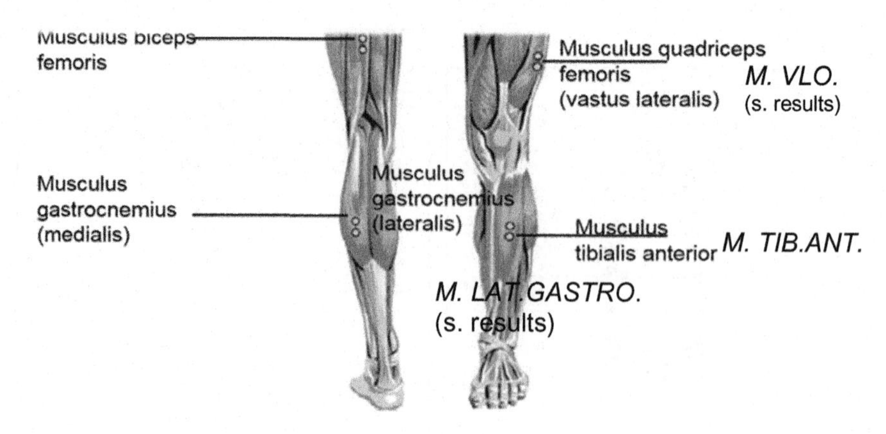

Fig. 3. Places for electrodes for measuring electrical activity (EA) in 3 leg muscles (EMG Fibel, Konrad (2005)

and analyzed by surface electrodes in three muscles of the left and right legs (Fig. 3):

- m. quadriceps femoris (vastus lateralis VLO)
- m. gastrocnemius lateralis and
- m. gastrocnemius medialis

The selection of these muscles based on the previous studies performed at IAD in Darmstadt (Wakula et al. 2016, 2017a).

Before and after the measurements, resting activities as well as MVC (Maximal Voluntary Contraction) measurements of the respective muscle in the legs were recorded in each test person. For resting measurement, the subject was in a lying position, while the leg muscles relaxed. The MVC measurements were carried out on the basis of the literature (AWMF online 2013). Six male young test persons without experience in assembly activities participated in the measurements. Data on the test persons are included in Table 1.

Table 1. Data related to the test persons, participated in the lab study

Test person	Age	Weight [kg]	Height [cm]	BMI-index
Person 1	29	72	1,85	21,0
Person 2	19	60	1,80	18,5
Person 3	19	85	1,87	24,3
Person 4	30	87	1,85	25,4
Person 5	29	85	1,87	24,3
Person 6	27	83	1,87	23,7

4 Results

The muscular strain data recorded using the Noraxon system was processed using the Myo Research XP Master Edition software version 1.08. The values of all test persons were transferred to Excel file and the last six working cycles from measured eight cycles were evaluated. Based on these results, diagrams were drawn up, in which the static EA fractions, dynamic EA-value and mean EA values for each measurement series are shown (eg. Wakula et al. 2016). The strain in the individual muscles was first analyzed separately in each test person. After this step an analysis of the results of each muscle with respect to all analyzed test persons was performed.

4.1 EMG Results Regarding the Muscle m. quadriceps femoris (vastus lateralis - VLO)

Figures 4a and b show the strain results (EA-mean, EA-dynamic and EA-static) related to the muscle m. quadriceps femoris (vastus lateralis) in the left leg (LT VLO-Fig. 4a) and in right leg (RT VLO, Fig. 4b) related to the person 6 as example, averaged over 6 cycles. The person was walking “sideways” about 30 min in following directions – 10 min counterclockwise (CC1) in the first series of measurement (SM1), then 10 min - clockwise (C) after 5 min resting and after that 10 min counterclockwise (CC2) again.

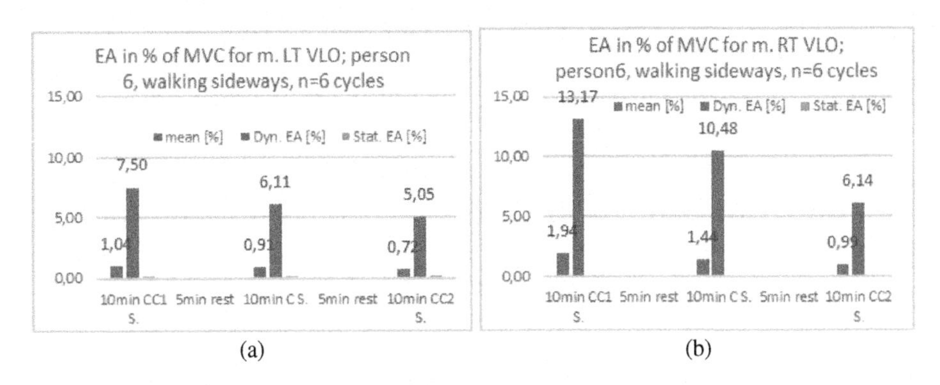

Fig. 4. a, b: EA in [%] of MVC of m. quadriceps femoris (vastus lateralis -VLO) in test person 6 during walking “sideways” in two directions counterclockwise-CC and clockwise-C

In the first series of measurements, during movement of subject in direction CC1 (scenario A), the dynamic EA-value and EA mean value are approximately twice higher in the right muscle as in the left one (Fig. 4a, b). In the next series, after 5 min resting, the subject was moved according scenario B in a clockwise direction (C), the muscular strain decreases slightly in the right muscle and remains at the same EA-level in the left one. In the last series (SM3), during 10 min walking “sideways” counter-clockwise (CC2, scenario A) the tendency of previous series is repeated. The static EA-values are very low in all three series.

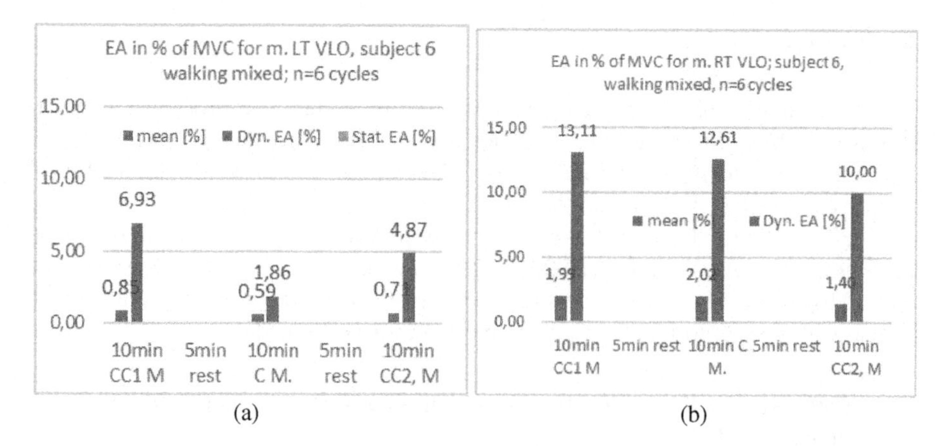

(a) (b)

Fig. 5. a, b: EA in [%] of MVC of m. quadriceps femoris (vastus lateralis-VLO) of subject 6 during "mixed" walking in two directions - counterclockwise-CC1, CC2 and clockwise-C)

The results related to similar three series of measurements during subject 6 performed "mixed" walking in different directions - counterclockwise-CC1 (scenario C) than clockwise – C (scenario D) and after that - counterclockwise-CC2 are presented in the Figs. 5a and b. These results show the same trends for the left and right leg as during the measurements in scenarios A and B by walking "sideways": The dynamic EA-values are in the left muscle significantly lower than in the right one. The static EA-values are very low also in all three series.

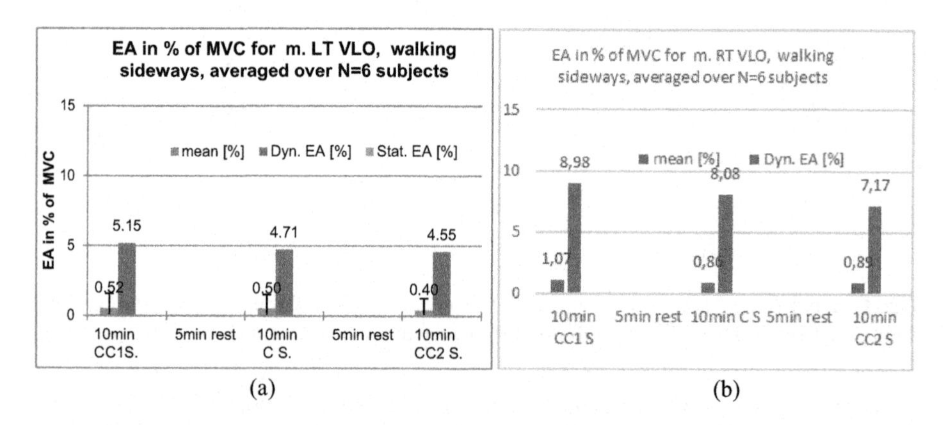

(a) (b)

Fig. 6. a, b: EA in [%] of MVC of m. quadriceps femoris (vastus lateralis) averaged over all subjects walking "sideways" in two directions CC1 → C → CC2 (LT VLO left, RT VLO right leg)

When comparing the two strategies of movement – walking "sideways" versus "mixed" walking - it is noticeable that the EA-values differ only slightly in the right leg. Another tendency is in the left muscle.

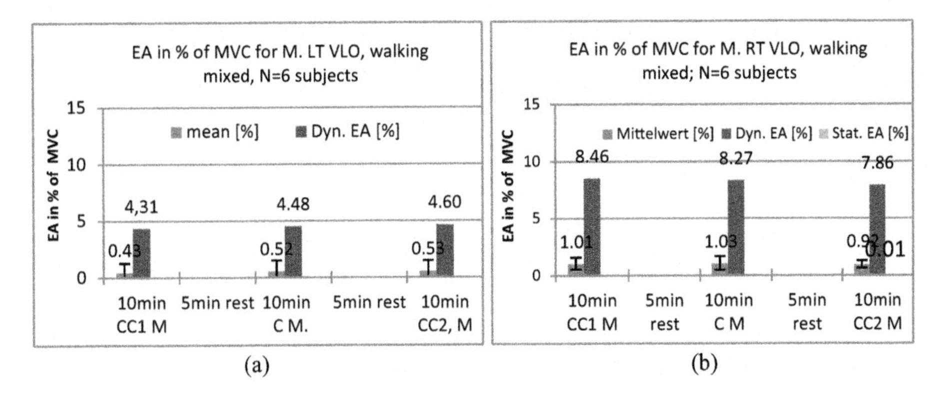

Fig. 7. a, b: EA in [%] of MVC of m. quadriceps femoris (vastus lateralis) averaged over all subjects during "mixed" walking in directions: CC1-C-CC2 (LT VLO left and RT VLO right leg)

In the next phase the EMG-results related to the muscle in left and right leg averaged over six test persons participated in the study were evaluated.

The results for the muscle m. quadriceps femoris (vastus lateralis) averaged over all test persons show a similar picture of muscular strain as in subject 6 – see Fig. 6a, b for walking "sideways" and Fig. 7a and b for "mixed walking". The dynamic EA-value as well as EA-mean have higher level in the right muscle as in the left one.

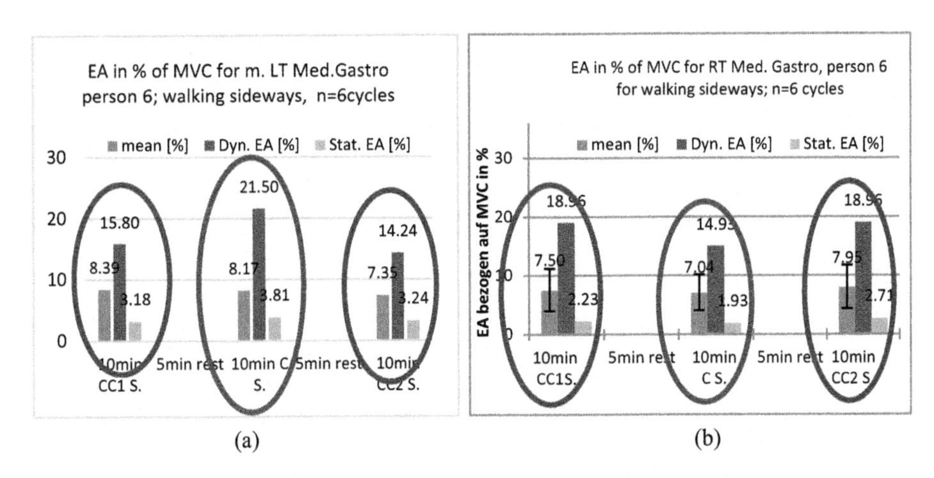

Fig. 8. a, b: EA in [%] of MVC for m. gastrocnemius medialis of subject 6 during "sideways" walking in different directions CC1 → C → CC2 (counterclockwise vs. clockwise)

If we compare two movements then the results of muscular strain (dynamic EA-value and EA-mean as well) show slightly higher EA-activities during walking "sideways" (Scenarios A and B) than those of "mixed" walking. It is noticeable that the EA-values are significantly higher in this muscle in the right leg comparing to left one.

4.2 EMG Results Regarding the Muscle *m. gastrocnemius medialis*

Now follows the analysis of the stress of the second muscle *m. gastrocnemius medialis*. Figures 8a and b present graphs of EA in [%] of MVC for the medial gastrocnemius muscle in "walking sideways" of subject 6.

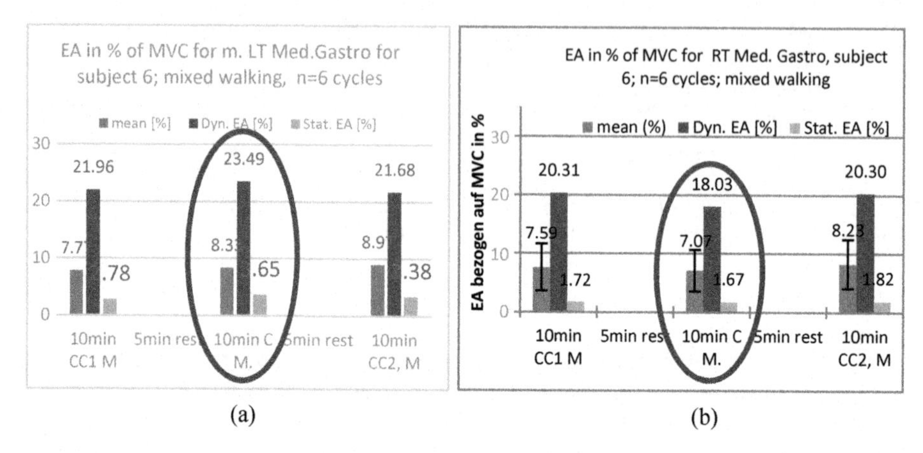

Fig. 9. a, b: EA in [%] of MVC for m. gastrocnemius medialis of test person 6 during "mixed" walking in different directions CC1 → C → CC2 (counterclockwise vs. clockwise)

Tendencies to unilateral asymmetric muscular stresses can be seen in this muscle and test person as well as the constant lateral walking in one direction. When looking at the diagrams, it is noticeable that when walking counterclockwise (CC, measurement series 1 and 3), the muscle of the right leg of subject 6 is subjected to somewhat more stress than this muscle in the left leg. Conversely, when walking in a clockwise direction (C, measurement series 2), this muscle shows significantly higher stress values in the left leg than in the right one (difference about 6%). In "mixed walking" in two directions of movement - CC and C-, overall slightly higher muscular demands were noted than when walking sideways (Fig. 9a, b).

In these scenarios, higher strain in % of the MVC was measured on the left (Fig. 9a) in the second series of measurements (walking in a clockwise direction, Fig. 9b). The standard deviations from the EA-value were approximately 4% of the MVC.

Overall, the EA values in the muscle m. gastrocnemius medialis shows higher strain levels than the first analyzed muscle. These tendencies can also be seen in the consideration of EA activities in muscle m. gastrocnemius medialis averaged over the six analyzed subjects (see Figs. 10a, and b).

Figure 11a and b present the muscular strain in m. gastrocnemius medialis over all subjects during "mixed" walking in different directions CC1 → C → CC2 (counterclockwise vs. clockwise)

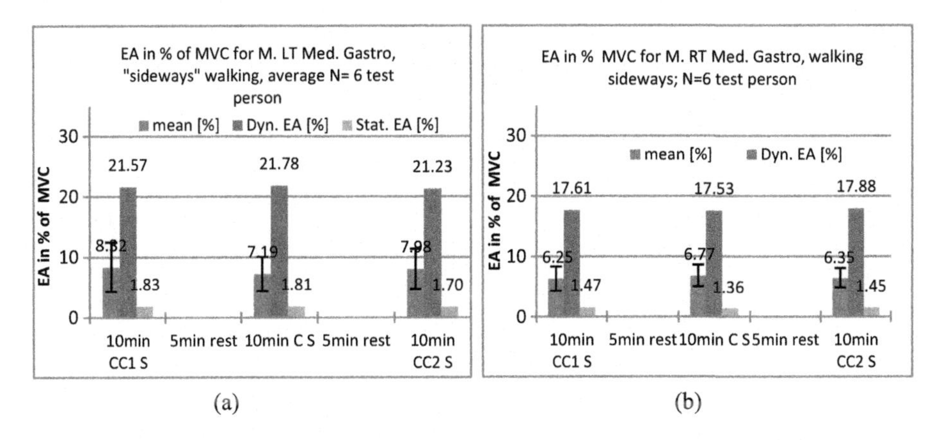

Fig. 10. a, b: EA in [%] of MVC for m. gastrocnemius medialis average above all test persons during "sideways" walking in directions CC1 → C → CC2 (counterclockwise vs. clockwise)

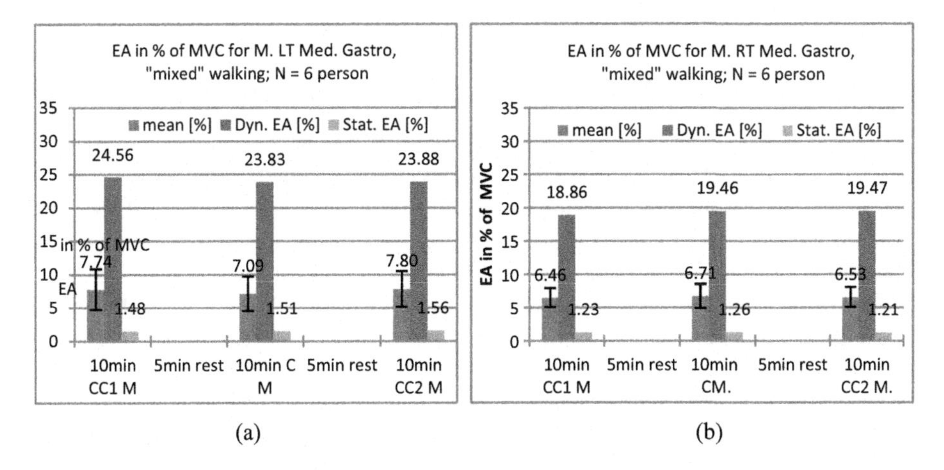

Fig. 11. a, b: EA in [%] of MVC for m. gastrocnemius medialis over all subject during "mixed" walking in different directions CC1 → C → CC2 (counterclockwise vs. clockwise)

4.3 EMG Results Regarding the Muscle *m. gastrocnemius lateralis*

Quite different results have been achieved to the third analyzed muscle - m. gastrocnemius lateralis (LAT. Gastro).

The strain results for this muscle averaged over all 6 subjects walking sideways are shown in Fig. 12a and b. Dynamic EA portion and also EA averages are slightly higher in the left muscle than in the right one.

In this muscle, as in the first muscle in the second series of measurements, the clockwise movement (middle block) also showed an increase in electrical activity in the left muscle and EA depression in the right muscle (see Figs. 12a and b).

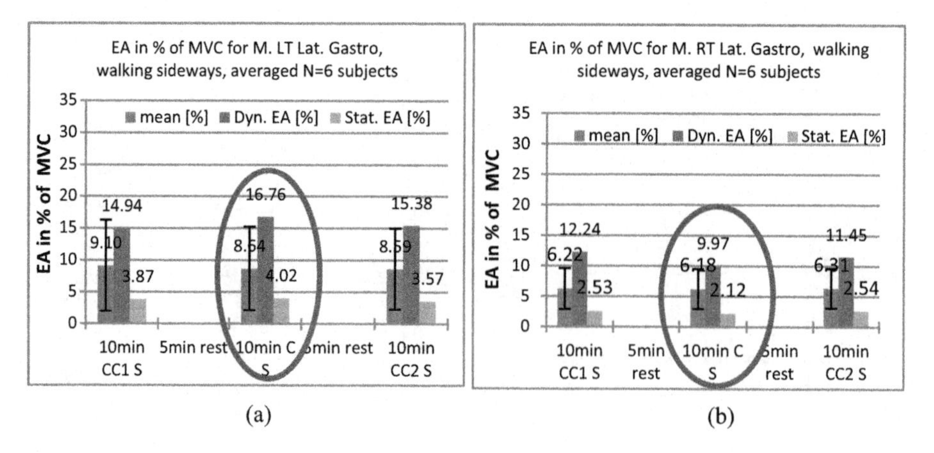

Fig. 12. a, b: EA in [%] of MVC of m. gastrocnemius lateralis averaged over all test persons walking "sideways" in three series in directions": CC1-C-CC2

5 Discussion of the Results and Conclusions

5.1 Discussion of Results Regarding Test Persons and EMG Data

Based on the analysis of results regarding muscular strain, the group of volunteers must first be considered. All test persons had no experiences concerning assembly work in U-shape especially with the walking sideways, typical for very narrow U-shapes. All person exercises regularly sport therefore the location of the surface electrodes on the muscles of legs were easy. In the present studies, as well as in Wakula et al. 2016 and Wakula et al. 2017a, the strain values were always considered over several 6 cycles with included standing and walking phases in the simulated U-shape. There are also some differences in the strain values in analyzed muscles in all test person. In all analyzed leg muscles high standard deviations in the EA-mean values have been obtained. This fact, as well as the short duration of assembly activities at the simulated U-shape (10 min per one series) with breaks (5 min) between each series of measurements, could be taken as a cause for the small variations in EA-mean values. Furthermore, this fact is also due to the subjectivity of the performed MVC measurement. As already discussed in the document related to Measurement and Analysis of surface Electromyography (3), the relation of the measured EA values to the MVC values depends on many boundary conditions, first of all on motivation of the subject to realize MVC during the testing. It is difficult for inexperienced persons in particular to retrieve a maximum isometric muscle contraction (MVC) intentionally during the respective MVC determination measurement, an explanation for the difference of the described stress values is given here. The MVC values are sometimes far apart from real values. According to the OEMG guideline (3), often only 60% to 80% of the maximum possible muscle contraction can be achieved by some test person.

5.2 Discussion of Results Regarding Variation of the Direction of Rotation as Well as Strategy of Movement

When looking at the movement strategies (walking "sideways" (scenarios A, B versus "mixed" walking-scenarios C, D), the EMG values of subjects and the individual muscles show slight EA differences. All EMG values for walking "sideways" are at a similar or slightly higher level than EA-values for "mixed" walking". It has to be considered here that the simulated U-shape layout has a slightly different for "mixed" working (Fig. 2b). Due to the increased distance between work station 1 and 2 and between station 4 and 5 a higher orbital distance in the U-shape was realized. This resulted in the greater number of steps required - walking "sideways" corresponds to approx. 20 steps and approx. 22 steps during "mixed" walking. According this fact "mixed" walking takes 12 additional steps to complete six cycles (10% more steps) This also means a higher work load and realizes in higher muscular strain for the analyzed scenarios C and D with "mixed" walking (see results in Sects. 4.1 and 4.2).

However, since the muscular strain values are not greater despite higher walking load, it can be concluded that the "mixed" walking has some advantages in terms of biomechanical-physiological criteria compared to purely "sideways" walking. This statement also coincides with the statements of the test subjects who perceive a lateral movement (with sidesteps) as unnatural. Test persons reported about higher subjective stress/strain in the legs and in the back as well.

A closer look at the musculoskeletal system of humans can allow following conclusions. The knee is realized better for "normal walking" and not for lateral walking with sidesteps. Therefore, the muscles of the lower extremities are loaded differently during the different strategies of walking. A rolling and dampening foot movement during "normal" walking also does not take place during "sideways" one. The musculature (muscle exertions) alone as well as the musculoskeletal system could thus be optimal used during acceleration and deceleration of the body when walking sideways.

Furthermore, during a purely lateral walking with sidesteps certain muscles are stressed unilaterally only. It should be remembered that during lateral walking adductor and abductor muscles are a greater involved in the generation of movement.

Regarding to the variation of the change of directions of movements (CC versus C), the following conclusions could be made. The EMG-results show small variation of the EA values in the analyzed muscles. This suggests that the changes/variation in the direction of movements - CC1 (about 10 min) to C (about 10 min) then to CC2 (about 10 min) causes compensation and regeneration effects in the muscles. This reduce unilaterally stress caused by walking in one direction only.

Looking at the achieved EMG-results, it is to notice that several muscles in the right leg of the test person are slightly more stressed than the left one during walking counterclockwise (CC). Conversely, when walking clockwise (C) some muscles show higher stress levels in the left leg than in the right one.

6 Conclusion

Based on the results of the study, the following conclusions can be drawn for the work design related to the U-shape:

- Trends towards unilateral muscular stresses/strains are evident in the constant movement of working person in one direction - counterclockwise or clockwise only;
- Change of direction on the U-shape: CC → C → CC → C is positive for the muscular strain situation, this brings mostly the balance of EA values in the right and left leg muscles;
- Varying the direction of movement strategy and rotation and is recommended so an ergonomic workload compensation will be achieved.
- "Mixed" walking in working process has some advantages compared to purely lateral/sideways walking in terms of biomechanical-physiological criteria of humans;
- Walking with "mixed" (lateral and two-three normal) steps in the analyzed U-shape did not reduce muscular strains in the legs compared to walking with "sidesteps" only.

These findings are based on a laboratory study with a simulated U-shape and six young subjects without practical experience in assembly work. Further studies in practice with a larger number of different test/working persons would be recommended in order to test statistically the statements listed above.

Acknowledgment. The authors would like to thank the professional association for wood and metal (Berufsgenossenschaft Holz und Metall - BGHM) and the professional association for electrical engineering (Berufsgenossenschaft Energie Textil Elektro Medienerzeugnisse - BG ETEM) for the support of the project "U-line assembly systems"/U-shape in the context of which this study was realized.

References

Lotter B, Wiendahl H-P (2006) Montage in der industriellen Produktion: Ein Handbuch für die Praxis, Auflage 2006. Ein Handbuch für Die Praxis. Springer, New York

Günther H-O, Mattfeld DC, Suhl L (2005) Supply chain management und logistik: optimierung, simulation, decision support. Springer Science & Business Media, Heidelberg

AWMF online (2013) Oberflächen-Elektromyographie in der Arbeitsmedizin, Arbeitsphysiologie und Arbeitswissenschaft- Arbeitsmedizinische S2 k-Leitlinie der Deutsche Gesellschaft für Arbeitsmedizin und Umweltmedizin (DGAUM) und der Gesellschaft für Arbeitswissenschaft (GfA). AWMF online-Das Portal der wissenschaftlichen Medizin

Wakula J, Fichtner K, Bruder R (2016) Analyse der physischen Belastungen und der muskulären Beanspruchungen an der simulierten U-Montagelinie mit unterschiedlichen Mechanisierungsgrad in der Prozesslernfabrik der TU Darmstadt. In: Gesellschaft für Arbeitswissenschaft (Hrsg.) Arbeit in komplexen Systemen. Digital, vernetzt, human?! GfA-Press, Dortmund

Wakula J, Müglich D, Bruder R (2017a) Walking "Normally" vs. "Sideways" in simulated, simple assembly operations: analysis of muscular strain in the legs. In: Schlick CM et al (eds) Advances in ergonomic design of systems, products and processes

Wakula J, Bauer S, Spindler S, Bruder, R (2017b) Analyse von Belastungen und muskulären Beanspruchungen der unteren Extremitäten beim „seitlichen" Gehen an einer simulierten Montagelinie mit unterschiedlichen Laufrichtungen und kurzen Taktzeiten. In: Dortmund (Hrsg.) Frühjahrs-kongress 2017, Brugg und Zürich: Soziotechnische Gestaltung des digitalen Wandels – kreativ, innovativ, sinnhaft. Gesellschaft für Arbeitswissenschaft e.V

Risk Assessment of Repetitive Movements of the Upper Limbs in a Chicken Slaughterhouse

Diogo Cunha dos Reis[1,2](✉), Adriana Seara Tirloni[1], Eliane Ramos[2], Natália Fonseca Dias[1], and Antônio Renato Pereira Moro[1,2]

[1] Technological Center, Federal University of Santa Catarina, Florianópolis, SC, Brazil
diogo.biomecanica@gmail.com
[2] Biomechanics Laboratory, CDS, Federal University of Santa Catarina, Florianópolis, SC, Brazil

Abstract. Brazil is one of the main exporters of chicken meat since 2004. However, improvements in the working conditions in this sector have not expand at the same rate as production grows. Thus, the aim of this study was to evaluate the risks associated with repetitive movements of the upper limbs in different meat processing tasks in a poultry slaughterhouse. The study was conducted in a poultry slaughterhouse with 3,500 workers, in which 300,000 chickens were slaughtered daily. In order to evaluate the risks associated with repetitive movements of the upper limbs, 10% of the workforce was evaluated while carrying out their work tasks, using the Checklist proposed by OCRA method. Descriptive statistics was used, as well as the Student t-test (SPSS 17.0) to compare the risk between left and right side of the body ($p \leq 0.05$). The average of occupational repetitive actions performed by workers was 64.7 ± 13.3 per minute, representing 9 points in the OCRA scale (0 to 10 points scale). The average score of OCRA Checklist was 19.5 ± 2.5 (moderate risk). The scores for the right upper limb (20.0 ± 3.0 - moderate risk) were significantly higher ($p = 0.024$) than the contralateral limb (17.7 ± 2.8 - moderate risk). Considering the five risk categories proposed by the OCRA method, one work task was considered high risk (7%) and 14 presented moderate risk (93%). Through simulations, it was possible to reduce the risk of UL-WMSDs to very low levels by reducing only the activity working pace ($-48.5 \pm 11.8\%$).

Keywords: UL-WMSDs · Poultry slaughterhouse · Risk assessment Ergonomics · OCRA

1 Introduction

Several factors have raised Brazil to a privileged position in the chicken meat market [1]. These factors include the abundance of land, fertile soil to produce feedstuff, favorable climate, and innovativeness of the companies in overcoming challenges [1]. Thus, Brazil became the leader in chicken meat exportation and the second largest

S. Bagnara et al. (Eds.): IEA 2018, AISC 825, pp. 323–329, 2019.
https://doi.org/10.1007/978-3-319-96068-5_35

producer following the USA [2]. Most Brazilian chicken meat exported is in the form of cuts (59%) [2], which are processed by means of hand tools (knives) or specific machines (chainsaws). However, improvements in the working conditions in this sector have not expand at the same rate as production grows [3]. Consequently, the slaughterhouse workers are exposed to high loading intensities and cyclic repetitive muscle actions of the upper limb contributing to an elevated risk of work-related musculoskeletal disorders [4–6]. According to Occupational Safety and Health Administration (OSHA) [7], the risk factors contributing to the development of musculoskeletal disorders (MSDs) in meatpacking plants are the repetitive and/or prolonged activities, forceful exertion, awkward and static postures, continued physical contact with work surfaces (edges), excessive vibration from powered tools, cold temperatures and inadequate hand tools.

Within this framework, the OCRA Checklist was developed to analyze the workers' exposure to various risks of developing upper limb work-related musculoskeletal disorders (UL-WMSDs) (repetitiveness, lack of recovery periods, strength demand, inappropriate posture and inadequate movements, and others defined as "complementary") related to activities performed [8, 9]. The Checklist was based on a consensus document of a technical committee in musculoskeletal injuries of the International Ergonomics Association (IEA) [8, 9].

Despite a series of studies evaluating the risks associated with the repetitive movements of upper limbs in poultry slaughterhouses have been carried out recently [10–13], the relevance of the present study is justified, given that each plant produces a different type of product, in order to meet the demands of the consumer market. The differentiation contributes to the unique characteristics of the work tasks. Therefore, it is evident the importance of surveying different slaughter and meat processing plants.

Thus, the aim of this study was to evaluate the risks associated with the repetitive movements of upper limbs in different tasks performed in a poultry slaughterhouse.

2 Method

The local Ethics Committee in Research with Human Beings, in accordance with the Helsinki Declaration, approved the procedures performed in this study.

The study was conducted in a poultry slaughterhouse with 3,500 workers in which 300,000 chickens were slaughtered daily, divided into two shifts. In order to evaluate the risks associated with repetitive movements of the upper limbs, 10% of the workforce was evaluated while carrying out their work tasks, using the Checklist proposed by OCRA method (Table 1) [14]. The tasks were analyzed individually, disregarding possible task rotations determined by the company and not regularly scheduled. Videos of 10 task cycles were recorded for each worker during their task routine, for subsequent analysis.

Descriptive statistics was used (mean, standard deviation and percentage), as well as the Student t-test (SPSS 17.0), in order to compare the risk between left and right side of the body ($p \leq 0.05$).

Table 1. Score description of the OCRA Checklist, risk classification and respective incidence of UL-WMSDs [14].

Color	Risk	Checklist score	Incidence of UL-WMSDs
Green	Acceptable	<7.5	<5.26
Yellow	Borderline or very low	7.6–11.0	5.27–8.35
Light red	Low	11.1–14.0	8.36–10.75
Dark red	Moderate	14.1–22.5	10.76–21.51
Purple	High	>22.5	>21.51

3 Results and Discussion

The 15 work activities analyzed were from the following sectors: cutting (7); packing (3); freezing tunnels (2); reception (1) and scalding (2) (Table 2).

Table 2. Risk assessment for repetitive movements of the upper limbs performed by the slaughterhouse workers, and the simulations for reducing the risk of UL-WMSDs by reducing the working pace.

Tasks - department	Actual situation observed				Simulations for minimizing the risk			
	Units/min	TA/min	OCRA	Risk	Units/min	TA/min	OCRA	Risk
Re-hanging chicken – Stork/cone	23.0	69.0	23.0	H	10.0	30.0	11.0	VL
Trimming chicken breast	13.0	93.0	22.0	M	5.0	30.0	11.0	VL
Trimming boneless leg	8.6	85.0	22.0	M	5.0	35.0	11.0	VL
Splitting thigh and drumstick	15.0	75.0	22.0	M	7.5	30.0	11.0	VL
Remove separators between packages	20.0	60.0	22.0	M	10.0	30.0	11.0	VL
Re-hanging chicken - chiller	20.0	60.0	21.0	M	10.0	30.0	11.0	VL
Placing wings at portioning machine	30.0	60.0	20.0	M	15.0	30.0	11.0	VL
Live chicken hanging	15.0	45.0	19.0	M	7.0	20.0	11.0	VL
Seal giblets packaging	4.0	76.0	19.0	M	2.1	40.0	11.0	VL
Chicken feet screening	–	70.0	19.0	M	–	35.0	11.0	VL
Inserting packaging plates	3.5	60.0	18.5	M	2.0	35.0	11.0	VL
Hanging chicken legs – Auto deboning	30.0	60.0	18.0	M	15.0	30.0	11.0	VL

(continued)

Table 2. (*continued*)

Tasks - department	Actual situation observed				Simulations for minimizing the risk			
	Units/min	TA/min	OCRA	Risk	Units/min	TA/min	OCRA	Risk
Primary packaging – deboning leg	2.7	55.0	16.5	M	2.2	40.0	11.0	VL
Meat transfer - bowl to box	4.6	50.0	16.5	M	3.0	30.0	11.0	VL
Seal deboned leg packaging	3.5	52.0	14.5	M	2.7	40.0	11.0	VL
Average	13.8	57.3	20.6	M	6.9	32.3	11.0	VL
Standard-deviation	9.7	12.3	5.8		4.6	5.3	0.0	

Risks: H-high; M-moderate; VL-very low; TA-technical actions.

Considering that a shift totaled 08 h 48 min of work with three rest breaks of 20 min, the net time of repetitive work was in the range of 421 to 480 min (multiplier factor 1.0). Three points were assigned to the "Recovery" risk factor (on a scale of 0 to 10). Regarding the other risk factors considered by the OCRA method (frequency of technical actions, force applied, posture and complementary factors), scores were assigned according to the characteristics of each task.

The average of occupational repetitive actions performed by the workers was 64.7 ± 13.3 per minute (Table 2), representing 9 points in the OCRA scale (0 to 10 points scale). Several studies in other poultry slaughterhouses have reported similar results, with averages of technical actions ranging from 59.1 to 64.4 [11–13, 15]. Kilbon [16] recommends that worker ought to not exceed 25–33 actions per minute, considering that frequencies above these values interrupt the physiological recovery mechanisms from operating efficiently, increasing the incidence of tendon injury.

Poultry processing includes strenuous and repetitive work, with workers at risk for overuse injuries. Upon entering the plant, the living chickens are received and then passed through a production line that requires workers to hang, kill, pluck, clean, eviscerate, cut, pack, and box poultry parts at a rapid pace. In addition, clean and repair equipment, assemble boxes, and move pallets of poultry packaged. Thus, potential risk for overuse injuries exists with each of these occupational duties [17].

Regarding the average score of OCRA Checklist was 19.5 ± 2.5 (moderate risk). The scores for the right upper limb (20.0 ± 3.0 - moderate risk) were significantly higher ($p = 0.024$) than the contralateral limb (17.7 ± 2.8 - moderate risk). In a poultry slaughtering industry, Colombini, Occhipinti and Fanti [14] revealed an average OCRA of 20 points (moderate risk) where 22.4% of the workers were affected by UL-WMSDs (confirmed by clinical examination and complementary testing). Previous studies also found moderate risk in another Brazilian poultry slaughterhouses [10–13].

Considering the five risk categories proposed by the OCRA method, one work task was considered high risk (7%) and 14 presented moderate risk (93%). Some previous studies have reported that there is a predominance of high-risk tasks in Italy poultry slaughterhouses (90%) [15], Iran (67%) [18], and in Brazil (56.5% - data collected before 2013, with a shorter duration for daily rest break - 10 min) [10]. On the other hand, recent Brazilian studies have shown lower percentages of high-risk tasks in poultry slaughterhouses (8%, 9% and 37%, respectively) [11–13]. It is hypothesized that the decrease in the number of high-risk tasks, observed in these studies, may have occurred due to the implementation of the Standard Regulatory Norm 36 (NR-36) [19].

The NR-36 establishes the minimum requirements for evaluation, control and monitoring risks in activities performed at meat processing industries [19]. Among the requirements of the NR-36 that directly influence the results of the OCRA Checklist, there is the minimum duration of the psychophysiological recovery periods (20 min, 45 min or 60 min). This requirement should be adopted in the productive sectors of the industries, according to the duration of the shift (up to 6 h, up to 7 h 20 min or 8 h 48 min, respectively). In addition, the recovery periods should be at least 10 min and at most 20 min, distributed in a fashion as not to be included in the first hour of work, contiguous with the meal interval and at the end of the last hour of the working day. When adopting these measures, there is a reduction of the time of exposure to repetitive activities and/or muscular overload, reducing the score of the variable "recovery" of the OCRA Checklist.

Based on epidemiological data, and starting from statistical procedures (regression analysis), the precursors of the OCRA method [14], developed hypotheses of disease prevalence expected in a particular occupational setting. Possible percentages were defined for each level of incidence of UL-WMSDs in the Checklist, as described in Table 1. In meat deboning activity, Colombini et al. [14] found an incidence of UL-WMSDs of 47.7% of workers from workstations classified with 28 points in the OCRA Checklist (high risk). Thus, it is possible to affirm that the individuals analyzed in the present study who performed activities of moderate risk have a probability of developing UL-WMSDs between 10.76 and 21.51%, whereas in high-risk activities the probability is >21.51%.

Due to predominantly highly repetitive movements of the upper limbs in poultry slaughterhouses [20], and previous studies suggesting the reduction in working pace to prevent UL-WMSDs [10–13], simulations of reduced working pace to achieve considerably low risk levels were performed, utilizing the OCRA Checklist. When performing these simulations by reducing only the working pace, it was possible to observe a reduction of the risk of developing UL-WMSDs to very low levels in all tasks (–48.5 ± 11.8%). Reis et al. [11–13] also carried out simulations aiming to lower the risk for UL-WMSDs in poultry slaughterhouses. When the pace was reduced to –42.1 ± 14.5%; –38.8 ± 13.3% and –44.9 ± 13.7%, very low risk levels were achieved for most of the tasks analyzed (24/26, 20/22 and 28/30, respectively), with the exception of those tasks requiring significant force exertion.

4 Conclusion

Given the results of this study and the postulation of the literature, the following can be concluded:

- Most of the tasks carried out by the workers were classified as presenting a moderate risk, increasing two to four times the chances of developing UL-WMSDs compared to the population that was not exposed;
- The risk of developing UL-WMSDs in the analyzed tasks was higher for the right side of the body;
- Simulations of reducing the working pace showed the effectiveness of this organizational measure to reduce the risk of developing UL-WMSDs in all analyzed tasks.

Future studies are necessary to determine whether the present findings can be generalized to other poultry slaughterhouses.

Organizational measures should be considered to reduce the risk of UL-WMSDs. Some of the suggestions are: reducing the work pace, adopting regular job rotation between tasks with different biomechanical requirements, adopting rest breaks well distributed throughout the workday (preferably hourly), increase the number of employees, use of sharp knives to reduce the effort required to perform cutting task, and monitor the risk level of work tasks through objective tools such as the OCRA Checklist.

References

1. Nääs IA, Mollo Neto M, Canuto SA, Walker R, Oliveira DRMS, Vendramento O (2015) Brazilian chicken meat production chain: a 10-year overview. Rev Bras Cienc Avic 17 (1):87–93
2. Brazilian Association of Animal Protein: Annual Report (2017). http://www.abpa-br.com.br. Accessed 20 May 2018
3. Sardá S, Ruiz RC, Kirtschig G (2009) Juridical tutelage concerning the health of meat packing workers: public service considerations. Acta Fisiatrica 16:59–65
4. Quandt SA, Grzywacz JG, Marin A, Carrillo L, Coates ML, Burke B, Arcury TA (2006) Illnesses and injuries reported by Latino poultry workers in western North Carolina. Am J Ind Med 49(5):343–351
5. Lipscomb H, Kucera K, Epling C, Dement J (2008) Upper extremity musculoskeletal symptoms and disorders among a cohort of women employed in poultry processing. Am J Ind Med 51(1):24–36
6. Punnett L, Wegman DH (2004) Work-related musculoskeletal disorders: the epidemiologic evidence and the debate. J Electromyogr Kinesiol 14:13–23
7. OSHA (2013) Occupational Safety and Health Administration: Prevention of Musculoskeletal Injuries in Poultry Processing. https://www.osha.gov/Publications/OSHA3213.pdf . Accessed 20 May 2018
8. Colombini D, Occhipinti E (2006) Preventing upper limb work-related musculoskeletal disorders (UL-WMSDs): new approaches in job (re)design and current trends in standardization. Appl Ergon 37:441–450

9. Occhipinti E, Colombini D, Cairoli S, Baracco A (2000) Proposta e validazione preliminare di una checklist per la stima dell'esposizione lavorativa a movimenti e sforzi ripetuti degli arti superiori. Med Lavoro 91:470–485
10. Reis DC, Reis PF., Moro ARP (2015) Assessment of risk factors of musculoskeletal disorders in poultry slaughterhouse. In: Azeres P, Baptista JS, Barroso MP, Carneiro P, Cordeiro P, Costa N, Melo R, Miguel AS, Perestrelo G (Org) proceedings book of the international symposium on occupational safety and hygiene – SHO 2015, 1 edn, vol 1, pp 294–296. Sociedade Portuguesa de Segurança e Higiene Ocupacionais, Guimarães - Portugal
11. Reis DC, Ramos E, Reis PF, Hembecker PK, Gontijo LA, Moro ARP (2015) Assessment of risk factors of upper-limb musculoskeletal disorders in poultry slaughterhouse. Procedia Manuf 3:4309–4314
12. Reis D, Moro A, Ramos E, Reis P (2016) Upper limbs exposure to biomechanical overload: occupational risk assessment in a poultry slaughterhouse. In: Goonetilleke R, Karwowski W (eds) advances in physical ergonomics and human factors, advances in intelligent systems and computing, vol 489, pp 275–282. Springer International Publishing Switzerland
13. Reis DC, Tirloni AS, Ramos E, Moro ARP (2017) Assessment of risk factors of upper-limb musculoskeletal disorders in a chicken slaughterhouse. Jpn J Ergon 53:458–461
14. Colombini D, Occhipinti E, Fanti M (2008) Método Ocra para análise e a prevenção do risco por movimentos repetitivos: manual para a avaliação e a gestão do risco. LTr, São Paulo
15. Colombini D, Occhipinti E (2004) Risultati della valutazione del rischio e del danno in gruppi dilavoratori esposti, in diversi comparti lavorativi, a movimenti e sforzi ripetuti degli arti superiori. Med Lavoro 95:233–246
16. Kilbom A (1994) Repetitive work of the upper extremity: part II—the scientific basis (knowledgebase) for the guide. Int J Ind Ergon 14:59–86
17. Cartwright MS, Walker FO, Blocker JN, Schulz MR, Arcury TA, Grzywacz JG, Mora D, Chen H, Marín AJ, Quandt SA (2012) The prevalence of carpal tunnel syndrome in Latino poultry processing workers and other Latino manual workers. J Occup Environ Med 54 (2):198–201
18. Mohammadi G (2012) Risk factors for the prevalence of the upper limb and neck work-related musculoskeletal disorders among poultry slaughter workers. J Musculoskelet Res 15 (1):1–8
19. Brasil (2013) Norma Regulamentadora 36. Segurança e Saúde no Trabalho em Empresas de Abate e Processamento de Carnes e Derivados, Ministério do Trabalho e Emprego, Portaria MTE n. 555, de 18 de abril de 2013, Diário Oficial da União
20. Sundstrup E, Jakobsen MD, Jay K, Brandt M, Andersen LL (2014) High intensity physical exercise and pain in the neck and upper limb among slaughterhouse workers: cross-sectional study. BioMed Res Int 218546

Ergonomics Now Has a Place in the Arkema Group as Part of a Permanent Improvement Initiative

Dominque Massoni[1] and Raphaële Grivel[2(✉)]

[1] Human Resources Development and Internal Communication, Arkema, Colombes, France
[2] HR and Social Development, Arkema, Colombes, France
raphaele.grivel@arkema.com

Abstract. The rollout of an initiative focusing on ergonomics came about as a result of the negotiation of an agreement on the prevention of harsh working conditions. The initiative has however been extended to be part of a continuous progress approach. It applies to all our production facilities.

The initiative entails a systematic diagnosis of working situations and the creation of a network of representatives: areas of application include analysis and improvement of existing situations as well as design of new plants.

This network of representatives, who come from a variety of professions and activities, is responsible for overseeing the initiative in the field. They have been trained in analyzing the operations, and meet on average twice a year to discuss and share their practices and continue their training.

This initiative has been communicated in the Group's Annual Report.

Keywords: Applied ergonomics · MSD prevention · Physical workload

1 Starting Point: The Introduction of an Ergonomics Initiative When an Agreement on the Prevention of Work Arduousness was Negotiated

1.1 Agreement on Prevention of Work Arduousness

– Reminder of legal background and law on pensions
– Initial negotiation in 2012. New agreement renegotiated in 2016 with more clear-cut affirmation of the place of ergonomics.
  • 2012–2015 agreement: prevention of work arduousness
  • 2016–2018 agreement: prevention of work arduousness and integration of ergonomics.

S. Bagnara et al. (Eds.): IEA 2018, AISC 825, pp. 330–332, 2019.
https://doi.org/10.1007/978-3-319-96068-5_36

1.2 Approach as Part of Social Dialog Policy

- Three trade union signatories to the last agreement
- Creation of a local joint body: the work arduousness working party comprising representatives of the signatory trade union, the HR manager, the HSE manager and the occupational physician.

1.3 Pre-diagnosis

1. Rollout with pre-diagnosis and subsequent diagnosis across all sites, as starting point: following the pre-diagnoses, joint working parties were set up on every site on the prevention of work arduousness, at-risk situations were identified, and action plans were drawn up. In-depth diagnoses were then carried out on identified at-risk situations. At the same time, action learning has helped train working parties in the ergonomics study.

2 Very Quickly, a Need to Dig Deeper

1. Move away from an intervention logic based on work arduousness thresholds to aim for an individual progress initiative.
   - Thresholds defined internally
   - Work arduousness thresholds or criteria relevant to our activity
   - Actions conducted below the thresholds: specific actions were conducted on the alternating successive periods (methodology guide for organization of shiftwork, internal awareness campaigns, field studies)
2. Capitalize on field experiences initiated.
   - Creation of a situations library (construction of library/structure): this library helps capitalize on the diagnoses carried out on the sites and thus facilitate the sharing of best practice.
3. Roll out a network of ergonomics officers to establish the initiative at individual site level.
   - Officers with different profiles have been identified on every site: technical design and development departments, HSE, site projects, HR
   - They were given 6 days' training in ergonomics: analysis of the activity, integration of ergonomics standards, integration of ergonomics in design
   - Running of network including Yammer and social network.

3 Concrete Achievements

1. In production: many examples can be given: improvement of raw material loading stations, improvement of maintenance and laboratory workstations, design and layout of stores and warehouses, etc.
2. Regarding the design of new plants (example of Honfleur: film)
3. Ergonomics as a key to understanding the activity for actions beyond working conditions.

   - Transfer of know-how: film Balan
   - Supply Chain and Supply Chain SAP
   - and soon Digital.

References

1. Zana JP, Victor B, Gravier E, Bulquin Y, Katlama Rouzet V, Burgaud M (2009) Démarche de prevention durable des TMS – engagement de trois organisations professionnelles
2. Daniellou F, Simard M, Boissières Y (2016) Human and Organizational Factors in industrial safety, les specifications of industrial safety, 2009-04, ICSI editor LNCS Homepage. http://www.springer.com/lncs. Accessed 21 Nov 2016
3. Prévenir les risques professionnels: un enjeu économique pour l'entreprise - EUROGIP - 124/F Paris, Février (2017)

Prevalence of Musculoskeletal Disorders and Posture Assessment by QEC and Inter-rater Agreement in This Method in an Automobile Assembly Factory: Iran-2016

Akram Sadat Jafari Roodbandi[1(✉)], Forough Ekhlaspour[2], Maryam Naseri Takaloo[2], and Samira Farokhipour[2]

[1] Research Center for Health Sciences, Institute of Health, Shiraz University of Medical Sciences, Shiraz, Iran
ergonomic.jafari@gmail.com

[2] School of Public Health, Bam Medical University, Bam, Iran
forough.ekhlasepoor71@gmail.com

Abstract. Background: Workers in the automotive industry due to the nature of the jobs are prone to musculoskeletal disorders (MSDs). The aim of this study was evaluating the prevalence of MSDs and inter-rater agreement in posture assessment with QEC method.

Methods: In a cross-sectional, descriptive and analytical in an automobile assembly plant in the Bam city in 2016, 148 people have completed MSDs Nordic questionnaire with census method sampling. Posture analysis was performed using the QEC. Two evaluators had experience of assessment in similar jobs at least 50 times before carrying out the assessment in this study. They independently evaluated 31 job tasks that were the worst posture in each task. SPSS software was used for statistical analysis with a significance level of 0.05.

Results: 88.6% of the subjects were married and they were in range age of 21 to 42 years. More prevalent have been in waist 78.3% (n = 90), wrist/hand 59.5% (n = 66), shank/feet 57.7% (n = 64), Shoulder 56.1% (n = 60), knee 55.1% (n = 59), upper back 46.2% (n = 49), neck 39.9% (n = 59), forearm 17.9% (n = 19) and thighs 13.8% (n = 13) during the last 12 months.

The incidence of MSDs of the neck and upper back were significantly associated with weight in the last 12 months. Neck disorder was the statistically significant correlation with the height.

Pearson correlation test shows quantitative evaluation score with QEC method in both evaluators was highly correlated and acceptable (r = 0.91). The maximum difference score between two evaluators in posture assessment was 22 which was 12% of maximum total score in the QEC method.

Conclusion: The high prevalence of MSDs was consistent with the high score in the QEC method. Ergonomic interventions are essential to improve the ergonomic status of the car assembly plant.

Keywords: MSDs · Automotive · Assembly · Posture assessment Agreement · QEC

S. Bagnara et al. (Eds.): IEA 2018, AISC 825, pp. 333–339, 2019.
https://doi.org/10.1007/978-3-319-96068-5_37

1 Introduction

The automotive industry includes all parts of the design, development, production, market and sale of motor vehicles. It is one of the largest industries in Iran that has a large part of the workforce. The automobile assembly industry, like any other industry, the workforce because of the nature of the work, are prone to musculoskeletal disorders (MSDs). The speed of the production line, the inappropriate work station space, repetitive movements, awkward posture, lifting and carrying the heavy load, applying force and snap-fit, prolonged standing are important factors in the automobile assembly industry, which intensively increases MSDs when simultaneously these non-ergonomic factors exist in the workplace [1, 2].

Automobile industry workers, especially assembly lines, because of the high work speeds, have to maintain a particular physical condition for a long time. Therefore, they are always exposed to occupational stress, which are the main causes of pain and discomfort in various parts of the body [3, 4].

Research has shown that feelings of pain and discomfort in various parts of the musculoskeletal system are major problems in the workplace and cause more than half of absenteeism [5]. Among the risk factors for work-related musculoskeletal disorders (WMSDs), awkward posture is one of the most important of them [6]. The risk of musculoskeletal disorders is determined based on posture evaluation results. And it is also possible to prioritize interventions and evaluate the effectiveness of any interventions [7, 8].

In order to assess the risk exposure to MSDs and postural analysis in workplace, observational, direct measurement and subjective judgment methods are used, in which observational methods are easier, less costly and more practical [9]. Quick Exposure Check (QEC) method can be used for holistic assessment of all the elements of a work system within 10 min (8). Studies have shown that the QEC has 'fair to moderate' levels of inter- and intra-observer reliability [10].

The aim of this study was to investigate the prevalence of MSDs and matching results of the QEC method with prevalence of MSDs. The second aim of our study was to examine the agreement between the evaluator in the QEC method.

2 Methods

The present study is a descriptive-analytic cross-sectional study in 2016. In the study, the workers of the three lines of automobile assembly participated in the study with census method in Bam city.

This study had two parts. In the first section, the prevalence of MSDs was assessed using the Nordic standard questionnaire. Data were gathered using a demographic questionnaire and Nordic questionnaire. A standard Nordic questionnaire with a validity and reliability of over 70%, questions about pain and discomfort of MSDs in a past year, past week and at the time of completing the questionnaire by division of the body including neck, shoulders, upper back, upper arm, Lower back, forearms, wrists, buttocks, thighs, knees, lower legs. Nordic Questionnaire is a widely used questionnaire that has high efficiency due to repeatability capabilities and can be used with high confidence to investigate MSDs [2].

In the second part, job analysis was performed and the worst posture was chosen in each job for postural analysis using Quick Exposure Check (QEC) method. Two evaluators were under the same training. After consultation between the two evaluators, 31 posture assessment was performed by two evaluators who had previously had a prior assessment of at least 50 times in similar jobs with the QEC method. The QEC allows the four main body areas (Back, Shoulder/Arm, Wrist/Hand, Neck) to be assessed by evaluator and involves workers in the assessment.

In the analysis of the data, the results of postural assessment were compared with the prevalence of MSDs. It was also calculated the inter-rater agreement in the QEC method.

Work experience above 1 year was considered as include criteria and history of surgery or congenital defects as exclude criteria in study. Data was analyzed by SPSS software version 18 at a 0.05 of significant level.

3 Result

148 male assembler workers participated in our study. The average age, height, weight and work experience were 29.38 ± 4.1, 172.6 ± 8.1, 70.2 ± 11.2 and 4.2 ± 1.96, respectively. 96.6% (n = 142) of workers that participants in our study had diplomas, 89.2% (n = 132) of them were married and 52.8% (n = 76) of them had regular exercise during the week.

87.2% (n = 129) of the assembly line workers involved in the study had at least one MSDs. The highest prevalence of MSDs was in the three limbs, including shoulders, hands and wrists, and waist and hip, with a prevalence of 68.2% (n = 101), 69.6% (n = 103), 60.8% (n = 90), respectively. Period prevalence and point prevalence of MSDs show in Table 1.

Table 1. Period prevalence and point prevalence of MSDs separated from 9 parts of the body in body map.

9 parts of the body in body map	Period prevalence percentage (Frequency)		Point prevalence percentage (Frequency)
	Last week	Last year	When completing the questionnaire
Neck	46.1(59)	39.9(59)	39.8(47)
Shoulders	47.3(53)	56.1(60)	41.7(45)
Upper back	41.5(51)	46.2(49)	37.8(45)
Forearms	15.1(18)	17.9(19)	16.1(19)
Wrist & Hands	54(68)	59.5(66)	50(58)
Back & Hip	70.9(90)	78.3(90)	65.3(79)
Thighs	16.2(19)	13.8(13)	14.9(17)
Knee	51.2(56)	55.2(59)	43.1(53)
Shank & feet	46.3(57)	57.7(64)	42.7(50)

Table 2. Relationship with MSDs with quantitative variables

Quantitative variables	Max-Min	Mean (SD)	t	P-value
Age	21–41	29.38(4.1)	1.6	0.09
Height	150–193	172.6(8.1)	–2.6	0.009
Weight	50–115	70.2(11.2)	–2.8	0.007
Work experience	1–11	4.2(1.9)	–0.56	0.53

18.1% of workers referred to a physician over the past 12 months due to MSDs and 10.3% had to resort to medical rest. 11% of them used physiotherapy services in the past year of study. And 12.1% (n = 18) of the workers spent on average 95 USD for the treatment of MSDs, which was at least 7 USD, and a maximum of 715 USD.

62.7% (n = 89) of the workers participating in our study were interested in their job. 47.8% of workers said they would probably have to give up jobs due to MSDs in the future.

T-test showed that having at least one MSDs with height and weight had a significant relationship. But there was no significant relationship with age and work experience.

The pearson correlation of posture assessment score with QEC method between two evaluators was 0.91. (p-value = 0.0001). In 64.4% (20 evaluations), the differences

Table 3. The final corrective action level in the 31 postural evaluation separated by the evaluators

	Level 1 n(%)	Level 2 n(%)	Level 3 n(%)	Level 4 n(%)
Evaluator A	2(6.5)	18(58.1)	10(32.3)	1(3.2)
Evaluator B	1(3.2)	14(45.2)	15(48.4)	1(3.2)

in evaluation did not change the level of final action (Table 3).

The agreement of the two evaluators is shown in the Bland-Altman plot (Fig. 1). The Bland-Altman plot is plotted with the mean of two evaluators in the X-axis and the

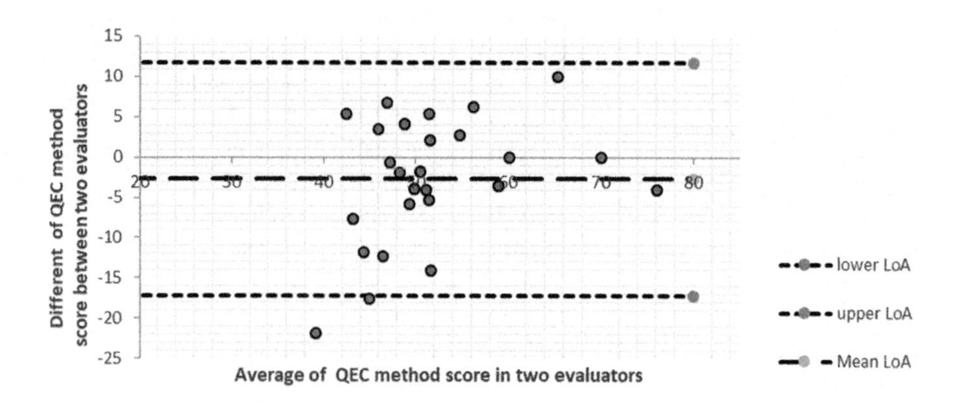

Fig. 1. Agreement of the two evaluators.

difference in the measurement of the two evaluators Y-axis. 30 of the assessments were in upper and lower of limit of agreement (LoA) rang, and only two assessments with 22 and 18 score difference was out of LoA range. Maximum difference between two evaluators was 12% of maximum total score (= 162) in the QEC method.

4 Discussion

In our study, there was a high prevalence of MSDs in automobile assembly workers. The limbs with the highest prevalence of MSDs were back/hip (78.3%), wrists/hands (59.5%), shank/feet and shoulders and knees (about 55%). 87.2% (n = 129) of the assembly line workers had at least one MSDs in their body part.

In the study of Nur et al. (2013), the most prevalent MSDs in the assembly workers were in the neck, shoulders, elbows, wrists/hands. And 76.9% of workers had experience pain or symptoms of MSDs at least in one part of their body [11]. In the study of Anita and et al. (2014), the prevalence of among the automotive assembly line workers was 78.2% and high prevalence body part were lower back, shoulder, wrist/hand, neck, upper back and knee [12]. In another study in Iran by Ghasemkhani et al. 2006, the commonest musculoskeletal symptoms were in feet (50%), back (47.4%), wrist/hand (30%). In the Ghasemkhani study 12-month period prevalence of MSDs in automobile assembly line workers was significantly more than MSDs symptoms in office worker in all of the 9 body part base on Nordic questionnaire [13].

Although the limbs with the highest prevalence of MSDs are in some studies similar to ours and in some studies are not. But in all previous studies similar to our study, reported a high prevalence of MSDs. The average prevalence of MSDs in our study is higher than the average of other studies that may be due to non-ergonomic conditions, such as inappropriate design of work stations, high work pace and the difference in workers' body size or other unknown causes.

In our study, height and weight were determined as significant factors associated with MSDs (Table 2). It increases significantly with increasing height and weight of MSDs. But there was no significant relationship between age and work experience with MSDs (Table 2).

In a study by Ohlander et al. (2016) was studied the effect of Snap-fits on upper limb functional limitations in German automotive worker. They stated that BMI and age had a significant relationship with the upper limb functional limitation in tasks with snap-fits. And a work experience of over 2 years is associated with a higher odds of the upper limb functional limitation [14]. Anita et al. (2014) reported that age, work history and awkward posture were three important and significant causes of musculoskeletal disorder. The age of more than 25 years and work experience over 3 years was significantly associated with MSDs [12]. Although in the study of Hollander (2016) and Anita (2014), significant relationship was found between age and work experience, but in our study, there was no relationship between MSDs and age and work experience. The meaningless association of age with MSDs can be due to the low average age in our study rather than Hollander's study. Considering that age and work experience are correlated, so justifiable is non-significant relationship between work experience and MSDs. In the study of Hollander BMI as an indicator of height and weight, correlated with MSDs similar our study.

According to Fig. 1, posture analysis only 2 of 31 items out of limit of agreement (LoA). In 64.4% (20 evaluations), the differences in evaluation did not change the level of final action. But this does not mean that necessarily the greatest difference in evaluations will necessarily lead to a change in the level. For example, in the same postural assessment, the greatest difference might be that the evaluator A has the lowest score in specific level and the evaluator B has the highest score in same level. On the other hand, the evaluators may be placed at a different level with a minor difference of 2 scores.

5 Conclusion

Musculoskeletal disorders are high prevalent in automobile assembly industries of Bam. The agreement between the evaluator is relatively good.

References

1. Eskandari D, Ghahri A, Gholamie A, Motalebi Kashani M, Mousavi GA (2011) Prevalence of musculoskeletal disorders and work-related risk factors among the employees of an automobile factory in Tehran during 2009–10. J Kashan Univ Med Sci 14(5):539–545 Feyz, (in Persian)
2. Roodbandi AJ, Choobineh A, Feyzi V (2017) The investigation of intra-rater and inter-rater agreement in Assessment of Repetitive Task (ART) as an ergonomic method. Occup Med Health Aff 3(5):1–5
3. Mazloumi A, Ghorbani M, Nasl Saraji G, Kazemi Z, Hosseini M (2014) Workload assessment of workers in the assembly line of a car manufacturing company. Iran Occup Health 11(4):44–55 (in Persian)
4. Kamalinia M, Nasl Saraji G, Choobine A, Hosseini M, Kee D (2009) Postural loading on upper limbs in workers of the assembly line of an Iranian Telecommunication Manufacturing Company using the LUBA technique. Sci J School Publ Health Inst Publ Health Res 6(3–4):49–60
5. Khanmohammadi E, Tabatabai Ghomsheh F, Osqueizadeh R (2017) Review the effectiveness of ergonomic interventions in reducing the incidence of musculoskeletal problems of workers in fatal truck assembly Hall. J Ergo 25(2):1–8 (in Persian)
6. Feyzi V, Mehdipoor S, Ghotbi Ravandi MR, Asadi M, Ghafori S (2015) Ergonomic assessment of workstations and musculoskeletal disorders risk assessment in the central oil refinery workshop of Hormozgan Province. J Health Dev 4(4):315–326 (in Persian)
7. Kohansal S, Koohpaei AR, Gharlipour Gharghani Z, Habibi P, Ziaei M, Gilasi H, Heidari Moghdam R, Ferasati F (2012) Ergonomic evaluation of musculoskeletal disorders among kitchen workers by QEC technique in the tehran university of medical sciences. Sci J Ilam Univ Med Sci 20(4):19–27 (in Persian)
8. David G, Woods V, Li G, Buckle P (2008) The development of the Quick Exposure Check (QEC) for assessing exposure to risk factors for work-related musculoskeletal disorders. Appl Ergon 39(1):57–69

9. David GC (2005) Ergonomic methods for assessing exposure to risk factors for work-related musculoskeletal disorders. Occup Med 55(3):190–199
10. Landis JR, Koch GG (1977) The measurement of observer agreement for categorical data. Biometrics 1:159–174
11. Nur NM, Dawal SZ, Dahari M (2014) The prevalence of work related musculoskeletal disorders among workers performing industrial repetitive tasks in the automotive manufacturing companies. In: Proceedings of the 2014 international conference on industrial engineering and operations management Bali, Indonesia, 7 Jan 2014, pp 7–9
12. Anita AR, Yazdani A, Hayati KS, Adon MY (2014) Association between awkward posture and musculoskeletal disorders (MSD) among assembly line workers in an automotive industry. Malays J Med Health Sci 10(1):23–28 ISSN 1675-8544
13. Ghasemkhani M, Aten S, Azam K (2006) Musculoskeletal symptoms among automobile assembly line workers. J Appl Sci 6(1):35–39
14. Ohlander J, Keskin MC, Weiler S, Stork J, Radon K (2016) Snap-fits and upper limb functional limitations in German automotive workers. Occup Med 66(6):471–477

Managing the Risk of Biomechanical Overload of the Upper Limbs in a Company that Produces High-End Clothing for Men

Nicola Schiavetti(✉) and Laura Bertazzoni(✉)

Corneliani Spa, via Panizza 5, 46100 Mantova, Italy
{nschiavetti, lbertazzoni}@corneliani.it

Abstract. The manufacturing facility that is the subject of this case study, where around 300 workers work, is organised by department (cutting, sewing, pressing, final testing) and has an incentive system based on the individual performance of each worker, which is calculated with respect to the standard timeframe needed to carry out operations.

The study carried out will analyse how risk indicators pertaining to biomechanical overload of the upper limbs vary in relation to the following factors:

- trends in the individual performance of workers with regard to individual operations
- rotation of personnel through various operations with differing risk indicators
- time that workers are exposed to different operations
- the inverse proportion between worker performance and times during which workers maintain inappropriate posture in each joint area

Due to the way in which all production cycle operations are mapped out, software has been implemented to enable users to manage risk in production departments in a more efficient and effective way.

In particular, via the integration of this system with the management system previously in use at the company, it is possible to obtain data relating to:

- the actual presence of workers in individual workplaces
- performance of workers in individual workplaces

As a result, it is possible to measure an individual worker's actual exposure to risk during an entire work shift.

Thanks to risk indicators being updated in real time, managing the rotation of personnel in production departments is improved. This also provides the versatility needed to balance times that workers spend carrying out operations with varying levels of risk.

Keywords: OCRA · Efficiency · Clothing

S. Bagnara et al. (Eds.): IEA 2018, AISC 825, pp. 340–354, 2019.
https://doi.org/10.1007/978-3-319-96068-5_38

1 Risk Assessment

1.1 The Organization of Manufacturing Departments

Corneliani Spa produces high-end men's clothing. In particular, in the manufacturing site under study there is the production of jackets and coats of the first formal line. Total annual production is 80 000 pieces with a workforce equal to 316 direct operators (310 women). Working content for the production of a jacket is on average 300 min and the work phases are about 150 for the production of the basic classic model.

There are different types of operations: fabric spreading, manual fabric cutting or with automatic machine, hand sewing or sewing with linear machine, sewing with automatic machine, manual ironing, ironing with press. All these operations are characterized by a high repetitiveness of the actions performed, and mainly involve the hand and wrist, elbow and shoulder districts.

The organization of production is structured by departments: cutting department, packaging department, ironing department and testing department. Each of these has families of operations typical of the department but in the sewing department for example, while being largely characterized by sewing operations, we also make trimmings of fabric with scissors or with punching machines, manual ironers, ironing with press.

The productivity of the departments is monitored through a management system that allows accurate information on the execution times of each operator in the various operations and the relative efficiency in a timely and instantaneous way. The remuneration of the persons is characterized by a variable part linked to the average monthly individual performance in the execution of the operations performed.

The corporate population consists of 98% female staff, with an average age of 46 (see Fig. 1) and an average working age of 18 (see Fig. 2).

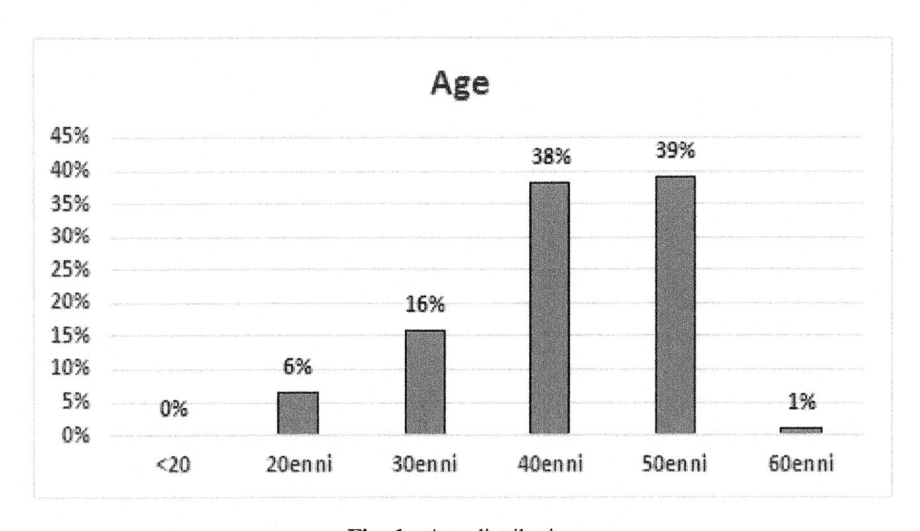

Fig. 1. Age distribution

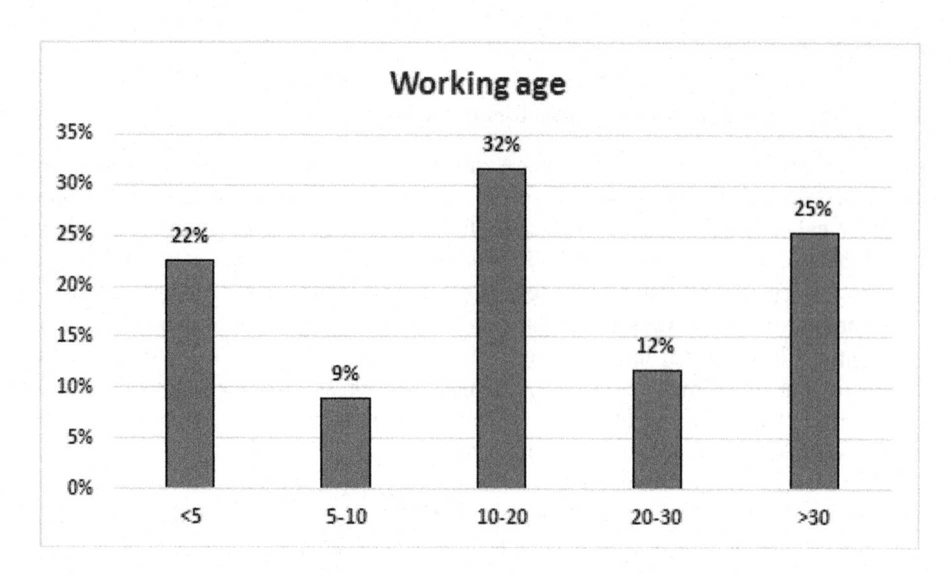

Fig. 2. Working age distribution

1.2 How to Hinder the Occupational Diseases

Since 2011, the demand for occupational diseases and job suitability limitations have increased. In particular since 2011 there has been an increase in occupational diseases and then a peak in the year 2015 (12), even if in reality the recognized diseases are about half of the requests (see Fig. 3).

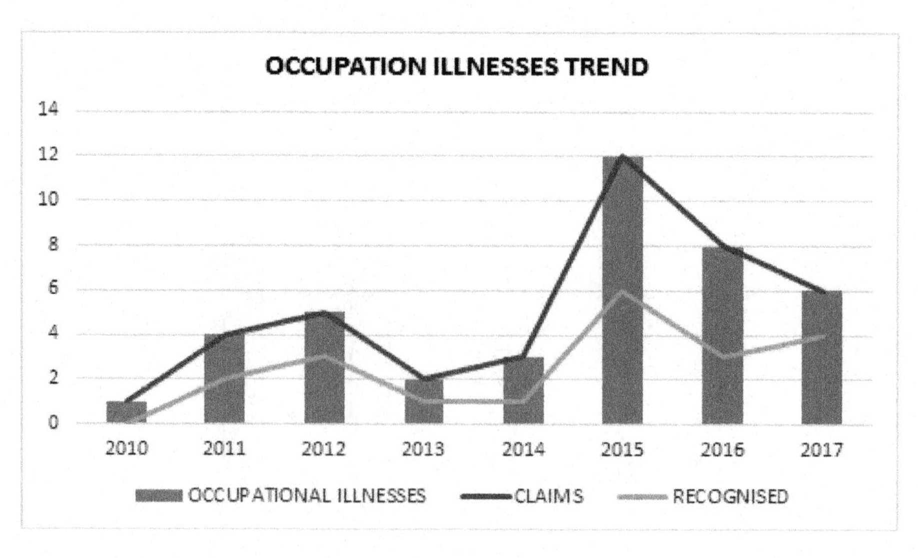

Fig. 3. Occupational illnesses trend

The increase in occupational limitations and required diseases led, on the one hand, to a lower flexibility of staff and, on the other, a decrease in the average efficiency of the departments due to displacements of specialized personnel no longer able to perform a given operation, replaced with other inexperienced staff.

This is the main reason that led the company management to take an interest in the topic of the biomechanical overload of the upper limbs in 2011, with the aim of understanding the real level of risk of the various tasks of the production departments and defining an action plan aimed at combating the onset of occupational diseases and physical problems limiting the performance of certain operations.

We contacted EPM and Dr. Colombini to share the situation of the company and then define an action plan that included:

- health mapping (first generalized screening - first level medical examination specialization in the strictly necessary cases);
- risk mapping
  - biomechanical overload risk of upper limbs from repetitive work (M RIP) with OCRA INDEX method [1];
  - risk of manual handling of loads (MMC) with the NIOSH method [2];
  - design of workstations and work methods (and subsequent verification of risk quantification);

In this way, the integration between the quali-quantitative mapping of the OCRA risk and the health mapping has enabled the identification of the critical areas and the lines of action necessary to draw up a program, aimed at ensuring the improvement over time of the levels of safety and health in the company, acting both at the level of ergonomic redesign of the critical positions, both at the level of work organization, and at the level of personnel management.

Health mapping was carried out through the administration of a specific questionnaire, developed by EPM and validated at international level and published in the scientific journals of the sector.

This questionnaire is aimed at highlighting the existence of work related diseases due to biomechanical overload of the arts higher from repetitive work.

It is also aimed at obtaining data that is significantly comparable with those available at national level both for the general population not exposed to the risk factor (generic comparison term) and for the working population exposed to risk (specific comparison term). So you have a clear idea of the global business situation.

The screening was conducted by the MC after an initial phase of coaching by the EPM/Clinical of Labor staff of Milan (University of Milan - Department of Occupational Medicine - "Laboratorio L. Devoto") and after specific path training at the EPM in Milan.

The company decided to proceed with the mapping of the risk following a specific training course and coaching by the EPM staff. It included 3 modules on the 3 main themes (risk assessment from repetitive movements, assessment of risk from manual handling of loads, the redesign of the work stations). It involved the staff of the time and methods office and production departments managers, raw materials warehouse and the shipping warehouse.

At the end of 2013, the administration of the anamnestic questionnaire was completed on the entire population of the production departments. We arrived at a first mapping of the health situation (people with a positive threshold and districts involved).

In addition, the time and method office has mapped the operations of the jacket packaging cycle in the basic model (about 70% of the total operations) in order to define the level of risk of the various tasks.

Repetitive Movement Risks Assessment

The first mapping, which began in 2012 and ended with the mapping of 145 operations in 2013, revealed that 22% of people were subjected to risk higher than 3.6 and in particular a 2% at intense risk (see Fig. 4).

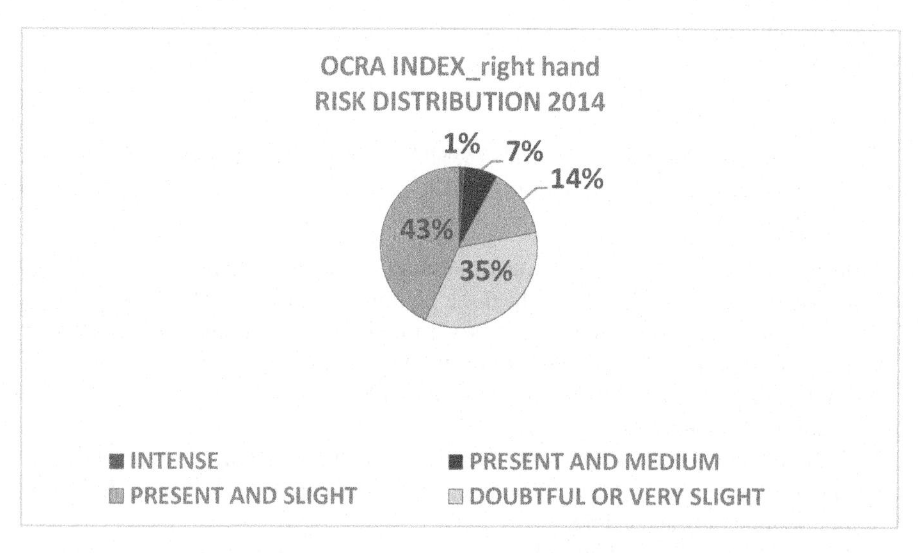

Fig. 4. Risk distribution in 2014 - right hand

For the left hand, however, the situation was less strongly as only 3% of workers had a risk greater than 3.6 (see Fig. 5).

The evaluation proceeded analyzing three different areas:

- Method evaluation for the execution of operations
- Promotion of versatility and job rotation
- Increase of rest periods

Method Evaluation

For high-risk tasks with an index of more than 3.6, we analyzed the districts most involved and a possible way to modifying the method of execution of the operations.

For example, for the operation of trimming lining shoulders and gluing, which provides the trimming of the front and back of the semi-finished jacket (fabric, lining and interior fabric) we reasoned the replacement of the scissors as a work tool that

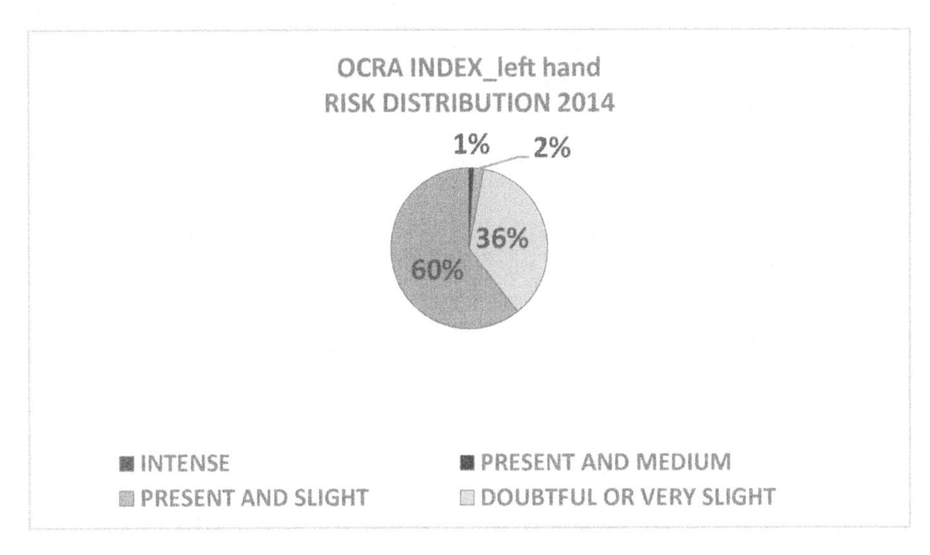

Fig. 5. Risk distribution in 2014 - left hand

imply a repetitiveness of high movement (high frequency) and high force usage for most of the cycle time. Tests were carried out to perform the cutting with electric shears and dies, but they did not give a positive result because they do not guarantee the necessary precision in the execution of the operation (it is 2–3 mm tolerance).

Job Rotation

In the past, the tasks known to people were on average 2, consequently the turnover of the staff was little practiced.

The management was more driven towards a maintenance of the fixed operator on a workstation, in the belief that this could lead to greater individual efficiency and therefore to greater productivity.

Over time, the different organizational context (production with smaller batches, less WIP, the need for narrower lead times) has led to a change in the type of work organization. As a result, the polyvalence has been proposed in favor of greater global efficiency and a better service level in terms of product delivery times.

Furthermore, the concept of job rotation was introduced. Explaining how this positively affects the efficiency not only of the system but also of the individual. Alleviating the negative effect of the repetitiveness of the actions during the execution of a single operation fatigue and increasing staff motivation.

After the training interventions, today each operator performs on average 3 operations, but about 10% still perform only one task.

Rest Periods

When technical measures and job rotation were not applicable, we decided to introduce time slots for the 6 operations with OCRA index above 9.

We obtained a reduction of 32% on the involved operators' index.

The details of the intervention are reported below.

It obviously had an impact in terms of additional labor costs. On the other hand we expected a reduction in the cost of occupational diseases (see Table 1).

Table 1. Rest period effects

Operations	No. of people	Index OCRA reduction	% of estimate sick workers reduction	Estimate of sick workers reduction
Soabaratura	4	32%	6,55%	0,3
Sbastitura finale	8	36%	14,32%	1,1
Ribattitura finale	8	31%	10,64%	0,9
Ribattitura angolini	4	26%	4,76%	0,2
Scarnitura	4	29%	7,84%	0,3
Attaccatura bottoni a mano	3	36%	7,50%	0,2
Total	31	31,67%	9,64%	3,0

At the end of 2017, after the actions described above, the distribution of the level of risk on the population is shown in Figs. 6 and 7.

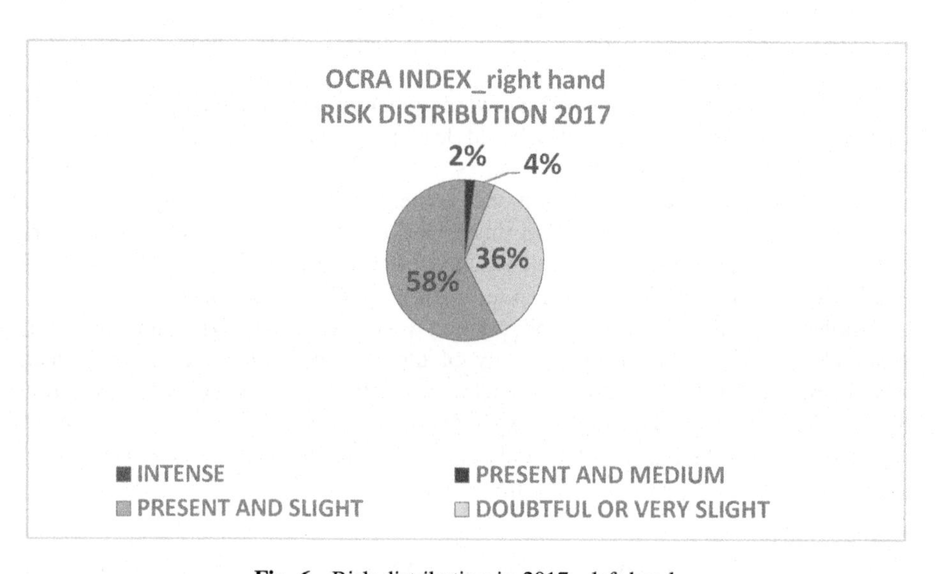

Fig. 6. Risk distribution in 2017 - left hand

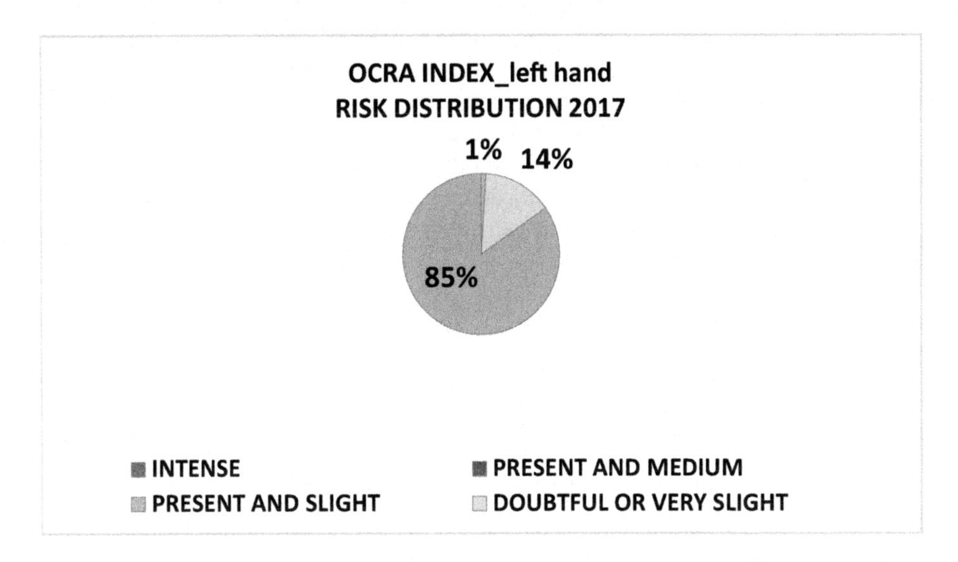

Fig. 7. Risk distribution in 2017 - left hand

2 Relationship of Direct Proportionality Between Efficiency and the OCRA Index

In order to obtain an assessment that is closer to reality, we must consider the mechanism for measuring individual efficiency and its effects on the real workers' risk exposure.

Despite the definition of the times and methods of work, the execution of the same operations by different workers will certainly change at different speeds of execution and consequently each worker will produce a different number of garments during the work shift.

For this reason efficiency is defined as the relationship between:

$$\text{efficiency} = n^{\circ}\,\text{pieces produced}/n^{\circ}\,\text{pieces with } 100\%\text{ efficiency} \quad (1)$$

The optimal number of pieces is equivalent to the number of pieces produced with 100% efficiency.

We can define:

- optimal efficiency (eff_{100}): the optimal efficiency is identified with the value 100%;
- real operation efficiency (eff_{ro}): it represents the average of the efficiencies of workers who perform the same operation.
- real worker's efficiency (eff_{rw}): is the average of the efficiency obtained by the person in a given period of time in a given operation.
- observed efficiency (eff_{ob}): it represents the operator's efficiency during the movie.

Because of the number of employees and the number of operations and the possible combinations at different efficiency values, it is basically impossible to assess the risk of biomechanical overload of the upper limbs.

The efficiency can vary for the single person during the day, weeks and months.

The efficiency certainly varies with the operator variation.

Is it possible to implement a simplification in order to facilitate risk management?

From the results of the mapping we can verify the following conditions and we can say for each operation that:

1. if the workers respect the working method, the number of technical actions in the cycle remains almost unchanged;
2. the efficiency is directly proportional to the number of pieces produced in the shift an so it is for the frequency;
3. the efficiency is inversely proportional to the cycle time.

It is also possible to make the following hypotheses:

- As the efficiency changes, the percentage of the individual postures on the cycle time remains unchanged;
- As the efficiency changes the percentage of the duration of the actions with force on the cycle time remains unchanged;
- The efficiency is inversely proportional to the duration of the static actions;
- As the efficiency changes, the stereotype score remains constant.

It is now possible to graph the OCRA index trend (see Fig. 8).

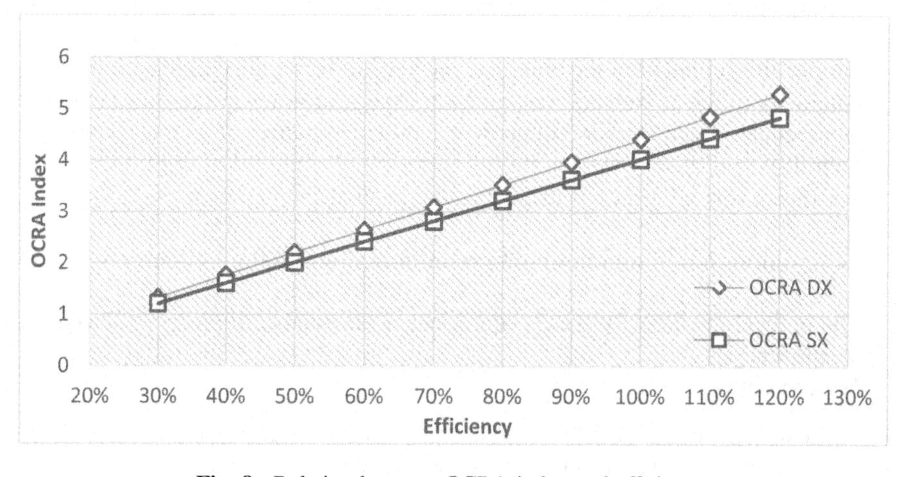

Fig. 8. Relation between OCRA index and efficiency

We can notice that there is a relationship of direct proportionality between efficiency and the OCRA index.

We could write:

$$\mathrm{OCRA}_{\mathrm{eff(X)}} = \mathrm{OCRA}_{\mathrm{ob}} * \mathrm{eff}_{(\mathrm{X})}/\mathrm{eff}_{\mathrm{ob}} \quad (2)$$

$$\mathrm{OCRA}_{\mathrm{eff(X)}} = \mathrm{OCRA}\ \mathrm{eff}_{100} * \mathrm{eff}_{(\mathrm{X})} \quad (3)$$

The slope of the line can change if there are static actions in the operation.

Especially when the efficiency decreases, the number of technical actions depends on the duration of static actions.

As the efficiency decreases, the duration of the static shares increases, therefore, by the comparison with the corresponding number of dynamic shares, the contribution of the static shares may be predominant.

In the next example (see Fig. 9) the trend of the OCRA index begins to diverge from the linearized OCRA for efficiency lower than 70%.

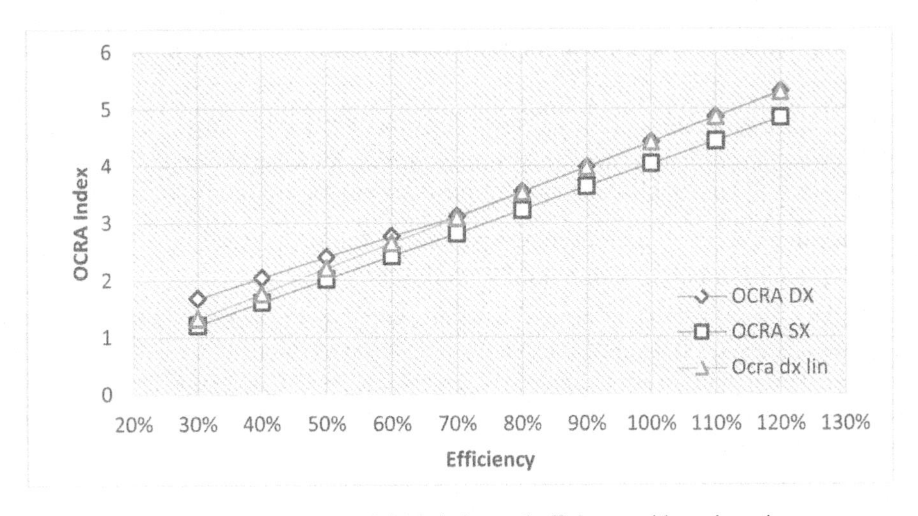

Fig. 9. Relation between OCRA index and efficiency with static actions

3 Automated Calculation of the Risk

We can apply the assumptions made for the OCRA index to the OCRA checklist.

Direct proportionality is not verified, however the trend of the OCRA checklist is comparable to the tendency of frequency factor score (see Fig. 10).

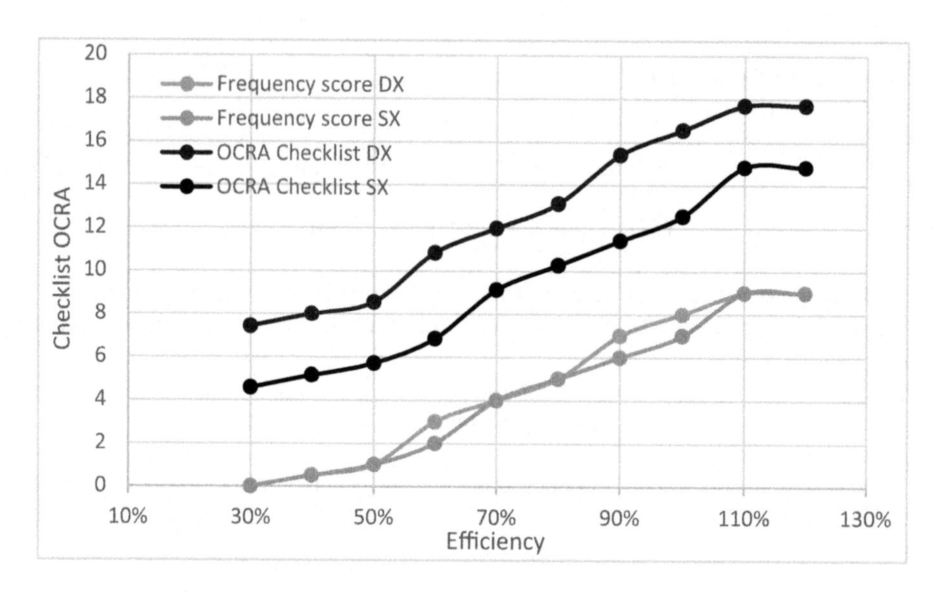

Fig. 10. Relation between OCRA checklist index, frequency score and efficiency

The direct proportionality is not verified for the OCRA checklist.

We have to recalculate the checklist value for each efficiency value.

To achieve the complete mapping of the departments, we need for each worker:

- Time of exposure to each operation performed during the day
- Working efficiency for each operation performed

Making the hypothesis of having 300 workers doing 3 operations each and with 3 different efficiencies it would be necessary to calculate 2700 daily checklists.

Therefore it's necessary to implement a software, Petra Automation [3], which can continually recalculate the checklist values while it interfaces with the production management system.

The input data for the development of the program are:

- Evaluation of the Ocra Index or Checklist with 100% efficiency;
- Minutes of work performed on each operation (data obtained from the production management);
- Efficiency recorded for each operation through the production management.

For the operation of Petra Automation it is necessary to have a mapping with OCRA Checklist.

It is necessary to transform all the OCRA index evaluations in high-precision OCRA checklists through a conversion table (see Table 2).

Table 2. _ Conversion table

Side	Organization Data						Frequency	Static actions	Force			Shoulder		Elbow	Wrist	Hand	Stereotype	Additional factors score
	ID	Number of hours without a recovery period	Multiplier for net duration	net duration of repetitive tasks (min)	n° of pieces at 100%	n° of piecdes observed efficiency	n° of actions	static actions score at observed efficiency	Borg 3-4	Borg 5-6-7	Borg ≥ 8	Hand over head	Hand not over the head	pronation - flexion/extension > 60°	flexion/extension >45° radio/ulnar deviation >45°	pinch		
DX	0	3,50	1,27	300,00	200,00	160,00	47,00	2,50	21,04	0,00	0,00	2,90	0,00	2,00	43,00	105,72	0,00	0,00
SX	0	3,50	1,27	300,00	200,00	160,00	65,00	0,00	21,04	0,00	0,00	44,90	0,00	2,00	42,00	68,00	0,00	0,00

The table data are:

- Operation ID.
- Multiplier for net duration.
- Number of hours without a recovery period (usually we have 30 min break a day in addition to the lunch).
- Net duration of repetitive tasks (min).
- n° pieces theoretically produced in the shift (8 h) at efficiency 100%.
- n° pieces theoretically produced in the shift (8 h) at the observed efficiency.
- n° technical actions (observed in the video).
- Static actions score at the observed efficiency.
- Borg “3–4” (s).
- Borg “5–6–7” (s).
- Borg “>8” (s).
- Postures (seconds).
- Stereotype (score).
- Additional factors score.

For the transition to automation it is necessary to interface the mapping with the production data extracted from the management production software, so we have the exposures of the workers to the various operations (see Fig. 11).

- Date.
- Worker’s ID.
- Worker’s name and surname.
- Operation ID.
- Duration (min).
- Produced pieces.
- OCRA index value.
- Pieces per shift.
- Duration of repetitive movements in a shift.
- Recalculated pieces per shift at real worker’s efficiency (eff_{rw}).
- Real worker’s efficiency (eff_{rw}).

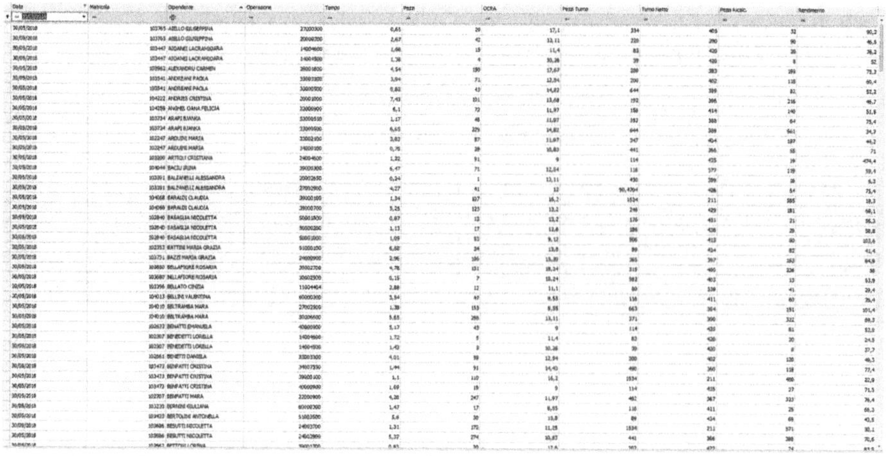

Data	Matricola	Dipendente	Operazione	Tempo	Pezzi	OCRA	Pezzi Turno	Turno Netto	Pezzi Ricalc.	Rendimento
30/05/2018	103765	AIELLO GIUSEPPINA	27000300	0,65	29	17,1	334	405	32	90,2
30/05/2018	103765	AIELLO GIUSEPPINA	20000700	2,67	42	13,11	220	290	90	46,5
30/05/2018	103447	AIDANEI LACRAMIOARA	14004600	1,66	15	11,4	83	420	26	76,2
30/05/2018	103447	AIDANEI LACRAMIOARA	14004500	1,38	4	10,26	39	420	8	52
30/05/2018	103962	ALEXANDRU CARMEN	39001600	4,54	150	17,67	200	383	199	71,3
30/05/2018	103541	ANDREANI PAOLA	33003300	3,94	71	12,94	200	402	118	60,4
30/05/2018	103541	ANDREANI PAOLA	33000500	0,82	43	14,82	644	389	82	52,2
30/05/2018	104222	ANDRIES CRISTINA	20001000	7,43	101	13,68	192	386	216	46,7
30/05/2018	104259	ANGHEL OANA FELICIA	22000900	6,1	72	11,97	158	414	140	51,6
30/05/2018	103734	ARAPI BIANKA	53000510	1,17	48	11,07	352	388	64	75,4
30/05/2018	103734	ARAPI BIANKA	33000500	6,65	229	14,82	644	389	561	34,7
30/05/2018	102247	ARDUINI MARIA	33002100	3,82	87	11,07	347	404	187	46,2
30/05/2018	102247	ARDUINI MARIA	34000100	0,75	28	10,83	441	366	55	71
30/05/2018	103200	ARTIOLI CRISTIANA	24004600	1,22	91	9	114	435	19	474,4
30/05/2018	104044	BACIU IRINA	39000300	6,47	71	12,04	118	577	139	59,4
30/05/2018	103391	BALZANELLI ALESSANDRA	20002650	0,24	1	13,11	430	399	16	6,3
30/05/2018	103391	BALZANELLI ALESSANDRA	27002900	4,27	41	12	90,4704	426	54	75,4
30/05/2018	104068	BARALDI CLAUDIA	39000100	1,34	107	16,2	1534	211	585	18,3
30/05/2018	104068	BARALDI CLAUDIA	39000700	5,25	123	13,2	240	429	181	68,1
30/05/2018	102840	BASAGLIA NICOLETTA	50001500	0,87	12	13,2	176	451	21	56,3
30/05/2018	102840	BASAGLIA NICOLETTA	50000200	1,13	17	12,6	186	436	29	58,8
30/05/2018	102840	BASAGLIA NICOLETTA	50001000	1,09	83	9,12	906	413	80	103,6
30/05/2018	102353	BATTINI MARIA GRAZIA	51000250	6,68	34	13,8	89	434	82	41,4
30/05/2018	103731	BAZZI MARIA GRAZIA	24000900	2,96	106	15,89	365	397	163	64,9
30/05/2018	103680	BELLAFIORE ROSARIA	20002700	4,78	131	18,24	319	400	226	58
30/05/2018	103687	BELLAFIORE ROSARIA	30602500	0,15	7	18,24	982	403	13	53,9
30/05/2018	103396	BELLATO CINZIA	11004404	2,88	12	11,1	80	339	41	29,4
30/05/2018	104013	BELLINI VALENTINA	60000300	3,54	47	8,58	116	411	60	78,4
30/05/2018	104010	BELTRAMBA MARA	27002500	1,38	153	8,55	663	304	151	101,4
30/05/2018	104010	BELTRAMBA MARA	30206600	5,65	286	13,11	371	390	322	88,3
30/05/2018	102673	BENATTI EMANUELA	40000900	5,17	43	9	114	435	81	52,9
30/05/2018	102307	BENEDETTI LORELLA	14004600	1,72	5	11,4	83	420	20	24,5
30/05/2018	102307	BENEDETTI LORELLA	14004500	1,43	3	10,26	39	420	8	37,7
30/05/2018	102661	BENETTI DANIELA	33003300	4,01	59	12,94	200	402	120	49,3
30/05/2018	103473	BENFATTI CRISTINA	34007350	1,44	91	14,43	480	360	118	77,4
30/05/2018	103473	BENFATTI CRISTINA	39000100	1,1	110	16,2	1534	211	480	22,9
30/05/2018	103473	BENFATTI CRISTINA	40000900	1,69	19	9	114	435	27	71,5
30/05/2018	102707	BENFATTI MARA	22000900	4,28	247	11,97	482	367	323	76,4
30/05/2018	103230	BERNINI GIULIANA	60000300	1,47	17	8,55	116	411	25	68,3
30/05/2018	103423	BERTOLINI ANTONELLA	51003500	5,6	30	13,8	89	434	69	43,5
30/05/2018	103686	BESUTTI NICOLETTA	24003700	1,31	172	11,25	1534	211	571	30,1
30/05/2018	103686	BESUTTI NICOLETTA	24502900	5,37	274	10,87	441	366	388	70,6
[illegible]	[illegible]	[illegible]	[illegible]	[illegible]	30	17,6	[illegible]	[illegible]	24	[illegible]

Fig. 11. Interface between production software and automation software

With real exposures, in minutes, it is possible to reconstruct the "real" day of the operator and calculate a very precise estimation of the daily risk index of each operator (Fig. 12).

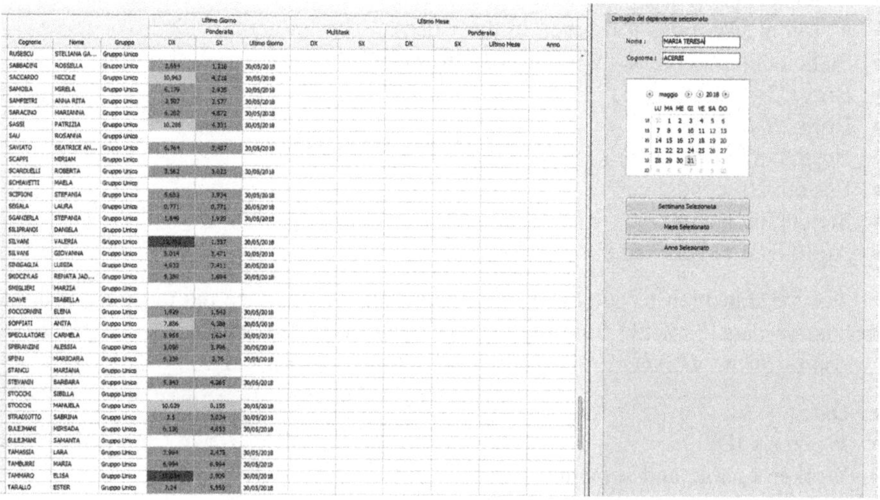

			Ultimo Giorno Ponderata			Ultimo Mese Multitask		Ultimo Mese Ponderata			
Cognome	Nome	Gruppo	DX	SX	Ultimo Giorno	DX	SX	DX	SX	Ultimo Mese	Anno
RUSESCU	STELIANA GA...	Gruppo Unico									
SABBADINI	ROSSELLA	Gruppo Unico	2,654	1,216	30/05/2018						
SACCARDO	NICOLE	Gruppo Unico	10,963	4,216	30/05/2018						
SAMOILA	MIRELA	Gruppo Unico	6,179	2,935	30/05/2018						
SAMPIETRI	ANNA RITA	Gruppo Unico	3,507	2,577	30/05/2018						
SARACINO	MARIANNA	Gruppo Unico	6,202	4,872	30/05/2018						
SASSI	PATRIZIA	Gruppo Unico	10,286	4,331	30/05/2018						
SAU	ROSANNA	Gruppo Unico									
SAVIATO	BEATRICE AN...	Gruppo Unico	6,764	2,487	30/05/2018						
SCAPPI	MIRIAM	Gruppo Unico									
SCARDUELLI	ROBERTA	Gruppo Unico	3,562	3,023	30/05/2018						
SCHIAVETTI	MAELA	Gruppo Unico									
SCIPIONI	STEFANIA	Gruppo Unico	5,653	3,934	30/05/2018						
SEGALA	LAURA	Gruppo Unico	0,771	0,771	30/05/2018						
SGANZERLA	STEFANIA	Gruppo Unico	1,846	1,929	30/05/2018						
SILIPRANDI	DANIELA	Gruppo Unico									
SILVANI	VALERIA	Gruppo Unico	[illegible]	1,337	30/05/2018						
SILVANI	GIOVANNA	Gruppo Unico	5,014	3,471	30/05/2018						
SINIGAGLIA	LUIGIA	Gruppo Unico	4,632	7,411	30/05/2018						
SKOCZYLAS	RENATA JAD...	Gruppo Unico	5,389	1,694	30/05/2018						
SMEGLIERI	MARZIA	Gruppo Unico									
SOAVE	ISABELLA	Gruppo Unico									
SOCCORNINI	ELENA	Gruppo Unico	1,929	1,543	30/05/2018						
SOFFIATI	ANITA	Gruppo Unico	7,806	4,388	30/05/2018						
SPECULATORE	CARMELA	Gruppo Unico	5,956	1,624	30/05/2018						
SPERANZINI	ALESSIA	Gruppo Unico	3,096	2,796	30/05/2018						
SPINU	MARIOARA	Gruppo Unico	6,239	2,76	30/05/2018						
STANCU	MARIANA	Gruppo Unico									
STEVANIN	BARBARA	Gruppo Unico	5,340	4,265	30/05/2018						
STOCCHI	SIBILLA	Gruppo Unico									
STOCCHI	MANUELA	Gruppo Unico	10,029	8,155	30/05/2018						
STRADIOTTO	SABRINA	Gruppo Unico	2,5	3,034	30/05/2018						
SULEJMANI	MIRSADA	Gruppo Unico	6,136	4,653	30/05/2018						
SULEJMANI	SAMANTA	Gruppo Unico									
TAMASSIA	LARA	Gruppo Unico	3,964	2,475	30/05/2018						
TAMBURRI	MARIA	Gruppo Unico	6,994	6,994	30/05/2018						
TAMMARO	ELISA	Gruppo Unico	13,034	3,909	30/05/2018						
TARALLO	ESTER	Gruppo Unico	7,24	5,950	30/05/2018						

Dettaglio del dipendente selezionato
Nome: MARIA TERESA
Cognome: ACERBI
maggio 2018
LU MA ME GI VE SA DO
Settimana Selezionata
Mese Selezionato
Anno Selezionato

Fig. 12. Daily pondered OCRA checklist

We can plot the weekly OCRA checklist trend for the right side (see Fig. 13) and the left side (see Fig. 14).

At the moment we are working to obtain similar monthly and annual trends.

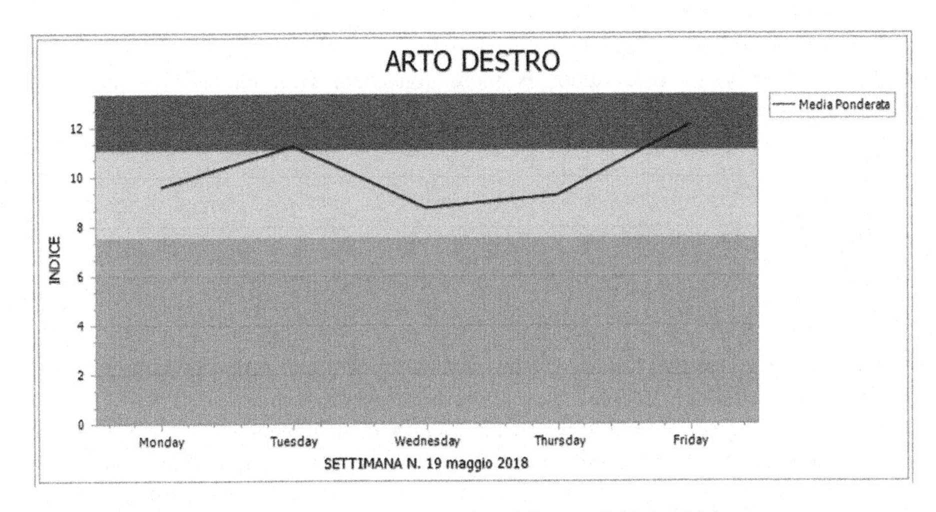

Fig. 13. Weekly OCRA Checklist trend (right side)

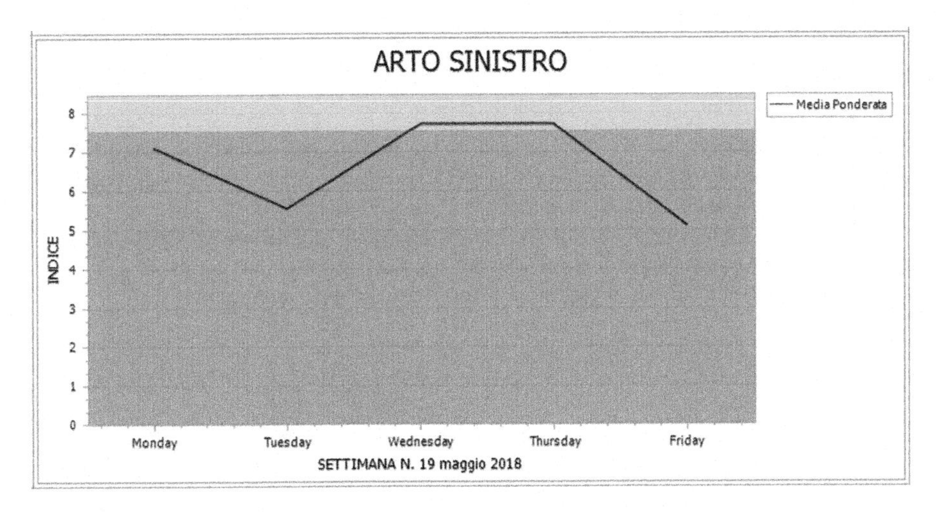

Fig. 14. Weekly OCRA Checklist trend (left side)

4 Conclusion

The proposed methodology highlighted the following strengths and weaknesses:

Advantages:

- Simplification in the calculation of risk indices
- The calculation of risk values is more consistent with changes in individual workers' performances.
- Ability to represent the trend of the risk level over time with good accuracy

Disadvantages:

- The workers must respect the working method
- No consideration of the anthropometric data of the personnel

The next steps to be taken will concern:

- Further checks on the validity of the correlation of direct proportionality between OCRA and efficiency (consistency of the posture score).
- Recording of work breaks through production management software in order to know the duration and the position of the pauses in the shift. In this way it would be possible to verify the variation of the risk when the work breaks change.
- Implement the multitask function.
- Integration with TACOs method.

References

1. Colombini D, Occhipinti E (2014) L'analisi e la gestione del rischio nel lavoro manuale ripetitivo. Manuale per l'uso del sistema OCRA per la gestione del rischio da sovraccarico biomeccanico in lavori semplici e complessi. Franco Angeli Editore
2. Colombini D, Occhpinti E, Battevi N, Cerbai M, Fanti M, Menoni O, Placci M (2010) Movimentazione manuale dei carichi, Dossier Ambiente n°89
3. Petra Automation Software. https://www.petrasoftware.it/automation-e-rpl

Analysis of Ergonomic Risk Factors in a Pharmaceutical Manufacturing Company

Jerrish A. Jose[1], Deepak Sharan[2(✉)], and Joshua Samuel Rajkumar[1]

[1] Department of Physiotherapy, RECOUP Neuromusculoskeletal Rehabilitation Centre, 312, 10th Block, Anjanapura, Bangalore 560108, KA, India
{jerrish, joshua.samuel}@recoup.in

[2] Department of Orthopedics and Rehabilitation, RECOUP Neuromusculoskeletal Rehabilitation Centre, 312, 10th Block, Anjanapura, Bangalore 560108, KA, India
deepak.sharan@recoup.in

Abstract. Ergonomic workplace analysis (EWA) in manufacturing industries have been carried out in various industrial sectors to evaluate the risk of Work Related Musculoskeletal Disorders (WRMSD). A prospective EWA was conducted by a team of ergonomists in a pharmaceutical manufacturing company located in an Industrially Developing Country. Initially, an onsite inspection of workstation was done to identify ergonomic risk factors that might be considered to predispose the workers to WRMSD. Borg Rating of Perceived Exertion (BRPE), Rapid Upper Limb Assessment (RULA), Rapid Entire Body Assessment (REBA) and Rodgers Muscle Fatigue Analysis (RMFA) were used to identify risks in the respective tasks. All the 11 tasks involved high risk of WRMSD involving neck, back, knees, ankle, arm, elbow, hands and fingers and some task involving most body parts on RMFA. Scores of REBA also showed a very high postural risk. The top 3 tasks that had highest ergonomic risk scores evaluated were: 1. filling powder bottle (Task 4), 2. primary packing of tablets (Task 8) and 3. Loading (Task 11).

Keywords: Ergonomic workplace analysis · Pharmaceutical company
Risk assessment

1 Introduction

Ergonomic workplace analysis (EWA) in manufacturing industries have been carried out in various industrial sectors to evaluate the risk of Work Related Musculoskeletal Disorders (WRMSD). However, no studies are available reporting the results of an EWA in the pharmaceutical manufacturing industry. The aim of this study was to estimate the ergonomic risk factors for WRMSD among a group of workers in a pharmaceutical manufacturing company and to provide appropriate recommendations to prevent WRMSD.

S. Bagnara et al. (Eds.): IEA 2018, AISC 825, pp. 355–356, 2019.
https://doi.org/10.1007/978-3-319-96068-5_39

2 Methodology

A prospective EWA was conducted by a team of ergonomists in a pharmaceutical manufacturing company located in an Industrially Developing Country. Initially, an onsite inspection of workstation was done to identify ergonomic risk factors that might be considered to predispose the workers to WRMSD. The task areas were visited and a detailed workplace assessment of the 11 tasks (jet printing, packing, packing of liquid medicine, filling powder bottle, powder packing to carton, changing of cap sealing, sealing of bottle, primary packing of tablets, secondary packing of tablets, packing and stapling, loading) was performed. Work-station information consisted of duration of work, intensity of work, breaks, working postures, working surface, weight of the object, force excretion, etc. Photographs and video recording of the tasks where obtained. Borg Rating of Perceived Exertion (BRPE), Rapid Upper Limb Assessment (RULA), Rapid Entire Body Assessment (REBA) and Rodgers Muscle Fatigue Analysis (RMFA) were used to identify risks in the respective tasks.

3 Results

The results of the study are as follows. Workers performing all 11 tasks were found to adopt hazardous posture and had improper work stations. All 11 tasks involved high risk of WRMSD involving neck, back, knees, ankle, arm, elbow, hands and fingers and some task involving most body parts on RMFA. Scores of REBA also showed a very high postural risk. The top 3 tasks that had highest ergonomic risk scores evaluated were: 1. filling powder bottle (Task 4), 2. primary packing of tablets (Task 8) and 3. Loading (Task 11). The highest risk scores reported in the evaluated scales of BRPE, RULA and REBA were 16, 7 and 12 for the tasks filling powder bottle (Task 4), primary packing of tablets/secondary packing of tablets (Task 8,9) and loading (Task 11) respectively.

4 Conclusions

A high prevalence of ergonomic risk factors for WRMSD was reported among workers in the pharmaceutical manufacturing company based on the ergonomic workplace analysis. Appropriate recommendations were provided to the company to help in prevention of WRMSD and a follow up study is planned to assess the effectiveness of the recommendations.

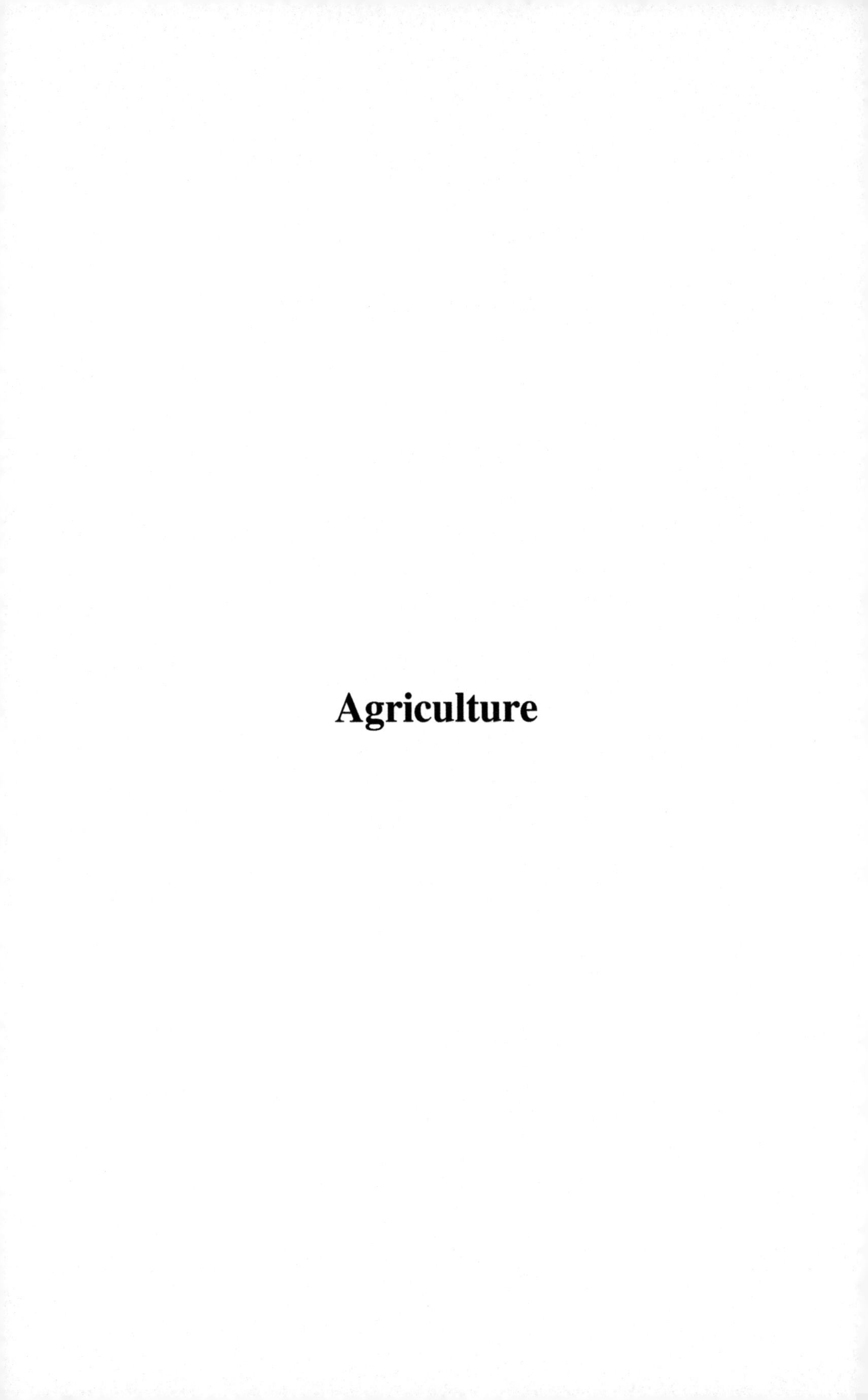

Agriculture

The Work of the Agricultural Pilot from an Ergonomic Perspective

Juliana Alves Faria, Mauro José Andrade Tereso Tereso, and Roberto Funes Abrahão(✉)

School of Agricultural Engineering, University of Campinas, Campinas, SP 13083875, Brazil
roberto@feagri.unicamp.br

Abstract. Brazil has the second largest agricultural aircraft fleet of the world, and more than 70 million hectares are sprayed annually by national agricultural aviation. Previous studies indicate that the work of agricultural pilots involves a series of risks factors and difficulties of physical, cognitive and organizational nature. Although agricultural aviation in Brazil represents only 5% of the national fleet, it accounts for 25% of all air accidents. This study aims to identify the main determinants of agricultural pilot activity, responsible for problems in the field of ergonomics and work safety. The method adopted is based on the application of semi-structured interviews, based on questionnaires, to the main actors involved in the problem: pilots, aircraft designers and air accident investigators. The main results allowed the characterization of the agricultural pilots according to demographic and labor relations criteria. The main ergonomic and occupational safety problems pointed out by the actors refer to aircraft design deficiencies and organizational issues. The conclusion points to the need for adjustments and improvements of several elements of the aircraft cabin, especially the flap lever, seat and joystick. Also, the environmental cabin conditions and organizational issues should also be improved.

Keywords: Agricultural aviation · Ergonomics · Engineering design

1 Introduction

In countries such as Brazil, where agribusiness is one of the main foundations of the economy, representing 21% of the GDP, agricultural aviation stands out as an important production factor.

Brazil has the second largest agricultural aircraft fleet in the world. Each year, 72 million hectares are sprayed by the national agricultural aviation. According to the Brazilian Aeronautical Registry (RAB), the country had 1,560 aircraft in 2010; that number jumped to 2035 in 2015, representing an increase of 23.3% in five years. The number of trained pilots also grows: in 2010, 98 agricultural pilots were licensed; in 2015, 191 pilots, i.e., an increase of 95% of qualified professionals in this segment (Simão 2010).

Agricultural aviation has several particular characteristics and an operating environment significantly different from other aviation segments. Pilots usually operate a

S. Bagnara et al. (Eds.): IEA 2018, AISC 825, pp. 359–366, 2019.
https://doi.org/10.1007/978-3-319-96068-5_40

single-engined aircraft, with minimal infrastructure, in unpaved runways, always taking off with as much cargo as possible, in the warmest periods (summer crops). They make dozens of landings per day, low maneuvers, and handle and apply pesticides and other inputs with low support infrastructure.

During the pilotage, agricultural pilots make maneuvers that require, in addition to great skill, a lot of attention during the entire flight trajectory when spraying. Besides focusing on the operating devices and equipment in the cabin, they need to be aware of several factors: weather conditions, obstacles, production calculations, the available time for the activities, aircraft maintenance, among others.

The work of the agricultural pilot is concentrated in 4 to 5 months of the year, beginning before sunrise and ending with the sunset. Pilots can make up to 90 takeoffs and landings in a single day's work. In the peak period of a crop, they fly up to 12 h a day. In the four or five months of crop, they fly approximately 400 h and make 2,000 takeoffs on average. The strain in these takeoffs is quite high, because the pilot must have the skill of taking off on a reduced runway, with total loading capacity and enough attention to avoid accidents (Prado 2002).

All these factors make the risks associated with the operation higher than those recorded for other aviation segments. Although agricultural aviation represents 5% of the national fleet, it is responsible for approximately 25% of all accidents in the Brazilian civil aviation (ANAC 2015a). In Brazil, between 2008 and 2014, these episodes grew by 60%, along with the growth of agricultural aircraft fleet in 72%, according to data from the Brazilian Aeronautical Registry (ANAC 2015b). The main causes pointed out by investigations are: loss of control in flight (LOC-I, human factor/pilotage); in-flight collision with obstacles (human factor and planning); engine failure in flight (maintenance factor); and loss of control on the ground (human factor/pilotage/infrastructure).

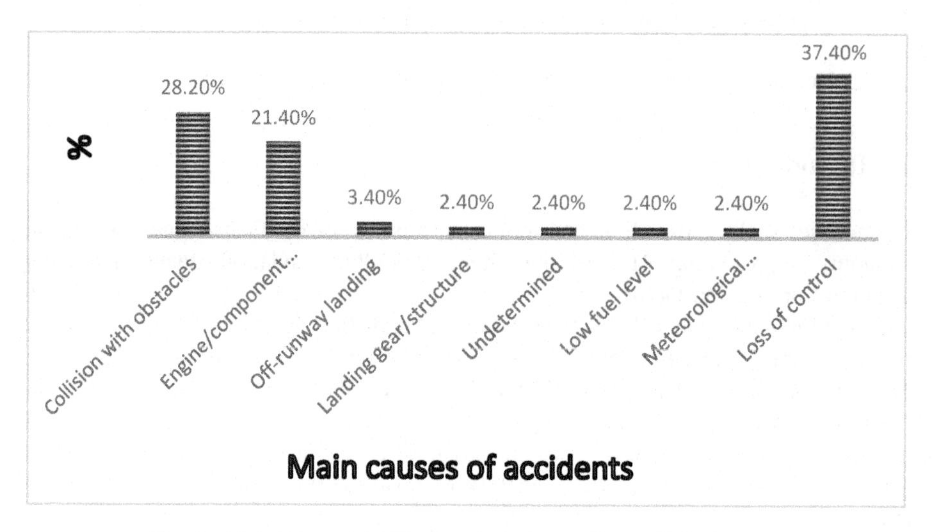

Fig. 1. Main causes of agricultural aircraft accidents (2007–2012).

Zanatta and Amaral (2015) compares the most common causes of accidents in agricultural aviation (Fig. 1), showing that the pilot's loss of control has the highest incidence. This phenomenon raises several questions: what happened at that moment for such an event to happen? Can the cabin size, in relation to the user, disturb and affect the pilot's performance and safety? Is the reach of the devices favorable and does it allow good visualization and user interface? Is the thermal sensation inside the cabin comfortable? These and many other factors mentioned by pilots may also be factors for these statistics of agricultural aviation accidents.

The uniform growth of the two factors – accidents and aircraft – may also suggest failures in the aircraft design, indicating that the industry and supervisory agencies need to accurately investigate the quality and operational condition of these equipment, in addition to the working conditions and its determinants.

2 Methodological Procedures

This research had the participation of 43 agricultural aviation pilots, two agricultural aircraft designers who work in construction companies, and one air accident investigator from the Aeronautical Accidents Investigation and Prevention Regional Service (SERIPA).

Initially, we prepared an interview guide, based on a pilot questionnaire, aiming to understand the context in which the activities of the agricultural pilot are developed. This guide consisted of the following elements: personal information; professional performance; aircraft information (model, maintenance); tasks performed; conditions in the support area (infrastructure and personnel); evaluation of the activity and difficulties in fulfilling the tasks (organizational aspects, equipment evaluation); training; considerations on safety and accidents; health conditions; and poisoning.

We seized the occasion of the National Agricultural Aviation Seminar, held in Primavera do Leste, Mato Grosso (Midwest region of Brazil) between August 13 and 14 of 2015, to apply the final questionnaire among those attending it. We handed out 36 questionnaires and there were 21 respondents. Subsequently, the definitive questionnaire was made available for six months on digital platform. We contacted 31 more pilots to participate in the research.

We conducted semi-structured interviews (based on the questionnaire) with two agricultural aircraft designers who work in construction companies (the one that produces the most widely used model in Brazil and another that produces a more sophisticated model). Finally, we conducted a semi-structured interview with one of the SERIPA VI investigators, to identify aspects regarding safety and causes of accidents in agricultural aviation.

3 Results and Discussion

Forty-three male individuals answered the questionnaire. They had an average age of 38 years, ranging from 24 to 60 years. The average time of experience with agricultural aviation was more than three years, ranging from three months to 25 years of professional experience. Official data show an increasing number of agricultural pilots licensed annually between the 2000 and 2013 (Simão 2010).

Thirty pilots are officially registered with formal contracts. Seven are self-employed and have their own aircraft, and the other six are service providers without aircraft. The six service providers declared working less than six months during the year. The other pilots go beyond this period, working from 10 to 11 months a year. Thirty-one pilots have working hours limited to 44 h per week, and the other twelve pilots work more than this. In this last group, six claimed working weekly journeys exceeding 80 h, which clearly violates the law and may have a direct impact on the health of these workers. The little rest and fatigue recovery time may increase the risk of accidents by affecting their attention and concentration. Most of them highlight their concern with the employment relationship, since the instabilities of the demands cause anxiety and insecurity about the future.

Pilots normally sleep and have their meals in the region where they perform the aerial spraying. After breakfast, they start their activities early in the morning. They make sure of the runway conditions for takeoffs and landings, which constantly change during a workday. These runways are generally precarious and often improvised, nothing more than simple roads in the properties.

The pilots conduct an area recognition flight, which allows the calculation of strips of land and attack area for the aircraft scheduling, supply, and calibration, in addition to the programming of the Differential Global Positioning System (DGPS), to make the work more accurate. They check the spray nozzles and calibrate the discharge.

The pilots' recognition regarding the benefits provided by the use of DGPS during aerial applications is clear. The advantages they mention include the ease for area registration and ease in operations. The speed of real-time information transfer gives it the first place in the ranking of the most used equipment in agricultural aviation. It gives pilots the advantage of staying oriented during flight, with sub-metric precision, resulting in maximum efficiency in air applications of pesticides, avoiding failures by overlaps of strips of land (Manso 1998; Schmidt 2006).

The journey can last for more than 8 h, depending on the extension of the crop area. On average, they carry out 40 takeoffs and landings in a single day of work, after which they refuel the aircraft and the spray tank.

During the activity, the pilots must be attentive to the fuel levels in the tanks and to the product to be applied. They must always be alert to unpredictable obstacles present in this type of operation, such as electric power transmission wires, which are the main cause of aircraft crashes.

During the flight, they monitor the DGPS to control the pulverized strips of land. Weather conditions determine the moment of application and/or impediment to the activity on that day. Wind direction, relative humidity, and light are some of the factors that must be monitored to evaluate and establish procedures.

They are exposed to vibration and high temperatures, especially in aircraft with not air-conditioned cabins.

Most aircraft have no air-conditioning system, forcing the pilots to open small hatches (located on the sides of the cabin or in the aircraft windows) in flight, which can cause contact with particles of the applied products.

The moments of greatest difficulty during the flight are the maneuvers required to achieve the contour of the spray strips and the flight at low altitudes at the time of product application, in which they fly from 3 to 4 meters above the crops. Figure 2 shows the most common formats of maneuvers adopted to cover the area to be sprayed. In the classic turn, the maneuvers are more abrupt, but scanning is continuous. In the racetrack, the maneuvers are smoother, but require greater precision.

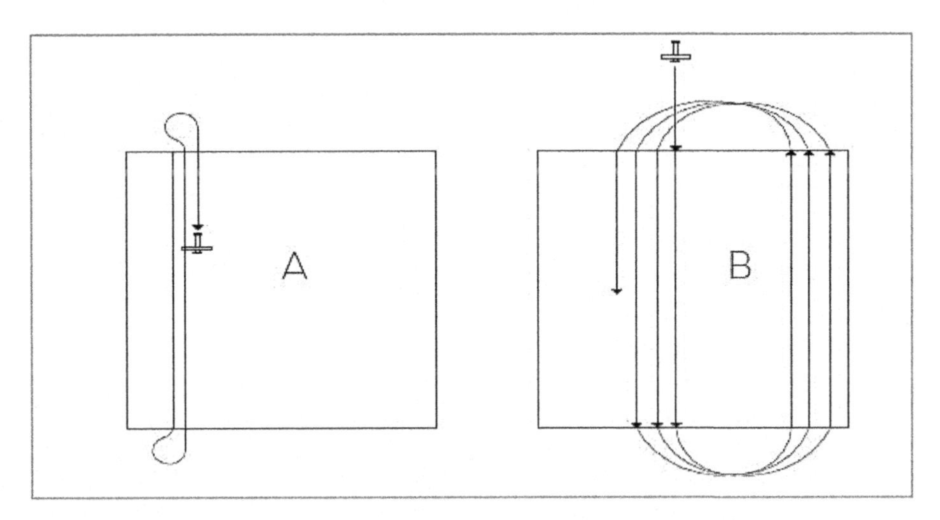

Fig. 2. Maneuvers during operations: (A) Classic; (B) Racetrack. Source: Rasi (2008)

For most pilots, the equipment used for solid or liquid application affect significantly the flight of the aircraft, increasing drag and stall speed. To compensate, they use the strategy of reducing the load on the tank of the product to be sprayed. However, this strategy requires pilots to carry out more landings, decreasing the productivity of their work. When they do not adopt this strategy, risk increases and aircraft performance decreases.

The pilots report that their activities are almost never followed by occupational safety professionals, as is usually seen in other segments of industrial production. Figure 3 summarizes the main issues and difficulties pointed out by the pilots during task fulfillment. Note that the lack of electrical network signaling is the most mentioned problem.

One of the risk factors pointed by the pilots is the proximity to pesticides. Direct contact with these products was reported by 16 pilots; 33 claimed receiving the adequate Personal Protection Equipment (PPE) from the companies.

Chatuverdi (2011), recognizing in aerial spraying activities the risks of chemical contamination to pilots and other workers, calls attention to the importance of occupational health programs for surveillance, detection, and control of the exposure to these products. Many compounds derived from organophosphorus and carbamates are naturally cholinesterase inhibitors. These enzymes, which act in red blood cells,

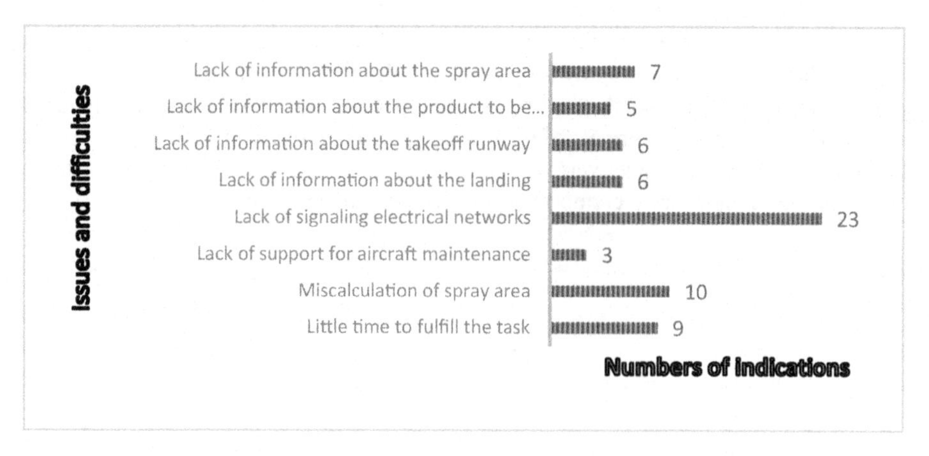

Fig. 3. Issues and difficulties found during operations.

plasma, or the whole blood (30–50% inhibition) can be routinely measured to monitor the exposure to these pesticides.

Foltz et al. (2010) studied the results of audiometric examinations of 41 agricultural pilots in Rio Grande do Sul, Brazil, and showed that more than half of the pilots reported frequent tinnitus and prevalence of hearing loss caused by the noise to which they are exposed during their work. Indeed, Kiefer et al. (1996) evaluated the equivalent noise level produced by two Rockwell Thrush SR2 agricultural aircraft, finding values between 103 and 104 dB(A), which go beyond the tolerance limits set by the regulatory standard NR15.

Regarding the hypotheses for causes of accidents and incidents, 29 pilots consider that they occur because of noncompliance with the norms and/or errors committed in the weight calculation or in the aircraft calibration: *"The pilot is responsible for his actions, and a failed planning increases his risk of falling"; "Pilots in the first crops want to show more than the plane can do and follow the more experienced ones, who no longer have to prove anything to the company, which requires productivity"; "In general, by the lack of proper training... And often because of lack of standardization"; "I need to earn money, because the crop period is short. The excess of pilots in the market leads to the prostitution of the profession."*. Other causes of accidents mentioned by pilots were: low number of practice hours, poor planning of operations; bad weather conditions; delays in equipment maintenance; lack of supervision; and deficiencies in training schools.

The professional responsible for the agricultural aviation accidents investigation department (SERIPA VI) points out the lack of aircraft maintenance, the application of wrong commands, and the incomplete mapping of the obstacles in the application area as the main contributing factors to the occurrence of fatalities. Other mentioned factors are the quantity of work, the long off-season periods, the precarious conditions of their rest hours, the flight at low height, the maneuvers too close to the operating limit of the aircraft, and the direct contact with pesticides. For him, the equipment is currently extremely reliable and mechanical failures rarely occur, and the decisions taken by pilots correspond to 80% of the causes of accidents. Regarding the ergonomic issues, he stated that, for a condition of pilot safety and comfort, the cabin must have an internal space compatible with their anthropometric characteristics. Furthermore, he said that they should be air-conditioned and provide an appropriate range of visual field for the pilot, which would ensure greater safety in the short-term decisions during the maneuvers during the spraying activities. He pointed to the need for changes in: the height of the pedals; the instrument panel to improve the field of vision; the positioning of the flap lever; the fuel tank replacement system; the seat in relation to the lever and panel; the availability of air-conditioning system.

Gasperini and Silva (2006) present, in Fig. 4, a cabin diagram of the most used agricultural aircraft model in Brazil, interacting with a mannequin, whose anthropometric data were provided by the aircraft manufacturer. The diagram shows, in the sagittal plane, the areas of easy (a), moderate (b), and difficult (c) access. The authors show that the flap lever is not even located in the area of difficult access, highlighting the need for a design intervention. In the past 30 years, this aircraft model had little evolution in aspects related to the user's comfort. The modifications were restricted to increasing the loading capacity and replacing some systems.

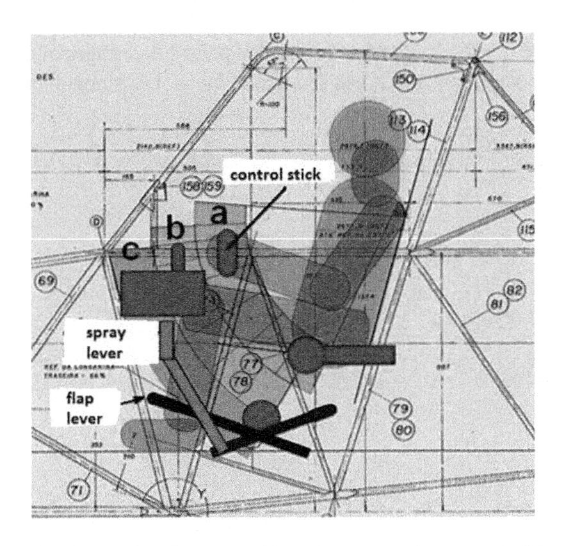

Fig. 4. Cabin diagram with anthropometric mannequin.

From this study it can be concluded that there are various aspects of the aircraft's design (mainly flap lever, seat, joystick), besides cabin's environmental conditions (noise, temperature, ambient air), that should be improved to enhance the safety and comfort of the pilots. Also, work organization issues play an important role in the pilots' unsafety and should be improved too. A greater involvement of the pilots in the process of aircraft improvement would be essential to achieve more effective results.

References

ANAC (2015a) RASO - Relatório Anual de Segurança Operacional. Agência Nacional de Aviação Civil, Brasília

ANAC (2015b) R.A.B. - Registro Aeronáutico Brasileiro (2008). www.anac.gov.br. Accessed 1 Nov 2015

Chatuverdi AK (2011) Aerospace toxicology overview: aerial application and cabin air quality. Rev Environ Contam Toxicol 9(1):15–40

Foltz L, Soares CD, Reimchembach MAK (2010) Perfil Audiológico de Pilotos Agrícolas. Arquivos Internacionais de Otorrinolaringologia 14(3):322–330

Gasperini R, Silva JCP (2006) Desenvolvimento de um novo conceito de cabine ergonômica para linha aeronáutica agrícola. In: SIMPÓSIO DE ENGENHARIA DE PRODUÇÃO (SIMPE, vol 13, pp 1–13. UNESP

Manso JA (1998) GPS uma abordagem prática. Bagaço, Recife

Prado FRW (2002) Investigação de Acidentes na Aviação Agrícola. CENIPA - Centro Nacional de Investigação de Prevenção de Acidentes Aeronáuticos, Brasília

Schimidt F (2006) A aviação agrícola no Brasil: um modelo para seleção de aviões com o uso da programação linear. Dissertation, Universidade Estadual do Oeste do Paraná

Simão AC (2010) Acidentes nas Operações Aeroagrícolas: Análise do Fator Humano. Revista Conexão Sipaer 1(3):130–148

Zanatta M, Amaral FG (2015) Problemas suportados pelos pilotos agrícolas e suas relações com segurança e saúde ocupacional. Revista Produção Online, Florianópolis, SC 15(4):1195–1223

How to Improve Farmers' Work Ability

Merja Perkiö-Mäkelä and Maria Hirvonen(✉)

Finnish Institute of Occupational Health, Kuopio, Finland
maria.hirvonen@ttl.fi

Abstract. The aim of this study was to identify the factors contributing to the improvement of farmers' work ability. The study sample comprised the full-time farmers (n = 2169) of the 'Occupational health and agriculture in Finland' study. The research material was analysed using cross tabulation and logistic regression methods, and we examined the relationship between perceived work ability and variables related to farmers' work, management, working conditions, workload, job satisfaction, health, and health-related behaviour. The 'good work ability' group (8–10 on a scale of 0–10) mostly included those under 64 years of age; those who placed a high value on managerial duties; those who felt that their work was not so physically strenuous; those who experienced work engagement more frequently; those who felt that their health was good; farmers with no debilitating long-term illness or injury diagnosed by a doctor; farmers who had not had, during the 30 days prior to the study, both musculoskeletal and mental symptoms, either for a prolonged period or recurrently; and farmers of normal weight.

Keywords: Work ability · Health · Agriculture

1 Introduction

In the last decades, Finnish agriculture has undergone great changes. Field areas and the sizes of herds per farm have increased, whereas the number of actual farms has decreased. Mechanization and automation have taken over.

In 2017, Finland had 48 562 agricultural and horticultural enterprises. The number of farms dropped by approximately1000 in 2016. The number of farmers and farms has decreased because many small farms have stopped production. The average agricultural area utilized by farms was 47 hectares. The average age of farmers on privately owned farms was 52 years. About 86% of farms were family run and 9% cent were farming syndicates; less than 3% were run by successors; and less than 2% were limited companies. Approximately 9% of all farmers on privately owned farms were under the age of 35, and 27% were over the age of 60. In 2017, cereals production was the most common farm operation in Finland (34% of farms); second, other plant production (30% of farms); and third, milk production (14% of farms) (Luke, Statistics database 2018).

According to a broad definition of work ability, humans and work make up a whole, which is affected by human resources, doing, the ability to function, knowing, working conditions, contents of work, work community, and organization. (Gould et al.

S. Bagnara et al. (Eds.): IEA 2018, AISC 825, pp. 367–374, 2019.
https://doi.org/10.1007/978-3-319-96068-5_41

2006). The Work Ability Index is a broadly used method for analysing work ability (Tuomi et al. 1997). It has been found that the response to the first question of the index (Let us assume that your best possible work ability is rated 10 points on a ratings scale. How would you rate your work ability today on a scale of 0–10?) correlates strongly with the whole work ability index scale (Ahlström et al. 2010).

The aim of this study was to identify the factors that contribute to the improvement of farmers' work ability.

2 Material and Methods

2.1 Materials

The sample in this study comprised the full-time farmers (n = 2169) of the 'Occupational health and agriculture in Finland' study (MTH2014). The aim of MTH2014 was to produce follow-up information on farmers' well-being at work; on work, working conditions and health. In addition, the study examined farmers' satisfaction with and their perceived benefits of occupational health services (OHS), and the areas that needed development (Perkiö–Mäkelä et al. 2016).

The sampling frame of the study comprised farms that submitted a subsidy application in 2014. The final sample consisted of 5774 farmers. Of these, 3117 (54%) were interviewed, 1370 (24%) refused to participate in the study and 1287 (22%) could not be reached. The response rate was 54% for the entire final sample and 69% among those reached. The data are suitably representative of Finnish-speaking farmers aged 18–68 years (Perkiö–Mäkelä et al. 2016).

A total of 2169 full-time and 948 part-time farmers were interviewed for the study. The results were weighted as equivalent to the production sector distribution of the entire random sample. The weighted number of full-time farmers was 1921 (62%) and that of part-time farmers 1196 (38%) (Perkiö–Mäkelä et al. 2016).

The mean age of full-time farmers was 50 and 71% had a vocational qualification in agriculture. Nearly nine out of ten (89%) were men and 11% women. Plant production represented 54% of farms, and domestic animal production and livestock production 46% of farms (Perkiö–Mäkelä et al. 2016).

2.2 Methods

MTH2014 was descriptive and cross-sectional. Most of the data were collected through telephone interviews, which were conducted between October 2014 and January 2015. The average interview length was 46 min for full-time farmers and 29 min for part-time farmers. The thematic areas of the telephone interview were background information related to farmers and their work, working conditions and working methods, mental well-being, management, accidents, physical workload, perceived work ability, health, lifestyle, continuing at work, and OHS (Perkiö–Mäkelä et al. 2016).

We analysed the research material using cross tabulation and logistic regression methods, and we studied the relationship between perceived work ability and the variables related to farmers' work, management, working conditions, workload, job satisfaction, health, and health-related behaviour. We used SAS (statistical analysis system) software (version 12.3) for the statistical analysis.

The background variables used were: production sector, age, gender, marital status, vocational qualification in agriculture, number of employees, and utilized agricultural area.

Work ability as a dependent variable was based on the following question:

- Let us assume that your best possible work ability is 10 points on a ratings scale. How would you rate your work ability today on a scale of 0–10? A rating of 0 would mean that you are unable to work. (0–7/8–10)

The independent variables in the different categories were:

- Farm work: development and changes in farm work, and successor of the farm
- Work outside the farm: side-line work, contractor work (selling and purchasing) and neighbourly help (sum variable)
- Management: meaningfulness, importance and strain
- Working conditions: perceived harm factors (noise, vibration, lighting, gas, air in production building, dust, mould, poisonous or irritating chemicals), accidents, safety at work, and risk management
- Physical workload: perceived physical strain of work, perceived harm of work postures and lifting
- Mental workload: perceived mental strain of work and the sum of mental strain factors at work
- Job satisfaction: job satisfaction in farm work, importance and meaningfulness of work, and work engagement
- Recovery from work strain
- Support for mental well-being
- Health: perceived health, chronic diseases diagnosed by doctor, musculoskeletal and mental symptoms, perceived stress
- Health-related behaviour: leisure-time physical activity, smoking, sleeping, alcohol consumption, and body-mass index
- OHS

Statistical analysis proceeded as follows

1. We tested the relationship between work ability and the single independent variable with crosstabs. Statistical significance was tested using the $\chi 2$ test ($p < 0.05$).
2. Single independent variables were put into the logistic regression analysis according to category, and statistically significant variables were selected for the final analysis. Statistical significance was tested using odds ratios and their 95% confidence intervals.
3. Finally, every statistical significant variable from the different categories were put in the same logistic regression analysis. Statistical significance was tested using odds ratios and their 95% confidence intervals.

3 Results

3.1 First Step of Analysis

The full-time farmers gave their work ability an average rating of 7.8 on a scale of 0–10. One in three (33%) full-time farmers rated their work ability as 0–7 on a scale of 0–10 and two thirds (66%) rated it as 8–10 (good work ability). Age, vocational qualification, number of employees, and utilized agricultural area were statistically significantly associated with work ability, tested using the $\chi 2$ test ($p < 0.05$) (Table 1).

Table 1. Perceived work ability on a scale of 0–10. Full-time farmers, n = 2164. Statistical significance, $\chi 2$ test (p), %

Perceived work ability	Points 0–7	8–10	N
All	33	67	2164
Production sector		p = 0.5831	
Dairy farming	32	68	798
Other cattle farming	34	66	271
Other livestock farming	29	71	216
Cereals production	34	66	558
Other plant production	36	65	321
Age, years		p = 0.0001	
<35	12	88	198
35–44	19	81	416
45–54	30	70	786
55–64	47	53	671
64>	56	44	93
Gender		p = 0.9395	
Male	33	67	1923
Female	33	67	241
Marital status		p = 0.6873	
Married/unmarried cohabiting	33	67	1715
Unmarried/divorced/widow	32	68	454
Vocational qualification		p < 0.0001	
No	38	62	1159
Yes	28	72	1005
Number of employees		p = 0.0002	
None	35	65	1613
At least one	26	74	551
Utilized agricultural area, hectares		p < 0.0001	
0–49	38	62	1137
50–	28	72	1032

Eighty per cent (80%) of full-time farmers received the majority (75–100%) of their income from agriculture and forestry. Food processing, tourism or care services and other contracting generated only a small share of income. Farmers considered management-related tasks quite meaningful and important for the farm's success, but also a source of additional workload. All the responses to the questions on farm work (development and changes in farm work and successor of the farm), work outside the farm [side-line work, contractor work (selling and purchase) and neighbourly help (sum variable)], and management (meaningfulness, importance and strain) were statistically significantly associated with work ability, tested using the $\chi 2$ test ($p < 0.05$).

Of the full-time farmers, 19% regarded dust, 15% noise, 8% lighting, 7% vibration, 4% mould, 4% poisonous or irritating chemicals, and 3% the gas and air in production buildings as quite or very harmful. Altogether 13% of the full-time farmers had suffered an occupational accident requiring a doctor's treatment. Accidents most commonly occurred when tending to cattle, carrying out forestry work, or maintaining and repairing machinery. The most common causes of occupational accidents were injury inflicted by an object, falling, or contact with an animal. All responses to the questions concerning working conditions: perceived harm factors (noise, vibration, lighting, gas, air in production building, dust, mould, poisonous or irritating chemicals), accidents, safety at work, and risk management were statistically significantly associated with work ability, tested using the $\chi 2$ test ($p < 0.05$).

Nearly every other full-time farmer perceived their work to be physically quite or very straining (47%). Every fifth full-time farmer considered awkward working postures (21%) and heavy lifting and carrying (21%) quite or very harmful. All responses to the questions on physical workload (perceived physical strain of work, work postures and lifting) were statistically significantly associated with work ability, tested using the $\chi 2$ test ($p < 0.05$)

Two thirds (66%) of the full-time farmers were quite or very satisfied with their work. Nevertheless, constant rushing, tight schedules and the farms' economic uncertainty troubled farmers. Two thirds (65%) of full-time respondents experienced little or no stress. A total of 40% of full-time respondents felt that their work was mentally quite or very straining. Nine out of ten (91%) full-time farmers felt that they recovered well or relatively well after the working day. All responses to the questions on job satisfaction (job satisfaction with farm work, importance and meaningfulness of work, and work engagement) and mental workload (perceived mental strain of work and the sum of mental strain factors at work), recovery, and support were statistically significantly associated with work ability, tested using the $\chi 2$ test ($p < 0.05$).

Two thirds (63%) of the full-time farmers considered their own health to be good in comparison to others their age. Approximately every fourth (23%) had a chronic disease that interfered with their work. The most common of these diseases were musculoskeletal disorders (12%). Seven out of ten (71%) full-time farmers had suffered a long-term or recurring musculoskeletal symptom during the past month. The most common musculoskeletal symptoms were trouble in the neck and shoulder region (42%), aches in the upper arms (38%) and pain in the lumbar region (38%). All responses to the questions on health (perceived health, chronic diseases diagnosed by doctor, musculoskeletal and mental symptoms, perceived stress) were statistically significantly associated with work ability, tested using the $\chi 2$ test ($p < 0.05$).

A total of 42% of the full-time farmers exercised at least twice a week. Every third (35%) farmer was of normal weight. Two out of three full-time farmers (65%) had never smoked, and 86% were not in the alcohol consumption risk group. Nearly three quarters (79%) of the full-time farmers slept sufficiently (6.5–8.5 h). Every fifth (18%) slept less than the amount recommended (under 6.5 h). Leisure-time physical activity, sleeping and body mass index were statistically significantly associated with work ability tested, using the $\chi 2$ test ($p < 0.05$).

A total of 61% of full-time farmers had purchased farmer-specific OHS. These farmers perceived their work ability as good (71%) more often than those who had no OHS (60%) ($\chi 2$ test $p < 0.0001$).

3.2 Second Step of Analysis

According the second step of the analysis, the background variable of age and following independent variables were included in final analysis:

- Farm work: development and changes in farm work and successor of the farm
- Work outside the farm, sum variable
- Management: meaningfulness, importance and strain
- Working conditions: perceived harm of vibration, dust, mould and accidents, safety at work and risk management,
- Physical workload: perceived physical strain of work, perceived harm of work postures and lifting
- Mental workload: perceived mental strain of work and sum of mental strain factors at work
- Recovery
- Job satisfaction: job satisfaction in farm work, work engagement
- Support: support from relatives or other entrepreneurs
- Health: perceived health, chronic diseases diagnosed by doctor, musculoskeletal and mental symptoms, perceived stress
- Health-related behaviour: leisure-time physical activity, sleeping, body mass index and
- OHS

3.3 Third Step of Analysis

According to the final step of analysis, the ‘good work ability’ (8–10 on a scale of 0–10) group most included those under 64 years of age; those who placed a high value on managerial duties; those who felt that their work was physically no more than somewhat strenuous; those who felt that heavy lifting and carrying was no more than somewhat bothersome; those who experienced work engagement more frequently; those who felt that their health was good; farmers who had no debilitating long-term illness or injury diagnosed by a doctor; farmers who had not had, during the 30 days prior to the study, both musculoskeletal and mental symptoms, either for a prolonged period or recurrently, and farmers of normal weight (Table 2).

Table 2. Factors associated with good perceived work ability (score 8–10). Full-time farmers, n = 2164. Odds ratios (OR) and their 95% confidence intervals (95% CI).

Good perceived work ability	OR	95% CI
Age, years		
<35	13.9	5.5–35.1
35–44	14.6	6.7–31.5
45–54	6.1	3.1–12.0
55–64	2.4	1.3–4.5
64>	1	
Importance of management		
Quite or very important	2.3	1.1–4.7
Not so important	1	
Physical strain at work		
Light or some strain	1.4	1.1–2.0
Quite a lot or very much strain	1	
Heavy lifting		
Not at all or somewhat harmful	2.0	1.3–3.0
Quite or very harmful	1	
Work engagement		
More	2.2	1.6–3.2
Less	1	
Perceived health		
Quite or very good	5.5	4.0–7.5
Poor or average	1	
Chronic diseases diagnosed by doctor		
No	1.9	1.3–2.7
Yes	1.8	1.2–2.8
Harm in agricultural work	1	
Musculoskeletal and mental symptoms		
Neither	3.2	2.0–5.1
Only musculoskeletal symptoms	1.9	1.3–2.7
Only mental symptoms	3.7	1.9–7.2
Both	1	
Body mass index		
<25	1.4	1.0–1.2
≥25	1	

4 Conclusions

All independent variables, except smoking and alcohol consumption, were statistically significantly associated with work ability, tested using the $\chi 2$ test ($p < 0.05$). Statistical analysis showed that the 'good work ability' group (8–10 on a scale of 0–10) mostly included younger farmers, those who placed a high value on managerial duties, those who felt that their work was not so physically strenuous, those who were more engaged in their work, and those who were in better health.

Good health is an important basis on which farmers perceive their work ability as good. If our aim is to prevent a reduction in farmers' work ability, we need health-promoting measures, good medical care, and alleviation of musculoskeletal and psychiatric symptoms.

There was a clear correlation between physical workload and perceived work ability. All possible ergonomic measures, particularly those related to handling heavy loads must be taken to reduce the physical workload of agricultural work. Such means and measures include, for example, development of the work environment and working methods, the use of auxiliary devices, appropriate tools, and the planning of working practices. Assistance from OHS specialists, especially in connection with farm visits, is an immense help.

Feelings of work engagement can also significantly contribute to perceived work ability among agricultural entrepreneurs. The opportunity to benefit from one's experience and skills and to improve challenging work, the content of which one can modify and develop, as well as the opportunity to see the results of one's work, are important factors contributing to work engagement.

References

Ahlström L, Grimby-Ekman A, Hagberg M, Dellve L (2010) The work ability index and single–item question: associations with sick leave, symptoms, and health – a prospective study of women on long–term sick leave. Scand J Work Environ Health 36(404):12

Gould R, Polvinen A (2006) Työkyky työuran loppupuolella. Kirjassa R Gould ym. (toim.) Työkyvyn ulottuvuudet. Terveys 2000–tutkimuksen tuloksia. (in Finnish) Eläketurvakeskus, Kansaneläkelaitos, Kansanterveyslaitos, Työterveyslaitos, Helsinki

Luke [Natural Resources Institute Finland]. Agricultural statistics. http://stat.luke.fi/en/maatalous

Perkiö-Mäkelä M, Hirvonen M, Kinnunen B, Koponen M, Louhelainen K, Mäittälä J, Sipponen J, Torpström A (2016) Occupational health and safety in agriculture in Finland 2014. Finnish Institute of Occupational Health, Helsinki. (in Finnish). http://www.julkari.fi/handle/10024/130362. Summary in English www.ttl.fi/maatalous

Tuomi K, Ilmarinen J, Jahkola L, Tulkki A (1997) Work ability index, Occupational health services 19. Finnish Institute of Occupational Health, Helsinki (in Finnish)

Tuomi K, Seitsamo J, Ilmarinen J (2006) Työkyvyn moninaisuus ja työkykyindeksi. Kirjassa R Gould, Ilmarinen J, Järvisalo J, Koskinen S. (toim.) Työkyvyn ulottuvuudet. (in Finnish) Terveys 2000 -tutkimuksen tuloksia. Eläketurvakeskus, Kansaneläkelaitos, Kansanterveyslaitos, Työterveyslaitos, Helsinki

A Study to Develop the Framework of Estimating the Cost of Replacing Labor Due to Job-Loss Caused by Injuries Based on the Results from Time Study in Agriculture of Korea

Hee-Sok Park[1(✉)], Yun-Keun Lee[2], Yuncheol Kang[1], Kyung-Suk Lee[3], Kyung-Ran Kim[3], and Hyocher Kim[3]

[1] Hongik University, Seoul, Korea
hspark720@naver.com
[2] Wonjin Institute for Occupational and Environmental Health, Seoul, Korea
[3] National Institute of Agricultural Sciences, Jeonju, Korea

Abstract. Measuring the time for human-driven agricultural work and establishing the standard time for the work are essential for estimating compensation or insurance cost when work-related damage and/or disaster are occurred. Our study aims to investigate the steps required for cultivating crops, and try to estimate the time and amount of job-loss due to the injuries occurring in the agricultural environment. From the analysis we found the work time recorded in the diary was greater than the time measured using GPS by 7% on average. Additionally, we develop the Excel-based macro application that can calculate the costs for the replacing labor forces based upon the data obtained from the time study.

Keywords: Agriculture · Work time · Replacing labor

1 Introduction

The rapid development of the Korean economy is reflected strongly in the changing role of agriculture. Until the 1960s, agriculture generated almost half of Korea's GDP and in 1970, agricultural production continued to contribute 25.5% of GDP and the labor force employed in the agricultural sector accounted for 50.5% of the country's total labor force. As the industrialization process progressed, however, the share of agriculture in the national economy declined sharply. In 2016, the share of agricultural production in GDP was 2.0% and the agricultural population accounted for 5.1% of the total population. Nevertheless, despite this decline agriculture continues to play an important role in the Korean national economy, accounting for a relatively large share of GDP, and with a large rural population and employment as compared to other OECD countries.

In terms of occupational health and safety, agriculture is considered as one of the most hazardous industries for workers all over the world. Farmers have a higher risk of

S. Bagnara et al. (Eds.): IEA 2018, AISC 825, pp. 375–379, 2019.
https://doi.org/10.1007/978-3-319-96068-5_42

work-related injury and illness than workers of most other occupations [1]. Farmers in Korea are facing a difficult situation of increasing longer work time and work intensity due to shortage of workforce by aging and empty of rural community. Besides, though dependency on the farming equipment and pesticides are increasing, and they are seriously exposed to musculoskeletal disorders, a compensation related with the accidents and disasters is meager.

In Korea, almost all businesses and workplaces with one or more workers have a comprehensive occupational safety and health system for preventing industrial accidents and establishing a healthy and safe work environment, based on the Labor Standards Act, the Industrial Safety and Health Act, and the Industrial Accident Compensation Insurance Act. These laws have benefited workers through improved regulation of the working environment, monitoring of work-related injuries and diseases, and ensuring regular medical examinations and compensation for industrial accidents [2].

Unfortunately, some workers have been excluded from eligibility under the laws. For example, the workers employed by businesses in the fields of agriculture that are run by non-corporate entities with less than five full-time workers are not covered by these laws. In addition, self-employed farmers and unpaid family workers are not included in either act [2].

Therefore, efforts to arrange for farmers to engage in farming stably by providing a proper compensation for the safety related incidents during farming work have been pursued by government and political circles (Ministry of Agriculture, Food and Rural Affairs, 2016), resulting in the Act on Safety Insurance for Farmers and Fishers and Prevention of Work Accidents [3].

Measuring the time for human-driven agricultural work and establishing the standard time for the work are essential for estimating compensation or insurance cost when work-related damage and/or disaster are occurred. To measure the work time, several methods have been utilized, including time diary, work sampling, and self-reported questionnaire. There is no consistent conclusion which method provides the most accurate result. Time diary is a popular approach since it is easy and convenient to use. In terms of accuracy, however, this method is prone to recall error, and it is also often cumbersome for the farmers working outdoors to fill out the requested form. To minimize the recall error, it is necessary to record the work time whenever each step of the work is finished, which is another work burden for the farmers. Therefore, a supplementary approach is needed to enhance the accuracy of the time-diary method.

Our study aims to investigate the steps required for cultivating crops, and try to estimate the time and amount of job-loss due to the injuries occurring in the agricultural environment. Additionally, we develop the Excel-based macro application that can calculate the costs for the replacing labor forces. This study will be helpful for coming up with fair compensation policies, which would be beneficial to both policy makers and injured workers who claims the compensation.

2 Methods

Time diaries and the data measured using GPS device were gathered from Kyeong-gi province, South Korea in 2015. 40 farmers agreed to participate in this study (all male, average age 58.4 years). Although the GPS provides correct information on the location at specific time, it cannot supply the information on the contents of the work done at the location. Therefore, the time-diary method needs also to be accompanied to extract the information. To minimize the recall error, the farmers were asked to record the start, end times and the contents of the works in the diary right after completion of each work step, not after all daily works are done. A GPS device (model ASCEN GPS850: 72 g, 81.7 mm × 54 mm × 22 mm) was attached to the farmers. It was attached lest it should interfere the farmer's work. The geographical information from GPS was overlaid onto the Google map to determine the location of the farmer at the specific time (Fig. 1). The work times for 92 work steps of 5 crops (apple, peach, pear, grape, and persimmon) were recorded.

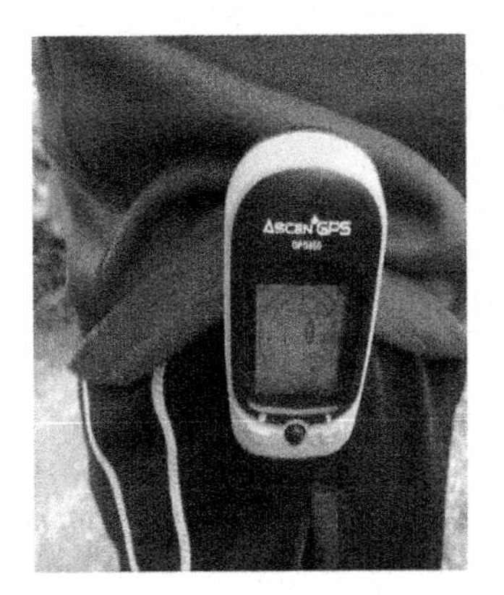

Fig. 1. A GPS device attached (left), Traces of a farmer overlaid onto Google map (right)

3 Results

3.1 Analysis of Work Time

Work time which was recorded in the diary is arranged according to the major work stage as in Table 1. Time for harvest for peaches and pears occupied 34.2% and 43.6%, respectively among the total work time, while weeding work for apples, grapes, and sweet persimmons occupied more than 20%.

The weight of indirect work time by crop showed that pears were the highest with 14.6% followed by peaches at 8.5%, grapes at 7.5%, apples at 4.7%, and sweet persimmons at 3.5%. Indirect work-time for five fruit crops was 7.1% among the total.

Higher indirect work for the grapes was due to higher processing work as compared with other fruit crops. In the indirect work, sales were included. In apples, peaches, and pears, time related with orchard management/maintenance was the highest, followed by

Table 1. Time for major steps (h/10a/year) (%)

Crop	Total (h)	Harvesting	Disinfection	Cutting off sprouts	Weeding	Others
Apple	87.39	8.20 (9.4)	6.93 (7.9)	11.79 (13.5)	20.40 (23.3)	40.07 (45.9)
Pear	26.66	11.63 (43.6)	5.36 (20.1)	0.00 (0.0)	1.03 (3.9)	8.64 (32.4)
Grape	111.55	18.54 (16.6)	5.22 (4.7)	21.45 (19.2)	27.19 (24.4)	39.15 (35.1)
Persimmon	29.38	8.19 (27.9)	4.88 (16.6)	3.72 (12.7)	8.58 (29.2)	4.01 (13.7)
Peach	50.46	17.28 (34.2)	4.49 (8.9)	3.37 (6.7)	7.57 (15.0)	17.75 (35.2)
Total	61.09	12.77 (20.9)	5.37 (8.8)	8.07 (13.2)	12.95 (21.2)	21.93 (35.9)

shipping related works, work preparation (packing and harvest), maintenance and repair of the facilities and machines, and agricultural items purchase (pesticides and package related items).

Process related work-time was the largest in the pears, while harvest related work-times were the highest in apples, pears, grapes, and peaches. Generally, processing took the highest work-time, followed by outgoing, maintenance and repair of facility/equipment.

3.2 Microsoft Excel Program

We developed a replacement labor cost calculator which is an excel program that can calculate the costs for the replacing labor forces. Input information includes the gender of the worker, type of crop, area of cultivation, market cost of a typical worker, and the activities that the replacing workers should carry out. Using the data from the time study, the excel program produces the cost of the replacing worker.

4 Conclusion

It was found that even though the farmers were subject to record the start and end times of the works in the diary right after completion of each work step, they overestimated their work times. From the analysis we found the work time recorded in the diary was greater than the time measured using GPS by 7% on average.

Time diary method, experimental sampling method, and questionnaire method, though have limits, these are recommended as recording methods for work time [4]. It has been reported that with daily diary method, time was tended to be overestimated as compared with the questionnaire method [5], and accuracy was more or less low as compared with the direct observation [6]. If continuous observation equipment like

GPS which is adopted in this study is used, reliability for measurement and data collection would be raised.

If we assume that the location and time information from GPS device is accurate, the information could be used to supplement the time-dairy method. GPS, however, provides only location and time information, therefore the contents of the work need to be identified from the time-diary. In this pilot study, it was shown that GPS could supplement time diary method when measuring work time. To our best knowledge, estimating work time using a GPS device is the first attempt in agricultural environment. A limitation of this study is the small size of study sample, and further studies are in progress.

Acknowledgement. This study was supported by Rural Development Administration (PJ010079052018).

References

1. Frank AL, McKnight R, Kirkhorn SR, Gunderson P (2004) Issues of agricultural safety and health. Annu Rev Publ Health 25:225–245. https://doi.org/10.1146/annurev.publhealth.25.101802.123007
2. Kwon YJ, Lee SJ (2014) Compensation for occupational injuries and diseases in special populations: farmers and soldiers. J Korean Med Sci 29(Suppl):S24–S31. https://doi.org/10.3346/jkms.2014.29.S.S24
3. Ministry of Agriculture, Food and Rural Affairs of Korea (2016) Law on the safety insurance and disaster prevention
4. Jacobs J (December 1998) Measuring time at work: are self-reports accurate? Monthly Labor Rev
5. Koning I, Harakeh Z, Rutger C, Engels R, Vollerbergh W (2010) A comparison of self-reported alcohol use measures by early adolescents: questionnaires versus diary. J Subst Use 15:166–173. https://doi.org/10.3109/14659890903013091
6. Prince S, Adamo K, Hamel M, Hardt J, Gorbe S, Tremblay M (2008) A comparison of direct versus self-report measures for assessing physical activity in adults: a systematic review. Int J Behav Nutr Phys Act 5. https://doi.org/10.1186/1479-5868-5-56

The Gap to Achieve the Sustainability of the Workforce in the Chilean Forestry Sector and the Consequences over the Productivity of System

Felipe Meyer(✉)

University of Concepcion, Concepcion, Chile
fmeyer@udec.cl

Abstract. The forestry sector is facing a number of challenges that having impact of the sustainability of the workforce and over productivity of the system. This study examines the relationship between the working conditions in the forestry sector and the workers. In addition examines the strategies the Chilean forestry companies are using. Finally, using two examples identify the consequence over the productivity of the system. The overall objective of this study was to investigate the impact of working conditions on the workforce in the Chilean forestry sector and over the productivity of the system. The study involved data collection from 350 forestry workers, both Chilean forest companies and contractors companies along with interviews with managers and experts in the area of forestry, ergonomics, working conditions and health and safety and the study cases. The findings indicate that even though working conditions in the Chilean forestry sector have been improve, they continue to have a negative impact on workers in terms of occupational health, market attractiveness and ageing population. Strategies that forestry organizations have implemented these are not enough to solve the problems mentioned before. The strategies are focuses on the prevention of accidents rather the OH problems and none of the strategies pays attention to the wellbeing of the workforce and the development of resources, aspects the workers demand. Finally, due the problem mentioned above, a negative impact over the productivity of the systems, based on examples related with pruning and harvesting.

Keywords: Sustainability workforce · Forestry sector · Working condition

1 Introduction

Sustainability principles have been growing significance in the forestry sector in particular [1–3]. The idea of using the concept of sustainability in business management is based on the strong belief that in order to be successful in the long term an organisation should have solid foundations, socially, economically and environmentally [4].

The empirical domain of this research is the forestry sector in Chile. The main aim is to explore the sustainability of the workforce in the forestry sector, a topic related to the social aspects of sustainability. This is a fundamental aspect of the production

S. Bagnara et al. (Eds.): IEA 2018, AISC 825, pp. 380–391, 2019.
https://doi.org/10.1007/978-3-319-96068-5_43

system in Chile is based mainly on workforce and their capacity. Social sustainability encompasses the impact of products or operations on human rights, labour, health, safety, regional development and other community concerns [5, 6] This research is informed by the incremental number of problems linked with social sustainability of the forestry workers internationally [7]. It is argued that a significant reason for the increase in problems in this area is due to the poor relationship between workers' capacity and working conditions [8].

However, if one looks at what is happening in the forestry sector worldwide it is possible to find other items that affect the sustainability of the workforce in the forestry sector. First, the attraction of the forestry sector to workers and people seeking employment has decreased [9] while the demand in some countries for workers has increased [10–12]. Employing and retaining people to work in the forestry sector was identified as the second biggest issue facing forestry businesses after environmental aspects [9]. The forestry sector is also facing the problem of recruiting and training fast enough to keep pace with business opportunities [7].

In Canada, a report steered by a committee of industry experts and British Columbian government representatives indicated that industry demand for skilled workers will increase by 26% in the next 10 years, while the occupational supply of labour is predicted to grow by only 8% [11]. According to the report, 75% of jobs needed are in categories already faced with high vacancy rates, including hand fellers (17%), forestry workers (13%) and logging machine operators (7%) [11]. It can be assumed, therefore, that there will be a lack of workers in the forestry sector over the next few years [10, 13].

In the particular case of the Chilean forestry sector, CORMA, a Chilean forestry association, developed a study in 2015 that concluded that this sector would need more than 10,000 forestry workers by 2030, with the workers in logging activities the most demanding positions to fill [10]. A second factor affecting the sustainability of the workforce in the forestry sector is that the forestry workforce is getting older [7, 10] and its capacity for work is decreasing at the same time [14]. This is an important issue since most of the activities in the forestry sector are physically very demanding [15].

These issues, the ageing of the workforce and the lack of interest in working in the forestry sector combined with incremental risk of OH problems among an older workforce, could be a threat to forestry organizations from a business sustainability point of view [16, 17]. Since it is widely perceived that the workforce is an increasingly critical factor in improving and sustaining organisational health in general [18] and the forestry sector in particular [9]. The workforce in the particular case of Chilean forestry is vital for the continuing development of this sector as, due mainly to the rugged landscape in which commercial forests are located, mechanized activities are not always possible [10]. Therefore, any aspects that could affect them, such as OH issues, an ageing population, and low employment attractions, will have an important impact from both an economic and social point of view in the Chilean forestry sector [10, 14].

Finally in terms of importance for the economy, the Chilean forestry sector, including silviculture, logging and industrial activities like wood elaboration, cellulose and paper production, is the second-largest exporting industry in Chile, behind large-scale mining. In 2017, forestry-related exports reached USD $5.271 million, 8.7% of the total goods exported by Chile. In the last 8 years, the average number of people employed in the sector was 300,000, both directly and indirectly employed in forestry-related sectors [19].

The objective of this paper were, for one side know what and how elements in the forestry sector are affecting the sustainability of the workforce. After that identify what activities, strategies have been implemented by both forestry companies and forestry contractors companies in this area. Finally investigating if as a consequence of the elements that were affecting the sustainability of the workforce, productivity of the system is affected, through two different study developed on the field.

2 Methods

This paper is based on two investigation. The first one, was a cross-sectional investigation that was undertaken in the Chilean forestry sector, between 2011 and 2015, more details on Meyer [7]. The second part was developed 2017–2018 and the objective was investigating the impact of the elements detected in the first part of the study in the productivity of the system through two different study developed on the field and interviews with expert. For these studies the methodology suggested by Apud et al. [20] was used.

3 Results and Discussion

The results of the first study are discussed in this section. Then a summary of the second part of the study are presented.

3.1 Demographic Data on Chilean Forestry Workers

The average age of the Chilean forestry workers who participated in the study was 37.8 years old, with the youngest 18 years and the oldest 61 years. In relation to the age distribution of the Chilean forestry workforce, shows the largest percentage of workers (32.15%) in the workforce today are aged between 30 and 39 years. However, in 1985 the largest percentage of workers (52%) was between 20 and 29 years [21]. In 1999 the age group between 20 and 29 years old made up 44% of the total workforce, still the largest [14]. In this study, the 20–29 age group drops to 24.1%.

The implication of the ageing of the population has a direct link with OH problems [22] and it is expected that as the workforce ages OH problems will increase, according to the results of the study by Ackerknecht [23]. Also initiatives to reduce the physical load of the work – for example, rotation between different positions in the crew – was not possible in some Forestry Contractor Companies because the oldest workers were not able to carry out activities with higher physical demands, such as the work done by chokers. The incapacity of older workers to carry out certain physically demanding tasks implies an extra load for younger workers, which in itself could affect the health of those younger workers and their willingness to remain in the sector. The physical capacity of workers remains a critical topic and is strongly related to system productivity, so this topic is a relevant issue and is further discussed in the next section.

Another significant finding is the low proportion of workers over 60 years old, 1.3% of the sample, considering the age of retirement in Chile is 65 years old. This is an interesting fact since a study in Chile in 2014/2015 found that by 2030 the Chilean forestry sector will need about 10,000 new workers [10]. The criteria to estimate the age of retirement for workers would be 65 years; however, the data in this research shows the actual retirement age is lower than 65 years old.

Finally, the activities that workers were engaged in at the time the data were collected showed that manual activities occupied 40.1% of the workforce; 27.7% were engaged in the operation of machinery, 29.4% worked in semi-mechanized activities - mainly as chainsaw operators, and 2.8% were involved in supervisory activities. This information shows that manual and semi-mechanised activities still involve the largest number of workers in the Chilean forestry sector and that the most common method of logging continues to be largely manual and semi-mechanized, and therefore a large workforce remains necessary for the development of the sector.

The demographic data on Chilean forestry workers confirms that there are elements, which threaten the sustainability of the workforce. Although some factors are related to the rise of OH problems, such as the incremental ageing of the population, the number of year's workers spend doing the same activities and the lack of formal training also present as significant elements.

3.2 The Impact of the Working Conditions on the Workers' Sustainability

The purpose of this section is to analyze the impact of working conditions base on worker's opinion, with a focus on logging operations. The elements, which influence that relationship, and how those elements are related to Safety and Health, productivity or Market attractions. As was mentioned before this part of the study was an investigation that was carry out in the Chilean forestry sector, between 2011 and 2015. In both Meyer, [7], Meyer and Tappin [24] is possible find more detail of the study. In Table 1, is presented a summary of the main results, with focus on workers' positive and negative perceptions of the Chile Forestry Sector in areas that historically has been critical in the Chilean Forestry sector, like physical environment, organizational and economics issue and individual growth and skill development

The first area to discuss is the result related to the physical demands of the work. Findings associated with the physical environment and physical demands of the task can have a strong influence on negative perception and low level of satisfaction mentioned by workers [25]. In this study, both areas have the strongest links with safety and health, productivity issues and low market attractiveness, according to both workers' opinions and the literature review. This negative perception is based on two main aspects – the physical environment and the physical demands of the task. The concerns of the workers associated with the physical demands of the work including the lack of attention paid to the source of the OH issues, or to the comfort and wellbeing of the workers associated with exposure to the working environment. Those are factors that are notably absent in

Table 1. Chilean worker's opinion about working conditions, with focus on Physical Environment, Organizational and Economics issue and Individual Growth and Skill Development aspects.

Worker´s opinion of how the element impacts over Physical Environment:
The negative impact of environ-mental issues, like noise, vibrations, high and low temperature, and dust. Very dangerous work to life and health. Inadequate personal protective equipment and Low maintenance in element related to the comfort of the machinery operator
Safety and health: Ninety-five percent of workers mentioned that their work was extremely dangerous in term of accidents, and 86% mentioned their work is link to health problems. Workers (77%) mentioned the very hard work during the winter season, since they did not get the opportunity to change their wet clothes, and the quality of clothing was not good enough to keep them dry and working properly. The same problems occurred in summer when the quality of clothing did not accord with environmental conditions. For 76% of the workers, their work demands moving heavy objects; for 74 % demand demands bad postures; for 72% work demands fixed positions72% and for 52% work demands repetitive movements 52%
Productivity: Over 90% of the workers mentioned that their productivity decrease due extremely demand during summer/winter. Over 85% of the workers mentioned that working physical demands are related with low productivity.
Market attractions: Physical environment and the physical demands of the work, are the worst aspects of working here, according to 92% of the workers and 74% of them would prefer change to another job if they have the chance.
Worker´s opinion of the element over Organizational and Economics issue
Working longer hours per week, bad system of days of work and rest, high pressure, inadequate rest allowances, low importance of the task, inadequacy of departmental structure for job performance, low influence within the organization, low chance to determine their own work schedule or procedures, inadequacy of tools, equipment, machinery, and technical support, for job performance, inadequacy of work flow input for job performance and low work pay and benefits and job security
Safety and health: For 66% of the workers the relation between working/resting days were associated with OH issues such as musculoskeletal disorders. For 76% of the workers the payment system were related with their physical health.
Productivity: For 81% of the workers the productive issue are due machinery problems, ageing and poor maintenance. For 64% of the workers the productivity issue are associated with lack of planning and logistics and 63% due lack of personnel in key positions.
Market attractions: For 77% of the workers the shift system-days away from the family-has a negative impact over them. For 87% of the workers the relationship between their effort and their salary was not enough and they don't have any work-benefits
Worker´s opinion of how the element impacts over Individual Growth and Skill Development
Market attractions: Positive initial chance to advance beyond their lack of studies or technical preparations, however low advancement opportunities beyond machine operator.

strategies the forestry sector has developed (Apud et al. 2014b; Mylek and Schirmer 2015). These strategies have concentrated on complying with legal aspects and certifications processes concerning safety, without incorporating OH issues or the improvement of the comfort of the workers (FSC 2009b).

Along with the impact of working conditions on the workers, 91% of those in this study consider that elements of logging activities, such as the physical environment and physical tasks, have affected the sector's market attractiveness. These elements can be considered as significant barriers to the recruitment of potential new workers and to retaining workers already in the sector, as it is well known that forestry involves very physical, demanding work carried out under sometimes very harsh environmental conditions and with a high accident ratio. This is an important point, as 90% of the workers in this study were motivated by relatives or friends to work in the sector and, of that percentage, 93% said they did not want their children or relatives to work in forestry. The remaining 7% said they would like their children to work in the forestry sector as long as they began by operating a machine and not as a manual worker, as the work was very hard.

These findings are consistent with information presented by the Government of Alberta in 2009, in their report they stated that the image the forestry sector had among young people was poor due to the working conditions and physical demands of the tasks and that this was the key reason young people were not motivated to work in that sector.

The second area is organisational and economics aspects of the forestry sector. These aspects are associated with the lack of workers in some key positions, problems with old or faulty machinery, lack support for solving mechanical problems and logistics. The lack of workers in some key positions, has a consequence in that the productivity of the system, as will be mentioned later. If key positions are not filled, the physical demands of the labour increase because workers must cover the missing workers, putting the rest of the crews' work under greater pressure and causing early fatigue (Apud and Valdes 2000). An interesting result is that 77% of the workers said that in the last couple of years the number of workers available for some key positions had decreased, which implies that the settings of the crew were adjusted. This aspect could be related to one of the aspects mentioned in the introduction; that is, a lack of interesting work in the forestry sector and therefore a lack of workers.

Finally, the third area is associated with individual growth and skill development; workers mentioned they developed skills for operating different tools, equipment and machinery. However, questions related to training, adequacy of technical expertise and adequacy of job training yielded highly debatable results, despite the positive tone of the previous paragraph, as there is no formal training in the Chilean forestry sector and the learning process is based on experience developed on the job throughout their professional life, as well as from mentoring. This is an important point since, according to Axelsson and Pontén [26] and Ackerknecht [23], the development of formal technical training is strongly related to reducing accidents and OH issues as well as to increasing system productivity.

Workers were positive about opportunities to advance and for individual professional growth. For example, 96% of those in the study began as manual workers on activities that require a minimum level of skill; now 40% of these workers are machine operators, chainsaw workers or supervisors. This sector has given these workers an opportunity to progress in their career into logging activities. This aspect influences positively on attracting people to work in the forestry sector. The sector does not, however offer many options beyond that, according to 74% of the workers.

According to the workers, the skills that they develop in the forestry sector, excluding machine operators, are not compatible with any other job apart from those related to logging. The chance to develop a career was the first and the third priority for forestry workers, according to studies by Lagos [27]. The forestry career is limited in opportunities, according to the workers, since it offered three options only: become a chainsaw operator, a machine operator or a supervisor. The absence of career development in the forestry sector, plus years spent doing the same work, can be related to the development of occupational illness and the decreasing of productivity, according to Slot and Dumas [28].

After describing the findings, it is possible to conclude that workers recognize the Chilean forestry sector has been working seriously on some aspects to do with working conditions. However, 83% of the workers believe the industry has a long way to go to improve its OH record, in terms of the comfort and wellbeing of its workers or which way this topic is linking with productive issue. For example, the introduction of technology to reduce the physical demands of the work or the exposure to the physical environment has been minimal, according to 85% of the workers.

So with regard to OH problems, the source of OH problems associated with the physical environment are still present in the Chilean forestry sector, based on the opinions of the sample of workers involved in this study. Workers' opinions coincide with information provided by Wästerlund (1998) and Poschen (2001), who both mentioned that workers are exposed to a series of elements and factors, including infrastructure, climate, technology, work methods, work organisation, economic aspects, contracting arrangements, worker accommodation and education and training, that could affect not only their health but also their comfort, wellbeing and performance.

3.3 Chilean Forestry Strategies Associated with the Working Conditions

This section focuses on what are the activities that Chilean forestry companies have implemented associated with working conditions. This section goes further and pursues knowledge about the activities which both the Chilean forestry companies (FC's) and Chilean forestry contractor companies (FCC's) have been working on in relation to this topic. The discussion of the findings is based on interviews conducted with people who work for both the Chilean forestry companies and the Chilean forestry contractor companies, Chilean experts from the area of ergonomics, safety and health and the forestry sector were interviewed.

The Chilean FC's, FCC's and experts in the field agreed about the link between working conditions and OH problems in the forestry sector. However, the FC's and the FCC's did not agree about the consequences those conditions have had on the OH of the workers. In this regard, the FC's and the FCC's have been developing strategies to improve working conditions, but not only for the factors linked to OH problems. These strategies include legislative improvements associated with new safety regulations, better control and supervision, and a certification process that Chilean forestry companies have implemented. As a consequence of those strategies aspects such as living conditions, shift design and payment systems - all aspects that have characterized the forest industry in a negative way in the past - have been improved. This was recognized by both workers and experts. However further developments were needed to improve working conditions.

Strategies about safety and health were discussed. FC's and FCC's are still focused on the prevention of accidents without paying much attention to the cause of OH problems, and practically no attention is paid to the wellbeing of the workers. This finding is consistent with the workers' opinions as previously mentioned. Also FC's and Fcc's does not link this elements, OH and wellbeing, with productivity and efficiency related. Consequently, improvement of the main factors that impact the OH of the workforce has been minimal. These factors related to the causes of OH problems have not been identified as an issue by either FC's or FCC's; to the point where forestry organizations do not fully recognize that forestry workers have OH problems related to their work. Associated with this is a lack of data about the OH of forestry workers. Finally, there is a lack of integration and collaborative work between FC's and FCC's to address this issue.

The last part of the discussion was about recruitment, training and individual growth, associated with the opportunity to develop a career in the forestry sector. These aspects are strongly related, based on the findings described in previous section, and have a robust and negative relationship with both OH and low market attractiveness and, consequently, the ageing of the workforce. Based on the findings and experts' opinions, these areas are also not seen as a priority by either the FC's or FCC's.

The lasts paragraphs discussed above reinforces some issue related with the limited attention that Chilean forestry companies have paid in the area of workforce sustainability [23]. Consequently, it is necessary that organizations improve their strategies to manage OH, based on workers needs and capacities. From the literature, it is known that OH strategies can have a direct impact on the sustainability of the workforce [29, 30]. This is particularly critical in the forestry sector where forestry workers have been considered a vulnerable class of workers, especially those who do not work directly for the main companies [1], an occurrence which is common in the Chilean forestry sector [31].

3.4 The Consequences Over the Productivity of System

In Table 1, was describe the Chilean worker's opinion about working conditions, with focus on Physical Environment, organizational and economics issue, and individual growth and skill development aspects. Also the impact of those elements over the productivity of the system. Therefore the objective of this section is discussed the consequences over the productivity of system due problems of workforce sustainability. Two cases developed between 2016 and 2018. The first one is an example base on a pine pruning and the second one is based on harvesting example.

Output in Pruning Activities

The first case was the study of pruning developed in pines three during 2017. Prunings are important activities for obtaining products of good quality. This activity has the purpose to eliminate mainly branches, cones and epicornics to obtain wood free of knots. To make this job efficient, the worker must have a good technique. If a branch is not properly cut, the occlusion will be delayed and the amount of clear wood will decrease, therefore the price and quality of that piece of wood will not be the possible highest.

Not all enterprises use the same scheme for pruning. In some cases, four treatments are made namely: First pruning from 0 to 2 m; Second pruning from 2 to 4 m; Third pruning from 4 to 6 m, and Fourth pruning from 6 to 8.3 m. This example was carry-on in a second pruning. First pruning are done with the subject standing on the ground. For second pruning and the next ones, tree ladders are used. The main demands of this technique are that the workers have to climb up and down the ladders, which means more physical load [14]. Mechanization is a feasible solution, however pruning trees is very difficult, mainly because operational issues and costs. Therefore still are cheaper use manual tools to do it.

Each year the company, where the study was conducted needs to prune thousands of hectares; however due the lack of workers, a percentage of hectares remains unpruned. The next year when those remaining hectares are pruning each branch has increased its diameter and their wood is harder since are one year older. What are impact of this late activity over the workers when its compare when is done in the right time? Three main and direct impact over the workers were identifies. Highest level of effort over shoulder, elbow and wrist of the pruners, since on average they are pruning around 7200 thicker branches per day. In addition, that means a higher cardiovascular load, 5% more when it compares with similar activities with thin branches. A third impact is this group reach 25% less output per day. The problem of the reduction of their productivity is that part of their salary; around 40–50% are based on what they are capable to done. However the payment, the price ($/pruned tree) is the same, if the work was done in a normal situation, with thinner branches and softer wood. The results of that could be 20 to 25% less salary for the workers.

Other impact is related with the mechanical quality of that wood, that are lower since the knots of the wood is bigger, and the amount of wood free of knots are less. Therefore, the prices of that piece of wood are cheaper.

Output in Logging Activities

The second case arises from the need of a Chilean forestry company that showed increasing concern because they observed a high turnover of forest chokers, indicating that it amounted to 70% of a month to another. Such rotation implies for this company the need to train new workers who develop this activity, as well as the loss of skilled labour, also exist the uncertainty of whether this staff will exist in future times [32].

Work-study was carry-on to see the output of a crew. The tower used for logging was URUS 2. The productive capacity of the tower was 150 m^3/day for a logging distance between 200–300 m with a slope of 50%. Each trees had a volume 0.7 to 0.8 m^3. The production goal established by the contractor company, for the logging distance and slope, was 53 cycles per day, with four trees per cycle. The chokers crew was composed by three workers, two working through the slope, to leash the threes, and one at the end of the slope to unleash it. The workers were certified in their mental, technical and physical capacities to develop their duties under CORMA criteria. The evaluated work begins its work at 08:00 in the morning, preparing equipment and moving to the specific place of work, the productive work begin at approximately

09:00 h. Lunch is served from 1:00 p.m. to 2:00 p.m, near 4,000 kcal/day as is suggested by Apud et al. (1999). The average temperature was 18.5 °C.

The results shows that during the three days that the data was collected, an average of 37 cycles was done, this means that they were 30% below the goal established for the logging crew. The time dedicated to the main activities, was near the 80% of the total, with 16% dedicated for lunch and 4% for pauses. The cardiovascular load of the day, reaches similar levels for the three evaluated chokers, which is around 37.5%. These cardiovascular loads qualify the work as almost physically heavy; being heavy over 40%, according to the criteria established Apud and Valdes [21]. The last two results, the time dedicated to the main activities and cardiovascular load, are indicates that the crew was working very close to their maximum capacities.

Some issue to highlight in this example related with the first section on this chapter.

- The average age of the workers was 45.7 years, with one workers over 55 years old, and the other two were 41 years old each one, ageing population.
- The youngest workers were working over the slope since the rotation with oldest one, was impossible due his age and physical capacities.
- For this conditions the ideal number of the crew are four, three working over the slope and one to unleash the trees, however due the lack of workers the crew was compose by only workers, with the consequence over the productivity, 30% below their goal.

The fact to not reach the estimate output has a consequence over the next step of the productivity chain, the transport of the tree from the forest to either the saw mill or the pulp mill. Fifty cycles per day, with four trees per cycle and each tree had a volume of 0.8 m^3, means a production of 160 m^3/day/crew. However due the reason mentioned before the output reached during the three days of the study was 118,4 m^3/day/crew. Each truck has the capacity to load on average 25 m^3, which means a daily difference of one to two trucks that couldn't be loaded due the lack of wood. This truck has the alternative to wait for the next day or be redirection it to other place, with the consequence over the cost of that, that cost are paid by the contractor.

There are chances to mechanization, using a system call "Steep slope logging winch assisted", however due the lack of preparation of the machine operator and the lower cost to do it using forestry tower, the necessary output to make the operation economically viable are not reach it and is an process that it is not consistently in its operation.

4 Conclusion

The purpose of this paper was to find out, based on the opinions of workers in the sector, if a relationship exists between working conditions in the forestry sector and the sustainability of the workforce, and to discover the impact working conditions have on the sustainability of the workforce with a focus on OH issues. This aspect was confirmed by the workers in this study, as the market attractiveness of forestry in Chile is affected by actual working conditions, and is therefore discussed as part of the findings. Both the working conditions and market attractiveness are also related to another important element threatening the sustainability of the forestry sector, which is the associated ageing of the population.

From FC's and FCC's initiatives to solve the issues associated with the OH of workers have been minimal. This confirms the limited view forestry organisations have about this topic.

The sustainability of the workforce is a fundamental aspect of the production system in Chile. The workforce in the particular case of Chilean forestry is vital for the continuing development of this sector. Therefore, any aspects that could affect them, such as OH issues, an ageing population, and low employment attractions, will have an important impact from both an economic and social point of view in the Chilean forestry sector.

References

1. Bolis I, Brunoro C, Sznelwar L (2014) Mapping the relationships between work and sustainability and the opportunities for ergonomic action. Appl Ergon 45(4):1225–1239
2. Dyllick T, Hockerts K (2002) Beyond the business case for corporate sustainability. Bus Strategy Environ 11(2):130–141
3. Vidal N, Kozak R (2008) The recent evolution of corporate responsibility practices in the forestry sector. Int Forest Rev 10(1):13
4. Longoni A, Cagliano R (2015) Environmental and social sustainability priorities. Int J Oper Prod Manag 35(2):216–245
5. Baumgartner RJ, Ebner D (2010) Corporate sustainability strategies: sustainability profiles and maturity levels. Sustain Dev 18(2):76–89
6. Mani V et al (2016) Supply chain social sustainability for developing nations: evidence from India. Resour Conserv Recycl 111:42–52
7. Meyer F (2017) Evaluation of workforce sustainability in the Chilean logging sector using an ergonomics approach. In: School of Management. Massey University, Auckland
8. Docherty P, Kira M, Shani AB (2008) Creating Sustainable Work Systems. Developing Social Sustainability, 2nd edn. Routledge, London
9. Brizay A (2014) The Rovanieni action plan for the forest sector in a green economy. In: Forest Europe Workshop on Green Economy and Social Aspects of SFM, Santander, Spain, p 46
10. Corma (2015) Fuerza Laboral de la Industria Forestal 2015–2030. Corma, Concepcion
11. DeVries P (2014) Wanted: skilled employees to work in B.C.'s woods. Business Vancouver, British Columbia
12. Ligaya A (2013) Forestry workers suddenly back in demand as average pay rises 11%. In: Financial Post, Canada
13. Östberg K (2014) The Forest Kingdom, a vision for jobs and growth in a green economy. In: Forest Europe Workshop on Green Economy and Social Aspects of SFM, Santander, Spain. Summaries of presentation
14. Apud E et al (1999) Manual de Ergonomia Forestal [Manual of Ergonomics in Forestry]. Proyecto FONDEF D96I1108: Desarrollo y Transferencia de Tecnologías Ergonómicamente Adaptadas para el Aumento de la Productividad del Trabajo Forestal, Chile [D96I1108 FONDEF project: Development and Technology Transfer Ergonomically Adapted for Increased Productivity of Forest Work, Chile

15. Enez K, Topbas M, Acar HH (2014) An evaluation of the occupational accidents among logging workers within the boundaries of Trabzon Forestry Directorate, Turkey. Int J Ind Ergon 44(5):621–628
16. Garland J (2008) Sustainable forestry? Only with a sustainable workforce: the idaho timber workforce development project. In: Council on Forest Engineering (COFE) Conference Proceedings: Addressing Forest Engineering Challenges for the Future
17. Forest Europe Workshop on Green Economy (2014) Forest Europe workshop on green economy and social aspect of SFM (Draft report), Santander, 29–30 April 2014
18. Genaidy A et al (2010) The role of human-at-work systems in business sustainability: perspectives based on expert and qualified production workers in a manufacturing enterprise. Ergonomics 53(4):559–585
19. Corma (2018) Exportaciones. http://www.corma.cl/perfil-del-sector/aportes-a-la-economia
20. Apud E et al (1989) Guide-lines on ergonomic study in forestry. Prepared for Research Workers in Developing Countries. ILO
21. Apud E, Valdes S (1995) Ergonomics in forestry: the Chilean case, E.I.L. Office. ILO, Geneva
22. Loeppke RR et al (2013) Advancing workplace health protection and promotion for an aging workforce. J Occup Environ Med 55(5):500–506
23. Ackerknecht C (2010) Relación Edad y Accidentalidad en Trabajadores del Sector Forestal en Chile. [Relationship between age and accidents in workers in rhe Forestry Sector in Chile. Ciencia y Trabajo 12(38):414–422
24. Meyer F, Tappin D (2014) Social Sustainability in the Chilean Logging Sector
25. Pontén B (2011) Physical safety hazards. In: Stellman JM (ed) Encyclopedia of Occupational Health and Safety. International Labor Organization, Geneva, vol 68
26. Axelsson S-Å, Pontén B (1990) New ergonomic problems in mechanized logging operations. Int J Ind Ergon 5(3):267–273
27. Lagos S (2006) Calidad de Vida Laboral en el contexto en el contexto de responsabilidad social empresarial del sector forestal chileno. [Quality of Working Life in the context in the context of corporate social responsibility of the Chilean forestry sector]. In: Ciencias Ambientales, Centro EULA. Universidad de Concepcion, Concepcion
28. Slot T, Dumas G (2010) Musculoskeletal symptoms in tree planters in Ontario, Canada. Work 36(36):67–75
29. Docherty P, Forslin J, Shani A (2002) Creating Sustainable Work Systems: Emerging Perspectives and Practice. Psychology Press, Abingdon
30. Kira M, Van Eijnatten FM, Balkin D (2010) Crafting sustainable work: development of personal resources. J Organ Change Manag 23(5):616–632
31. ILO (2011) Industria Forestal. [Forestry Industrie]. In: Encyclopedia of Occupational Health and Safety. International Labor Organization, Geneva, Switzerland
32. Toledo F (2017) Study about turnovers in chokers in forestry activities in Ergonomics. Universidad de Concepcion, Concepcion, Chile

Ergonomic Practices in Africa: Date Palm Agriculture in Algeria as an Example

Mohamed Mokdad[1(✉)], Mebarki Bouhafs[2], Bouabdallah Lahcene[3], and Ibrahim Mokdad[4]

[1] University of Bahrain, Sakhir, Bahrain
mokdad@hotmail.com
[2] University of Oran, Oran, Algeria
[3] University of Setif, Setif, Algeria
[4] Exa. Co., Manama, Bahrain

Abstract. For more than fifty years, early ergonomists such as [1–4] called for the application of ergonomics in developing countries to expand its application rather than its confinement to developed countries. Indeed, ergonomists from developing countries applied ergonomics to many traditional workplaces, machines, and jobs in the field of what was called at that time, traditional ergonomics. Despite what has been done, the African share of ergonomic studies is modest when compared to the ergonomic work done in other continents. Date palm farming is considered as one of the most important economic resources especially in hot and dry areas in Africa. In Algeria, according to Bouguedoura et al. [5], the number of date palms is in millions. The number of people who work in date palm industry is also very great. The majority of date palm work is carried out after the farmer climbs the trunk to reach the crown. The worker climbs the palm tree, which may be about 21–23 m in height, barefooted and in rare cases uses a harness or a rope for support. The work is insecure and associated with significantly higher rates of work related musculoskeletal disorders. This paper aims to answer the following questions:

- What attempts can be made to solve the problem of falling from the date palm crown?
- What attempts can be made to fight WRMSDs?

Keywords: Ergonomics · Africa · Date palm agriculture
Developing countries · WRMSDs

1 Introduction

For more than fifty years, early ergonomists either from industrially developed countries [2, 6–9], or from industrially developing countries [1, 3, 4] called for the application of ergonomics in industrially developing countries to assist them in their development processes.

Indeed, ergonomists from industrially developing countries applied ergonomics to many workplaces, traditional machines, and jobs in the field of what was called at that

S. Bagnara et al. (Eds.): IEA 2018, AISC 825, pp. 392–403, 2019.
https://doi.org/10.1007/978-3-319-96068-5_44

time, traditional ergonomics [10–18]. But despite these actions, the African share of ergonomic studies is modest when compared to the ergonomic work done in other continents.

African countries are predominantly agricultural countries [19, 20]. Their economies are dependent on agriculture. In addition, a large portion of the population lives and works in rural areas. Work in these areas is still done traditionally like their ancestors [21, 22].

Farming work seems to be a safe job, but it is actually fraught with many farming hazards and injuries. Muscles and tendons are stretched or over-used beyond their capabilities [23–26]. These hazards and injuries may lead to work related muscular-skeletal disorders (WRMSDs). Workers in a variety of farming types have been found to be affected. Some of those include oil palm farmers [27], wheat harvesters [28], tea pickers [29, 30], apple harvesters [31], self-employed farmers [32], livestock workers [33], and tomato farmers [34].

Various efforts have been attempted to develop agricultural work in Africa [22, 35–39]. But most or all of them dealt with the subject from the economic and technology transfer approaches. There is no doubt that these approaches helped to develop African countries, but it is necessary to give an opportunity to ergonomics, which has contributed significantly to the development of the economies of industrialized countries.

Date palm (Phoenix dactylifera) farming is considered one of the most important economic resources especially in hot and dry areas in Africa. In Algeria, according to Bouguedoura et al. [5], the number of date palms is in the millions. The number of people who work in the date palm industry is also very large. The majority of date palm work (climbing, dethroning, pollinating, bunch thinning, dry leaves pruning, bunch covering and date harvesting) is carried out after the farmer climbs the trunk to reach the crown. The worker climbs the palm tree, which may be about 21–23 m in height, barefooted, and rarely uses a harness or a rope for support.

The work is insecure and associated with significantly higher rates of work related musculoskeletal disorders (WRMSDs) [40–42]. The major source of insecurity is falling which can result in death or disability. The farmer may fall as he climbs the trunk of the palm, especially when he catches an old leaf base or when he puts his foot on a cracked old leaf base. He may also fall while connecting the rope to the belt. Likewise, he may fall when he accidentally cuts off the rope during the cutting of dry leaves.

There have been numerous incidents of workers falling from the tree trunk. In August 2012, a farmer fell and broke his spine when he accidentally cut his cord while cutting the dry leaves. In another accident in July 2015, the farmer fall from a date palm trunk because the knot of the rope he held opened without being noticed. This incident resulted in the farmer breaking his spine. In a third accident in 1986, the farmer did not press the harness very well, and he came down because it was not tied to his body. When he came back a little bit as he was cutting the dry leaves, his back was not raised, and he fell. He is now paralyzed and in a wheelchair. With this method, farmers keep the harness on their back by the pressure of their legs on the trunk of the palm at work. If the farmer forgets to press, or loosen the pressure, the harness descends, and the farmer falls.

However, the major source of WRMSDs is that date palm farmers have to do various tasks and operations on the palm crown with a large amount of static work (gripping tools for long time, holding the arms out or up to perform tasks, or standing on unstable surface {stalk of the date palm} for prolonged periods…), bending, and adopting awkward postures.

To address these problems, it is necessary to take advantage of the ergonomic interventions which make the workers work comfortably, safely and effectively, while at the same time leading to increased production quantitatively and qualitatively. Therefore, this paper aims to answer the following questions:

(1) How climbing the date palm trunk is done?
(2) What ergonomic attempts are made to fight WRMSDs in date palm work?

2 Methodology

2.1 Method

To achieve the aims of this study, the descriptive research method, was used as follows:

(1) **Library Research**. A literature survey was conducted as follows: Initially, an electronic search in the following data bases: Ergonomics Abstracts, Scopus, and Science Direct, was conducted using a variety of key words (date palm agriculture, mechanization of date palms, occupational risk factors among date farmers). In addition, manual searches were carried out in the following journals: Accident Analysis and Prevention, Ergonomics, Applied Ergonomics, Human Factors, International Journal of Industrial Ergonomics, Reviews of Human Factors and Ergonomics, Theoretical Issues in Ergonomics Science, Human Factors and Ergonomics in Manufacturing, Ergonomics in Design, and Le Travail Humain. The result of this survey was the collection of more than 400 items, mainly articles. Second, the articles obtained were then classified into two categories: climbing date palms and falling, and occupational health and WRMSDs of date-palm farmers.
(2) **Field Research.** To ensure the spread of WRMSDs among date palm farmers, the Nordic Musculoskeletal Questionnaire [43] was used. Before its use, the questionnaire was translated into Arabic, the language of farmers in Algeria. 100 questionnaires were distributed to farmers in the region of Ziban (Algeria). The farmers who were able to read were asked to answer the questionnaire themselves. For non-reading farmers, the questionnaire was applied to them in the form of a structured interview. Regarding the validity and reliability of the questionnaire, the researchers relied on the studies of other researchers who used it, especially the studies on farmers [44–51].

2.2 Study Sample

One of the researchers (MM) requested the help of the Directorate of Agriculture in the governorate of Ouled Djellal to distribute the questionnaire to the farmers working in the date palm cultivation. A total of 100 questionnaires were distributed on a randomly selected sample of farmers. The response rate was 63%. After examining the retrieved responses, seven of them were found to be incomplete and were excluded. The analysis of the results is based on 56 questionnaires. The average age of these farmers was 42 years with a standard deviation of 5.3.

3 Results and Discussion

3.1 How Climbing the Date Palm Trunk Is Done?

It has been found that climbing the date palm trunk is done in two ways: the traditional and the modern ways.

In the Traditional Way. We can distinguish the following methods:

- **Free climbing:** In this way, the farmer climbs the trunk of the palm using his hands and legs with no aid (see Fig. 1). He places his two feet on the base of the leaf, then raises his trunk with his hands to maintain his balance and prevent him from falling. He continues this way up to the trunk of the palm.

Fig. 1. A farmer climbing date palm barefooted (the Monkey method).

  We can call this method "the Monkey method". In this method work is done with one hand while the other hand holds the worker's body to the trunk of the palm. The farmers used this method for years, but they realized that it is very tiring as a lot of static work is done. While the farmer does his work in this position, he exerts a lot of static effort in an awkward position that prevents blood from reaching the muscles bringing with it nutrition and removing wastes. For this, the farmer is forced to change the hand he works with constantly. At the beginning, he works with the main hand, and when fatigue reaches its range, he changes to the other hand which is not as skilled as the main hand to finish the work. This method is mainly used to accomplish the tasks just below the crown with no need to use two hands for dry leaves cutting, ripe dates picking, and bunch cutting of some date varieties.
- **Aided climbing:** Human beings have been in a continuous pursuit in improving working methods. According to Bhuiyan and Baghel [52], continuous improvement

is a "culture of sustained improvement targeting the elimination of waste in all systems and processes of an organization". In this context, the use of harness first appeared (see Fig. 2). The harness that is man-made - is a plexus made of thin ropes (10 to 15 mm) that combine to form a rectangular piece with a width of about 100 mm, and a length of about 800 mm. On either side of the harness, an overhand loop is stitched. If the farmer who uses the harness is right-handed, he will fix the working rope on the right side of the Harness, otherwise, it is fixed on the left-side. Farmers usually choose the rope based on their own experience or the experience of their colleagues. Therefore, they do not receive accurate information about the type of cord they should use. Even if they look for this information, they do not find it because it is not available. After all the falls that have occurred, and the painful consequences that have been obtained (disabilities, death, various psychological problems), the farmers are alerted to the need to connect the harness to the body before embarking on the climb. This simple action reduced the incidence of falls very dramatically. If the fall has been reduced, the fatigue caused by barefoot climbing still exists. In addition, the risk of falling and WRMSDs are still present.

Fig. 2. Aided climbing using a harness.

Modern Ways of Climbing Are. As date palm growers realized the difficulty of climbing and the danger of falling from the trunk of the palm, they began to search for safe alternatives that facilitate climbing and working on the trunk. Here are some solutions that have been introduced:

- **The use of ladder:** Despite the fact that there are many types of ladders, the only one that can be used is the extension ladder (see Fig. 3). It is a ladder that can be extended as needed. It is located at different lengths where the total length after the expansion is about 15 m. When used, it is leaned against the date palm trunk and connected to the trunk so that it does not move. This helps prevent the farmer from falling. The ladder is not widely used in the Algerian date palm orchards. The major reason may be the fact that orchards are not designed for using ladders. Date Palms water basins and canals may hinder the use of these ladders. In addition, the installation of ladder on a palm trunk is time consuming and tiring process. For this

Fig. 3. Climbing using an extension ladder

reason, in some date palm farms in Coachella valley in California (USA), a ladder has been installed on each palm trunk on the farm.

- **The use of climbing device:** At the Technical Institute for the Development of Saharan Agronomy, (Biskra, Algeria) some engineers, have developed a date palm climber (see Fig. 4). This device is made up of two pieces: upper piece and a bottom one. The two pieces are connected to the trunk of the palm in a safe manner which allows them to move upward or downward. The farmer sits on the upper piece and lifts the lower piece with his feet. Then stands on the lower piece and lifts the top piece with his hands to a higher level and so up to the crown of the palm. There he does the work for which he is climbing the date palm.

Fig. 4. Climbing using the climbing device.

- **The use of modern technology:** Many researchers have realized that climbing the trunk of a date palm to perform a certain task is tiring, difficult and very dangerous. They took upon themselves the task of facilitating the work. Some of the results are:

The Development of Elevating Farmers' Machines. Various machines for elevating workers to the date palm crown have been designed.

- Aly-Hassan, et al. [53] point out that the technology that was designed to work with palm trees does not exist. What exists is machinery for operation in the field of industry. Thus, they examined a large number of machines, to see to what extent they can be used in date palm fields. Two of which were selected: the BEN (10) which is a French hydraulic (10 m high) lifting machine and the aerial work platform which is an American machine. The goal of the researchers was first, to see whether the two machines could be used in date palm farming. Second, to know which parts or pieces had to be modified to make machines usable in date palm orchards. And finally, to what extent the modified machines were suitable and capable of helping date palm workers to do their work properly and safely. Finally, they were suggested for use in date palm farms.

- Al-Suhaibani et al. [54] first measured the length and size of date palms in different orchards. Then, the land of the palm plantations has been cleared to see the nature of the plant and the nature of the machine that can walk without difficulty. In light of these two variants, the initial shape of a machine that could be used to lift the peasants to the palm crown was designed.
- Moustafa [55] developed what he called a date palm tree service machine. This machine is fixed on the back of a traditional 80 horsepower agricultural tractor. Also, it is totally dependent on the hydraulic capacity of the tractor. The machine can carry the worker and the tools he needs to work on the service platform to a height of up to 4.5 m. In addition to the work he does, he can maneuver his workplace to move right and left, up and down. In order to protect both the worker and the tree from any failure, Moustafa mentioned that attention was paid to providing security means during design.
- Jahromi, et al. [56] proposed the Lifting Model for Gripper Date Palm Service Machines. The researchers claim that the gripper machine was manufactured in such a way as to be fixed the trunk of the date palm firmly without causing any damage to the trunk. Further, they claim that it can hold the trunk of the date palm no matter what the circumference of that trunk is. It machine consists of four grippers, two at the bottom and two at the top. When the machine is placed on the trunk, the worker sits in a chair and controls the lifting, lowering and rotation around the trunk in a way that enables him to carry out his various functions under or inside the crown. The researchers noted that during its conception, they took into consideration the level of security.
- Garbati-Pegna [57] designed the Self-moved ladder which is a self-contained telescopic ladder that can reach a height of 15 m. It was mounted on a small tracked dumper. The ladder is moved by the hydraulic system of the dumper, while its extension is carried out with a manual winch. The author mentions that the self-moved ladder is simple and does not require any effort for the worker who can then climb the trunk up to a height of 15 m. And that laying and extending the ladder to a maximum length takes about 60 s.
- **Robotization:** Throughout history, man has been trying to look for ways to minimize work and prevent occupational diseases and accidents. In recent years, humans have discovered that robots can achieve this. For this reason, robots have entered the field of farm work [58, 59]. Similar to other farming work, date palm farming has seen the introduction of robotics. In this regard, Shapiro et al. [60] built a robot for date palm work. Similarly, [61] designed the Mechanical Analysis of a

Robotic Date Harvesting Manipulator. Further, Asfar [62] developed a robot that can climb the date palm tree to accomplish some operations on the trunk.

3.2 What Ergonomic Attempts Are Made to Fight WRMSDs at Date Palm Work?

Prevalence of WRMSDs in date palm farming: It is seen in Table 1 that WRMSDs are prevalent among date palm farmers. The peasants complained about their presence in the shoulders, hands and wrists, lower back, hips knees and feet. All this in the course of the week. However, after twelve months, we can see some reduction in complaints, but the amount of WRMSDs is still high.

Table 1. Prevalence of WRMSDs in nine different body parts among date palm farmers (n = 56).

Body part	7-day prevalence		12-month prevalence	
	Number	%	Number	%
Neck	18	32.2	15	48
Shoulder	51	91	52	93
Elbows	19	34	14	25
Hands/Wrists	55	98	56	100
Upper back	49	87.5	39	69.5
Lower back	54	96.5	51	91
Hips	49	87.5	29	52
Knees	51	91	49	87.5
Feet	56	100	53	94.5

Preventing WRMSDs: It has been found that exerted efforts to reduce WRMSDs were at both administrative (intensive work in the early morning, a lot of breaks after that) and personal levels (wearing wrist supports and large back belts). It should also be noted that in the last years, some engineering controls have started to be used. They take the form of inventing some apparatus which reduces work pressure and risks.

The following are some examples:

- Harvesting: Nourani and Garbati-Pegna [63] and Nourani et al. [64] designed a machine that can harvest date bunches while the farmer is on the ground. The machine consists of four parts: the base of the machine, the basket for collecting dates, the device of raising or lowering the basket, and the tool that cuts date stalks. After the base is installed on the trunk of the palm, the basket is lifted until it reaches the date bunch. The farmer adjusts the basket to be directly under the bunch, and then he lifts it up to enter the basket. Then the cutting device is operated to cut the bunch. The chopped dates are then brought down to ground.

- Pollinating: Mostaan, et al. [65], developed an electrical apparatus for pollinating date palms. They claim that the new apparatus is much better than the other traditional and mechanical tools in terms of size, weight, operation time and efficiency. The researchers showed that the first experiments of the device with some types of dates showed that its effectiveness was manifested significantly in increasing the amount of dates obtained from its use compared with the production of previous manual or mechanical pollinations.
- Dry Leaves cutting (Pruning): Ismail and Al-Gaadi [66] developed a portable AC machine. The aim was to automate the process of pruning the dry date palm leaves. While developing the machine, researchers considered three variables: the energy (the machine consumes little energy), the force (the machine can cut all types of dry and green leaves), and the time at work (the leaves are cut in as little time as possible). The researchers have shown that the machine achieved all the desired goals.

All of these efforts aim to reduce WRMSDs and thus make work easier, less tiring and more secure than before.

4 Conclusion

Current research sheds some light on the cultivation of palm trees in Algeria, especially on how to climb the palm and the musculoskeletal disorders experienced by the workers in this agriculture. It has been shown that date palm climbing is still done in traditional ways where the worker climbs the trunk of the palm barefooted either with no aid or using a harness that helps him to perform his duties when reaching the date palm crown. It was also been found that workers in this type of farming are exposed to musculoskeletal problems.

Despite the emergence of some technologies that facilitate the cultivation of date palm trees, it has been shown that the use of these technologies, is still very limited for a number of reasons, most notably financial and structural ones. The current research considered that the technological efforts used in the cultivation of date palms are in fact ergonomic efforts as long as they sought to make the work comfortable and secure. Ergonomic is not widely used in date palm farming, not only in Algeria, but in all the countries that are famous for date palm farming. Unless doors are opened for ergonomics, this agriculture will not experience the desired development.

References

1. Sen RN (1984) The Ergonomics Society: the Society's Lecture 1983. Ergonomics 27:1021–1032
2. Wisner A (1985) Ergonomics in industrially developing countries. Ergonomics 28(8):1213–1224

3. Abeysekera JDA (1987) A comparative study of body size variability between people in industrialized countries and industrially developing countries, its impact on the use of imported goods. In: Proceedings of international symposium on ergonomics in developing countries, 18–21 November 1985, Jakarta, Indonesia. International Labor Office, Geneva, pp 65–100
4. Shahnavaz H (1987) Workplace injuries in the developing countries. Ergonomics 30:397–404
5. Bouguedoura N, Bennaceur M, Babahani S, Benziouche SE (2015) Date palm status and perspective in Algeria. In: Al-Khayri JM, Jain SM, Johnson D (eds) Date palm genetic resources and utilization, vol 2. Springer, Netherlands, pp 125–168
6. Chapanis A (1975) Ethnic variables in human factors engineering. John Hopkins University Press, Baltimore
7. Kogi K, Sen RN (1987) Third world ergonomics. Int Rev Ergon 1:77–118
8. Wisner A (1997) Anthropotechnologie. Vers un monde industriel pluricentrique. Octarès, Toulouse
9. O'Neill DH (2000) Ergonomics in industrially developing countries: does its application differ from that in industrially advanced countries? Appl Ergon 31:631–640
10. Nag PK, Goswami A, Ashtekar SP, Pradhan CK (1988) Ergonomics in sickle operation. Appl Ergon 19(3):233–239
11. Sen RN, Sahu S (1996) Ergonomics evaluation of a multipurpose shovel-cum-hoe for manual material handling. Int J Ind Ergon 17:53–58
12. Chang SP, Park S, Freivalds A (1999) Ergonomic evaluation of the effects of handle types on garden tools. Int J Ind Ergon 24:99–105
13. McNeill M, Westby A (1999) Ergonomics evaluation of a manually operated cassava chipping machine. Appl Ergon 30:565–570
14. Sen RN (1997) Ergonomic modifications of shovels in India. Environ Manag Health 8 (5):173–174
15. Wu SP, Hsieh CS (1997) Ergonomics study on the handle length and lift angle for the culinary spatula. Appl Ergon 33:493–501
16. Kumar A, Mohan D, Patel R, Varghese M (2002) Development of grain threshers based on ergonomic design criteria. Appl Ergon 33:503–508
17. Junga HS, Jungb HS (2003) Development and ergonomic evaluation of polypropylene laminated bags with carrying handles. Int J Ind Ergon 31:223–234
18. Choobineha A, Tosianb R, Alhamdic Z, Davarzanieb MH (2004) Ergonomic intervention in carpet mending operation. Appl Ergon 35:493–496
19. Bilsborrow RE (1987) Population pressures and agricultural development in developing countries: a conceptual framework and recent evidence. World Dev 15(2):183–203
20. Cotula L, Vermeulen S, Leonard R, Keeley J (2009) Land grab or development opportunity? Agricultural investment and international land deals in Africa. IIED/FAO/IFAD, London
21. Nwuba EIU, Kanl RN (1986) The effect of working posture on the Nigerian hoe farmer. J Agric Eng Res 33:179–185
22. Obi OF (2015) The role of ergonomics in sustainable agricultural development in Nigeria. Ergon SA 27(1):33–45
23. Hoppin JA, Valcin M, Henneberger PK, Kullman GJ, Umbach DM, London SJ, Alavanja MCR, Sandler DP (2007) Pesticide use and chronic bronchitis among farmers in the agricultural health study. Am J Ind Med 50(12):969–979
24. Geng Q, Stuthridge RW, Field WE (2013) Hazards for farmers with disabilities: working in cold environments. J Agromed 18:140–150
25. Stoleski S, Minov J, Mijakoski D (2014) Bronchial hyper-responsiveness in farmers: severity and work-relatedness. Open Access Maced J Med Sci 2(3):536–543

26. Cha ES, Jeong M, Lee WJ (2014) Agricultural pesticide usage and prioritization in South Korea. J Agromed 19:281–293
27. Mohd Nizam J, Rampal KG (2014) Study of back pain and factors associated with it among oil palm plantation workers in Selangor. J Occup Saf Health 2(2):36–41
28. Chandra N, Parvez R (2016) Musculoskeletal disorders among farm women engaged in agricultural tasks. Int J Home Sci 2(2):166–167
29. Hazaarika M, Bhatt Achorya N (2001) Postures adopted and musculoskeletal problems faced by women workers in tea industry. J Agric Sci 14(1):88–92
30. Karukunchit U, Puntumetakul R, Swangnetr M, Boucaut R (2015) Prevalence and risk factor analysis of lower extremity abnormal alignment characteristics among rice farmers. Patient Prefer Adherence 9:785–795
31. Earle-Richardsona G, Jenkinsb P, Fulmerc S, Masona C, Burdickb P, May J (2005) An ergonomic intervention to reduce back strain among apple harvest workers in New York State. Appl Ergon 36(3):327–334
32. Hartman E, Oude Vrielink HH, Huirne RB, Metz JH (2006) Risk factors for sick leave due to musculoskeletal disorders among self-employed Dutch farmers: a case–control study. Am J Ind Med 49:204–214
33. Kolstrup CL (2012) Work-related musculoskeletal discomfort of dairy farmers and employed workers. J Occup Med Toxicol 7(23):1–9
34. Palmer KT (1996) Musculoskeletal problems in the tomato growing industry: tomato trainer's shoulder. Occup Med 46:428–431
35. Doss CR (2001) Designing agricultural technology for African women farmers: lessons from 25 years of experience. World Dev 29(12):2075–2092
36. Kleaver KM (1993) A strategy to develop agriculture in Sub-Saharan Africa and a focus for the World Bank. The World Bank, Washington, DC
37. Kelly V, Adesina AA, Gordon A (2003) Expanding access to agricultural inputs in Africa: a review of recent market development experience. Food Policy 28:379–404
38. Gabre-Madhin EZ, Haggblade S (2004) Successes in African agriculture: results of an expert survey. World Dev 32(5):745–766
39. Salami A, Kamara AB, Brixiova Z (2010) Smallholder agriculture in east Africa: trends, constraints and opportunities. Working Papers Series No 105. African Development Bank, Tunis, Tunisia
40. Grivna M, Aw TC, El-Sadeg M, Loney T, Sharif AA, Thomsen J, Mauzi M, Abu-Zidan FM (2012) The legal framework and initiatives for promoting safety in the United Arab Emirates. Int J Inj Control Saf Promot 19:278–289
41. Marzban A, Hayati A (2014) Ergonomic assessment of manual harvesting of date palm. In: 1st international scientific conference on applied sciences and engineering, 20–21 December. Pak Publishing Group, Kuala Lumpur, p 8
42. Grivna M, Eid HO, Abu-Zidan FM (2015) Epidemiology of spinal injuries in the United Arab Emirates. World J Emerg Surg 10(20):01–07
43. Kuorinka I, Jonsson B, Kilbom A, Vinterberg H, Biering-Sørensen F, Andersson G, Jørgensen K (1987) Standardized Nordic questionnaires for the analysis of musculoskeletal symptoms. Appl Ergon 8:233–237
44. Miranda H, Viikari-Juntura E, Martikainen R, Riihimaki H (2002) A prospective study on knee pain and its risk factors. Osteoarthr Cartil 10:623–630
45. Gallis C (2006) Work-related prevalence of musculoskeletal symptoms among Greek forest workers. Int J Ind Ergon 36:731–736
46. Ostensvik T, Veiersted KB, Nilsen P (2009) Association between numbers of long periods with sustained low-level trapezius muscle activity and neck pain. Ergonomics 52:1556–1567

47. Scuffham AM, Legg SJ, Firth EC, Stevenson MA (2010) Prevalence and risk factors associated with musculoskeletal discomfort in New Zealand veterinarians. Appl Ergon 41:444–453
48. Bernard C, Courouve L, Bouee S, Adjemian A, Chretien JC, Niedhammer I (2011) Biomechanical and psychosocial work exposures and musculoskeletal symptoms among vineyard workers. J Occup Health 53:297–311
49. Das B (2015) Gender differences in prevalence of musculoskeletal disorders among the rice farmers of West Bengal, India. Work 50:229–240
50. Kolstrup CL, Jakob M (2016) Epidemiology of musculoskeletal symptoms among milkers and dairy farm characteristics in Sweden and Germany. J Agromed 21:43–55
51. Douphrate DI, Nonnenmann MW, Hagevoort R, de Porras DGR (2016) Work-related musculoskeletal symptoms and job factors among large-herd dairy milkers. J Agromed 21:224–233
52. Bhuiyan N, Baghel A (2005) An overview of continuous improvement: from the past to the present. Manag Decis 43(5):761–771
53. Aly-Hassan OS, Sial FS, Ahmed AE, Abdalla KN (1986) Modification of some industrial equipment to suit date palm orchards mechanization. In: The 2nd date palm symposium, 3–6 March. King Feisal University, Al-Hassa, KSA, pp 543–554
54. Al-Suhaibani SA, Babeir AS, Kilgour J, Flynn JC (1988) The design of a date palm service machine. J Agric Eng Res 40:143–157
55. Moustafa AF (2005) Development of a tractor-mounted date palm tree service machine. Emir J Food Agric 17(2):30–40
56. Jahromi MK, Mirasheh R, Jafari A (2008) Proposed lifting model for gripper date palm service machines. Agric Eng Int CIGR E-J X(October):01–10
57. Garbati-Pegna F (2008) Self-moved ladder for date palm cultivation. In: Innovation technology to empower safety, health and welfare in agriculture and agro-food systems international conference, 15–17 September, Ragusa, Italy
58. Brynjolfsson E, Mcafee A (2014) The second machine age: work, progress, and prosperity in a time of brilliant technologies. W. W. Norton & Company, New York
59. Graetz G, Michaels G (2017) Is modern technology responsible for jobless recoveries? Am Econ Rev 107(5):168–173
60. Shapiro A, Korkidi E, Demri A, Edan Y (2009) Toward elevated agrobotics: development of a scaled-down prototype for visually guided date palm tree sprayer. J Field Rob 26(6–7): 572–590
61. Razzaghi E, Massah J, Asefpour-Vakilian K (2015) Mechanical analysis of a robotic date harvesting manipulator. Russ Agric Sci 41(1):80–85
62. Asfar KR (2016) Palm tree climbing robot. J Autom Control Eng 4(3):220–224
63. Nourani A, Garbati Pegna F (2014) Proposed harvester model for palm date fruit. J Agric Technol 10(4):817–822
64. Nourani A, Kaci F, Garbati Pegna F, Kadri A (2017) Design of a portable dates cluster harvesting machine. Agric Mech Asia Afr Lat Am 48(1):18–21
65. Mostaan A, Marashi S, Ahmadizadeh S (2010) Development of a new date palm pollinator. In: The 4th international date palm conference, 15–17 March, Abu Dhabi, UAE
66. Ismail KM, Al-Gaadi KA (2006) Development and testing of a portable palm tree pruning machine. Int J Agric Res 1(3):226–233

Agriculture into the Future: New Technology, New Organisation and New Occupational Health and Safety Risks?

Kari Anne Holte[1(✉)], Gro Follo[3], Kari Kjestveit[1,2], and Egil Petter Stræte[3]

[1] International Research Institute of Stavanger, Stavanger, Norway
Kaho@Norceresearch.no
[2] University of Stavanger, Stavanger, Norway
[3] Ruralis, Trondheim, Norway

Abstract. Agriculture is a hazardous industry, with a high frequency of injuries. As working life has changed over the last decades, so has also agriculture. In Norway, farm size has increased, and agriculture has become technology intensive with a high amount of automated milking systems (AMS) and is now more dependent on hired help. The aim of the study is by sociotechnical system theory to explore how a new generation of farmers describe their work organisation in relation to occupational health and safety. The study is an explorative interview study at five farms having implemented AMS. An open interview guide was used. The interviews were recorded and thereafter transcribed. Analyses were based on the balance-theory with the domains technology, organisation, physical environment, task design, and individual characteristics. The results show that AMS changes the farm as a sociotechnical work system. AMS is considered a relief with regards how tasks become less physically demanding, less time consuming, and with less animal contact. On the other hand, cognitive demands increase. The results indicate that the technology increases both complexity and vulnerability, these factors being less considered by the farmers. The findings underline the importance of farmers' increasing awareness of their role as a manager and for an increased system perspective.

Keywords: Agriculture · Work organisation · Sociotechnical system theory Injuries

1 Introduction

Agriculture is a hazardous industry, with a high frequency of injuries [1, 2]. A Norwegian study concludes that farmers constitute a heterogenous group considering occupational health and safety (OHS) risk, asking for addressing specific groups of farmers [3]. Increased injury risk is found among dairy farmers [4]. Dairy farming is the major agricultural production in Norway. By the beginning of 2017, there were 8,326 dairy farms with an average of 27 dairy cows and 5,143 suckler cow farms with an average of 16 cows. Average numbers of cows on both types of farms have increased in recent years [5]. Dairy farming has undergone major technological developments past decades, one of the most

S. Bagnara et al. (Eds.): IEA 2018, AISC 825, pp. 404–413, 2019.
https://doi.org/10.1007/978-3-319-96068-5_45

important innovations being the introduction of milking robots or automatic milking systems (AMS). Implementation of modern technology alters work activities, behavior, and how work is organised, which in turn affect workplace safety [6]. Implementing AMS may therefore change the risk picture for dairy farmers. The aim of this study is by sociotechnical system theory to explore how a new generation of farmers, having implemented AMS, describe their work organisation in relation to OHS-risks.

1.1 Automated Milking Systems; Organisational Aspects

It is estimated that more than 35,000 AMS operate around the world [7]. Norway has one of the highest relative numbers of AMS: 44.5% of the milk produced by end 2016 was milked by a robot [8]. The first AMS was installed in Norway in 2000 [9]. In 2006, there were 170 robots, and by end 2016 there were 1,726 dairy farms with robots (total number of dairy farms by end 2016 was 7,880). The number of AMS is rapidly increasing, and approximately 200–250 AMS are installed in Norway each year.

The milking robot is a device associated with increased efficiency and productivity, and it is often combined with other devices, like robot for feeding, activity measuring, robot for cleaning, etc. The milking robot requires loose-housing, which is a specific design of the cowshed, which enables the cows to enter the robot whenever they want to. Around the clock, cows can freely step into the robot, as well as freely move around in the cowshed, and the robot traffic becomes influenced by herd dynamics [10].

Arguments for investing in AMS involve reduced labor [11]. In a review of AMS studies, Jacobs and Siegford [12] reported a decrease in labor by as much as 18%. Others have found minor differences in labor use, but rather differences in task and work flexibility [13]. Similarly, Butler et al. [14] found that although AMS reduced the need for labor in the milking parlor, farmers' labor changed rather than decreased. This is firstly related to the maximum capacity of one milking robot (i.e. approximately 70 dairy cows). This maximum capacity can be viewed as a target to make optimum use of the robots from both a technical and an economic perspective, as higher milk volume reduces the fixed costs per liter. However, the higher volume of milk requires more fodder, which means need for more land and more transportation. All in all, more labor may be needed on the farm [15].

Moreover, farmers' main benefits of investing in AMS are; less time spent on milking, more interesting farming, more stable treatment of the cows, and less need for relief in the cow house [16]. To achieve the benefits, farmers must succeed in implementing AMS. Successful implementation depends on the motivation level of the farmers and whether they manage to adapt this technology to their specific needs (op. cit.). AMS may also have disadvantages for farmers, such as being constantly on call and information overload (op. cit.). A Norwegian study concludes that the primary motives for investing in milking robots include a more flexible work day, reduced physical work, and a desire to join what is regarded as the future standard of dairy farming. A motive is also increased flexibility and new opportunities and challenges related to the management of the herd [15]. In another Norwegian (interview) study, robot-farmers said they spend less time on milking and are less hands-on physically with each individual cow, particularly the udder, which changes the relationship with the animals, spending more time observing the "whole" cow and the herd. The authors point to farmers facing a transformation from emphasis on tacit knowledge, towards an increased importance of codified knowledge [15]. A biannual trend survey, conducted by a representative sample of Norwegian farmers in 2016, showed robot farmers being more satisfied with their own health, work environment and occupational safety compared to dairy farmers without a robot. However, standard of the operational building better explained the variation in satisfaction with the work environment and occupational safety than did the milking robot [17]. Considering the substantial changes in farming entailed by the implementation of AMS, there is a need for studies exploring *how* OHS risks may change due to the technology itself and the altered conditions brought along.

1.2 Sociotechnical Perspectives Applied on Agriculture

The above studies argue for using a sociotechnical perspective when exploring how implementation of AMS may alter OHS risks, as sociotechnical system theory emphasizes the organisation's interdependence of both the technical and the social system [18]. Moreover, a questionnaire study among a representative sample of Norwegian farmers found factors like workplace design, production type, work organisation, income, and size as risk factors for injuries [3], pointing to a complex risk picture. A relevant model is the "balance theory" [6, 19], where the work system is described by the domains *technology and tools* (design and usability), *tasks* (job content, workload, control, repetitiveness, learning abilities, feedback, etc.), *organisational conditions* (social and organisational support, role ambiguity and conflict, job security, culture, work scheduling, etc.), *physical environment* (workplace design, noise, temperature, climate, lightening), and the *individual* (cognitive and physical characteristics, needs and abilities, experience, demographic characteristics), and where these domains interact with each other [19]. These domains together constitute and represent work activities, highlighting the sharp end, with the human in the center of the work system [6]. Moreover, the work system influence processes, which in turn affect certain outcomes, like risk [20]. The aim of this study is therefore, by means of the "balance theory", to explore how a new generation of farmers, having implemented AMS, describe their work organisation in relation to OHS risks. Research questions are:

- How do dairy farmers having implemented AMS reflect upon OHS risks?
- How may OHS risks change due to alterations in the work system caused by implementation of AMS?

2 Methods

The study is an explorative interview study, at five farms located in three areas of Norway. Common for these farms were dairy production by means of AMS. They otherwise constitute a variety of production combinations, topography/climate, and different organisational forms (lone farmers, family farms, and joint venture). The latter caused variations in number of informants at each farm, and the study includes a total of 12 informants.

The interviews were based on an open interview guide, covering topics like the farm as a work place and a home, accidents at the farm and their consequences, risk attitudes, and safety culture. The interviews lasted from 90 to 120 min. The interviews were audiotaped and thereafter transcribed verbatim and anonymized.

In the initial analysis, the text was searched for descriptions of accidents, injuries and OHS risk factors that were associated with the robot. Thereafter, these risk factors were associated with processes being influenced by the robot and assigned to the different domains of the model: task design, technology, organisation, individual, and environment, in accordance with how aspects of work are allocated to the different domains [19], as described in the former section. For each farm, we drew a picture of the actual configurations describing the dynamics between processes, domains and OHS risk. Then these pictures were aggregated to identify overall patterns of processes and associated system configurations.

3 Results

The farmers gave rich descriptions of their daily work, enabling us to explore how the milking robot changes activities and OHS risks. In the analysis we identified different processes, described by a specific work system configuration, that could be associated with changes in OHS risk. In this paper we will restrict our results to two of them; the *man-animal-robot interaction* and *animal-flow in the cowshed.*

3.1 The Man – Animal – Robot Interaction

Considering the milking process, implementation of AMS (*technology*) interacts highly with and completely alters the *task design,* furthermore also having impact on the *organisational domain.*

Changes in Task Design

The robot entails a complete change in the *job content* associated with milking, including exposures to risk factors like *physical demands* and *animal contact*. When installing AMS, cows are trained to enter the robot themselves and the robot takes care of cleaning and milking. This is seen as a positive outcome by the interviewed farmers, as they perceive their *total work load* as being reduced. On the other hand, the robot introduces new tasks, increasing the *cognitive demands*. One farmer says: *"(...) and then I go to the cow house. And there it's the daily routine; see through stuff with the robot, is there anything...? There are standard reports I must check. It takes about 5–10 min."* This description points to *routines* of checking the computer daily. However, the robot monitors many different variables for each animal. This provides the farmer with a huge amount of data to analyse and to use as a background for decisions. Another farmer states: *"(...) I would claim that the use of data and the... The information on the robot... I don't know if maybe 70% of those who have a robot, don't make use of it at all, so to say. They only learn what they must to make it work."* In this farmer's opinion, most of the information goes to waste, because farmers do not know how to utilise it. If used in an efficient way, there is a large potential using this information for improving the production. Tasks may therefore become more interesting and challenging and allow for increased *possibilities for learning.*

Changes in task design due to implementing AMS lead to new risks. Two incidents are described involving the robot. An incident happened when the robot was new to the farmer and the herd. The farmer tried to prevent a cow from entering the robot, but he was not well located. As the cow was persistent, he was pushed into the robot along with the cow and thereafter squeezed between the cow and the device. Those witnessing could not assist. The incident was painful and gave the farmer bruises, but he did not see a doctor. Lessons were learned regarding how to place and protect oneself in the cow house when directing the herd. The other case was retold by our informants and was about a farmer being trapped in the robot with a cow. This robot was designed in a way that when opening the robot door, it stopped working and the entire system shut down. This made it impossible to identify what was wrong. The farmer may therefore have tried to step into the robot while it was running, and then squeezed by the robot arms and pushed underneath the cow.

Other situations were also associated with risks, for instance how cows had to be taught to use the robot. One farmer stayed in the cow house all day, pulling unwilling animals through the robot. Cows that do not want to enter the robot, do not fit in the robot, and even destroying elements of the robot by kicking, were described. Another potential risk situation mentioned was when having to help a cow: *"(...) with which the robot has a problem, for example."* Common for these situations was unpredictability in animal behavior, and situations arising when being outside of what the algorithms of the robot could handle. This required performing tasks that could be characterized as *non-routinised tasks,* leading to potential risk situations.

Changes in Organisational Domain

While regular tasks associated with traditional milking disappear, milking becomes a continuous around the clock process, and non-routinised tasks may show up at any time. This has impact on the *organisational domain,* particularly *working hours* and *training of employees.* When having technological breakdowns, the system itself sends a message to the farmer by means of the phone, and the farmer can be interrupted 24/7 irrespective of where he or she is. One farm was organised as a joint venture with five equal farmers and an established *shift schedule.* Thus, all farmers were allowed periods with relief from responsibility and unpredictable incidents that needed to be solved. Another farmer, operating the farm alone, was required to handle these interruptions himself. Farmers with employees can choose to leave responsibility for the robot to employees. This requires training, and there were different views considering *training of employees,* regarding how they approach skills, type of training, and experiences from allowing employees working with the robot. Several of the visited farms had foreign employees. When a farmer was asked for his view on involving employees within the robot, this entailed this reflection:

> *"Yes, you can easily switch the robot over to Estonian, of course. (I: Yes, that's what we've heard. And it's actually true?) That's not a problem, no. No, no, it is build-in. You just have to remember to switch it back again. But I know (...) how to go and change the language."*

This farmer seems to accept that just switching the language is enough, indicating that he understands the robot as logic and rational. He also trusts that the employees, as they are educated within agriculture, can handle the robot in a proper way, not reflecting upon the need for experience and training on the robot. At another farm the language issue entails a somewhat different reflection, when the farmers were asked about changing language at the robot:

> *"Husband: Yes, I think you can do that.*
> *Wife: But it has to be taught anyway...*
> *Husband: But we have needed long time too, to figure out the robot, so it's not "just"...*
> *I: It's not as easy as that...?*
> *Husband: But I think that if... Like, he who has been here for ten years... We could easily have taught him the robot, if that was what we wanted. I don't think that's the problem. However, they would have to be there on a regular basis, and not just on day now and then."*

This couple talked about the robot as having certain qualities entailing a need for understanding the robot. This requires daily experience, which make them reluctant to let employees work with the robot. Yet another farmer reflected on involving a potential new employee: "(…) *regarding training him to take responsibility for the daily care, it is how he must respond when things stop. Because some things you may fix in a straightforward way, but other things you should not try to repair, because it can be dangerous*". This also address how they evaluate having the skills of knowing when *not to do* certain activities, like not to try to repair or fix the technology, especially where these tasks are non-routinised, as explained earlier.

3.2 The Animal-Flow in the Cowshed

For implementing AMS, loose housing is a premise. Therefore, technology interacts and alters the *physical work environment*, influences *the tasks*, and impacts overall *organisational decisions* at the farm.

Changes in Physical Work Environment
Workplace design impacts the flow of cows in the part of the cowshed where the animals choose themselves where to go. The farmers reflected upon a design that considers cows' behavior and the animal hierarchy. Does the chief cow hamper others to approach the robot, the feeder for grain concentrates, the huge drinking trough, or the brush? As said at one farm, it is important that the cowshed has a design that spread these areas, so the chief cow is unable to be at all these locations. When solutions are optimal, the result is a continuous robot traffic and increased amount of milk. Moreover, loose housing is more on the cows 'premises' compared to a tie-stall, giving better animal welfare.

Changes in Task Design
Removing physical barriers between animals and between the animals and the farmer may change certain exposures, changing injury risk when performing tasks within this area. Such tasks are cleaning of manure, clearing, and visual observations of animal health. When the physical barriers are removed, farmers must be observant in another way compared to the traditional tie-stalls. They talked about the need for having *"eyes in their neck"*, as these tasks demand them *to be attentive*. In the interviews, stories

were told about cows almost and in fact running humans down, and of dangerous *"trains of cows"*. One farmer with 100 milking cows and two robots claimed: *"Most of the cows are ok, but they are loose, and we never know"*. On the other hand, when animals are well accustomed to the robot (if not adapted to the robot; sold or slaughtered), there is less need for chasing or physically moving animals as opposed to in the traditional tie-stalls. According to the farmers, this is perceived as reducing the risk for injuries.

Changes in Organisational Domain
When talking about the changes in physical work environment, the farmers also talk about how this interacts with what we could called *strategic decisions,* as well as *risk management*. This is underlined in one farmer's statement: *"Yes, but we have very calm cattle. It is very... Like... You should never say that nothing can happen, but the chances for injury caused by handling cattle, with the breed that we have, are very, very much smaller than with regular (...) cows"*.

Decisions about breed are part of strategic decisions considering how to improve the quality of the milk and increase the production at the farm. Decisions about breed is also about animal behavior within the cowshed, as farmers anticipate that a calmer breed reduces risk for injuries due to animal contact. The ability to take strategic decisions is further supported by and closely connected to the increased ability for learning, as the robot produce a lot of information, as described under the heading: *the man – animal – robot interaction.* When properly utilised, this give farmers increased opportunities for professionalising the organisation and increasing their production.

4 Discussion

The results show that implementing AMS changes the farm as a sociotechnical work system and alters OHS risk in both positive and negative terms. Despite the small number of informants, the results clearly point to changes that allow us to reflect upon arising OHS challenges. In the discussion we will firstly discuss the perceived changes in OHS risks. Thereafter, we will reflect around overall work system changes, and how it might entail risks in the longer term.

4.1 Perceived Changes in OHS Risks

Farmers who implement AMS experience a huge change in content and structure of work, changing OHS risks in both positive and negative directions. They experience an overall reduction in total work load. This is in line with a biannual trend survey conducted by a representative sample of Norwegian farmers. The survey showed in 2016 that robot-farmers were more satisfied with their own health, work environment and occupational safety compared to dairy farmers without a robot [17]. Figures from the Department of Occupational Health Surveillance (NOA) show that farmers are among the occupational groups with the highest proportions of self-reported work-related pain [21]. As physical demands reduce, we could assume that risk for musculoskeletal disorders decrease. Our results indicate a shift from physical demands to cognitive demands. One of the reasons why cognitive demands increase is the amount of data generated by the robot. A Norwegian study concludes that farmers face new opportunities and challenges related to data and computer work and in management of the herd [15]. In this way, increased cognitive demands become positive. In the demand-control model by Karasek and Theorell, high demands and high control constitute the active job situation, which may lead to increased motivation and learning opportunities [22].

A considerable risk factor for injuries among farmers is animal contact [23, 24]. The robot requires a design where physical barriers in the cowshed are removed. Interestingly, the farmers perceived this as a risk-reducing element, as animals are calmer in loose-housing system compared to a tie-stall. Despite few incidents to report, new risks seem to be unpredictable situations involving the robot, for instance being trapped inside the device. The descriptions point to situations that are extraordinary or unusual, as they are non-routinised tasks for the farmers and outside of algorithms for the robot. This is not surprising, considering repair and maintenance of farm machinery being activities related to injury risk [23]. However, there is lack of knowledge regards how OHS risks change due to implementation of AMS.

4.2 Changes in the Work System – Latent Conditions

Sociotechnical system theory provides a holistic perspective allowing for exploring complexity and dynamics within work systems, as well as reflecting upon risk reductions through redesign [19]. In the results, we described two processes and their specific configurations. Both processes depend on the technology, making the processes tightly coupled and vulnerable for technological breakdowns [25]. The couplings are tighter than for a tie-stall, because of the build-in assumption that AMS increases the number of animals, which increases the vulnerability in case of a breakdown. This vulnerability is to a minor degree addressed among the farmers. As the robot is still new to many Norwegian farmers, there is lack of experiences and stories to learn from. However, those who in one way or another had experienced a robot incident, showed less trust in the technology compared to those without such experiences. We may therefore speculate if the work organisation is still being immature considering how risk introduced by the technology is understood and handled. Based on lack of experience, the technology is trusted and perceived as rational. The vulnerability of the technology itself and how it changes the overall organisation may therefore be overseen.

While the capacity of a milking robot is about 70 milking cows a day, utilisation of this capacity makes investment in a robot more efficient. Dairy farms often double their production when installing AMS, which in the next step requires more fodder and increases the amount of work load at the farm [15]. This is more or less the case for our farmers as well, and requires increased professionality, paying attention to new and upcoming tasks, reliance on external factors/suppliers, as well as on hired labor. All these factors may challenge the farmer as a manager.

The increased cognitive demands and potentially being accessible 24/7, are not addressed as a negative outcome of implementing AMS. We have previously referred to the learning potential and more interesting farming. However, increased cognitive demands may also be potential stressors. It may be difficult to discover the negative long-term effects of always being accessible in case of interruptions and breakdowns. Increased economic constraints and need for effectively utilising the AMS may be another cognitive demand that is overlooked, and which is a latent condition for OHS risk. Glasscock and coworkers [26] indicates that stressors and stress symptoms, like work overload/time pressure, role conflict, economic concerns, administrative burden, and unpredictability are risk factors for occupational injuries in agriculture. Several studies point to organisational complexity due to size, income level and number of employees [23, 24, 27, 28]. The immediate positive OHS outcomes from implementing AMS may therefore be replaced or balanced out by other, underlying outcomes, which over time may increase OHS risk in other ways than the farmers are able to discover after a few years with the new system. Studying the same farmers after some years might shed light on some of these elements.

References

1. Jadhav R et al (2015) Risk factors for agricultural injury: a systematic review and meta-analysis. J Agromed 20(4):434–449
2. Jadhav R et al (2016) Review and meta-analysis of emerging risk factors for agricultural injury. J Agromed 21(3):284–297
3. Kjestveit K, Aas O, Holte KA. Can organizational aspects serve as latent conditions for occupational injury rates among Norwegian Farmers? (manuscript)
4. Hartman E et al (2004) Risk factors associated with sick leave due to work-related injuries in Dutch farmers: an exploratory case-control study. Saf Sci 42(9):807–823
5. Norwegian Agriculture Agency (2017) KU - Foretak med felles melkeproduksjon 2016, fylkesfordeling
6. Carayon P et al (2015) Advancing a sociotechnical systems approach to workplace safety - developing the conceptual framework. Ergonomics 58(4):548–564
7. Salfer J et al (2017) Dairy robotic milking systems – what are the economics? https://articles.extension.org/pages/73995/dairy-robotic-milking-systems-what-are-the-economics. Accessed 25 Jan 2017
8. Norsk landbruk. http://www.norsklandbruk.no/nyhet/antallet-melkeroboter-oker-i-Norge. Accessed 6 Dec 2017
9. Kjesbu E, Flaten O, Knutsen H (2006) Automatiske melkingssystemer - en gjennomgang av internasjonal forskning og status i Norge. NILF, Oslo
10. John AJ et al (2016) Review: milking robot utilization, a successful precision livestock farming evolution. Anim Int J Anim Biosci 10(9):1484–1492
11. Drach U et al (2017) Automatic herding reduces labour and increases milking frequency in robotic milking. Biosys Eng 155:134–141
12. Jacobs J, Siegford J (2012) The impact of automatic milking systems on dairy cow management, behavior, health, and welfare. J Dairy Sci 95(5):2227–2247
13. Steeneveld W et al (2012) Comparing technical efficiency of farms with an automatic milking system and a conventional milking system. J Dairy Sci 95(12):7391–7398
14. Butler D, Holloway L, Bear C (2012) The impact of technological change in dairy farming: robotic milking systems and the changing role of the stockperson. R Agric Soc Engl 173:1–6

15. Stræte EP, Vik J, Hansen BG (2017) The social robot: a study of the social and political aspects of automatic milking systems. In: System dynamics and innovation in food networks. International Journal on Food System Dynamics, Innsbruck
16. Hansen BG (2015) Robotic milking-farmer experiences and adoption rate in Jæren, Norway. J Rural Stud 41:109–117
17. Hårstad RB, Stræte EP (2017) Melkerobotbønder mer tilfreds med HMS, ferie og fritid, in Faktaark 7/17, Ruralis, Editor. Ruralis, Trondheim
18. Davis MC et al (2014) Advancing socio-technical systems thinking: a call for bravery. Appl Ergon 45(2, Part A):171–180
19. Carayon P (2009) The balance theory and the work system model... Twenty years later. Int J Hum Comput Interact 25(5):313–327
20. Carayon P et al (2006) Work system design for patient safety: the SEIPS model. Qual Saf Health Care 15:I50–I58
21. Tynes T et al (2015) Faktabok om arbeidsmiljø og helse 2015 - status og utviklingstrekk. STAMI - rapport, Årg 16, Nr 3. STAMI, Oslo
22. Theorell T, Karasek R (1996) Current issues relating to psychosocial job strain and cardiovascular disease research. J Occup Health Psychol 1(1):9–26
23. Karttunen JP, Rautiainen RH (2013) Distribution and characteristics of occupational injuries and diseases among farmers: a retrospective analysis of workers' compensation claims. Am J Ind Med 56(8):856–869
24. Taattola K et al (2012) Risk factors for occupational injuries among full-time farmers in Finland. J Agric Saf Health 18(2):83–93
25. Perrow P (1984) Normal accidents. Basic Books, New York
26. Glasscock DJ et al (2006) Psychosocial factors and safety behaviour as predictors of accidental work injuries in farming. Work Stress 20(2):173–189
27. Rautiainen RH et al (2009) Risk Factors for serious injury in finnish agriculture. Am J Ind Med 52(5):419–428
28. Van den Broucke S, Colemont A (2011) Behavioral and nonbehavioral risk factors for occupational injuries and health problems among Belgian farmers. J Agromed 16(4):299–310

Estimation of Output in Manual Labor Activities: The Forestry Sector as Example

Felipe Meyer(✉) and Elias Apud

University of Concepcion, Concepcion, Chile
{fmeyer, eapud}@udec.cl

1 Introduction

Apud and Valdes [1] and Apud, Gutierrez [2] developed an area of research were they proposes study different activities in forestry sector and determine which element were related with the output of the workers. Planting, pruning, site clearing, commercial harvesting and harvesting were the main activities that were studies. The main objective was set up standard references for output in forestry work, since the output depends on the workload that a worker can reach in a sustainable way without fatigue or other risks and on the difficulties, they can found to carry out his job, mainly related to the type of trees, climate and ground. In other words, in forestry there is no chance to demand of a worker always the same amount of work. This puts a difficulty to calculate incentives and salaries, especially when workers are paid by piece rate [3]. Therefore is not enough to carry out work-studies, having only subjective criteria to choose the workers and to decide after if they are doing the job at the right pace.

The idea of estimate output from the work load and characteristics of the site was to transferred to the operation and used for the scheming of expected output and for the calculation of basic salaries or basic salaries plus bonus. The output, in any forest operation tends to be broad. Therefore, with predictions, managers and planners may have an objective basis to calculate salaries. For example, if a stimulus for production is going to be paid, this should be based on the difficulty of the job. However, it is important to highlight that these equations should not be used rigidly because they propose an estimation of yields per hour, but at least allowances for rest pauses and lunch should be considered to avoid fatigue. The objective is discuss over reference formulas for calculation of both and incomes in manual forest workers, using an example of pruning activities.

1.1 Pruning in Forestry

Pruning are important activities for obtaining products of good quality. This activity has the purpose to eliminate mainly branches, cones and epicornics to obtain wood free of knots. To make this job efficient, the worker must have a good technique. If a branch is not properly cut, the occlusion will be delayed and the amount of clear wood will decrease, therefore the price and quality of that piece of wood will not be the possible highest.

S. Bagnara et al. (Eds.): IEA 2018, AISC 825, pp. 414–419, 2019.
https://doi.org/10.1007/978-3-319-96068-5_46

Not all enterprises use the same scheme for pruning. In some cases, four treatments are made namely: First pruning from 0 to 2 m; Second pruning from 2 to 4 m; Third pruning from 4 to 6 m, and Fourth pruning from 6 to 8.3 m. This example was carry-on in a second pruning. First pruning are done with the subject standing on the ground. For second pruning and the next ones, tree ladders are used. The main demands of this technique are that the workers have to climb up and down the ladders, which means more physical load, therefore is a very fatiguing job. The workers complain of neck, shoulders, arms and low back pain. Mechanization is a feasible solution, however pruning trees is very difficult, mainly because operational issues and costs. Therefore still are cheaper use manual tools to do it. This activity is important and related with the mechanical quality of the wood. If the pruning were not done, the wood would fall in quality are, since the knots of the wood would be bigger, and the amount of wood free of knots would less. Therefore, the prices of that piece of wood would be cheaper.

The following study was an attempt to discuss standard references of second pruning, based on a recently study; were big difference in output and income were detected between crews that also produce discussion between, workers, contractors company and main forestry company. Also have the objective to discuss the importance of this topics, since the forestry sector need to revise all the elements that are threatening the incorporation new workers to this sector, and the salary and payment system is one of the key area beyond many positive chances has been incorporated in last years [4].

2 Methods

The case that will explain in this paper is based on a pruning developed in pines three during 2017–2018.

Methods Follow-ups in the field were carried out using techniques proposed by Apud, Bostrand [5]. In addition statistical procedures were applied to the data and equations were proposed to calculate output.

3 Results and Discussion

Each year the company, where the study was conducted needs to prune thousands of hectares; however, due the lack of workers, a percentage of hectares remains unpruned. The next year when those remaining hectares are pruning each branch has increased its diameter and their wood is harder since are one year older. Associated with that Apud and Valdes [1] mentioned, after carrying out an evaluation of different variables from the trees, ground and environment it was found that pruning output was negatively correlated with average branch diameter ($r = -0.82$), and to the number of verticil ($r = -0.51$). Therefore What are impact of this late activity over the workers when its compare when is done in the right time?

In Table 1, show variables of the stand one and two, characteristics of the terrain and ambient temperature in the premises in which the evaluations of the second pruning were carried out.

Table 1. .

Variables	Unit	Stand 1 (n = 9)		Stand 2 (n = 9)	
		Average	S.E.	Average	D.E.
DBH (Diameter at breast height)	cm	18.7	0.54	19.3	1.05
DB (Diameter of branch)	cm	4.7	0.14	6.1	0.17
Number of verticil	n	5.2	0.47	5.4	0.43
Number of branches prune	n	22.5	3.03	25.7	5.99
Slope	%	5.3	15.8	8.3	3.11
Average temperature	%	15.3	3.1	16.2	3.1

As can be seen in Table 1, the trees pruned on site 1 were smaller than those processed on site 2, lower diameter at breast height, lower diameter of branch, less number of verticil and less number of branches to prune. No differences in the slope of the pruned stands in either studies or the temperature that workers were exposes.

Table 2, summarizes the distribution of times during the study. Nine workers were evaluated in both cases.

Table 2. Main time, secondary and rest times expressed as a percentage of the total of the workday during second pruning work.

Time	Site 1 (n = 9)		Site 2 (n = 9)	
	Average	D.E.	Average	D.E.
Pruning	73.0	5.1	74.9	5.2
Secondary	9.2	4.6	8.7	4.6
Total principal time	82.2%		83.6%	
Resting and lunch	17.8		16.4	
Total resting times	17.8%		16.4%	

As it stands out, there were no great differences in the distribution of times during the pruning carried out in the two farms studied. The workers devoted an average of 83% of the workday to the main activities, which reveals an adequate use of time. It should be noted that the secondary times include scheduled breaks of 15 min for each hour of work, which reveals that virtually all secondary activities, including tool maintenance, were done during this period.

In Table 3 includes the output, per day and per hour. Also the cardiovascular load.

Table 3. Performance, physical load of work and duration of the days in the first pruning tasks studied

Variable	Unit	Site 1 (n = 9)		Site 2 (n = 9)	
		Average	D.E.	Average	D.E.
Output					
Pruning trees per day	n	301	21	244	45.1
Pruning trees per hour	n	41	3.7	34.9	5.5
Carga cardiovascular	%	37.3	5.6	36.2	4.5

The price that forestry contractor companies paid to the workers per each pruned tree is US$ 0.16. On average workers expend 21 days per month working, in a scheme of 5 days working and 2 days resting. The salary that is based on their productivity, for the workers that pruned site 1, receive US$ 1021 per month (US$ 0.16*301*21), instead workers that worked en site 2 receive US$ 830 (0.16*244*21) per month, an 82% of the salary of the first group.

Therefore, answer the initial questions about *What are impact of this late activity over the workers when its compare when is done in the right time?* Three main and direct impact over the workers were identifies. Highest level of effort over shoulder, elbow and wrist of the pruners, since on average they are pruning around 6200 thicker branches per day. In addition, that means a higher cardiovascular load, 5% more when it compares with similar activities with thin branches. A third impact is this group, as was mentioned before, they reach between 15–25% less output per day. The problem of the reduction of their productivity is that part of their salary; around 40–50% are based on what they are capable to done. However the payment, the price ($/pruned tree) is the same, if the work was done in a normal situation, with thinner branches and softer wood. The results of that could be between 17 to 25% less salary for the workers. Therefore, what would happen if the salary were estimate using a standard reference formula that considered the natural difference of each forestry site?

A systematic regression was carried out to evaluate the possibility of estimating the yield based on simple terrain, stand and climate variables, also including the physical load reached as a criterion during the evaluation days. The following equation was obtained:

$$EO = 52.3149 - 3.518 * NV - 1.1107 * BD + 0.322\,CL$$

where EO, is expected output, NV is number of vertical, BD is branch diameter, and CL is cardiovascular load. The statistical indicators were $R = 0.81$; R squared $= 0.7$; Standard Error = 2.9 trees per hour; $p < 0.002$.

In Table 4, a comparison between the real output and the estimate output using the formula above.

Table 4. Performance, physical load of work and duration of the days in the first pruning tasks studied

Real measure (prunned tree/hour)		Estimation using formula (prunned tree/hour)	
Site 1	Site 2	Site 1	Site 2
43	34.9	41.78	36.65

Looking at the Table 4, one can see that the range of expected output is very close. Therefore, the users of the table, including managers and planners, may have an objective basis to calculate salaries. For example, if a stimulus for production is going to be paid, this should be based on the difficulty of the job but, at the same time, it has to be thought in such a way that the workers can obtain it, without the harmful effects of fatigue. In other words, such a stimulus will not be motivating if the workers, to get their normal salaries, have to work at or over 40% cardiovascular load. As a matter of fact, if such a system is to be applied, we have recommended to calculate the basic salaries considering a work load of no more than 30% cardiac cost. In such a way, if a bonus is to be paid, the workers still can increase their cardiovascular load around 10%, allowing them to earn the stimulus, but without fatigue.

There are risks if tables like the one under discussion are used too rigidly. The estimation is for the output expressed as trees pruned per hour. In Chile a normal shift in the forest last nine hours: eight hours work and one hour at mid-day for lunch. Our observations in the field have shown that it is highly advisable to provide two rest pauses of 15 min each at mid-morning and mid-afternoon. The workers to have a snack and to break the monotony of this repetitive task can use that time. In such a case, working time is reduced to 7.5 h. If the hourly output is multiplied by 7.5 and by the number of workers in the crew, daily production can be calculated.

In addition, the monthly output can be estimated since it is considered that 21 days are worked in summer and in winter 18. It is very likely that in summer the estimation of 21 working days is correct because the weather is not an impediment. However, in winter the situation is different. During the rainy winter months, it is not always possible to work 22 days. Then, when time is lost and the contractors have to finish a job, they put pressure on workers making them to work longer shifts or sometimes two weeks without resting days in between. There is no doubt that, when those things happened, any criteria will fail if planners do not consider the unpredictable factors which in forestry are very common to occur.

The equation obtained is significant and the variables selected in the function explain around 80% of the variation in the performance achieved in this task. The variables are simple to measure and the number of whorls, the ruggedness of the terrain and the actual height can be estimated before the start of the work. For its part, the cardiovascular load should be used as a reference variable.

4 Conclusion

The approach to obtain information on standard performance is laborious. Material and human resources are needed to spend time observing and measuring work. However, it is difficult but not impossible. Ergonomics specialists should struggle to develop research programs more than isolated studies with the aim to make forest work fair aiming to keep balance between wages and reasonable effort.

References

1. Apud E, Valdes S (1995) Ergonomics in forestry: the chilean case. E.I.L. Office, ILO, Geneva
2. Apud E et al (1999) Manual de Ergonomia Forestal. [Manual of Ergonomics in Forestry]. Proyecto FONDEF D96I1108 "Desarrollo y Transferencia de Tecnologías Ergonómicamente Adaptadas para el Aumento de la Productividad del Trabajo Forestal" Chile [D96I1108 FONDEF project "Development and Technology Transfer Ergonomically Adapted for Increased Productivity of Forest Work" Chile]
3. Apud E, Meyer F (2004) Ergonomics. In: Burley J (ed) Encyclopedia of forest sciences. Elsevier, Londres, pp 639–645
4. Meyer F (2017) Evaluation of workforce sustainability in the Chilean logging sector using an ergonomics approach. School of Management, Massey University, Auckland, New Zealand
5. Apud E et al (1989) Guide-lines on ergonomic study in forestry. Prepared for research workers in developing countries. ILO, Geneva

Comparison of Ergonomic Training and Knee Pad Using Effects on the Saffron Pickers Musculoskeletal Disorders

Nasrin Sadeghi[1](✉) and Mojtaba Emkani[2]

[1] Department of Occupational Health Engineering, Kerman University of Medical Sciences, Kerman, Iran
na.sadeghi@gmail.com

[2] Department of Occupational Health Engineering, Gonabad University of Medical Sciences, Gonabad, Iran

Abstract. Saffron picking is the job that capable to create a field for musculoskeletal disorders due to inappropriate working positions such as bending, kneeling and pressure that imposed to the body in these positions. The purpose of this study was compare two methods contain knee pad and training pamphlet application to prevent and reduce musculoskeletal disorders in saffron pickers. This survey was randomized clinical trial with register number IRCT2012100811043N1. Samples included 60 persons of saffron picker workers. Sampling was random & easy and used tools were including REBA, Body Discomfort Chart (BDC). During the saffron harvest before intervention take code to parts of workers bodies based on the REBA method. Workers discomfort levels were determined by BDC questionnaire. These steps are repeated after the intervention. Then the data entered into SPSS software version 16 and data analysis were performed with statistical tests (Mann–Whitney and t-test). BDC questionnaire analysis indicated a significant effect on the knee discomfort after knee pad using ($P = 0.0004$). But in the upper back ($P = 0.27$) and lower back discomfort ($P = 0.12$) had no significant effect. Also, REBA analysis results indicate a significant impact of the training pamphlet intervention on the scores: A ($P = 0.025$), C ($P = 0.027$) and final ($P = 0.028$). But no significant effect on B table scores ($P = 1.000$). According to the results, the effect of training pamphlets was higher than kneepad using. Probably, both interventional methods can be appropriate methods to prevent and reduce knee and trunk discomforts of saffron pickers.

Keywords: Musculoskeletal disorders · REBA · BDC · Saffron pickers
Ergonomic training · Knee pad

1 Introduction

Work-related musculoskeletal disorders (MSDs) are a major issue with important economic consequences. These disorders are manifested with varying degrees of severity with regard to the type of occupation and are common causes of occupational injury and disability in industrialized and developing countries [1]. Work-related

S. Bagnara et al. (Eds.): IEA 2018, AISC 825, pp. 420–431, 2019.
https://doi.org/10.1007/978-3-319-96068-5_47

diseases are harmful realizations of technology advancement in work environments that threaten the health of a huge body of workforce. According to International Labor Organization (ILO), 160 million cases of occupational disease annually occur worldwide, causing the death of 1.1 million people and imposing irreparable costs on economy of societies [2]. According to ACGIH (American Conference of Governmental Industrial Hygienists) report, musculoskeletal disorders rank second among work-related diseases in terms of importance, frequency, severity, and probability of progression. Contrary to the expansion of automated processes, work-induced musculoskeletal disorders are a major factor for loss of working time, increase in costs, and human damages of the workforce [3]. There have been few studies in our country on musculoskeletal disorders and their direct and indirect outcomes. According to the report of medical committee of Social Security Organization in Tehran Province, musculoskeletal disorders account for 14.4% of diseases and disabilities. Lower back pain is the second cause of absenteeism, the third cause of medical visits, and the fifth cause of hospitalization in this country [4].

Farming is a difficult occupation, although it is depicted in a beautiful and poetic manner [5]. As an important production sector in the country, agriculture has an effective role in job creation and prosperity. Therefore, safeguarding occupational health and safety can improve quality management in this sector [6]. Agriculture is one of the most widespread and most dangerous occupational activities, and nearly 63% of population in developing countries is engaged in farming activities [7].

The physical working conditions of farm workers show a potential for susceptibility to musculoskeletal disorders such as osteoarthritis of hip and knee joints, lower back pain, upper extremity disorders, and trauma. In addition, some pieces of evidence suggest that farmers may be more vulnerable to other rheumatologic conditions, including rheumatoid arthritis [5]. Due to its high frequency, musculoskeletal pain has become a stressful factor for farm workers [8]. Performing the tasks of agricultural occupation causes an inappropriate physical state in the body. Kneeling, bending, lifting and carrying heavy objects, vibrations of tractor driving, and working in inappropriate postures are associated with pelvic and knee pain. Musculoskeletal disorders are an integral part of this occupation due to improper conditions of the body while performing the duties, and almost all farmers suffer from these disorders [8, 9]. The physical condition of most saffron pickers is at a high ergonomic risk, and it is necessary to take immediate action to correct their body condition; otherwise, the risk of musculoskeletal disorders will be high in them [10]. In today's industrial world, especially in developing societies, due to inappropriateness of the environment and tools relative to human physical and mental conditions, workers are forced to tolerate these inappropriate conditions of the working environment and tools, which will have serious consequences and can have an adverse effect on the quality of life of individuals and their health. Persistence of these conditions can lead to musculoskeletal disorders among these workers. These disorders are one of the main causes of human workforce damage, loss of working time, and increase in costs, and according to research, musculoskeletal disorders account for over half of the absence cases from workplace and cause one-third of work-related absenteeism [3, 11, 12]. Recently, a comprehensive report by National Academy of Sciences has been presented, which indicates that ergonomic risks in the workplace lead to musculoskeletal disorders, but appropriate interventions will prevent from them [13]. Hong Xiao points out in his study that further research should focus on the development of preventive interventions to carry out tasks related to increased risk of pain. These interventions should specifically target different types of agricultural tasks [14, 15].

The above statements indicate that saffron harvesting is a stressful physical process given the inappropriate conditions of picking saffron flowers. This flower grows in a low height on the ground, causing the farmer to be exposed to inappropriate physical conditions such as sitting on two knees, bending, swinging of back, and doing severe monotonous work. Also, the harvesting period of this flower is limited, which causes high psychological pressure on farmers. Our aim in this study was to determine the risk of musculoskeletal disorders in saffron pickers and the application of two methods of ergonomics training and knee pad application to solve these problems and compare the effects of these two methods on musculoskeletal disorders.

2 Research Method

This research is a randomized clinical trial with register number IRCT20121008110 43N1. The sample consisted of 105 people who were recruited into the study by simple non-probability sampling. The inclusion criteria were as follows: (a) age range of 20–50 years, (b) kneeling during work, (c) no history of musculoskeletal disorders unrelated to saffron harvesting. Eventually, after applying the inclusion and exclusion criteria, the number of samples decreased to 60. Then, they were randomly divided into two groups of knee pad and training pamphlets. In the knee pad group, 30 knee pads were distributed among the farmers and the method of its application was trained to them. In the pamphlet group, 30 educational pamphlets were distributed among farmers and relevant explanations were provided.

Body Discomfort Chart (BDC) and Rapid Entire Body Assessment (REBA) were used to assess musculoskeletal disorders and body position, respectively. The research method was as follows. Before intervention, when the worker was picking saffron, pictures were taken from certain body angles according to REBA method. Then, based on the pictures, various parts of the body (trunk, neck, arm, forearm, leg, wrists, and hands) were encoded with respect to their condition. The codes were then entered into relevant tables, and A, B, C, and final scores were calculated. Also, the BDC charts were completed by farmers. In ergonomics, Rapid Entire Body Assessment (REBA) is a method to assess the physical body condition during occupational activity. Determining the risk of musculoskeletal disorders is one of the bases to determine whether there is a need for corrective action in the workplace. Rapid entire body assessment is an approach designed and presented to achieve this goal and to assess the physical body during activities with varying physical body conditions [11, 16].

REBA method was the most appropriate approach for assessing farmers due to grip in hands and fingers as well as the whole body involvement while performing the occupational tasks. In this method, different parts of the body were divided into two groups for analysis: A and B. The group A organs included the trunk, neck, and legs and group B organs were arms, forearms, and wrists, which produced a total of 36 combined physical conditions. A and B scores were calculated using rapid entire body assessment, and the C score was determined after combining them. Then, the activity score was added to C score based on rapid entire body assessment, and the final score was obtained. After determining the final score, the risk level and priorities for corrective actions were determined [17].

Body disorder chart (BDC) was another method used in this study. BDC is a mental assessment technique for individual discomfort, which is performed using experience [11]. This form consists of two parts. In the first part, questions about age, work experience, previous occupations, and the history of some discomforts are asked from the subject, and in the next part, he is asked to determine the severity of each disorder in a subjective manner as follows: without discomfort, mild discomfort, moderate discomfort, severe discomfort, and very severe discomfort. Musculoskeletal disorders considered in this form include discomfort in the following parts: neck, shoulder, upper and lower back, arm, elbow, forearm, wrist, hand and finger, buttocks, thigh, knee, leg, and ankle. To evaluate the effectiveness of interventions, Rapid Entire Body Assessment (REBA) and Body Disorder Chart (BDC) were again completed for all participants after two weeks. Then, the collected data entered SPSS version 16. The data were analyzed by Mann–Whitney and t-test to determine if the interventions reduced the level of pain in different parts of the farmers' bodies and whether the intervention was successful.

3 Results

In training pamphlet group, 10 samples (33.3%) were male and 20 samples (66.7%) were female, and in the kneepad group, there were 9 men (30.0%) and 21 women (70.0%). Statistical data on age, weight, work experience, and daily working time are given in Table 1. Statistical data of demographic characteristics of samples is also shown in Table 2. Frequency distribution of BDC form data before/After intervention in educational pamphlet and Kneepad groups showed in Tables 3 and 4 respectively. Also frequency distribution of REBA Scores results before/After intervention in educational pamphlet and Kneepad groups showed in Tables 5 and 6 respectively.

Table 1. Frequency distribution of age, weight, work experience and daily work time in samples

Variable	Pamphlet group		Kneepad group	
	$\mu \pm SD$	min–max	$\mu \pm SD$	min–max
Age (year)	9.50 ± 38	20–50	10.20 ± 36.6	17–50
Weight (kg)	12.33 ± 65.3	45–90	12.78 ± 59.7	42–90
Work history (year)	11.47 ± 14.7	1–40	13.44 ± 16.7	1–45
Working Time (hr)	1.68 ± 5	2–8	5.1 ± 1.98	3–12

According to the statistical data, the significant difference between two groups was the following: neck (p = 0.34), shoulder (p = 0.38), upper back (p = 0.61), arm (p = 0.31) (P = 0.32), elbows (p = 0.32), lower back (p = 0.25), forearm (p = 0.31),

Table 2. Frequency distribution of demographic characteristics in samples

Variable	Pamphlet group		Kneepad group	
	Yes n (%)	No n (%)	Yes n (%)	No n (%)
Previous job	28 (93.3)	2 (6.7)	29 (96.7)	1 (3.3)
Exercise	4 (13.3)	26 (86.7)	7 (23.3)	23 (76.7)
Low back pain	17 (56.7)	13 (43.3)	16 (53.3)	14 (46.7)
Smoking	1 (3.3)	29 (96.7)	2 (6.7)	28 (93.3)
Drinking	0 (0)	30 (100)	0 (0)	30 (100)
Accident history	0 (0)	30 (100)	1 (3.3)	29 (96.7)
Musculoskeletal surgery	2 (6.7)	28 (93.3)	1 (3.3)	29 (96.7)
Vertebral disc surgery	1 (3.3)	29 (96.7)	0 (0)	30 (100)
Knee discomfort	9 (30.0)	21 (70.0)	11 (36.7)	19 (63.3)
Other discomfort	2 (6.7)	28 (93.3)	2 (6.7)	28 (93.3)

wrist ($p = 0.3$), hands and fingers ($p = 0.56$), hips ($p = 0.7$), femur ($p = 0.43$), knee ($p = 0.031$), leg ($p = 0.36$) and ankle ($p = 0.7$). According to the statistical data, the significant difference between the two groups was the following: neck ($p = 0.31$), shoulder ($p = 0.65$), upper back ($p = 0.27$), arm ($p = 0.15$), elbow ($P = 0.12$), forearm ($p = 0.12$), forearm ($p = 0.32$), wrist ($p = 0.56$), hands and fingers ($p = 0.56$), hips ($p = 0.73$), femur ($p = 0.4$), knees ($p = 0.0003$), legs ($p = 0.13$) and ankle ($p = 0.68$). The statistical information related to the REBA form before and after the intervention is presented in Tables 5 and 6. According to the statistical data, the significant difference between the two groups of pamphlets and kneepads before intervention was: neck ($P = 0.49$), foot ($P = 0.42$), trunk ($P = 0.06$), A ($P = 0.1$), forearm ($P = 1.0$), wrist ($P = 1.0$), arm ($P = 1.0$), B ($P = 1.0$), C ($P = 0.1$), activity ($P = 1.0$) and final ($P = 0.1$). According to the statistical data, the significant difference between the two groups was the following: neck ($P = 0.09$), foot ($P = 0.6$), trunk ($P = 0.008$), A ($P = 0.025$), forearm ($P = 1.0$), wrist ($P = 1.0$), arm ($P = 1.0$), B ($P = 1.0$), C ($P = 0.027$), activity ($P = 1.0$) and final ($P = 0.028$). The results of this study indicate that education has a greater impact on the prevention and reduction of musculoskeletal disorders.

4 Discussion

The study of musculoskeletal disorders before and after intervention in the two groups of pamphlet and knee pad showed that feeling of pain in knee region was significantly reduced in the knee pad group, but did not change significantly in the pamphlet group.

When asked to specify the effectiveness of knee pads on reducing knee pain and discomfort, farmers stated two reasons: (a) reduced pressure on knee when kneeling in farmland; (b) warming of knee region due to use of kneepad. From the time knee pad was used as a means to prevent and reduce knee disorders, several studies have been conducted to determine whether the use of knee pads is suitable for this purpose. In

Table 3. Frequency distribution of BDC form data before/after intervention in educational pamphlet group

Body part	Very severe n (%)		Severe n (%)		Medium n (%)		Mild n (%)		No pain n (%)		p-value
	Before	After	Before	After	Before	After	Before	After	Before	After	
Neck	8 (26.7)	5 (16.7)	18 (60)	13 (43.3)	3 (10)	5 (16.7)	1 (3.3)	7 (23.3)	0 (0)	0 (0)	0.34
Shoulder	0 (0)	0 (0)	0 (0)	1 (3.3)	2 (6.7)	2 (6.7)	1 (3.3)	1 (3.3)	27 (90)	26 (86.7)	0.38
Upper back	3 (10)	2 (6.7)	1 (3.3)	1 (3.3)	14 (46.7)	13 (43.3)	6 (20)	5 (16.7)	6 (20)	9 (30)	0.61
Arm	0 (0)	0 (0)	0 (0)	0 (0)	1 (3.3)	1 (3.3)	0 (0)	1 (3.3)	29 (96.7)	28 (93.3)	0.31
Elbow	0 (0)	0 (0)	0 (0)	0 (0)	0 (0)	1 (3.3)	0 (0)	0 (0)	30 (100)	29 (96.7)	0.32
Lower back	7 (23.3)	6 (20)	5 (16.7)	4 (13.3)	16 (53.3)	18 (60)	1 (3.3)	1 (3.3)	1 (3.3)	1 (3.3)	0.25
Forearm	0 (0)	0 (0)	0 (0)	1 (3.3)	0 (0)	0 (0)	1 (3.3)	0 (0)	29 (96.7)	29 (96.7)	0.31
Wrist	0 (0)	0 (0)	0 (0)	0 (0)	3 (10)	2 (6.7)	0 (0)	0 (0)	27 (90)	28 (93.3)	0.3
Hand & Fingers	0 (0)	0 (0)	0 (0)	0 (0)	2 (6.7)	2 (6.7)	0 (0)	0 (0)	28 (93.3)	28 (93.3)	0.56
Hip	0 (0)	0 (0)	0 (0)	1 (3.3)	3 (10)	2 (6.7)	0 (0)	0 (0)	27 (90)	27 (90)	0.7
Femur	0 (0)	0 (0)	0 (0)	0 (0)	5 (16.7)	4 (13.3)	1 (3.3)	2 (6.7)	24 (80)	24 (80)	0.43
Knee	3 (10)	2 (6.7)	8 (26.7)	6 (20)	16 (53.3)	17 (56.7)	3 (10)	5 (16.7)	0 (0)	0 (0)	0.031
Leg	0 (0)	0 (0)	0 (0)	0 (0)	3 (10)	3 (10)	2 (6.7)	0 (0)	25 (83.3)	27 (90)	0.36
Ankle	1 (3.3)	0 (0)	0 (0)	1 (3.3)	4 (13.3)	4 (13.3)	0 (0)	0 (0)	25 (83.3)	25 (83.3)	0.7

Table 4. Frequency distribution of BDC form data before/after intervention in Kneepad group

Body Part	Very severe n (%)		Severe n (%)		Medium n (%)		Mild n (%)		No pain n (%)		p-value
	Before	After	Before	After	Before	After	Before	After	Before	After	
Neck	8 (26.7)	8 (26.7)	13 (43.3)	13 (43.3)	5 (16.7)	5 (16.7)	2 (6.7)	2 (6.7)	2 (6.7)	2 (6.7)	0.31
Shoulder	2 (6.7)	2 (6.7)	1 (3.3)	0 (0)	2 (6.7)	3 (10)	0 (0)	0 (0)	25 (83.3)	25 (83.3)	0.65
Upper Back	6 (20)	6 (20)	0 (0)	0 (0)	11 (36.7)	11 (36.7)	9 (30)	9 (30)	4 (13.3)	4 (13.3)	0.27
Arm	0 (0)	0 (0)	0 (0)	0 (0)	0 (0)	0 (0)	0 (0)	0 (0)	30 (100)	30 (100)	0.15
Elbow	0 (0)	0 (0)	0 (0)	0 (0)	1 (3.3)	1 (3.3)	0 (0)	0 (0)	29 (96.7)	29 (96.7)	1.0
Lower Back	8 (26.7)	9 (30)	8 (26.7)	9 (30)	12 (40)	12 (40)	0 (0)	0 (0)	1 (3.3)	1 (3.3)	0.12
Forearm	0 (0)	0 (0)	0 (0)	0 (0)	0 (0)	0 (0)	0 (0)	0 (0)	30 (100)	30 (100)	0.31
Wrist	0 (0)	0 (0)	0 (0)	0 (0)	1 (3.3)	1 (3.3)	0 (0)	0 (0)	29 (96.7)	29 (96.7)	0.56
Hand & Fingers	0 (0)	0 (0)	0 (0)	0 (0)	1 (3.3)	1 (3.3)	0 (0)	0 (0)	29 (96.7)	29 (96.7)	0.55
Hip	0 (0)	0 (0)	0 (0)	0 (0)	0 (0)	4 (13.3)	4 (13.3)	0 (0)	26 (86.7)	26 (86.7)	0.73
Femur	2 (6.7)	2 (6.7)	0 (0)	0 (0)	6 (20)	6 (20)	0 (0)	0 (0)	22 (73.3)	22 (73.3)	0.4
Knee	0 (0)	11 (36.7)	7 (23.3)	1 (3.3)	10 (33.3)	10 (33.3)	2 (6.7)	19 (63.3)	0 (0)	0 (0)	0.0003
Leg	4 (13.3)	4 (13.3)	0 (0)	0 (0)	3 (10)	3 (10)	0 (0)	0 (0)	23 (76.7)	23 (76.7)	0.13
Ankle	3 (10)	3 (10)	0 (0)	0 (0)	2 (6.7)	2 (6.7)	1 (3.3)	1 (3.3)	24 (80)	24 (80)	0.68

Table 5. Frequency distribution of REBA Scores before/after intervention in educational pamphlet group

Score		1	2	3	4	5	6	7	8	9
Neck	Before	0 (0%)	4 (13.3%)	26 (86.7%)	0 (0%)	0 (0%)	0 (0%)	0 (0%)	0 (0%)	0 (0%)
	After	0 (0%)	12 (40%)	18 (60%)	0 (0%)	0 (0%)	0 (0%)	0 (0%)	0 (0%)	0 (0%)
Leg	Before	0 (0%)	12 (40%)	18 (60%)	0 (0%)	0 (0%)	0 (0%)	0 (0%)	0 (0%)	0 (0%)
	After	0 (0%)	15 (50%)	15 (50%)	0 (0%)	0 (0%)	0 (0%)	0 (0%)	0 (0%)	0 (0%)
Trunk	Before	0 (0%)	1 (3.3%)	14 (46.7%)	15 (50%)	0 (0%)	0 (0%)	0 (0%)	0 (0%)	0 (0%)
	After	0 (0%)	1 (3.3%)	17 (56.7%)	12 (40%)	0 (0%)	0 (0%)	0 (0%)	0 (0%)	0 (0%)
A-Score	Before	0 (0%)	0 (0%)	0 (0%)	0 (0%)	0 (0%)	9 (30%)	13 (43.3%)	8 (26.7%)	0 (0%)
	After	0 (0%)	0 (0%)	0 (0%)	1 (3.3%)	5 (16.7%)	9 (30%)	9 (30%)	6 (20%)	0 (0%)
Forearm	Before	0 (0%)	30 (100%)	0 (0%)	0 (0%)	0 (0%)	0 (0%)	0 (0%)	0 (0%)	0 (0%)
	After	0 (0%)	30 (100%)	0 (0%)	0 (0%)	0 (0%)	0 (0%)	0 (0%)	0 (0%)	0 (0%)
Wrist	Before	30 (100%)	0 (0%)	0 (0%)	0 (0%)	0 (0%)	0 (0%)	0 (0%)	0 (0%)	0 (0%)
	After	30 (100%)	0 (0%)	0 (0%)	0 (0%)	0 (0%)	0 (0%)	0 (0%)	0 (0%)	0 (0%)
Arm	Before	0 (0%)	30 (100%)	0 (0%)	0 (0%)	0 (0%)	0 (0%)	0 (0%)	0 (0%)	0 (0%)
	After	0 (0%)	30 (100%)	0 (0%)	0 (0%)	0 (0%)	0 (0%)	0 (0%)	0 (0%)	0 (0%)
B-Score	Before	0 (0%)	30 (100%)	0 (0%)	0 (0%)	0 (0%)	0 (0%)	0 (0%)	0 (0%)	0 (0%)
	After	0 (0%)	30 (100%)	0 (0%)	0 (0%)	0 (0%)	0 (0%)	0 (0%)	0 (0%)	0 (0%)
C-Score	Before	0 (0%)	0 (0%)	0 (0%)	0 (0%)	0 (0%)	9 (30%)	13 (43.3%)	8 (26.7%)	0 (0%)
	After	0 (0%)	0 (0%)	0 (0%)	6 (20%)	0 (0%)	9 (30%)	9 (30%)	6 (20%)	0 (0%)
Activity	Before	30 (100%)	0 (0%)	0 (0%)	0 (0%)	0 (0%)	0 (0%)	0 (0%)	0 (0%)	0 (0%)
	After	30 (100%)	0 (0%)	0 (0%)	0 (0%)	0 (0%)	0 (0%)	0 (0%)	0 (0%)	0 (0%)
Final-Score	Before	0 (0%)	0 (0%)	0 (0%)	0 (0%)	0 (0%)	0 (0%)	9 (30%)	13 (43.3%)	8 (26.7%)
	After	0 (0%)	0 (0%)	0 (0%)	0 (0%)	6 (20%)	0 (0%)	9 (30%)	9 (30%)	6 (20%)

Table 6. Frequency distribution of REBA Scores before/after intervention in Kneepad group

Score		1	2	3	4	5	6	7	8	9
Neck	Before	0 (0%)	6 (20%)	24 (80%)	0 (0%)	0 (0%)	0 (0%)	0 (0%)	0 (0%)	0 (0%)
	After	0 (0%)	6 (20%)	24 (80%)	0 (0%)	0 (0%)	0 (0%)	0 (0%)	0 (0%)	0 (0%)
Leg	Before	0 (0%)	9 (30%)	21 (70%)	0 (0%)	0 (0%)	0 (0%)	0 (0%)	0 (0%)	0 (0%)
	After	0 (0%)	13 (43.3%)	17 (56.7%)	0 (0%)	0 (0%)	0 (0%)	0 (0%)	0 (0%)	0 (0%)
Trunk	Before	0 (0%)	0 (0%)	8 (26.7%)	22 (73.3%)	0 (0%)	0 (0%)	0 (0%)	0 (0%)	0 (0%)
	After	0 (0%)	0 (0%)	8 (26.7%)	22 (73.3%)	0 (0%)	0 (0%)	0 (0%)	0 (0%)	0 (0%)
A-Score	Before	0 (0%)	0 (0%)	0 (0%)	0 (0%)	3 (10%)	5 (16.7%)	4 (13.3%)	18 (60%)	0 (0%)
	After	0 (0%)	0 (0%)	0 (0%)	0 (0%)	3 (10%)	6 (20%)	6 (20%)	15 (50%)	0 (0%)
Forearm	Before	0 (0%)	30 (100%)	0 (0%)	0 (0%)	0 (0%)	0 (0%)	0 (0%)	0 (0%)	0 (0%)
	After	0 (0%)	30 (100%)	0 (0%)	0 (0%)	0 (0%)	0 (0%)	0 (0%)	0 (0%)	0 (0%)
Wrist	Before	30 (100%)	0 (0%)	0 (0%)	0 (0%)	0 (0%)	0 (0%)	0 (0%)	0 (0%)	0 (0%)
	After	30 (100%)	0 (0%)	0 (0%)	0 (0%)	0 (0%)	0 (0%)	0 (0%)	0 (0%)	0 (0%)
Arm	Before	0 (0%)	30 (100%)	0 (0%)	0 (0%)	0 (0%)	0 (0%)	0 (0%)	0 (0%)	0 (0%)
	After	0 (0%)	30 (100%)	0 (0%)	0 (0%)	0 (0%)	0 (0%)	0 (0%)	0 (0%)	0 (0%)
B-Score	Before	0 (0%)	30 (100%)	0 (0%)	0 (0%)	0 (0%)	0 (0%)	0 (0%)	0 (0%)	0 (0%)
	After	0 (0%)	30 (100%)	0 (0%)	0 (0%)	0 (0%)	0 (0%)	0 (0%)	0 (0%)	0 (0%)
C-Score	Before	0 (0%)	0 (0%)	0 (0%)	3 (10%)	0 (0%)	5 (16.7%)	4 (13.3%)	18 (60%)	0 (0%)
	After	0 (0%)	0 (0%)	0 (0%)	3 (10%)	0 (0%)	6 (20%)	6 (20%)	15 (50%)	0 (0%)
Activity	Before	30 (100%)	0 (0%)	0 (0%)	0 (0%)	0 (0%)	0 (0%)	0 (0%)	0 (0%)	0 (0%)
	After	30 (100%)	0 (0%)	0 (0%)	0 (0%)	0 (0%)	0 (0%)	0 (0%)	0 (0%)	0 (0%)
Final-Score	Before	0 (0%)	0 (0%)	0 (0%)	0 (0%)	3 (10%)	0 (0%)	5 (16.7%)	4 (13.3%)	18 (60%)
	After	0 (0%)	0 (0%)	0 (0%)	0 (0%)	3 (10%)	0 (0%)	6 (20%)	6 (20%)	15 (50%)

some of these studies, positive effects were observed when knee bands were applied but there were no such effects in a number of other studies [18, 19].

Several biomechanical studies have also been conducted to assess the effects of knee pads in preventing knee injury; however, these studies are not able to conclusively document the positive or negative effects of using kneepads [20, 21].

A study in the United States on athletes found that the use of knee pad had a positive effect on knee protection [22]. The results of posture evaluation by REBA method before and after intervention in the two groups of pamphlet and knee pad showed a significant difference in trunk, A, C, and final score in the pamphlet group, which was somehow indicative of a significant effect of training on farmers' postures. In REBA form, the final score indicates the degree of interventions that need to be taken. Score 1 means no need for corrective action. Score 2 or 3 means a low risk and the probable need to change the posture. The scores 4–7 represent a moderate risk, requirement for further studies, and imminent posture change. The scores 8–10 represent a high risk, requirement of more studies, and a rapid change in posture. The +11 score represents a very high risk and a rapid change in posture.

As it is clear from the tables of results section, a significant change has occurred in the number of subjects with high risk; while there have been few changes in this respect in the knee pad group. Beach and colleagues showed that in a majority of the world's most prestigious dental schools, ergonomic training is essential for students to prevent musculoskeletal disorders. In this study, there has been an emphasis on continuous and long-term training of individuals on the principles of ergonomics, and it has been argued that cross-sectional and short-term training can reduce the level of risk but have not been sufficient [23]. Obviously, in studies conducted by Derebery and colleagues, it was found that the use of instructional booklets does not affect the reduction in neck discomfort [24]. Another study recommends to pay more attention to training and to do assignments of training pamphlets over a longer period [25]. In a study in Kermanshah on students, it was found that the use of training pamphlets had a positive effect on raising awareness [26]. In the study of Momeni et al., the results also indicate the effect of education on the awareness of health volunteers, and it is noted that if the pamphlet is properly designed, it can be used as an effective educational tool [27].

Comparison of the results of knee pad intervention shows a positive effect, but other studies on the impact of knee pad cannot indicate its positive effect with certainty. Also, the results of educational pamphlet intervention are indicative of a positive effect, which confirms other studies in this regard.

5 Conclusion

According to results, both knee band and educational pamphlet methods can be effective in preventing and reducing musculoskeletal disorders. However, comparison of the two methods showed that knee pad intervention only reduced knee discomfort but educational pamphlet was more effective on inappropriate posture to decrease and prevent musculoskeletal disorders. As a result, training has been more effective on the prevention and reduction of musculoskeletal disorders.

Acknowledgment. We appreciate the research deputy of GMU who sponsored the project, as well as diligent farmers who participated in this project.

References

1. Motamedzade M, Saedpanah K, Salimi K, Eskandari T (2016) Risk assessment of musculoskeletal disorders by muscle fatigue assessment method and implementation of an ergonomic intervention in assembly industry. J Occup Health Eng 38(1)
2. Jalali Naeini SGh (2002) Ergonomics place in Iran and world. In: National conference on ergonomics in industry and production, 1st edn. Tehran, pp 25–34
3. Mouodi MA, Hasanzadeh H (2004) CTD from ergonomics and occupational medicine viewpoint (along with postures analysis techniques), vol 150. Abasaleh-Hayan Publication, Tehran [Persian book]
4. Nasl Saraji J, Ghaffari M, Shahtaheri S (2006) Survey of correlation between two evaluation method of work related musculoskeletal disorders risk factors REBA & RULA. Iran Occup Health J 7(2)
5. Walker-Bone K, Palmer K (2002) Musculoskeletal disorders in farmers and farm workers. Occup Med 52(8):441–450
6. Razavi Asl S, Ezzatian R (2006) Occupational health in agricultural sector, 1st edn. Andishe Mandegar Publication, Qom
7. Levy BS, Wegman DH (2000) Occupational health recognizing and preventing work-related disease, 4th edn. Little Brown, Boston
8. Davis KG, Kotowski SE (2007) Understanding the ergonomic risk for musculoskeletal disorders in the United States agricultural sector. Am J Ind Med 50(7):501–511
9. Osborne A, Blake C, McNamara J, Meredith D, Phelan J, Cunningham C (2010) Musculoskeletal disorders among Irish farmers. Occup Med 60:598–603
10. Sadeghi N, Delshad A, Fani MJ (2010) REBA method posture analysis in Saffron pickers in Gonabad. Horiz Med Sci 15(4):47–53 [Persian]
11. Choobineh A (2004) Posture assessment methods in occupational ergonomics. Fanavaran Publication, Hamedan [Persian book]
12. Molteni G, De Vito G, Sias N, Grieco A (1995) Epidemiology of musculoskeletal disorders caused by biomechanical overload (WMSDs). La Medicina del Lavoro 87(6):469–481
13. Bernard BP (1997) Musculoskeletal disorders and workplace factors: a critical review of epidemiologic evidence for work-related musculoskeletal disorders of the neck, upper extremity, and low back. NIOSH
14. Xiao H, McCurdy SA, Stoecklin Marois MT, Li CS, Schenker MB (2013) Agricultural work and chronic musculoskeletal pain among Latino farm workers: the MICASA study. Am J Ind Med 56(2):216–225
15. Allen KD, Chen J-C, Callahan LF, Golightly YM, Helmick CG, Renner JB et al (2010) Associations of occupational tasks with knee and hip osteoarthritis. Johnston county osteoarthritis project. J Rheumatol 37(4):842–845
16. McAtamney L, Hignett SREBA (2000) Rapid entire body assessment. Appl Ergon 31:201–205
17. Mouodi M, Hasanzadeh H (2004) CTD from ergonomics and occupational medicine viewpoint. Hayyan 9–15
18. Hansen BL, Ward JC, Diehl RC Jr (1985) The preventive use of the Anderson knee stabler in football. Physician Sportsmed 13(9):75–81

19. Hewson JR, George F, Mendini RA, Wang JB (1986) Prophylactic knee bracing in college football. Am J Sports Med 14(4):262–266
20. Griffin LY, Albohm MJ, Arendt EA, Bahr R, Beynnon BD, DeMaio M et al (2006) Understanding and preventing noncontact anterior cruciate ligament injuries: a review of the Hunt Valley II meeting, January 2005. Am J sports Med 34(9):1512–1532
21. Paulos LE, France EP, Rosenberg TD, Jayaraman G, Abbott PJ, Jaen J (1987) The biomechanics of lateral knee bracing: part I: response of the valgus restraints to loading. Am J Sports Med 15(5):419–429
22. Sitler M, Ryan CJ, Hopkinson LW, Wheeler LJ, Santomier J, Kolb LR et al (1990) The efficacy of a prophylactic knee brace to reduce knee injuries in football: a prospective, randomized study at West Point. Am J Sports Med 18(3):310–315
23. Beach J, DeBiase C (1998) Assessment of ergonomic education in dental hygiene curricula. J Dent Educ 62(6):421–425
24. Derebery J, Giang GM, Gatchel RJ, Erickson K, Fogarty TW (2009) Efficacy of a patient-educational booklet for neck-pain patients with workers' compensation: a randomized controlled trial. Spine 34(2):206–213
25. Karimi A, Mozafari F, kamaledini H, Jokar S (2011) The effectiveness of therapeutic exercises and general care by educational booklet on reduction of neck pain. Res Rehabilitation Sci 6(2)
26. Azizi A, Amirian F, Amirian M (2008) To compare the evaluation of HIV/AIDS female high school peer education with lecture by physician and pamphlets in Kermanshah. Iran J Epidemiol 3-4(4):71–76 [Persian]
27. Mouemeni E, Malekzadeh JM (2000) Comparing the effect of pamphlet versus lecture on the nutritional knowledge of health communicators. Yasuj J Med Sci 19–20(5):49–54

Building and Construction

Evaluation of Participatory Strategies on the Use of Ergonomic Measures and Costs

Steven Visser, Henk F. van der Molen(✉), Judith K. Sluiter, and Monique H. W. Frings-Dresen

Coronel Institute of Occupational Health, Amsterdam Public Health Research Institute, Academic Medical Center, University of Amsterdam, PO Box 22700, 1100 DE Amsterdam, The Netherlands
h.f.vandermolen@amc.nl

Abstract. The implementation and use of ergonomic measures is dependent on behavioural changes of both employers and employees. In these stakeholder-groups different barriers could emerge with respect to using ergonomic measures. Participatory strategies – guided by professional ergonomic consultants – are thought to stimulate behavioural change of the stakeholders and a better chance for starting interventions on barriers in order to increase the use of ergonomic measures. In the present cluster randomized controlled trial, the effect of two participatory guidance strategies – a face-to-face or an e-mail guidance strategy - on the use of ergonomic measures among construction workers and related company costs were studied. Only five out of twelve companies actually implemented ergonomic measures. Within these five companies, both participatory guidance strategies are thought to be capable of improving the actual use of ergonomic measures by workers. The face-to-face guidance strategy, however, may be more suitable in a company context where lack of insight in relevant work-related risk factors exists. Costs were determined by guidance costs in the F2F group and purchasing costs in the EG group. The cost analysis provided insight into the financial consequences of the ergonomic measures to the companies, but the large variety in purchasing costs made a comparison between the two guidance strategies in this study irrelevant.

Keywords: Participatory ergonomics · Ergonomic measures · Costs
Construction industry · Prevention · Cluster randomized controlled trial

1 Introduction

The implementation and use of ergonomic measures is dependent on behavioural changes of both employers and employees. In these stakeholder groups different barriers could emerge with respect to using ergonomic measures. Barriers for using ergonomic measures could be: lack of information, no availability, and having no test opportunities. In addition, the use of new tools requires training and therefore also time and money on the part of the employers. When the benefits of the new measures are unclear in the short term, employers are not easily motivated to invest in new ergonomic measures. Participatory strategies – guided by professional ergonomic consultants – are thought to

S. Bagnara et al. (Eds.): IEA 2018, AISC 825, pp. 435–437, 2019.
https://doi.org/10.1007/978-3-319-96068-5_48

stimulate behavioural change of the stakeholders and a better chance for starting interventions on barriers in order to increase the use of ergonomic measures. In the present study [1], the effect of two participatory guidance strategies – a face-to-face or an e-mail guidance strategy - on the use of ergonomic measures among construction workers and related company costs were studied.

2 Methods

Twelve construction companies were randomly assigned to a structured step by step face-to-face guidance strategy (F2F; N = 6) and a comparable e-guidance strategy (EG; N = 6). An ergonomic consultant applied either a face-to-face or an e-mail guidance in a participatory ergonomics strategy. The percentage of workers using ergonomic measures, as primary outcome measure were assessed using questionnaires at baseline and after six months. A comparison was made on individual level between both strategies. Costs were divided into three items: the costs of the guidance strategy, purchasing the ergonomic measures, and costs of training the workers. Guidance costs were calculated by multiplying the hourly rate of the ergonomic consultants by the number of hours they spend for guiding the F2F or the EG. For the costs regarding purchasing the implemented ergonomic measure, suppliers of the ergonomic measures were asked to supply information about the purchase costs, depreciation time, maintenance costs of the ergonomic measure, and additional energy costs for using the ergonomic measure. Costs for each worker given training were calculated by multiplying the number of hours of the training (obtained from interviews with the employers) by their hourly costs.

3 Results

Five companies – two in the F2F and three in the EG – implemented new ergonomic measures during the intervention. For the F2F, the percentage of workers using a newly-implemented ergonomic measure after the PE intervention was 23% (11 out of 48 workers) and 42% (13 out of 31 workers) for the EG. This difference was not statistically significant ($p = 0.271$). Only the increased use of ergonomic measures to adjust working height differed significantly different ($p = 0.001$) between F2F (+1%) and EG (+10%).

For the F2F, the total costs in the first year were between €3,294 and €5,781 per company and were mainly due to the guidance costs. The purchasing of the ergonomic measures accounted for 2% to 29% of the costs incurred. For the EG, the total costs were between €1,479 and €3,754 per company. The largest costs were guidance costs (82% of the total costs) for one company and the purchasing costs (83% to 93% of the total costs) for two other companies.

4 Conclusions

Only five out of twelve companies actually implemented ergonomic measures. Within these five companies, both participatory guidance strategies are thought to be capable of improving the actual use of ergonomic measures by workers. The face-to-face guidance strategy, however, may be more suitable in a company context where lack of insight in relevant work-related risk factors exists. Costs were determined by guidance costs in the F2F group and purchasing costs in the EG group. The cost analysis provided insight into the financial consequences of the ergonomic measures to the companies, but the large variety in purchasing costs made a comparison between the two guidance strategies in this study irrelevant.

Reference

1. Visser S (2015) Ergonomic measures in construction work: enhancing evidence-based implementation. Thesis, University of Amsterdam. ISBN 978 94 6259 5057

Effectiveness of Interventions for Preventing Injuries in the Construction Industry: Results of an Updated Cochrane Systematic Review

Henk F. van der Molen[1(✉)], Prativa Basnet[2], Peter L. T. Hoonakker[3], Marika M. Lehtola[4], Jorma Lappalainen[1,2,3,4,5], Monique H. W. Frings-Dresen[1], Roger A. Haslam[5], and Jos H. Verbeek[2]

[1] Department: Coronel Institute of Occupational Health, Amsterdam Public Health Research Institute, Academic Medical Center, University of Amsterdam, P.O. Box 22700, 1100 DE Amsterdam, The Netherlands
h.f.vandermolen@amc.nl
[2] Cochrane Work Review Group, Finnish Institute of Occupational Health, Kuopio, Finland
[3] Center for Quality and Productivity Improvement, University of Wisconsin, Madison, Madison, WI, USA
[4] Promotion of Occupational Safety, Finnish Institute of Occupational Health, Kuopio, Finland
[5] Loughborough Design School, Loughborough University, Leicestershire, UK

Abstract. Various interventions to prevent occupational injuries in the construction industry have been proposed and studied. This continuing updated Cochrane review systematically summarizes the most current scientific evidence on the effectiveness of interventions to prevent injuries associated with construction work. Search terms that covered the concepts of 'construction workers', 'injury', 'safety' and 'study design' were used to identify intervention studies in five electronic databases up to April 2017. Acceptable study designs included randomized controlled trials (RCT), controlled before–after studies (CBA) and interrupted time series (ITS). In total 17 studies, 14 ITS and three CBA studies, from the US (6), UK (2), Italy (3), Denmark (1), Finland (1), Austria (1) Germany (1) Spain (1), Belgium (1) met the inclusion criteria. Most studies were at high risk of bias. There is very low-quality evidence that introducing regulations as such may or may not result in a decrease in fatal and non-fatal injuries. There is also very low-quality evidence that regionally oriented safety campaigns, training, inspections or the introduction of occupational health services may not reduce non-fatal injuries in construction companies. There is very low-quality evidence that company-oriented safety interventions such as a multifaceted safety campaign, a multifaceted drug workplace programme and subsidies for replacement of scaffoldings may reduce non-fatal injuries among construction workers.

Keywords: Occupational injury · Interventions · Prevention · Construction industry · Systematic review

S. Bagnara et al. (Eds.): IEA 2018, AISC 825, pp. 438–440, 2019.
https://doi.org/10.1007/978-3-319-96068-5_49

1 Introduction

Various interventions to prevent occupational injuries in the construction industry have been proposed and studied. This continuing updated Cochrane review [1] systematically summarizes the most current scientific evidence on the effectiveness of interventions to prevent injuries associated with construction work.

2 Methods

Search terms that covered the concepts of 'construction workers', 'injury', 'safety' and 'study design' were used to identify intervention studies in five electronic databases up to April 2017. Acceptable study designs included randomized controlled trials (RCT), controlled before–after studies (CBA) and interrupted time series (ITS). To obtain comparable and reliable results from included ITS studies, data from original papers were extracted and reanalyzed according to recommended methods for analysis of ITS designs for inclusion in systematic reviews. Re-analysis with autoregressive modelling made it possible to estimate regression coefficients corresponding to two standardized effect sizes for each study: change in level, and change in slope of the regression lines before and after the intervention. Data were standardised by dividing the outcome and standard error by the pre-intervention standard deviation as recommended by Ramsay 2001 [2] and entered into Review Manager 5 (RevMan 5) as effect sizes. An ITS study was eligible for inclusion when (i) there were at least three time points before and after the intervention, irrespective of the statistical analysis used, and (ii) the intervention occurred at a clearly defined point in time. CBA studies were eligible for inclusion when the outcome was measured in both the intervention and control group before and after the introduction of the intervention. We used the GRADE approach for assessing the evidence and results.

3 Results

In total 17 studies, 14 ITS and three CBA studies, from the US (6), UK (2), Italy (3), Denmark (1), Finland (1), Austria (1) Germany (1) Spain (1), Belgium (1) met the inclusion criteria. Most studies were at high risk of bias. The ITS studies evaluated the effects of the introduction or change of regulations which laid down safety and health requirements for the construction sites (N = 9), a safety campaign (N = 2), a drug-free workplace programme (N = 1), a training programme (N = 1), and safety inspections (N = 1) on fatal and non-fatal occupational injuries. One CBA study evaluated the introduction of occupational health services such as risk assessment and health surveillance, one evaluated an training programme and one evaluated subsidy for the replacement of scaffoldings.

The regulatory interventions at national or branch level may not have a considerable initial effect (effect size of −0.33; 95% confidence interval (CI) −2.08 to 1.41) and no sustained effect (effect size of −0.03; 95% CI −0.30 to 0.24) on fatal and nonfatal injuries (9 ITS studies). Inspections may not result in a considerable reduction (effect size of 0.07; 95% CI −2.83 to 2.97) of non-fatal injuries (one ITS study).

Introduction of occupational health services may not result in a decrease of fatal or non-fatal injuries (one CBA study). Safety training interventions may not result in a significant reduction of non-fatal injuries (one ITS study and one CBA study).

In companies that received subsidies non-fatal injuries from falls to a lower level may decrease more (risk ratio at follow-up: 0.93; 95% CI 0.30 to 2.91) than in companies that do not receive subsidies (1 CBA study). A multifaceted drug-free workplace programme at the company level may reduce non-fatal injuries in the years following implementation by −7.6 per 100 person-years (95% CI −11.2 to −4.0) and in the years thereafter by −2.0 per 100 person-years per year (95% CI −3.5 to −0.5) (one ITS study).

A safety campaign intervention may result in an initial and sustained decrease in injuries at the company level (one ITS study) but not at the regional level (one ITS study).

The quality of the evidence was rated as very low for all interventions.

4 Conclusions

There is very low-quality evidence that introducing regulations as such may or may not result in a decrease in fatal and non-fatal injuries. There is also very low-quality evidence that regionally oriented safety campaigns, training, inspections or the introduction of occupational health services may not reduce non-fatal injuries in construction companies.

There is very low-quality evidence that company-oriented safety interventions such as a multifaceted safety campaign, a multifaceted drug workplace programme and subsidies for replacement of scaffoldings may reduce non-fatal injuries among construction workers.

Additional strategies are needed to increase the compliance of employers and workers to the safety measures that are prescribed by regulation. An evidence base is needed for the vast majority of technical, human factors and organisational interventions that are recommended by standard texts of safety, consultants and safety courses.

References

1. van der Molen HF, Basnet P, Hoonakker PLT, Lehtola MM, Lappalainen J, Frings-Dresen MHW, Haslam R, Verbeek JH (2018) Interventions to prevent injuries in construction workers. Cochrane Database Syst Rev (2):CD006251. https://doi.org/10.1002/14651858.cd006251.pub4
2. Ramsay C, Grimshaw J, Grilli R (2001) Robust methods for analysis of interrupted time series. In: 9th annual cochrane colloquium, October 2001, Lyon, France

Co-design in Architectural Practice: Impact of Client Involvement During Self-construction Experiences

Pierre Schwaiger[1], Clémentine Schelings[1], Stéphane Safin[2], and Catherine Elsen[1(✉)]

[1] LUCID, Faculty of Applied Sciences, University of Liège, Liège, Belgium
catherine.elsen@uliege.be
[2] I3-SES, CNRS, Télécom ParisTech, 75013 Paris, France

Abstract. This paper investigates how self-construction processes, considered as the utmost form of clients' involvement in the realm of building a family house, impact clients' and architects' interactions. The study of four cases (two involving "traditional" processes, two involving "self-built" processes) and the drawing of Experience Maps for each of them nurture reflections about satisfaction assessment, perceived quality and clients' integration to the architectural design process (potentially including co-design attitudes).

Keywords: Self-construction design process · Co-creation
Satisfaction assessment · Architectural design

1 Clients and Architects Interactions in Architectural Design

It is widely accepted that designers and users are inextricably related in regard of both the design process and output. Designers have major impacts on the quality of the built environment, i.e. on the quality of life of many people. Designed artifacts, on the other hand, are meaningless unless endorsed by end-users (in power of taking ownership or rejecting them) [1]. These end-users are nowadays recognized as "owning the factual problem" [2] i.e. being experts of their own personal behaviors, experiences and issues. Research moreover points out that end-users are no longer willing to undergo the design process simply as external observers [3]: better informed, they expect to have their say all along the decision-making process, considering themselves as "part of the team" [4].

Acknowledging this evolution, disciplines such as product, service or software design progressively shifted from "usability" to "user-centered approaches" and eventually to "users-driven innovation" [5], while resources for participation such as "participatory design" or "co-creativity" and "co-design" also emerged, either in an institutionalized [6] or horizontal way [7]. In the field of architectural design, though, research shows that most architects rarely go beyond early conversational interactions to reach out to users' needs and expectations. Clients/architects' relationships have been investigated for decades [8, 9], and the analysis of their interactions offers provoking results: communication gaps largely subsist [4, 10, 11], limiting users' input to

S. Bagnara et al. (Eds.): IEA 2018, AISC 825, pp. 441–452, 2019.
https://doi.org/10.1007/978-3-319-96068-5_50

functional and structural recommendations, with very rare attempts to integrate users into the design process.

It's a fact that architectural design processes entail numerous intricate, co-dependent constraints that might explain why architects are so reluctant to involve end-users as soon as preliminary design phases. But so is the case for closely related disciplines, such as product design or urban planning, where participation has yet been implemented (with various degrees of success) for several decades. This reluctance to involve users into the architectural process might therefore rather originate from some disciplinary tendency to consider architects as sole Masters of the design process, in control of prioritizing constraints in their own way. As discussed by Cole-Colander, architects indeed tend to focus on concepts, on powerful ideas that their achievements will convey; they try to persuade their clients to invest in a sustainable way of living, or in innovative building techniques and materials; they pursue recognition of their peers; … End-users, on the other hand, and even more specifically clients/future owners and occupants of the architectural artifact, are generally more attentive to other, complementary practical criteria such as cost, duration of the building process, return on investment, added value or future consumptions [12]. Expectations and definition of priorities, roles and missions therefore seem to quite drastically differ between architects and their clients. It is yet of crucial importance to re-align these perceptions, as the quality (of a service or a product) is defined as the intersection between the service/product provided and the initial expectations of the person served. Thus, if the provided service/product is below initial expectations, it is considered of poor quality; if the service/product meets the expectations, it is considered of acceptable quality; if it exceeds these expectations, it is considered of excellent quality [13].

Quality and satisfaction assessment in both the architectural field and construction industry have been the subject of research, but rather conducted in an attempt to develop processes such as "total quality management" and tools to manage customers' service expectations, post-process and post-occupancy assessments of such quality and satisfaction [13–17]. Those tools limit users' input to brief and design evaluation, neglecting the fact that their requirements constantly evolve through time, i.e. neglecting satisfaction towards the process and how it unfolds [18]. Satisfaction levels, as a consequence, remain under increased scrutiny and point towards delays, cost overruns and poor quality of products and services [19]. Since the 1960's government and industry reports have consistently warned about these low levels of users satisfaction, specifically within the architectural profession [1].

In this paper, we argue that architectural processes could benefit from the users' willingness for involvement and situated creativity. In order to do so, users should be considered as resources for the process, supportive of the architects who still have to deal with down-to-earth constraints (norms and regulations, timing, budget, …) and processes' uncertainties. When correctly facilitated, any level of co-design (even basic) might support this interaction, enabling clients to engage [20], learn and apprehend design values [1], and even reach problem and solution co-evolution [2]. On that basis, this research investigates how self-construction processes, where future occupants are largely involved all along the design process, impact end-users' and architects' interactions, satisfaction and assessment of the perceived quality (mainly in regard of the shared design experience).

2 Methodology

To answer this question, four case studies were chosen – two "traditional" design processes and two self-construction cases – all of these cases concerning new buildings (full-scale family houses or additions to existing buildings). We focused on such reduced-scale projects for two reasons: first because they constitute the larger part of the design activity of most Belgian architects [21], and therefore are representative of their daily realities, and second because clients, in the process of designing their own future dwelling, generally demonstrate more attachment and willingness to take part to the whole architectural experience. Additionally, according to Siva and London [1], the reduced scale does not diminish the relative complexity of the project in regard of the architect's involvement, in regard of stakeholders' respective definition of roles nor the multiplicity of constraints each has to adapt to. Cases were moreover chosen because they all still belonged to the early phases of the design process, understood here as the time interval between the first desire to build a new home, and the very first steps of the construction process.

Eleven in-depth interviews were conducted with the main stakeholders involved in those four cases, namely clients, architects and one carpenter with a recurrent expertise as "self-construction counselor" (two cases, one "traditional" and one "self-built", called on the same architect). Several themes were chosen to structure these semi-directive interviews: general presentation of the project (and its initial brief); criteria for architect's selection (or, when addressing to the architects: reasons why they agreed to accept the clients, their projects); questions about the perceived roles, missions and respective responsibilities of both clients and architects (as well as other trades, when necessary); description of the experience, the relationship and how it evolved/maintained through time; definition and description of the perceived "quality" all along the architectural processes.

Additionally, we took part to two work meetings (one concerning a "traditional" process, one concerning a "self-built" one) during which we conducted "fly on the wall" observations. This observation technique, after a short adaptation time, enables the observer to stay away from the stakeholders' activity and to interfere as little as possible with the ongoing exchanges and situations. If unnoticed enough, the researcher might this way observe the users, the contents and the contexts of a collaborative session in almost ecological conditions [22].

The contents of the in-depth interviews and fly-on-the-wall observations informed the drawing of Experience Maps for each case, illustrating evolution of actors' experiences and (dis)satisfaction levels with time. These Maps are built on basis of Adler's famous five-steps development model [23]:

- The **honeymoon step** refers to initial excitement and optimism, the individual feeling euphoric in regard of the new experience;
- During the **disintegration step**, or "culture shock", the individual goes through a period of confusion and disorientation, the differences (with the expected experience) becoming increasingly noticeable and cultural distinctions creating tensions and frustrations;

- The **reintegration step** is characterized by a strong rejection of the new culture; it demonstrates how the individual grows awareness for what really causes negative feelings and how he/she builds a basis for new cognitive experiences;
- The **autonomy step** is the moment when the individual feels comfortable with his/her status of both "insider-outsider" in two different cultures. This stage is characterized by more personal flexibility and by the development of "appropriate coping skills";
- In the fifth and last **independence step**, the individual is able to accept and draw benefits from cultural differences and similarities. As Adler underlines, "*he or she is capable of experiential learning that is holistically incorporated into identity, while at the same time capable of again having preconceptions, assumptions, values and attitudes challenged*" (p. 18).

The first, fourth and fifth stages are considered as "positive experiences", while the second and the third (disintegration and reintegration) are rather considered as "negative experiences" in this paper. The Experience Maps are designed to only document the most significant moments, as defined by the subjects themselves: they therefore have no ambition to illustrate the complete design flow. Although they do not aim for exhaustiveness, we argue that these Maps constitute an effective way to visually question how traditional vs. self-built processes may impact clients' overall experience and satisfaction in regard of early architectural design phases.

3 Results

3.1 Case #1, Traditional Process – "Perseverance in the Face of Disillusionments"

The case, the building of a family house, gathers a couple of clients (referred here as A & B, to respect anonymity) and the architect A, with 30 years of experience. The client B involved herself a lot in the design of her house, providing the architect with very precise requirements in regard of architectural desires, functional needs and spaces' articulations. The client A, on the other hand, got rather involved in the technical follow-up of the building site. The construction follows a "traditional" process, involving several, independent trades. The clients will themselves only be in charge of a few late finishing work.

Following in-depth interviews conducted with each stakeholder separately, we draw an Experience Map divided in 7 main phases, in accordance with the descriptions provided by each subject (Fig. 1). The first phase only involves the client B, who started to write down her ideas for her "future, ideal house" a decade ago: "*I observed everything, I collected a lot of inspiration material and images from other projects, in magazines, during my travels*" (Client B, translated from French by the authors). The client, thus, demonstrates a high level of involvement (even prior to any concrete design project) and progressively builds her expectations. After 5 years as married couple, the clients A & B decide to buy a piece of land and to build a house. The client B recalls she felt some apprehension in sharing her vision of her "ideal house" to her husband: "*I have a strong character, and I don't dare imagine what would have happened if my husband had refused my ideas*" (client B). Client A, on the other hand, at that time had no particular requirements and therefore really easily accepted client B's suggestions.

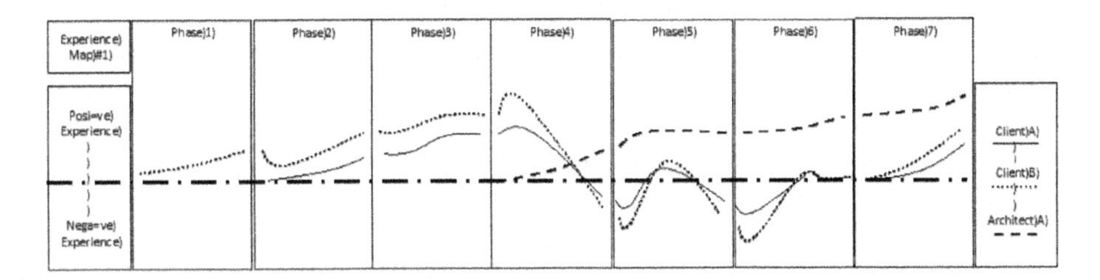

Fig. 1. Experience Map #1, traditional case.

The third phase corresponds to the search of an architect. Self-built process as well as turnkey solutions are rapidly dismissed: the first one necessitates time and skills the clients claim they don't possess; while the second one offers too limited architectural potentialities, according to the clients, to adequately meet their expectations. Both clients dreaded the search for an architect, especially client B who had at that time a strong opinion about architects in general: "*I had the image of an architect seen as someone cold, and uncommunicative*" (client B).

To the fourth phase corresponds the first acknowledged "negative experience" for both clients. The architect A has been chosen through word of mouth, and the very first meeting (on the building site) goes really well, as client B underlines: "*I was happily surprised to observe that the architect was looking for dialogue, instead of imposing his ideas*" (client B). The architect A is indeed very keen to listen to his clients' desires, but yet has to warn them that the single-story house they always dreamt of will not be technically possible, given the strong declivity of the land they just bought. This is experienced as a real shock for both clients, who imagined that somehow the land could be leveled to deal with this aspect. The client B, in particular, expresses difficulty in recovering from such a bad new, as she projected herself already a lot in her "perfect, future home". From the architect' point of view, it was no real surprise to deal with such disillusion and frustration; he'd rather work with clients that have a precise idea of their needs, even though it means help them face some discouragements from time to time.

The fifth phase constitutes an additional shock, as the architect presents an adapted project, well fitted to the clients' desires and to the land's slope, but way out of budget considering what the clients had in mind. The architect remains confident, as he's used to this kind of situations; he knows it will take time to work hand-in-hand with the clients to find compromises that will help decrease the total cost. During phase 6, the clients go back to a few turnkey contractors to evaluate how much a simpler project would cost. They rapidly realize that given the slope of their land, not even turnkey solutions will help drastically reduce the total cost. The architect A, meanwhile, remains optimistic and keeps submitting new ideas and solutions to the clients. Both of them will eventually admit that working with architect A remains the best solution if they want to build their "dream house". The end of the preliminary design phases does not generate any more disappointments according to the stakeholders, as clients and architect cooperate to find concrete solutions to adjust the project to the limited budget, and vice-versa. The construction experience is thus positively experienced by the two clients, who define it as "*laborious and energy-consuming*", but also instructive and exciting. The architect recalls a positive, demanding but exciting experience, given his interesting and courageous clients.

3.2 Case #2, from Self-built to Traditional Process – "About Keeping Control, Even During Resignation"

The second case involves the same architect (A) and one main client (C). The latter wished to double the surface area of his town house by self-building an extension. He decides to call on an architect but only to make sure that his project is in line with some urban regulations. From his point of view, and at the beginning of the project, he felt perfectly capable of dealing with such a construction project all by himself. At the beginning (phase 1, Fig. 2), the client is therefore quite euphoric and positive about his own capabilities to conduct the project successfully. The second phase marks a first disillusionment, as architect A underlines that the envisioned project actually infringes several urban regulations and that it won't be permitted to built it as such. The client C then admits the importance to call on the expertise of an architect, and asks architect A to re-draw a new design proposal for his extension: "*he opened my eyes on so many things in such a few time... it would have saved me a lot of time and effort if I had called on him at the very beginning*" (Client C). The architect, during the third phase, suggests three different design drafts to the client, who consider all of them "*completely off the mark*", but still interesting because they "*raise questions I never asked myself before*". The refusal of all three designs constitutes a disappointment for the architect, even though he realizes that the effort made at least had the added value to open up new perspectives. To the fourth phase corresponds the submission of a fourth sketch, this time considered as adequate by the client. Both stakeholders declare being enthusiastic at the end of this phase.

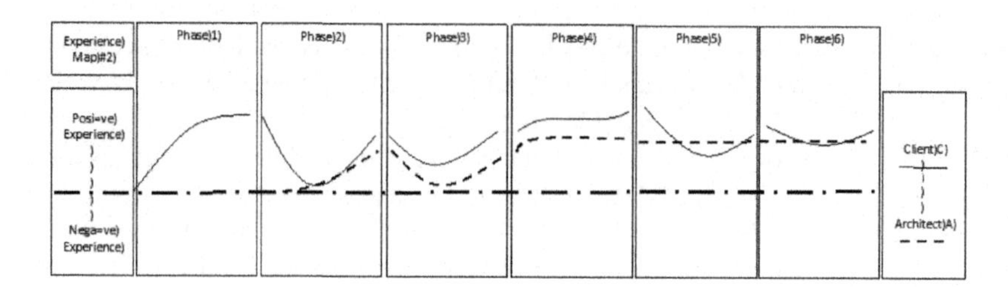

Fig. 2. Experience Map #2, from a "self-built" to a "traditional case".

While waiting for estimates from several sub-contractors, client C decides to undertake some preparatory work (such as, for instance, delimitating the layout of the premises). This work consumes way more time than he had expected: "*I didn't realize it would take me a whole week-end to delineate the land... I have the necessary equipment, though, but learning how to use it correctly is another story*" (client C). Following this realization, the client discusses his project with several sub-contractors and eventually admits that he won't be able to conduct a self-built construction site as initially expected: "*although my project is simple, the construction techniques are complicated. It's a full time job, that's all*" (client C). Phase 6 is a real disillusion phase, but qualified by the client himself as a "*positive, instructive experience*" as he remains in complete control of each decision and resignation, even this setback to a more traditional construction process. Architect A, meanwhile, describes the experience as globally positive as his 30 years experience made him accustomed to these kinds of uncertainties and hesitations.

3.3 Case #3, Self-built Process – "About Letting Things Go, and Dealing with Communication Gaps"

The third case is about the extension of a four-façade existing house, built on a sloping ground. The client (D) has made the decision to self-construct in order to control the costs, but also because he says he's "someone manual" who likes making things by himself in order to control the overall quality: "*when I do something myself, contrary to sub-contractors, I take the time to do it well*" (client D). He wants the extension to be modest, not some "*ostentatious project*" that would denature the perception of the surroundings. This simplicity will not be to the liking of the architect (B), who comments: "*it's a functional, unoriginal project that answers the needs of the client, nothing less, nothing more*" (architect B). At the beginning of the project (phase 1, Fig. 3), the client is reluctant to choose an architect; he says he never got good feedbacks about clients-architects relationships in general, and he dreads to enter in a potentially conflicting relation with someone he knows. He therefore selects architects he absolutely does not know; another selection criterion is that the architects have to agree with a self-built process involving timber structure. It's client's D partner who will at first meet the chosen architect (B; 20 years experience); she explains being very attracted to the "*modern-looking realizations*" that the architect presents on his website. The architect, after this first encounter, meets with the clients several times (phase 2), but admits "*having a hard time deciphering the clients' wishes; after two meetings, I still had no clear understanding of their program*" (architect B). He still decides to submit a first sketch, essentially to foster a more constructive dialogue in that regard. The client D considers this first sketch as unsatisfactory: "*he* [the architect] *didn't respect any of our wishes; the project was too modern with no specific opening towards the interesting view*" (client D). From the architect's point of view, the goal of this first design proposal was to "generate some debate". He adds having some difficulties in reconciling both clients' points of view: "*several times I had the feeling to act as marriage counselor*". Although this first submission is considered as disappointing by all main stakeholders (see phase 3), they decide to pursue their collaboration and to keep discussing other design drafts. According to the architect B, "*they* [the clients] *kept sending me my prints back with written comments and corrections that weren't at the right scale, as well as out-of-scale free-hand sketches that expressed furniture's arrangements simply impossible to implement*" (architect B). The client D, on the other hand, didn't understand why the architect kept "*showing us modified plans including several errors we had already pointed out*" (client D). Those multiple attempts to adapt the project constitute consecutive shocks for the stakeholders that express "exhaustion" in regard of a "negative experience".

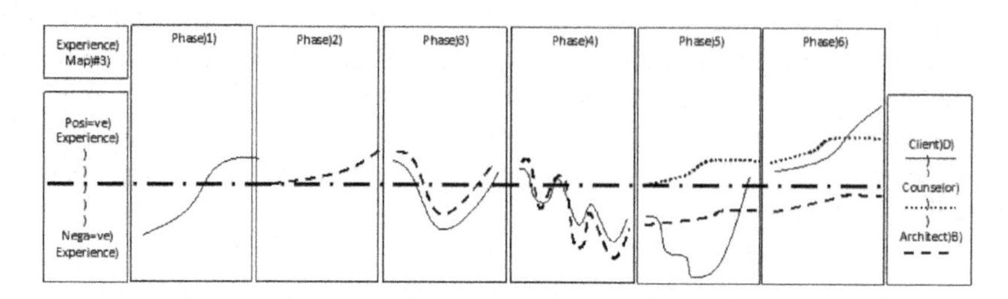

Fig. 3. Experience Map #3, self-built case.

They eventually reach an agreement, and require a building permit for the selected project. The subsequent exchanges with the urban department cause additional shocks for client D (phase 5): the architect B is required to adapt several aspects of the project, modifications that very badly experienced by the client. Additionally, client D complains about several mistakes that subsist through the file. For instance: "*all along the report, he* [the architect] *stipulates I only have two kids and that the project only contains three bedrooms. But that is completely incorrect, I have three kids, each one with their own bedroom !*" (client D). These mistakes and communication gaps exacerbate the negative feelings. Meanwhile, the client D decides to invite an external counselor, expert in self-construction, in order to prepare for the construction site and to receive some technical advices. The counselor, after looking through the project, reassures the client D: the timber structure is absolutely not a problem, nor will be the self-built process. This encounter helps client D gaining back some optimism: "*since we got the building permit, I started to prepare for the construction and I'm quite excited to start*" (client D).

Phase 6, eventually, relates to the very first work meeting gathering all stakeholders (client D, counselor and architect B). From the client's point of view, the counselor provides professional advice and "*produces, in a single meeting, as much work as I had expected from the architect in several months*" (client D). The architect B, on the other hand, remains divided: "*he* [the client] *made every choice I wouldn't have done. Perhaps I should have refused this client after the very first encounter*" (architect B). At the end of this meeting, the counselor remains optimistic, but also a little anxious in view of the difficult relation between the client and the architect.

3.4 Case #4, Self-built Process – "About Shared Involvement and Co-design"

The project is about extending an existing family house in order to build to client E, a professional caterer, an up-to-standards culinary shop. The architect C, 24 years experience, works with many clients interested in self-construction and finds the collaboration always fruitful: in his opinion, self-construction is like a "*rite of passage during which I like to see my customers transform and gain some experience*" (architect C). The client E, as independent caterer, argues having "some sympathy and compassion" for the job done by an architect, whose difficulties are quite similar to hers in her own words: "*services performed in tight budgets and whose clients will never be completely satisfied*" (client E). Client F (client E spouse) also gets involved in the project and holds a certain technical background in metal construction, acquired thanks to former activity in the industry.

The clients had not envisioned self-construction at first. But given their very tight budget and their willingness to build an "eco-friendly" extension, the self-built process suggested by the selected architect eventually convinced them. Client F quickly suggests integrating his expertise in metal construction to the project. Architect C is enthusiastic: "*it is very nice to meet clients with specific skills, that give us a place to pull ideas from*" (architect C). Client E, as for her, remains much more neutral and quiet: only the functionality of the final project really matters (phase 2).

Phase 3 relates the process of obtaining the permit. Although the urban department underlines the quality of the project (thanks to its apparent metallic structure, among other features), it requires the building to be slightly repositioned on the sloping ground. Both the architect and the client E have difficulties dealing with this demand, which, according to them, calls into question several qualities of the project. Client E remains again rather impassive, hoping that this difficulty won't slow down the process.

After obtaining the permit, and waiting for the construction to begin, client F starts studying the building plans with great attention. He detects some errors and incoherencies, and experiences them as a shock (phase 4): "*it would be better to avoid this kind of inaccuracy with auto-builders. Self construction itself is already pretty scary !*" (client F). The architect apologetically recognizes the errors, explains that several collaborators have reviewed the project (including interns) and that it is sometimes difficult to control a perfect transmission of information inside an architectural office.

Phase 5 starts with a first work meeting (including the clients and the architect) on the building site. Both the clients are really receptive and qualify this first "initiation to self construction" as very instructive and positive. Client E starts to really believe in the project's potential and its outcome: "*at first, I didn't think he* [her husband] *would be able to build all this by himself. But after a few weeks, I realized he could !*" (client E). Phase 6, yet, generates some disappointment as the implementation of the metal frame on site proves to be more complicated than expected. In view of the total weight of this frame, it seems practically impossible to put it in place without the intervention of some equipped sub-contractor. Client F has difficulties facing this new and comments: "*at that time, I had the impression to suffer some architect's cosmetic caprice*" (client F). The client F therefore asks the architect to find a solution, in order for him to be able to self-pose the metallic frame. Eventually, the frame will be divided in two sub-parts, which makes the manipulation hazardous but feasible. At that point of the construction process (phase 7), all stakeholders judge the experience as positive. The clients, even though they describe the self-construction process as being "exhausting", underline the excitement as well as the instructive aspect of it. As for the architect C, even though the project generated some problems of communication (in between his office and his clients), he insists on the pleasure he experienced in seeing his clients blossom and be proud of their achievement (Fig. 4).

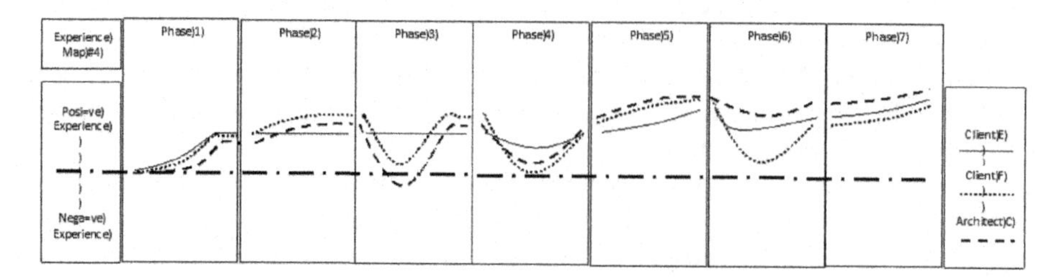

Fig. 4. Experience Map #4, self-built case.

4 Clients-Architects' Interactions in Regard of Traditional vs. Self-built Processes

The data collected through these four highly contextual cases have to be handled with care: future work will include a necessary step of Experience Maps' validation with the clients and architects themselves, and their limited representativeness should not lead to any kind of premature generalizations. We select here only some specific observations that might fuel some reflections about how traditional vs. self-built processes impact clients-architects' interactions during early architectural design phases.

From the first case, we recall that the perseverance of the clients and architect surpasses successive disappointments generated by a too large gap subsisting between early expectations and building realities, gap that generally causes disappointments in terms of perceived quality of a product or a service [13].

The second case, although not eventually conducted as a self-built process, is largely shaped by the strong character of a determined client who needs to keep control on every step of the process, including the resignation and decision to resort to a more traditional building process. This case demonstrates how crucial it is for an architect to remain focused, flexible and capable of adaptation when dealing with such uncertainty.

The third case illustrates the difficulty, for some architects, to "let things go". If the role of the client limits to "evaluating and detecting errors" from draft to draft, moments of interaction only resume in dealing with frustrations and tensions, whereas self-construction should instead generate constructive sharing of experiences/expertise. In this case, communication gaps largely subsist, and although the clients are eager to invest self-construction, they reveal unprepared to deal with the complexity of such a process. The integration of a "counselor", seen here as a new complementary expert, might help bridging this communication/preparation gap. Communicating regularly and explicitly also means enabling the clients to acquire enough knowledge (architectural "culture", awareness of projects' complexities, or even basic vocabulary) necessary to correctly express needs and requirements [1].

The fourth case, eventually, demonstrates how working hand in hand with informed clients (either because they hold some technical skills, or more simply hold some understanding of the architect's role and day-to-day difficulties) generates added value for the whole self-built process. Clients express satisfaction in view of an instructive process, as well as the architect who expresses satisfaction in view of generating what we would refer to as "architectural awareness". From this case we also recall that such clients-architects collaboration might generate creative insights and even, in some aspects, evolve through mutual learning and co-creative experiences typical of co-design processes.

Self-built processes, hypothesized in this paper as the utmost form of users' involvement, do not yet systematically translate into "positive" Experience Maps (i.e. experiences considered as fruitful and satisfying). Be they self-built or traditional, one should recall that architectural design and construction processes remain complex and ill-defined, full of uncertainties, incoherencies and unscheduled events expected to occur. After some "honeymoon phase", clients and architects will certainly experience some degree of "habitus shock", a concept we borrow from both Adler's five stage development model [23] and Bourdieu's sociological observations [in 24], and that explains why users experience disorientation, frustration and even stress as they face an unfamiliar (architectural) habitus. Self-built processes, we argue, might nevertheless ease the path towards Adler's autonomy step as the adjustment process necessary to deal with each of these shocks gets nurtured by a better understanding of the project's complexities and the realities of the world of construction.

Acknowledgments. Authors wish to thank all stakeholders (clients, architects, counselor) who accepted to take part to this research project.

References

1. Siva JPS, London K (2011) Investigating the role of client learning for successful architect-client relationships on private single dwelling projects. Archit Eng Des Manag 7(3):177–189
2. Reymen I, Dorst K, Smulders F (2009) Co-evolution in design practice. In: McDonnell J, Llyod P (eds) About: designing. Analysing design meetings. CRC Press, Taylor and Francis Group, p 434
3. Sanders EB-N (2005) Information, Inspiration and Co-creation. In: Proceedings of the 6th international conference of the European academy of design, University of the Arts, Bremen
4. Lawson B (2005) How designers think – the design process demystified, 4th edn. Architectural Press, Routledge, p 321
5. Barcenilla J, Bastien J-M-C (2009) L'acceptabilité des nouvelles technologies: quelles relations avec l'ergonomie, l'utilisabilité et l'expérience utilisateur? Le Travail Humain 2009/4 (vol 72), Presses Universitaires de France, pp 311–331
6. Chesbrough H, Vanhaverbeke W, West J (2006) Open innovation. researching a new paradigm. Oxford University Press, Oxford
7. Cardon D (2005) Innovation par l'usage. In: Ambrosi A, Peugeot V, Pimienta D (eds) Enjeux de mots. Regards multiculturels sur les sociétés de l'information, Caen, C&F Editions
8. Schön DA (1983) The reflective practitionner: how professionals think in action. Basic Books, NY
9. Cuff D (1991) Architecture: the story of practice. MIT Press, Cambridge, MA

10. Cairns GM (1996) User input to design: confirming the "user-needs gap" model. Environ Des 1(2):125–140
11. Olsson E (2004) What active users and designers contribute in the design process. Interact Comput 16:377–401
12. Cole-Colander C (2003) Designing the customer experience. Build Res Inf 31(5):357–366
13. Maloney WF (2002) Construction product/service and customer satisfaction. J Constr Eng Manag 128:522–529
14. Zeithaml VA, Berry LL, Parasuraman A (1993) The nature and determinants of customer expectations of service. J Acad Mark Sci 21(1):1–12
15. Love PED, Holt GD (2000) Construction business performance measurement: the SPM alternative. Bus Process Manag J 6(5):408–416
16. Ling FYY, Chong CLK (2005) Design-and-build contractors' service quality in public projects in Singapore. Build Environ 40(6):815–823
17. Ng ST, Palaneeswaran E, Kumaraswamy MM (2011) Satisfaction of residents on public housings built before and after implementation of ISO9000. Habitat Int 35(1):50–56
18. Ahmed SM, Kangari R (1995) Analysis of client-satisfaction factors in construction industry. J Manag Eng 11(2):36–42
19. Masrom MA, Skitmore M, Bridge A (2013) Determinants of contractor satisfaction. Constr Manag Econ 31(7):761–779
20. Luck R (2007) Learning to talk to users in participatory design situations. Des Stud 28:217–242
21. CAE - Conseil des Architectes d'Europe (2017) La profession d'architecte en Europe 2016 – Une étude du secteur. Mirza & Nacey Research Ltd, Royaume-Uni
22. Gronier G, Lallemand C (2016) Méthodes de design UX 30 méthodes fondamentales pour concevoir et évaluer les systèmes interactifs. Eyrolles, Paris
23. Adler P (1975) The transitional experience: an alternative view of culture shock. J Humanist Psychol 15(4):13–23
24. Siva JPS, London K (2009) Habitus shock: a model for architect-client relationships on house projects based on sociological and psychological perspectives. In: Proceedings of the future trends in architectural management international symposium, Taiwan, pp 209–220

Thermal Comfort Differences with Air Movement Between Students and Outdoor Blue-Collar Workers

Yu Ji and Hong Liu(✉)

Chongqing University,
No. 83 Shabei Street, Shapingba District, Chongqing, China
liuhong1865@163.com

Abstract. In order to explore the comfort zone of airflow and thermal comfort differences with air movement between students and outdoor blue-collar workers, the experiment was carried out in a climate chamber in Chongqing university. The research shows that the range of the outdoor blue-collar worker (0.7 m/s–2.89 m/s) is wider than that of the students (0.9 m/s–2.76 m/s). In addition outdoor blue-collar workers are not sensitive to changes of the temperature compared to the students. The upper threshold of the acceptable temperature and humidity is different between the two groups. The upper limit of the white-collar workers is 30 °C/70%, while the outdoor blue-collar workers could maintain their thermal comfort with air movement even in 32 °C/90%.

Keywords: Thermal comfort · Thermal sensation · Air movement

1 Introduction

Human thermal comfort is usually affected by ambient physical parameters and human physiological conditions. Difference of thermal comfort requirement between different human groups in non-neutral environment exist, due to different body physiological conditions and thermal experience. However, most previous thermal comfort studies were basic on the data from college students, whose results may different from the labors. Increasing air movement is an effective and low-cost method to improve thermal comfort in hot-humid environment, which is widely used by low-income labors (which is called "bangbang") in Chongqing, China. Thus, this study aim to evaluate outdoor blue-collar workers' thermal responses to air movement in warm and humid environment, and compare the responses to air movement between the two different groups.

S. Bagnara et al. (Eds.): IEA 2018, AISC 825, pp. 453–458, 2019.
https://doi.org/10.1007/978-3-319-96068-5_51

2 Method

The experiment was carried out in a climate chamber with 12 chosen healthy college students (6 females and 6 males) and 12 outdoor blue-collar workers. Nine conditions with air temperature at 28 °C, 30 °C and 32 °C, while relative humidity at 50%, 70% and 90% were conducted. They were healthy and had rested well the night before the experiment (Table 1).

Table 1. Basic information of the subjects.

Population	Age (year)	Height (cm)	Weight (kg)	Clothing description (clo)	Residence time in Chongqing (year)
Outdoor blue-collar workers	50 ± 5.23	168.25 ± 2.36	67.65 ± 6.83	0.32	20.5 ± 2.65
Student	23.67 ± 0.83	168.17 ± 3.71	56.58 ± 5.15	0.32	3.92 ± 1.65

3 Experimental Procedures

Experiments were conducted in a thermal comfort laboratory which was a room measuring 3 m (width) by 4 m (length) by 3 m (height). During the thermal comfort experiment, the subjects were asked to enter laboratory 20 min before the experiment for adaptation. For each thermal environment, the subjects were exposed to 3 patterns of airflows for 20 min, during which they filled in two questionnaire as shown in Fig. 1, and then another airflow pattern was provided. In total, the thermal comfort experiment for each subject lasted for 200 min. To avoid the effect of circadian rhythms, all subjects participated in the experiment at the same time of the day. An online questionnaire was used during the experiment to collect each subject's response for each airflow pattern. The online questionnaire included questions on thermal sensation (on a 7-point scale), temperature preference, air movement perception, thermal comfort and perceived air quality. The thermal sensation questions complied with ASHRAE Standard 55-2013 [18].

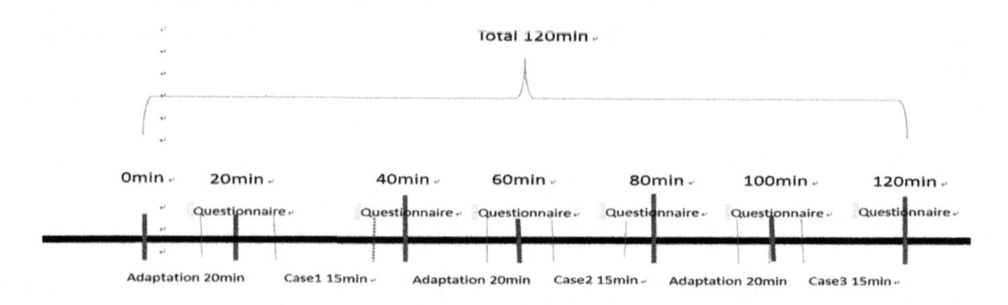

Fig. 1. Timeline for the thermal comfort experiment

4 Results and Discussion

4.1 Skin Temperature Differences Between Students and Outdoor Blue-collar Workers Before Opening the Fan

Figure 2 shows the overall skin temperature of students and outdoor blue-collar workers before the fan opens. Table 2 is a significant level test of the corresponding difference between the two groups of people. From Table 2, it can be seen that when the ambient temperature and humidity of environment are 28 °C/50%, 28 °C/70%, 28 °C/90%, and 30 °C/50%, the skin temperature differences of the two groups is not significant. When the ambient temperature and humidity are 30 °C/70%, the overall skin temperature of the students is significantly smaller than that of the outdoor blue-collar workers, while the ambient temperature and humidity are 30 °C/90%. At 32 °C/70%, the overall skin temperature of the students group was significantly higher than that of the outdoor blue-collar workers.

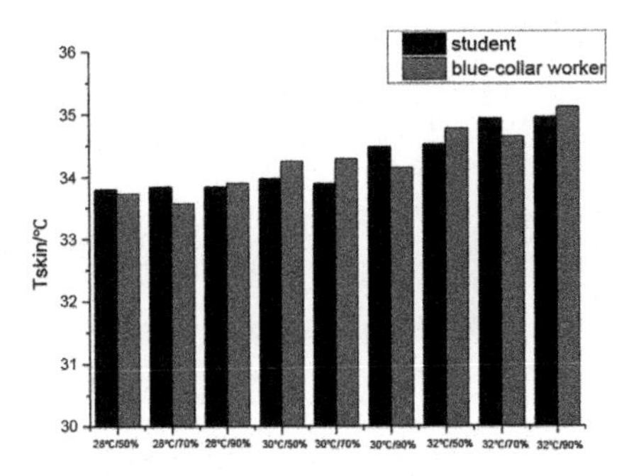

Fig. 2. The differences of average skin temperature between students and outdoor blue-collar workers

Table 2. The significance level of the average skin temperature difference between students and outdoor blue-collar workers

T/RH	Mean difference (students - outdoor blue-collar workers)	Std. error	Sig.	95% confidence interval	
				Lower bound	Upper bound
28 °C/50%	0.068	0.092281	0.464	−0.117974	0.254233
28 °C/70%	0.275	0.147089	0.244	−0.472424	0.124197
28 °C/90%	−0.053	0.131128	0.683	−0.317314	0.20971
30 °C/50%	−0.277	0.152867	0.081	−0.590229	0.036437
30 °C/70%	−0.4	0.133359	0.005	−0.671059	−0.128667
30 °C/90%	0.334	0.110526	0.005	0.109594	0.558352
32 °C/50%	−0.255	0.143115	0.085	−0.546956	0.037603
32 °C/70%	0.287	0.117031	0.017	0.053478	0.523843
32 °C/90%	−0.162	0.12757	0.211	−0.41947	0.095424

4.2 Analysis of Skin Temperature Changes of the Two Groups Under the Different Wind Speeds

Previous studies show that it is not appropriate to use the skin temperature to evaluate the blowing efficiency because of that the initial skin temperature of the subject is different. Therefore, it is better to use the changes of skin temperature before and after blowing to quantify the effect of blowing. Figure 3 shows the skin temperature changes before and after blowing of the two groups.

As can be seen from Fig. 3, as the temperature rises, the decrease in skin temperature decreases. When the ambient temperature is 28 °C and 32 °C, the difference is both about 0.8 °C–1.0 °C. Compared to 28 °C, the selection of wind speeds of 32 °C are larger. So the same conclusion can also be drawn. It can be clearly seen from the Fig. 3 that, the changes of skin temperature of manual workers are greater than that of students at 28 °C. While the ambient temperature are 30 °C and 32 °C, the changes are less than that of students. As the humidity increased, the changes of skin temperature decreases.

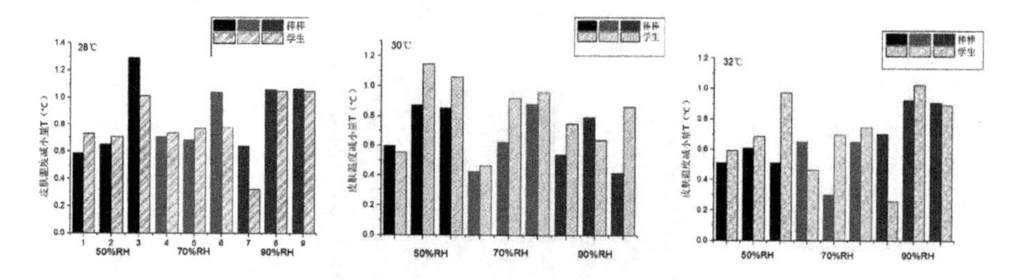

Fig. 3. The skin temperature changes before and after blowing of the two groups.

4.3 Thermal Sensation

Figure 4 shows the thermal sensation votes of students and outdoor blue-collar workers under the different conditions. As can be seen from Fig. 4, as the temperature rises, the thermal sensation of students increases. Besides, the thermal sensation increases obviously as the humidity rises at the same ambient temperature. The higher the ambient temperature, the more obvious the influence of relative humidity will be. This is similar to the results of other scholars' research that the effect of humidity on the thermal sensation is more significant under the hot environment. For example, at 28 °C, the thermal sensation difference is less than 0.5 under the different humidity, while at 32 °C, the thermal sensation difference exceeds 1. From the Fig. 4, we can see that airflow can significantly improve the thermal sensation. As the wind speed increases, the thermal sensation vote decreases. However, comparing the thermal sensation votes in different conditions, we can find that there was no airflow demand at 28 °C. But when the temperature is 30 °C, students can significantly reduce the thermal sensation through the airflow. The increase in wind speed at 32 °C could no longer meet the thermal comfort requirement (TSV > 0.5).

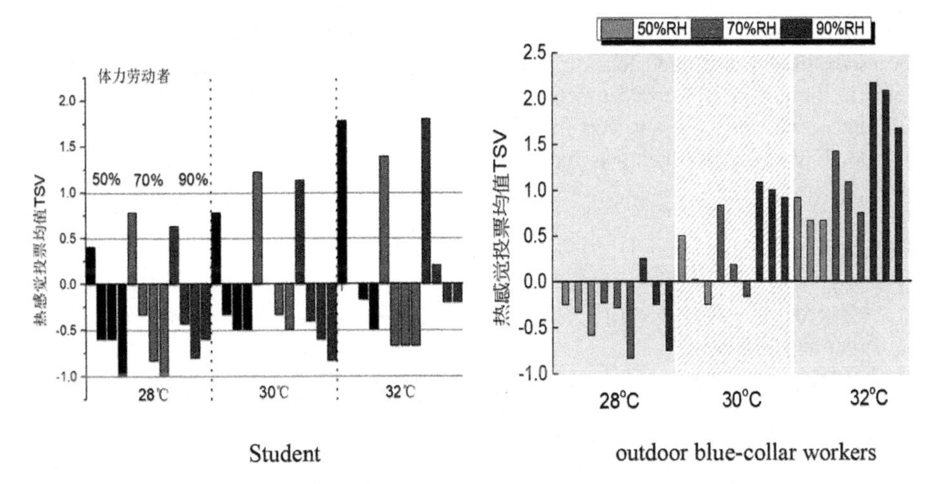

Fig. 4. Thermal sensation votes of students and outdoor blue-collar workers under the different conditions

For outdoor blue-collar workers, the increase of relative humidity under the same temperature will obviously increase the heat sensation of the subjects, which is the same as those of the students. As the ambient temperature rises, the effect of humidity on the heat sensation is not obvious, which is different from the students. Similar to mental workers, airflow can significantly improve the thermal sensation of manual workers. When the ambient temperature is 28 °C, the outdoor blue-collar workers have no need for airflow. While when the ambient temperature is 30 °C and 32 °C, airflow can obviously improve the thermal sensation of the outdoor blue-collar workers. Unlike students, outdoor blue-collar workers can improve thermal sensation through the airflow, which may be due to that outdoor blue-collar workers work outside for a long time who is well tolerated for high temperature and high humidity.

5 Conclusion

The research shows that air movement can significantly improve the thermal sensation in hot-humid environment. But the upper threshold of the acceptable temperature and humidity is different between the two groups. The upper limit of students is 30 °C/70%, while the outdoor blue-collar workers could maintain their thermal comfort with air movement even in 32 °C/90%. Outdoor blue-collar workers are not sensitive to changes of the temperature compared to the students. When the temperature raise from 28 °C to 32 °C, the thermal sensation vote of outdoor blue-collar workers increase about 0.5, while the value of students increase about 0.7. By analyzing the relationship between the draft vote and air velocity, we find that the draft vote is similar between the two groups. But the acceptable ranges of air velocity are different, the range of the outdoor blue-collar worker (0.7 m/s–2.89 m/s) is wider than the students (0.9 m/s–2.76 m/s).

The results show that outdoor blue-collar worker are not sensitive to the changes of the environment and need less air velocity to maintain thermal comfort than the students in the hot-humid environment. Besides, the ranges of comfort air velocity under different conditions are showed in this article and the thermal comfort zone of the outdoor blue-collar worker was built.

References

1. ASHRAE Standard 55-2013 (2013) Thermal environmental conditions for human occupancy, Atlanta
2. EN ISO 7730:2005 (2005) Ergonomics of the thermal environment - analytical determination and interpretation of thermal comfort using calculation of the PMV and PPD indices and local thermal comfort criteria, Geneva
3. Fanger PO (1970) Thermal comfort: analysis and applications in environmental engineering. Danish Technical Press, Copenhagen
4. Yang Y (2015) Indoor thermal response of human body in uniform environment (warm conditions). PhD Thesis, Chongqing University
5. Lind AR, Bass DE (1963) Optimal exposure time for development of acclimatization to heat. Fed Proc 22(22):704–708
6. Rohles FH, Woods JE, Nevins RG (1974) The effect of air movement and temperature on the thermal sensations of sedentary man. ASHRAE Trans 80(1):101–119
7. Rohles FH, Konz S, Jones B (1983) Ceiling fans as extenders of the summer comfort envelope. ASHRAE Trans 89(1):245–263
8. Tanabe S, Kimura K (1994) Effects of air temperature, humidity, and air movement on thermal comfort under hot and humid conditions. ASHRAE Trans 100(2):953–969
9. Mc Intyre DA (1978) Preferred air speed for comfort in warm conditions. ASHRAE Trans 84(2):263–277
10. Kubo H, Isoda N, Hikaru E-K (1997) Cooling effects of preferred air velocity in muggy conditions. Energy Build 32(3):211–218
11. ISO7730:2005 (2005) Ergonomics of the thermal environment—analytical determination and interpretation of thermal comfort using calculation of the PMV and PPD indices and local thermal comfort criteria. International Standard Organisation, Geneva
12. ASHRAE (2004) ANSI/ASHRAE 5-2004, thermal environmental conditions for human occupancy. American Society of Heating, Refrigerating and Air Conditioning Engineers, Inc., Atlanta
13. Gagge AP, Nishi Y (2010) Heat exchange between human skin surface and thermal environment. Comprehensive physiology. Wiley
14. ISO (1998) International Standard 7726: thermal environment-instruments and methods for measuring physical quantities. International Organization for Standardization, Geneva

Standardizing Human Abilities and Capabilities Swedish Standardization with a Design for All Approach

Jonas E. Andersson[1,2(✉)]

[1] Royal Institute of Technology, KTH, Stockholm, Sweden
jonas.andersson.arch@gmail.com
[2] Swedish Governmental Agency for Participation, Stockholm, Sweden

Abstract. Several standard works in Sweden from the period 2000–2017 have been focused on converting visionary welfare political goals into down-to-earth-oriented guidelines for subsequent realization and implementation. The present paper is focused on the conversion of general welfare goals into standards that apply to areas that require a trans-disciplinary approach to address accessibility issues in built environment, services and transportation. The study suggests that standardization with a design for all perspective becomes an interpretive work in which words and phrases are contemplated in relation to the ethical stance of the national disability policy. This framework is situated at the very interface between real-life settings and visionary thinking. Consequently, participants in standardization works revolving around design for all activate several individual knowledge fields of ethical, ideological, practical and theoretical nature. In communal discussions between the participants, the development of standards proceeds through an analytical work that is like an iterative creative process that uses concepts, phrases and words as instruments. The overall conclusion is that standardization with a design for all approach has left the strict focus on products and started to target the design process in view of a built environment, products or services that are centered on the fit between the design and a wide range of human abilities.

Keywords: Standardization · Design for all · Multi-disciplinary knowledge

1 Introduction

In 2017, just two days before Christmas eve, the governmental Committee for Modern Building Rules, in the following CMBR, appointed by the Swedish government, submitted its first report on the efficiency of national building standardization. The committee suggested a new take on building standardization to overcome a widening gap of mistrust between authorities, building industries and the national standardization organization, Swedish Standards Institute, SIS [1]. At the heart of the conflict lies a new high-profile management of the SIS that makes the building industries hesitant to engage in new standard project and prefer company agreements. In addition, this approach has also generated problems in the public governance, especially for the National board for housing, planning and building, in the following NBHPB,

S. Bagnara et al. (Eds.): IEA 2018, AISC 825, pp. 459–468, 2019.
https://doi.org/10.1007/978-3-319-96068-5_52

(in Swedish Boverket) in charge of accessibility and usability in the built environment. The Swedish legal tradition prescribes that any act or regulation must be accessible in Swedish, which creates a problem with standard copyrights. Currently, the conflict impedes effectively the engagement of the NBHPB in any standardization projects.

Despite the current heated conflict between the national standardization organization and different influential actors on the building market, standardization has played an important role in the construction of the modern Swedish welfare state (ibid). Sweden joined electrical standardization in 1907 while standardization for the building market grew strong during the 1920s. Paired with the architectural style of functionalism, standardization of various building elements and functions became an essential element for improving the national housing standard and realize new socio-political ideals for appropriate housing of the Swedish "folkhemmet," i.e. people's home in English. Standardization led to the development of a whole new knowledge on cooking and kitchen work and their requirements in terms of cabinet modules, length of worktops and appropriate square-meters for dining. The work also touched the design of bathrooms and hygiene installations. The standardization works involved both architects, designers and engineers. Minimum requirements were established, which until the beginning of the 1980s also supplied the threshold square-meters for state loans for housing construction.

Sweden was among the first countries to embrace the idea that the adaptation of existing and new built environment was a prerequisite for including people with disabilities on equal terms in the surrounding welfare society. During the 1960s, this ambition forged the Swedish concept of accessibility that today implies an equal access to the built environment for people with cognitive or locomotorary impairments. Over the years, physical requirements have been developed as minimum requirements to be met in architectural designs for existing or new buildings. Since, accessibility has become an integral part of other legal frameworks, i.e. the national building code, as well as acquiring the status of being a key concept in the Swedish disability policy. Driven by the essential mechanism for Swedish disability policy – the so-called disability perspective, in the following termed DOA (in Swedish funktionshinderperspektiv). The DOA is the national homologue of the international concepts of design for all, inclusive design or universal design – a universal level of accessibility in any type of built environment, service or transportation mode is promoted [2]. To this level, assistive technologies can be added according to personal needs to achieve the highest possible level of participation in the modern Swedish welfare society.

1.1 Aims and Purposes

Since the mid-1970s, when accessibility was propellered as the key concept for removing perceived and physical barriers for the full participation of people with disabilities in the surrounding Swedish welfare society [3], standardization has been used for this quest. The present study aims at framing Swedish standardization with a design for all approach and analyzing how mutual agreements can be used for converting broad disability policy goals into detailed guidelines for products and processes that are applicable in a series of trans-disciplinary areas that address accessibility issues in built environment, services and transportation. The study used a close reading process of archive materials [4]. The study is anchored in a self-reflective inquiry framework that moves freely between the inner and outer arcs of attention [5].

A critical analysis approach was used to create a narrative that shed light on how standardization with a design for all approach can benefit people with disabilities [6–10]. The research material was of a transdisciplinary nature and included research fields that focus on conditions for people with disabilities or accessibility issues concerning the built environment, e.g. architectural and building research, research on disabilities, and research on ergo-therapy and gerontology [11, 12].

2 Retrospective Backdrop to Standardization in Sweden

Before entering the topic, it is necessary to give a brief account of the context in which Swedish standardization with a design for all approach, in the following DFA standardization, takes place. Although Sweden joined electro-mechanical standardization already in 1907, DFA standardization materialized gradually for the 1950s until the mid-1970s. Associations and non-government organizations for people with disabilities were first founded at the end of the 19th century [3]. At that time, poliomyelitis was a considerable threat to public health and caused complex impairments among adults, but children. Following the polio epidemic in 1911, a national collection was opened that aimed at improving living conditions at large institutions that served as combined housing, education and work centers for people with disabilities. The Nobel prize-winner for literature in 1909, Selma Lagerlöf, was the main fundraiser and contributed by publishing a pamphlet in Swedish newspapers that stressed the need for helping the large group of young and older people afflicted by the epidemic. The campaign accumulated such generous contributions, above half a million Swedish crowns (SEK), that a national committee was founded. This was the Swedish Central Committee for Disabilities, in the following the SVCK after the Swedish acronym (ibid).

During the 1940s and until 1965, SVCK hosted different initiatives to break the isolating and uprooting effect of concentrating a large group of people with just one common denominator – a cognitive, physical or sensory deficiency – to large institutions on the city edge or in the countryside (ibid). During the 1960s, the normalization principle, which was promoted by the Swedish sports organization for people with disabilities, expanded into other socio-political sectors such as appropriate housing and work opportunities. SVCK emphasized that cognitive, physical or sensory disabilities were not a barrier to a full and rich life, but rather that attitudes and beliefs among the rest of the population were the main barrier, poignantly demonstrated by an inaccessible built environment. In 1965, the Swedish Board for Disabled People (SBDP) was created to disseminate knowledge about living conditions for people with disabilities on national, regional and local level. In 1968, SVCK was divided into two new organizations. The remainder of the large sums from 1911 was entrusted to a foundation for research and support for people with disabilities. In addition, a non-profit association was founded, the Swedish Institute for Handicapped people (SIHP). It focused on rehabilitative equipment, technology and work to integrate people with disabilities on equal terms into the welfare society. In 1999, the institute was reorganized and formed SIAT.

3 Assistive Technologies Promote Standardization

The SIAT was converted into an authority in 2014, the current Agency for participation, in the following MFD after the Swedish name Myndigheten för delaktighet. This conversion was motivated by the fact that the SIAT, along with its previous predecessors from the late 1960s, had become an influential player in the field of disability policy, both nationally and internationally. In line with this development, the SIAT with predecessors also started to issue guidelines for assistive technologies as well as for the built environment [13]. In the absence of other players in these field, these guidelines quickly became popular by both public and private contractors. They filled the void of missing knowledge on the interaction between human beings, disabilities and the appropriate design of assistive equipment and built environment. Being a non-for-profit association with guidelines with a similar impact as recommendations and regulations issued by civil administrations became an anomaly that called for a new legal stature of the institute. This was the main reason for dismantling the successful SIAT and replacing it with the MFD.

The key factor in this remarkable ascension of the SIAT dates to the late 1968, when the SIHP took over after the SVCK. Replacing previous random testing during the 1950s and early 1960s, regular and systematic testing of assistive equipment and adjustments of cars for individual needs became one of the assignments for the newly formed institute for people with disabilities, i.e. SIHP. The testing was directly linked to the procurement of assistive equipment for people with disabilities. Originally, this procurement was made by the state, but in 1976 this assignment was transferred to counties and regional authorities (ibid). Since Sweden belonged to the pioneer countries in the field of disability policy, this testing and evaluation also propellered the SIHP into becoming a renowned laboratory of various assistive products on both a European and international level. One special project that forged SIHP's fame during the 1980s, was an innovation and renewing project that pertained to the design of wheelchairs and walkers. The project abandoned the static prototype of a device that was supposed to be pushed by a helper or as a temporary support. Instead, the project stressed the direct focus on user needs and user-outcomes so that the wheelchair and the walker promoted independent living and participation in the surrounding society. The extensive work on assistive technologies, including the redevelopment of wheelchairs, also generated a Nordic interest, e.g. the Nordic network on sit comfort in wheelchairs.

The step from thorough commitment to evaluations, tests and rethinking designs for assistive equipment was not far from both national and international standardization projects. Towards the end of the 1970s the International Standardization Organization, ISO, vetted intensively the area of assistive equipment and started the first standardization projects that pertained to safety and security issues in relation to users of these products [14]. The new committee was equipped with a Swedish chair, and the SIHP formed the secretariat. The SIHP took part in the formulation of about 60 global standards [13]. During the 1990s, standardization projects in conjunction with European Union directives were initiated. Sweden joined the union after a referendum in 1995. Even in this context, the SIHP assumed the roles of chair and secretary, while the

Swedish Standards Institute, SIS, assisted with a secretariat. This work also involved mobility devices such as walkers and wheelchairs, but mostly medicine-technical equipment, for instance hygiene products for people with incontinence and stomia (Ibid). In a parallel track, the institute also participated in national standardization that concerned accessibility issues in the built environment.

4 A New Deal in Swedish Disability Policy and the CRPD

The first decade of the new millennium saw the birth of a 10-year action plan for realising goals of Swedish disability policy. The title of the action plan, 'From being a patient to achieving full citizenship,' encapsulated a fundamental shift and reorientation of national policy based on the assumption that once the built environment was made accessible on a general level, a full social inclusion of people with disabilities would follow [15]. Swedish state organisations were designated to be key players for the realisation of national goals concerning accessibility in the built environment and the full inclusion of people with disabilities in the surrounding welfare society [16].

The concept of accessibility was firmly integrated in a disabilities-obstacle approach, the DOA, that aimed at abolishing attitudes and cultural beliefs about people with disabilities and eradicating physical barriers in the built environment. With the DOA approach, additional trimming of the user-environment fit must rely on finely tuned assistive technologies that are based upon the individual needs of the disabled person. The DOA orientation displays similarities with the egalitarian and non-discriminatory goals with which universal design had been equipped since 1997 [17]. In 2003, a set of physical barriers to the public in the existing official buildings and open spaces, defined as obstacles of an easily removable character, was defined, and, therefore, targeted for adjusting measures. These can be summarised in the six bullet points [18]:

1. heavy doors that are difficult to open, therefore, requiring door-opening devices for easy access for people with disabilities;
2. high thresholds or levels, therefore, requiring level-adjusting measures like a ramp, inclined runway or elevator for improved access for people with disabilities;
3. insufficient signage cues for orientation or warning, therefore, requiring contrast or tactile markings for people with disabilities;
4. inappropriate design of handrails for stairways or handles that pose dangers, especially people with disabilities;
5. insufficient electric lighting that creates glare or provide poor illumination, especially for people with disabilities;
6. other building aspects like slippery flooring, over-compact hygiene spaces, or narrow passages.

In the following year, the first set of guidelines was published. It supplied guidelines to overcome physical obstacles in the built environment and immaterial barriers that people with disabilities might experience in relation to media and organisations [18]. The guidelines supplied a tripartite understanding of the notion accessibility:

- Guidelines on physical requirements for buildings and the built environment
- Guidelines on codes of conduct and strategic planning in work organisations
- Guidelines for information in books and other media, including the internet.

In 2007, a second edition of guidelines, entitled "Tear down the obstacles; guidelines for accessibility" (TDOGA) was published [19]. These guidelines are mandatory for state organisations, so that they provide exemplary models of accessible buildings, along with inclusive work organisations and accessible media, for the rest of the building sector to follow [20]. In 2008, Sweden signed and ratified the UN Convention 26 on the Rights for People with Disabilities (UN CRPD) which, when it entered into force in 2009, extended the understanding of the national tripartite understanding of accessibility to include two new concepts, both involving a focus on user needs and the necessary fit between architectural design and future users of the built space. The intention was to highlight even further the potential fit between architectural design and the various needs of a plethora of users. This concretisation of future users of the built environment had also resulted in changes to building regulations and in 2008, the concept of usability in close combination with accessibility was introduced to emphasize the fit between the built space and a large and varied group of users [21].

5 Designing Guidelines for All - Guide 6

Like most European countries, Sweden entered a still on-going demographic change in the composition of the population by the end of the 1980s, whose most prominent feature is an increasingly larger and older share of retired people, estimated to equal the number of children and adolescents, about 25%, around 2030 [22, 23]. Swedish representatives participated in the foundation of the European Institute for Design and Disability (EIDD) – Design for all Europe, founded in 1993. By the end of the 1990s, the EU also realized the need for adapting products and service to the approximately 120 million Europeans who continued to lead an independent life within the ordinary stock of housing, often elderly but also people with disabilities [24]. In 1999, the EU issued a mandate to the European standard organisation CEN for a guidance document in the field of safety and usability of products for people with special needs, e.g. elderly and disabled (ibid.). CEN asked the Swedish standards Institute, SIS, to form a secretariat, while the chair and the secretary was recruited from the newly formed SIAT [25].

The work on an EU guidance document soon merged with the one of the ISO/IEC standardisation work, since the third European organisation ETSI preferred to work within this context and with the same topic. The ISO Guide 71 was published in 2001. The guidance standard for the EU, Guide 6, was harmonised with this publication and issued a year later. In 2005, the first Swedish version was published in 2006. Until 2017, this guide has been downloaded sparsely. Revision works of both Guide 6 and Guide 71 were initiated in 2009, and the revised version of both guides were published in 2014. Even this new version of the guides has been downloaded sparsely. In 2015, the SIS started a translation work of Guide 6 into Swedish that continued until the fall of 2017.

Like previous guidelines, also this guide has not produced a boom in downloads and subsequent implementation. This is problematic, since 2017, the principle of universal design for built environment and products is a corner stone in the current disability policy [26].

In line with the Swedish disability policy, the two guides 6 and 71 emphasize a view on disability in which the design and the organisation of the surrounding society produce obstacles vis-à-vis the individual rather than that these are due to the individual's set of personal abilities and capabilities, i.e. the social model of disability [27, 28]. During the first decade of the new millennium, the SIAT initiated several standardization projects in a similar orientation. For instance, the SIAT initiated the development of a national standard on informative icons – i.e. SS 30600 Graphical symbols – which generated an in interest by the ISO and some icons were adopted as icons in the international standard ISO 21542 [13].

5.1 Expanding the Circle of Stakeholders

Since the early 1990s, established by the SIS, a board with consumer representatives, the so-called 'SKA-rådet' has guaranteed a continuous participation of end-user representatives in standardization projects to widen the group of stakeholders outside the prime target groups for establishing new standardized solutions [28]. From 2000, the consumer representatives also include end-user representatives from both disability organizations and work unions. The SKA is financed by the Swedish government with a yearly allowance of approximately 31 million SEK, of which, about 3,5 million SEK (2017) may be used to sponsor full membership of non-profit organisation, i.e. disability organisations, in various technical committees [1, Appendix 6, p. 255]. The broad and mixed participation in technical committees of authorities, manufacturers and end-user representatives is seen as essential in Swedish policy for standardization [29]. The envisioned outcome is a stronger Swedish turnover on the European and international market (ibid).

Concerning what in Sweden has become a twin concept, i.e. accessibility and usability, the matter of accessible and usable products and designs of buildings and built elements is present in several standardization projects. Approximately 1500 standards, national (SIS), European (CEN) or international (ISO), deal with the built environment and was under revision by July 1st, 2017 [1]. To address the severe national housing crisis, standardization is also suggested for physical planning in view of new exploitation. Not all these standards address accessibility and usability per se. Given the complexity of accessibility and usability, the SIS has formed a special technical committee, SIS/TC 536 Coordination group for accessibility issues, approximate translation of the Swedish name 'Samordningsgrupp för tillgänglighet'. This committee supervises on-going standardization projects with a potential effect on accessibility for disabled or older people. The committee has 15 members and is taking part of three ongoing European standardization projects, i.e. CEN/BT/WG 213, Strategic Advisory Group on Accessibility (SAGA), CEN/CLC/JTC 12, Design for All and CEN/TC 320, Transportation services.

6 Conclusions and Discussion

This paper has reflected upon Swedish standardization with the intent to harmonize human abilities and capabilities with requirements for assistive technologies, the built environment, products and services. The studied example lends support to five conclusions on how standardization with a design for all approach has developed: This type of standardization:

1. has evolved from specific requirements for assistive products and technologies;
2. is associated with a national prescription system of assistive products and technologies;
3. has a close relationship to awareness-raising measures of living conditions for people with disabilities and the promotion of inclusion and participation of every persons regardless of age or disability in a democratic and egalitarian welfare society;
4. presents a clear link to the development and confirmation of a national disability policy;
5. is promoted by the signature and ratification of the UN CRPD.

As such, this type of standardization has gone from clear cases of the appropriate fit between a multitude of users and assistive technology products to become standardization that aims at integrating a design for all approach already in design process. The period of standardization, which targeted mainly requirements for assistive products, spanned some 25 years. The period of standardization that pertains to processes has just begun. Hence, the design for all approach is slowly turning into becoming a decisive parameter in design and development processes of buildings, products and services. However, the poor spread of guidelines on safety and usability of products for people with special needs, e.g. elderly and disabled corroborates the conclusions of the committee report of CMBR [1]: there is a gap between standardisation and the subsequent uptake by stakeholders who can promote the principle of universal design.

Although outcomes of national disability policies often are difficult to evaluate in terms of measurable progress, the number of standardization projects that pertain to integrating a design for all approach in design and development processes of buildings, products and services can be seen as an indicator of the efficiency of such a policy. In the Swedish case, the abandonment of an obsolete view on people with disabilities as dependent and in need of constant assistance towards an increasingly stronger emphasis of the DOA and a view on people with disabilities equal to other citizens has created the need for tools for implementing the DOA concretely. Here, standardization of processes in view of producing built environment, products and service probably will gain a larger importance in a near future.

Currently, Swedish building standardization is under investigation. To gain momentum and infuse new power in this field of standardization, the governmental Committee for Modern Building Rules has suggested a stronger state supervision [1]. This could either be executed as a special two-year assignment for a negotiator or a permanent committee that monitor standardization of buildings, infrastructure and travel services. In any case, standardization is viewed as essential for promoting development and upholding fundaments for the Swedish welfare society. Standardization with a design for all approach will continue to develop and become an important factor for the realisation of the Swedish disability policy.

References

1. SOU 2017:106 (2017) Nystart för byggstandardiseringen genom stärkt samverkan [New take on building standardization through closer collaboration. White papers]. Statens Offentliga Utredningar, Stockholm
2. von Axelson H, Linden A, Andersson JE (2016) Equalization and participation for all: Swedish disability policy at a crossroads. In: Petrie H et al (eds) Learning from the past, designing for the future. IOS Press
3. Andersson J (2016) Improved Swedish accessibility hindered by a housing imbroglio. Nord J Archit Res 28(2):9–32
4. Brummett B (2010) Techniques of close reading. SAGE Publications, Los Angeles
5. Marshall J (2001) Self-reflective inquiry practices. In: Reason P, Bradbury H (eds) Handbook of action research. SAGE Publications, London
6. Blum-Kulka S (1997) Discourse Pragmatics. In: van Dijk TA (ed) Discourse as social interaction. SAGE Publications Inc., London, Thousands Oaks, New Delhi
7. Gunnarsson B (1998) Applied discourse analysis. In: van Dijk TA (ed) Discourse as social interaction. SAGE Publications, London, Thousands Oaks, pp 285–312
8. Mumby DK, Clair RP (1998) Organizational discourse. In: van Dijk TA (ed) Discourse as social interaction. SAGE Publications Ltd., London, Thousands Oaks, pp 181–205
9. Perren L, Sapsed J (2013) Innovation as politics: The rise and reshaping of innovation in UK parliamentary discourse 1960–2005. Res Policy 42:1815–1828
10. van Dijk TA (1998) Discourse as interaction in society. In: van Dijk TA (ed) Discourse as social interaction. SAGE Publications, London, Thousands Oaks, New Delhi
11. Iwarsson S, Ståhl A (2003) Accessibility, usability and universal design - positioning and definition of concepts describing person-environment relationships. Disabil Rehabil 25(2):57–66
12. Andersson JE, Rönn M (2014) Projektredovisning: Arkitektur för Bo bra på äldre dar [Architecture for Growing Old, Living Well]. Kungl. Tekniska Högskolan, Statens Byggeforskningsinstitut, SBi and Rio Kulturkooperativ, Stockholm
13. Sjöberg M (2017) Uppdraget slutförts. Hjälpmedelsinstitutets historia [Mission accomplished. The history of the Swedish Institute for Assistive Technologies, SIAT, Swedish Institute for Assistive Technologies]. Hjälpmedelsinstitutet, Stockholm
14. Michele Micheletti M, Dietlind S (2004) Politics, products, and markets: exploring political consumerism. Routledge, Cambridge
15. Swedish Government (2000) Från patient till medborgare. En nationell handlingsplan för handikappolitiken. 1999/2000:79 [From being a patient to achieving full citizenship. A national action plan for the disability policy]. Swedish Government Offices, Stockholm
16. SFS 2001:526 (2001) Om de statliga myndigheternas ansvar för genomförande av funktionshinderspolitiken [Ordonnance on the responsibility of state authorities to promote the realization of the national disability policy]. Statens Författningssamling Stockholm
17. Steinfeld E, Maisel J (2012) Universal design. Creating inclusive environments. Wiley, New York

18. NBHBP (2003) Boverkets föreskrifter och allmänna råd om undanröjande av enkelt avhjälpta hinder till och i lokaler dit allmänheten har tillträde och på allmänna platser [Guidelines and recommendations of the NBHBP, National Board of Housing, Building and Planning, concerning the removal of minor barriers in public buildings or public space]. Boverket [National Board of Housing, Building and Planning], Karlskrona
19. Disability Ombudsmannen: Riktlinjer för en tillgänglig statsförvaltning [Guidelines for an accessible civil administration]. Handikappsombudsmannen, Stockholm (2002)
20. Handisam (2007) Riv hindren - riktlinjer för tillgänglighet. Enligt förordning 2001:526 [Tear down obstacles, guidelines for accessibility according to ordonnance 2001:526]. Handisam, Stockholm
21. MFD (2015) Riktlinjer för tillgänglighet, riv hindren. Enligt förordning 2001:526 om statliga myndigheters ansvar för genomförande av handikappolitiken [Guidelines for accessibility, tear down obstacles]. Myndigheten för Delaktighet, Stockholm
22. NBHBP (2008) BBR 15. BFS2008:6 [Building regulations, BBR 15]. Boverket [National Board of Housing, Building and Planning], Karlskrona
23. Statistics Sweden (2017) Aktuell prognos över folkmängd. Folkmängd 31 dec efter ålder, kön och år [Prognostics for population. Population by 31 December according to age, gender and year]. Statistiska Centralbyrån, Stockholm
24. European Commission (1999) Mandate to the European standards bodies for a guidance document in the field of safety and usability of products by people with special needs (e.g. elderly and disabled). In: European_Commission (ed) Directorate-general XXIV consumer policy and consumer health protection. European Commission, Brussels
25. Swedish Standards Institute (2006) SIS-CEN/ CENELEC Guide 6:2006 utg 1 [SIS-CEN/CENELEC Guide 6:2006, 1st issue]. Swedish Standards Institute, Stockholm
26. Government bill 2016/17:188 (2017) Nationellt mål och inriktning för funktionshinderspolitiken [Goals and directions for the Swedish disability policy]. Swedish Government Offices, Stockholm
27. Kaplan D (2000) The definition of disability: perspective of the disability community. J Health Care Law Policy 3(2000):352–364
28. SOU 2004:54 (2004) Handikappolitisk samordning - organisation för strategi och genomförande [Coordination of national disability policy, organization, strategy and realization]. Statens Offentliga Utredningar, Stockholm
29. www.skaradet.se/Pdf/Swedish_Standards_Consumers_Workers_Council.pdf
30. SOU 2007:47 (2007) Den osynliga infrastrukturen. Om förbättrad samordning av offentlig IT-standardisering [The invisible infrastructure. On improved harmonization of public ICT standardization]. Statens Offentliga utredningar, Stockholm

Construction Ergonomics: A Support Work Manufacturer's Perceptions and Practices

John Smallwood(✉)

Nelson Mandela University, PO Box 77000, Port Elizabeth 6031, South Africa
john.smallwood@mandela.ac.za

Abstract. Temporary works designers influence construction ergonomics directly and indirectly. The direct influence is because of design, details, and method of connecting, and depending upon the type of procurement system, supervisory, and administrative interventions. The indirect influence is because of the type of procurement system used, pre-qualification, project duration, partnering, and the facilitating of pre-planning.

A questionnaire survey was administered among attendees attending an inhouse support work designer and supplier 'designing for construction ergonomics' workshop.

The following constitute the salient findings. A range of temporary works design related aspects impact on construction ergonomics, and the respondents' organisation considers/refers to such aspects frequently, and on a range of design, procurement, and construction occasions. Experience predominates in terms of how ergonomics knowledge was acquired.

The paper concludes that respondents contribute to construction ergonomics, but there is potential for enhanced contributions.

Recommendations include that tertiary-built environment education should address temporary works design and construction H&S and ergonomics, temporary works design standards should highlight designing for construction H&S and ergonomics, and practice notes, and continuing professional development (CPD) should be evolved.

Keywords: Construction · Ergonomics · Support work

1 Introduction

1.1 A Subsection Sample

The definition of 'designer' in the South African Construction Regulations [1] includes, inter alia, a competent person who designs temporary work, including its components. According to the South African Construction Regulations [1], designers must take cognisance of ergonomic design principles during the design stage to minimise ergonomic related hazards in all phases of the life cycle of a structure. This alludes to the term 'designing for safety', which Behm [2] defines as "The consideration of construction site safety in the preparation of plans and specifications for construction projects." Thorpe [3] in turn contends that design is an important stage of projects, as it is at this stage that conceptual ideas are ideally converted into constructable realities.

S. Bagnara et al. (Eds.): IEA 2018, AISC 825, pp. 469–476, 2019.
https://doi.org/10.1007/978-3-319-96068-5_53

Furthermore, 'designing for H&S' is one of the designing for constructability principles. Thorpe [3] further states that designing for safety is one of a range of considerations that need to be balanced simultaneously during design.

The Construction Regulations and international literature highlight the relevance of designing for H&S and ergonomics, which resulted in a study that was conducted among staff of a major multinational temporary works designer and supplier, the objectives being to determine relative to their organisation and personally:

- Importance of ergonomics during seven temporary works stages of projects;
- Frequency at which construction ergonomics is considered on various occasions and relative to various temporary works design related aspects;
- Extent to which various temporary works design related aspects impact on construction ergonomics, and
- Rating, and source of ergonomics knowledge.

2 Review of the Literature

2.1 Health and Safety Legislation and Recommendations Pertaining to Designers

Prior to the promulgation of the Construction Regulations, all designers were required to address H&S, as in terms of Sect. 10 of the Occupational H&S Act [4] designers are allocated the responsibility to ensure that any 'article' is safe and without risks when properly used. This includes buildings and structures. In terms of the South African Construction Regulations [1], clients and designers, including temporary works designers, have responsibilities with respect to construction H&S and ergonomics.

Clients are required to, inter alia, prepare an H&S specification based on their baseline risk assessment (BRA), which is then provided to designers. They must then ensure that the designer takes the H&S specification into account during design, and that the designers carry out their duties in terms of Regulation 6 'Duties of designers'. Thereafter, clients must include the H&S specification in the tender documentation, which in theory should have been revised to include any relevant H&S information included in the designer report as discussed below.

Designers in turn are required to, inter alia: consider the H&S specification; submit a report to the client before tender stage that includes all the relevant H&S information about the design that may affect the pricing of the work, the geotechnical-science aspects, and the loading that the structure is designed to withstand; inform the client of any known or anticipated dangers or hazards relating to the construction work, and make available all relevant information required for the safe execution of the work upon being designed or when the design is changed; modify the design or make use of substitute materials where the design necessitates the use of dangerous procedures or materials hazardous to H&S, and consider hazards relating to subsequent maintenance of the structure and make provision in the design for that work to be performed to minimise the risk. To mitigate design originated hazards, requires hazard identification and risk assessment (HIRA) and appropriate responses, which process should be structured and documented.

In terms of the Draft Ergonomics Regulations [5] '5.1 Designers of machinery, equipment or articles for use at work, must: eliminate ergonomic risk factors from the design, or where this is not reasonably practicable, must minimise the ergonomic risk factors that workers may be exposed to in each possible use of the items; provide information regarding the ergonomic risk factors identified and the controls to the manufacturer, so that the manufacturer may take action where reasonably practicable to eliminate or minimise residual ergonomic risk factors, and provide information to the manufacturer for potential users involved in each phase of the lifecycle regarding the ergonomic risk factors that he/she could not eliminate, and the conditions for safe use. Although these are draft regulations, they are not onerous, and merely require design HIRAs, and appropriate responses.

Furthermore, the International Labour Office (ILO) [6] as early as 1992 recommended that designers should: receive training in H&S; integrate the H&S of construction workers into the design and planning process, and not include anything in a design which would necessitate the use of dangerous structural or other procedures or hazardous materials which could be avoided by design modifications or by substitute materials.

3 Research

A questionnaire survey was administered among staff of a major international temporary works designer and supplier attending an in house 'designing for construction ergonomics' workshop presented by the author.

A previous study conducted among engineers in South Africa to determine their perceptions and practices with respect to construction H&S investigated the: frequency at which construction H&S is considered on various occasions and relative to various design related aspects; extent to which various design related aspects impact on construction H&S; source of H&S knowledge, and the potential of various aspects to contribute to an improvement in construction H&S [7]. The study reported on constitutes a replication of this prior study, which study in turn constitutes the origin of the occasions, aspects, and sources.

Table 1 indicates the importance of ergonomics to respondents' organisations during seven temporary works stages of projects in terms of percentage responses to a scale of 1 (not important) to 5 (very important), and a MS ranging between 1.00 and 5.00. It is notable that the MSs are all above the midpoint of 3.00, which indicates that in general the respondents can be deemed to perceive the parameters as important to their organisation. However, given that the MSs for all the parameters are $>4.20 \leq 5.00$, the respondents can be deemed to perceive them to be between more than important to very important/very important. It is notable that supply of equipment is ranked first and detailed design (Stage 3) second. Concept and feasibility (Stage 2), and project initiation and briefing (Stage 1) are ranked third and fourth respectively.

Table 1. Importance of ergonomics to respondents during seven temporary works stages of projects.

Stage	Response (%)						MS	R
	Unsure	Not.........................Very						
		1	2	3	4	5		
Supply of equipment	8.7	0.0	0.0	4.3	26.1	60.9	4.62	1
Detailed design	4.3	0.0	0.0	8.7	26.1	60.9	4.55	2
Concept and feasibility	13.0	0.0	0.0	13.0	17.4	56.5	4.50	3
Project initiation and briefing	17.4	0.0	0.0	8.7	26.1	47.8	4.47	4
Construction documentation and management	8.7	0.0	4.3	4.3	39.1	43.5	4.33	5
Project close out	13.0	0.0	4.3	17.4	13.0	52.2	4.30	6
Tender documentation and procurement	13.0	0.0	0.0	13.0	39.1	34.8	4.25	7

Table 2 presents the frequency at which the respondents' organisation considers or refers to construction ergonomics on fourteen occasions in terms of percentage responses to a frequency range, never to always, and a MS ranging between 1.00 and 5.00. It is notable that all the MSs are above the midpoint of the range, namely 3.00, which indicates the consideration of or reference to construction ergonomics on these occasions can be deemed to be prevalent.

Table 2. Frequency at which respondents' organisation considers/refers to construction ergonomics on various occasions.

Occasion	Response (%)						MS	Rank
	Unsure	Never	Rarely	Sometimes	Often	Always		
Design (U)	4.3	0.0	0.0	0.0	30.4	65.2	4.68	1
Detailed design (U)	4.3	0.0	0.0	0.0	39.1	56.5	4.59	2
Discussions with the principal contractor (U, M, D)	4.3	0.0	0.0	0.0	52.2	43.5	4.45	3
Site visits/inspections (D)	4.3	0.0	0.0	8.7	39.1	47.8	4.41	4
Working drawings (U)	4.3	0.0	4.3	8.7	34.8	47.8	4.32	5
Equipment delivery (M)	4.5	0.0	9.1	0.0	40.9	45.5	4.29	6
Client (Contractor) meetings (U, M, D)	8.7	0.0	4.3	13.0	34.8	39.1	4.19	7
Project progress meetings (D)	13.0	0.0	0.0	21.7	43.5	21.7	4.00	8
Design coordination meetings (U)	9.1	0.0	13.6	4.5	40.9	31.8	4.00	9
Project close out reports (D)	26.1	0.0	8.7	13.0	30.4	21.7	3.88	10
Deliberating project duration (U)	8.7	4.3	4.3	17.4	39.1	26.1	3.86	11
Preparing project documentation (M)	4.3	4.3	4.3	26.1	43.5	17.4	3.68	12
Constructability reviews (U)	27.3	0.0	4.5	40.9	9.1	18.2	3.56	13
Discussion of H&S plan (M)	14.3	4.8	19.0	28.6	9.5	23.8	3.33	14

It is notable that 6/14 (42.9%) occasions have MSs > 4.20 ≤ 5.00 – between often to always/always. 3 are upstream, 1 is midstream, 1 is downstream, and 1 is triple-stream. 7/14 (50%) of the occasions have MSs > 3.40 ≤ 4.20 – between sometimes to often/often. 3 are 'upstream', 1 is 'midstream', 2 are 'downstream', and 1 is triple-stream. Only one MSs is >2.60 ≤ 3.40 – between rarely to sometimes/sometimes, namely discussion of H&S plan, which is midstream.

Table 3 presents the frequency at which the respondents' organisation considers/refers to construction ergonomics relative to fifteen temporary works design related aspects, in terms of percentage responses to a frequency range, never to always, and a MS ranging between 1.00 and 5.00. It is notable that all the MSs are above the midpoint of 3.00, which indicates consideration of/reference to H&S relative to these temporary works design related aspects can be deemed to be prevalent.

It is notable that 5/15 (33.3%) MSs fall within the range >4.20 ≤ 5.00 – between often to always/always, and 9/15 (60%) MSs are >3.40 ≤ 4.20 – between sometimes to often/often. The remaining MS, which is virtually in the upper range, is >2.60 ≤ 3.40 – between rarely to sometimes/sometimes.

The top six ranked occasions predominate, namely method of connecting, method of fixing, details, specification, mass of components, and design of temporary works (general).

Table 3. Frequency at which respondents' organisation considers/refers to construction ergonomics relative to various temporary works design related aspects.

Problem	Response (%)						MS	Rank
	Unsure	Never	Rarely	Sometimes	Often	Always		
Method of connecting	8.7	0.0	0.0	8.7	43.5	39.1	4.33	1
Method of fixing	8.7	0.0	0.0	8.7	43.5	39.1	4.33	2
Details	4.3	0.0	4.3	13.0	26.1	52.2	4.32	3
Specification	8.7	0.0	4.3	13.0	30.4	43.5	4.24	4
Mass of components	8.7	0.0	4.3	17.4	21.7	47.8	4.24	5
Design of temporary works (general)	4.5	0.0	4.5	13.6	36.4	40.9	4.19	6
Surface area of components	8.7	0.0	8.7	13.0	34.8	34.8	4.05	7
Finish of components	4.3	0.0	4.3	17.4	43.5	30.4	4.05	8
Elevations	4.3	0.0	8.7	21.7	26.1	39.1	4.00	9
Position of components	8.7	0.0	8.7	17.4	30.4	34.8	4.00	10
Plan layout	4.3	0.0	17.4	8.7	26.1	43.5	4.00	11
Sectional area of components	13.0	0.0	8.7	17.4	30.4	30.4	3.95	12
Site location	8.7	4.3	4.3	26.1	34.8	21.7	3.71	13
Edge (s) of components	17.4	0.0	17.4	17.4	26.1	21.7	3.63	14
Texture of components	21.7	4.3	17.4	21.7	13.0	21.7	3.39	15

Table 4. Extent to which various temporary works design related aspects impact on construction ergonomics.

Aspect	Response (%)							MS	R
	Unsure	Does not	Minor......................... Major						
			1	2	3	4	5		
Details	4.3	0.0	0.0	4.3	4.3	30.4	56.5	4.45	1
Method of connecting	8.7	0.0	0.0	0.0	13.0	26.1	52.2	4.43	2
Method of fixing	8.7	0.0	0.0	0.0	13.0	26.1	52.2	4.43	3
Specification	8.7	0.0	0.0	0.0	17.4	30.4	43.5	4.29	4
Plan layout	4.3	0.0	0.0	4.3	13.0	30.4	47.8	4.27	5
Design of temporary works (general)	4.3	0.0	0.0	8.7	4.3	34.8	47.8	4.27	6
Mass of components	4.3	0.0	0.0	0.0	26.1	21.7	47.8	4.23	7
Elevations	8.7	0.0	0.0	4.3	17.4	30.4	39.1	4.14	8
Surface area of components	13.0	4.3	0.0	4.3	21.7	21.7	34.8	3.85	9
Finish of components	9.1	4.5	0.0	4.5	22.7	31.8	27.3	3.75	10
Site location	8.7	8.7	0.0	4.3	17.4	26.1	34.8	3.71	11
Position of components	13.0	4.3	8.7	0.0	17.4	26.1	30.4	3.65	12
Sectional area of components	17.4	4.3	8.7	0.0	17.4	26.1	26.1	3.58	13
Edge (s) of components	21.7	4.3	4.3	4.3	21.7	17.4	26.1	3.56	14
Texture of components	8.7	4.3	4.3	4.3	34.8	21.7	21.7	3.43	15

Table 4 indicates the perceived impact of fifteen temporary works design related aspects on construction ergonomics, in terms of percentage responses to 'does not' and a scale of 1 (minor) to 5 (major), and a MS ranging between 0.00 and 5.00. Given that a 'does not' option was provided, the scale effectively consists of six points, and hence the MS range. It is notable that all fifteen MSs are above the midpoint of 2.50, which indicates the respondents perceive the design related aspects to impact on construction ergonomics.

It is notable that 7/15 (46.7%) MSs are $>4.17 \leq 5.00$ – between a near major to major impact/major impact. The remaining 8/15 (53.3%) aspects' MSs are $>3.34 \leq 4.17$, which indicates that they have between an impact and a near major impact/near major impact on construction ergonomics.

Table 5. Respondents' self-rating of their knowledge with respect to ergonomics aspects.

Aspect	Response (%)						MS	R
	Unsure	Limited.......................Extensive						
		1	2	3	4	5		
Ergonomics	4.3	21.7	0.0	56.5	13.0	4.3	2.77	1
Designing for construction ergonomics	4.3	21.7	8.7	43.5	13.0	8.7	2.77	2
Construction ergonomics	4.3	17.4	8.7	56.5	8.7	4.3	2.73	3

Experience (39.1%) predominates in terms of respondents' source of ergonomics knowledge, followed by workshops (26.1%), tertiary education (17.4%), magazine articles (13.1%), and other (13.1%). Post graduate qualifications were only identified by 4.1% of respondents, and conference papers, CPD seminars, journal papers, and practice notes by no respondents. Respondents' source of ergonomics knowledge is informal as opposed to formal.

Table 5 indicates the respondents' self-rating of their knowledge of ergonomics, construction ergonomics, and 'designing for ergonomics' skills in terms of percentage responses to a scale of 1 (limited) to 5 (extensive), and a MS ranging between 1.00 and 5.00. Given that the MSs are ≤ 3.00, the knowledge can be deemed to be limited as opposed to extensive. However, all three MSs are $>2.60 \leq 3.40$ – less than average to average/average.

4 Conclusions

Construction ergonomics is more important during the supply of equipment, detailed design, concept and feasibility, and project initiation and briefing temporary works stages of projects, than the midstream and downstream stages of construction documentation and management, project close out, and tender documentation and procurement. Therefore, it can be concluded that the respondents' organisation understands and appreciates that ergonomics can be influenced more during the upstream than downstream stages. This is underscored by the extent the respondents' organisation could influence construction ergonomics during the detailed design, supply of equipment, and concept and feasibility temporary works stages of projects.

The respondents' organisation does consider construction ergonomics on various occasions, however, more so during upstream phases than mid-stream phases, design (upstream), detailed design (upstream), and discussions with the principal contractor (upstream, midstream, downstream). Therefore, it can be concluded that the cited importance thereof does manifest itself. Furthermore, the MSs relative to the top 6/14 (42.9%) indicate a frequency of often to always/always.

The respondents' organisation considers/refer to construction ergonomics relative to fifteen temporary works design related aspects. The top five (33.3%) aspects, namely method of connecting, method of fixing, details, specification, and mass of components, have MSs, which indicate a frequency of often to always/always. The frequency relative to mass of components is notable due to the manual handling of components, and relative to the other aspects, which impact on construction ergonomics.

Respondents do appreciate the extent to which various temporary works design related aspects impact on construction ergonomics in that they maintain 7/15 (46.7%) aspects have between a near major to major impact/major impact, and 8/15 (53.3%) an impact to near major impact/near major impact thereon.

Given the sources of respondents' ergonomics knowledge it can be concluded that the sources are more informal than formal. It can also be concluded that tertiary built environment education and the related professions are not addressing ergonomics to the extent that they should. These conclusions are reinforced by the respondents' 'below average' self-rating of their knowledge of 'ergonomics', 'designing for construction ergonomics', and 'construction ergonomics' skills.

5 Recommendations

Tertiary built environment education should address temporary works design and construction H&S and ergonomics, and highlight the role thereof in overall project performance. Furthermore, designing for construction H&S and ergonomics, temporary works design included, should be introduced and more importantly, embedded in tertiary built environment education programmes.

Temporary works design standards should highlight designing for construction H&S and ergonomics, and practice notes, and continuing professional development (CPD) should be evolved. The Ergonomics Regulations should be promulgated the soonest.

References

1. Republic of South Africa (2014) No. R. 84 Occupational Health and Safety Act, 1993 Construction Regulations 2014. Government Gazette No. 37305, Pretoria
2. Behm M (2006) An analysis of construction accidents from a design perspective. The Center to Protect Workers' Rights, Silver Spring
3. Thorpe B (2006) Health and safety in construction design. Gower Publishing Limited, Aldershot
4. Republic of South Africa (1993) Government Gazette No. 14918. Occupational Health & Safety Act: No. 85 of 1993, Pretoria
5. Republic of South Africa (2017) Draft Ergonomics Regulations No. 40578. Pretoria
6. International Labour Office (ILO) (1992) Safety and health in construction. ILO, Geneva
7. Smallwood JJ (2004) The influence of engineering designers on health and safety during construction. J S Afr Inst Civ Eng 46(1):2–8

Construction Ergonomics: Construction Health and Safety Agents' (CHSAs') Perceptions and Practices

John Smallwood(✉) and Claire Deacon

Nelson Mandela University, Port Elizabeth 6031, South Africa
john.smallwood@mandela.ac.za

Abstract. Construction entails exposure to a range of ergonomics hazards and risks: bending, lifting, repetitive movement, and vibration; environmental stresses such as heat, sun, noise, poor illumination, and wet or damp work; skin and respiratory exposure to chemicals and dust, as well as mental stress. In accordance with the South African Construction Regulations, clients may appoint CHSAs to fulfil their functions, which requires CHSAs to interface with and guide clients and designers, and conduct interventions during the construction process.

A self-administered questionnaire survey was conducted among CHSAs registered with the South African Council for the Construction and Project Management Professions (SACPCMP) to determine, inter alia, CHSAs' perceptions and practices, and CHSAs' source of knowledge.

Findings and conclusions include: CHSAs understand and appreciate that construction ergonomics can be influenced during all the stages of projects, and especially design development; CHSAs do consider/refer to construction ergonomics on various occasions, design included, and relative to various design related aspects, which indicates that it is important to CHSAs. CHSAs do have an understanding and appreciation of the impact of design related aspects on construction ergonomics. CHSAs are 'lacking' the requisite competencies, as they rate themselves as average in terms of their knowledge with respect to ergonomics, and their source of ergonomics knowledge being informal as opposed to formal.

Recommendations include: CHSAs should be appointed during the initial stages of projects; CHSAs should register for and complete formal tertiary education programmes, and appropriate continuing professional development (CPD) courses should be evolved.

Keywords: Construction · Ergonomics · Perceptions · Practices

1 Introduction

1.1 A Subsection Sample

The report 'Construction Health & Safety Status & Recommendations' highlighted the considerable number of accidents, fatalities, and other injuries that occur in the South African construction industry [1]. The report cited the extensive non-compliance with

S. Bagnara et al. (Eds.): IEA 2018, AISC 825, pp. 477–485, 2019.
https://doi.org/10.1007/978-3-319-96068-5_54

H&S legislative requirements, which is indicative of a deficiency of effective management and supervision of H&S on construction sites as well as planning from the inception/conception of projects within the context of project management. The report also cited a lack of sufficiently skilled, experienced, and knowledgeable persons to manage H&S on construction sites. The Council for the Built Environment (CBE) then mandated the South African Council for the Project and Construction Management Professions (SACPCMP) in terms of Act No. 48 [2] to register construction H&S professionals. This in turn led to the identification of three such categories of registration, namely Professional Construction Health and Safety Agent (Pr CHSA).

Given the function of CHSAs, the draft Ergonomics Regulations, the influence of construction project management, design, and procurement, in addition to construction, on ergonomics, an exploratory study was conducted to determine, inter alia, the:

- CHSAs' perceptions and practices relative to construction ergonomics;
- Potential of aspects/interventions to contribute to an improvement in construction ergonomics during the various project phases, and
- CHSAs' source of ergonomics knowledge.

2 Review of the Literature

2.1 Legislation

The amended Construction Regulations [3], lay down important requirements with respect to clients and designers, and contractors.

Clients are required to, inter alia: prepare a baseline risk assessment (BRA); prepare an H&S specification based on the BRA; provide the designer with the H&S specification; ensure that the designer takes the H&S specification into account during design; ensure that the designer carries out the duties in Regulation 6 'Duties of designers', and include the H&S specification in the tender documents. In theory, the H&S specification should schedule the residual hazards on projects. Contractor related client requirements include: ensure that potential PCs have made provision for the cost of H&S in their tenders; ensure that the PC to be appointed has the necessary competencies and resources; take reasonable steps to ensure cooperation between all contractors appointed by the client; ensure that every PC is registered for workers' compensation insurance cover and in good standing; appoint every PC in writing; discuss and negotiate with the PC the contents of the PC's H&S plan and thereafter approve it; ensure that a copy of the PC's H&S plan is available; take reasonable steps to ensure that each contractor's H&S plan is implemented and maintained; ensure that periodic H&S audits and documentation verification are conducted at agreed intervals, but at least once every 30 days; ensure that a copy of the H&S audit report is provided by the PC within seven days of the audit; stop any contractor from executing an activity which posed a threat to the H&S of persons, which is not in accordance with the H&S specification and H&S plan; when changes are made to the design or construction work provide sufficient H&S information and resources available to the PC; ensure that the H&S file is kept and maintained by the PC; when additional work is required the client must ensure that sufficient H&S information and appropriate additional resources are available to execute the work safely; in the case of a fatality or permanent disabling injury the client must ensure that the contractor provides the provincial director with a report that includes the measures that the contractor intends to implement to ensure a healthy and safe construction site, and must ensure co-operation between all principal contractors and contractors. Furthermore, where a construction work permit is required, a client must appoint a competent person in writing as an agent, and where notification of construction work is required the client may appoint a competent person in writing as an agent. However, an agent must manage the H&S on a construction project, and be registered with a statutory body. Clearly the requirements of clients are onerous, given that they are invariably not built environment professionals, or H&S professionals.

Designers are required to, inter alia: ensure that the H&S standards incorporated into the regulations are complied with in the design; take the H&S specification into consideration; include in a report to the client before tender stage all relevant H&S information about the design that may affect the pricing of the work, the geotechnical-science aspects, and the loading that the structure is designed to withstand; inform the client of any known or anticipated dangers or hazards relating to the construction work, and make available all relevant information required for the safe execution of the work upon being designed or when the design is changed; modify the design or make use of substitute materials where the design necessitates the use of dangerous procedures or materials hazardous to H&S, and consider hazards relating to subsequent maintenance of the structure and make provision in the design for that work to be performed to minimize the risk. Therefore, hazard identification (HIRA) is of relevance in terms of the requirement to modify the design or make use of substitute materials where the design necessitates the use of dangerous procedures or materials hazardous to H&S. Furthermore, the report that is required to be submitted to the client should schedule the residual hazards on projects, which in turn should be included in the H&S specification.

In terms of the Draft Ergonomics Regulations [4] '5.1 Designers of machinery, equipment or articles for use at work, must: eliminate ergonomic risk factors from the design, or where this is not reasonably practicable, must minimise the ergonomic risk factors that workers may be exposed to in each possible use of the items; provide information regarding the ergonomic risk factors identified and the controls to the manufacturer, so that the manufacturer may take action where reasonably practicable to eliminate or minimise residual ergonomic risk factors, and provide information to the manufacturer for potential users involved in each phase of the lifecycle regarding the ergonomic risk factors that he/she could not eliminate, and the conditions for safe use. Although these are draft regulations, they are not onerous, and merely require design HIRAs, and appropriate responses.

3 Research

A previous study conducted among engineers in South Africa to determine their perceptions and practices with respect to construction H&S investigated the: frequency at which construction H&S is considered on various occasions and relative to various design related aspects; extent to which various design related aspects impact on construction H&S, and source of H&S knowledge [5]. The study reported on constitutes a part-replication of this prior study, which study in turn constitutes the origin of the occasions, aspects, and sources. A self-administered questionnaire was circulated to 40 CHSAs registered with the SACPCMP. It consisted of 23 questions, 22 of which were close ended, one being open ended. 10 of the 22 close ended questions were five- or six-point Likert scale type questions. 14 Questionnaires were included in the analysis of the data, which equates to a response rate of 35%.

Industrial (28.1%), and infrastructure (27.9%) predominate in terms of the type of construction respondents provided CHSA services relative to, followed by commercial (20.4%), residential (12.1%), and other (11.5%).

50% of CHSAs reported to clients, 35.7% to construction project managers (CPMs), and 14.3% to other. However, 78.6% of CHSAs were paid by clients, 14.3% by principal agents, and 7.1% by CPMs.

Table 1 indicates the importance of ergonomics to respondents during the six stages of projects in terms of percentage responses to a scale of 1 (not important) to 5 (very important), a mean score (MS) ranging between 1.00 and 5.00, and a rank (R). It is notable that all the MSs are > 3.00, which indicates that in general, the respondents perceive ergonomics to be important during the six project stages.

Table 1. Importance of ergonomics to respondents during the six stages of projects.

Stage	Response (%)						MS	R
	Unsure	Not..........................			.Very			
		1	2	3	4	5		
Design development	0.0	0.0	0.0	7.1	21.4	71.4	4.64	1
Construction documentation and management	0.0	0.0	0.0	35.7	28.6	35.7	4.00	2
Tender documentation and management	0.0	0.0	7.1	50.0	28.6	14.3	3.50	3
Concept and feasibility	0.0	7.1	7.1	35.7	28.6	21.4	3.50	4
Project close out	0.0	14.3	7.1	28.6	21.4	28.6	3.43	5
Project initiation and briefing	0.0	14.3	21.4	14.3	28.6	21.4	3.21	6

Design development is the only stage that falls within the range > 4.20 $\leq$ 5.00 – between more than important to very important/very important. Construction documentation and management, tender documentation and management, concept and feasibility, and project close out in turn fall within the range > 3.40 $\leq$ 4.20 – between important to more than important/more than important. Project initiation and briefing falls within the range > 2.60 $\leq$ 3.40 – between less than important to important/important.

Table 2 presents the frequency at which respondents consider or refer to construction ergonomics on seventeen occasions in terms of a frequency range, never to always, and a MS ranging between 0.00 and 5.00. It is notable that all the MSs are > 3.00, which indicates the consideration of or reference to construction ergonomics on these occasions is frequent as opposed to infrequent.

Table 2. Frequency at which respondents consider/refer to construction ergonomics on various occasions.

Problem	Response (%)							MS	Rank
	Unsure	Cannot	Never	Rarely	Some-times	Often	Always		
Discussions with the principal contractor (D)	0.0	0.0	0.0	7.1	7.1	50.0	35.7	4.14	1
Site audits (D)	0.0	0.0	0.0	7.1	14.3	35.7	42.9	4.14	2
Discussion and approval of H&S plan (M)	0.0	0.0	0.0	14.3	7.1	42.9	35.7	4.00	3
Audit reports (D)	0.0	0.0	0.0	7.1	21.4	42.9	28.6	3.93	4
Design (U)	0.0	0.0	0.0	7.1	42.9	21.4	28.6	3.71	5
Detailed design (U)	0.0	0.0	0.0	14.3	21.4	42.9	21.4	3.71	6
Project progress meetings (D)	0.0	0.0	0.0	7.1	42.9	21.4	28.6	3.71	7
Design coordination meetings (U)	0.0	0.0	0.0	7.1	42.9	28.6	21.4	3.64	8
Site handover (M)	0.0	0.0	0.0	14.3	35.7	28.6	21.4	3.57	9
Preparing project documentation (M)	0.0	0.0	0.0	21.4	21.4	35.7	21.4	3.57	10
Constructability reviews (U)	0.0	0.0	0.0	7.1	42.9	42.9	7.1	3.50	11
Client meetings (U)	0.0	0.0	0.0	15.4	30.8	46.2	7.7	3.46	12
Evaluating tenders (M)	0.0	0.0	0.0	35.7	14.3	21.4	28.6	3.43	13
Pre-qualifying contractors (M)	0.0	0.0	7.1	28.6	21.4	28.6	14.3	3.14	14
Pre-tender meeting (M)	0.0	0.0	0.0	35.7	28.6	28.6	7.1	3.07	15
Working drawings (U)	0.0	7.1	0.0	14.3	57.1	14.3	7.1	2.93	16
Deliberating project duration (U)	0.0	7.1	7.1	14.3	42.9	28.6	0.0	2.79	17

It is notable that no occasions have MSs $> 4.17 \leq 5.00$ – between often to always/always. 13/17 (76.5%) MSs are $> 3.33 \leq 4.17$ – between sometimes to often/often. It is notable that 3/4 (75%) of the occasions that have MSs > 3.76 are 'downstream' occasions. 5/9 (55.5%) occasions that have MSs $> 3.33 \leq 3.76$ are 'upstream', 3/9 (33.3%) are 'midstream', and 1/9 (11.1%) is a 'downstream' occasion. The last four (23.5)% ranked 'occasions' MSs are $> 2.50 \leq 3.33$ – between rarely to sometimes/sometimes. Two (50%) are 'midstream' occasions, and two (50%) are 'upstream' occasions.

Table 3 indicates the perceived extent to which sixteen design related aspects impact on construction ergonomics, in terms of percentage responses to 'does not' and a scale of 1 (minor) to 5 (major), and a MS ranging between 0.00 and 5.00. It is notable that all sixteen MSs are above the midpoint of 2.50, which indicates the respondents perceive the design related aspects to impact on construction ergonomics.

Table 3. Extent to which various design related aspects impact on construction ergonomics.

Problem	Response (%)							MS	Rank
	Unsure	Cannot	Never	Rarely	Sometimes	Often	Always		
Method of fixing	0.0	0.0	0.0	0.0	21.4	35.7	42.9	4.21	1
Position of components	0.0	0.0	0.0	7.1	7.1	50.0	35.7	4.14	2
Edge of materials	7.1	0.0	0.0	0.0	28.6	28.6	35.7	4.08	3
Design (general)	0.0	0.0	0.0	0.0	28.6	35.7	35.7	4.07	4
Type of structural frame	0.0	0.0	7.1	0.0	14.3	42.9	35.7	4.00	5
Specification	0.0	0.0	0.0	0.0	28.6	50.0	21.4	3.93	6
Details	0.0	0.0	7.1	0.0	7.1	71.4	14.3	3.86	7
Elevations	0.0	0.0	0.0	7.1	28.6	42.9	21.4	3.79	8
Mass of materials	0.0	7.1	0.0	14.3	14.3	21.4	42.9	3.71	9
Surface area of materials	0.0	0.0	7.1	14.3	14.3	35.7	28.6	3.64	10
Content of material	0.0	7.1	0.0	14.3	14.3	28.6	35.7	3.64	11
Plan layout	0.0	7.1	0.0	0.0	28.6	50.0	14.3	3.57	12
Finishes	0.0	7.1	0.0	7.1	28.6	50.0	7.1	3.36	13
Schedule	0.0	7.1	0.0	7.1	35.7	50.0	0.0	3.21	14
Texture of materials	0.0	7.1	0.0	21.4	21.4	35.7	14.3	3.21	15
Site location	0.0	7.1	7.1	28.6	14.3	42.9	0.0	2.79	16

It is notable that only one MS, namely method of fixing, is > 4.17 $\leq$ 5.00 - between a near major to major impact/major impact. 12/16 (75%) aspects' MSs are > 3.33 $\leq$ 4.17, which indicates that they have between an impact and a near major impact/near major impact on construction ergonomics.

Notable rankings include method of fixing (first), design (general) (fourth), type of structural frame (fifth), finishes (eighth), mass of materials (ninth), and surface area of materials (tenth), as these have a major effect in terms of manual handling.

Table 4 presents the frequency at which respondents consider/refer to construction ergonomics relative to sixteen design related aspects, in terms of a frequency range, never to always, and a MS ranging between 0.00 and 5.00. It is notable that all the MSs

Table 4. Frequency at which respondents consider/refer to construction ergonomics relative to various design related aspects.

Problem	Response (%)							MS	Rank
	Unsure	Cannot	Never	Rarely	Some-times	Often	Always		
Specification	0.0	0.0	0.0	14.3	14.3	42.9	28.6	3.86	1
Method of fixing	0.0	0.0	0.0	21.4	7.1	42.9	28.6	3.79	2
Design (general)	0.0	0.0	0.0	7.1	35.7	42.9	14.3	3.64	3
Details	7.1	0.0	0.0	21.4	14.3	50.0	7.1	3.46	4
Plan layout	0.0	0.0	0.0	14.3	42.9	28.6	14.3	3.43	5
Mass of materials	0.0	0.0	7.1	21.4	21.4	21.4	28.6	3.43	6
Edge of materials	7.1	0.0	0.0	21.4	28.6	28.6	14.3	3.38	7
Position of components	0.0	0.0	7.1	14.3	28.6	35.7	14.3	3.36	8
Elevations	0.0	0.0	0.0	28.6	28.6	21.4	21.4	3.36	9
Type of structural frame	0.0	7.1	0.0	28.6	14.3	21.4	28.6	3.29	10
Surface area of materials	7.1	0.0	0.0	35.7	21.4	28.6	7.1	3.08	11
Site location	0.0	0.0	7.1	21.4	42.9	21.4	7.1	3.00	12
Content of material	0.0	0.0	14.3	21.4	28.6	21.4	14.3	3.00	13
Finishes	0.0	7.1	7.1	21.4	21.4	28.6	14.3	3.00	14
Schedule	0.0	7.1	0.0	42.9	21.4	21.4	7.1	2.71	15
Texture of materials	0.0	0.0	14.3	35.7	28.6	14.3	7.1	2.64	16

are above the midpoint of 2.50, which indicates consideration of/reference to H&S relative to these design related aspects can be deemed to be prevalent.

It is notable that no occasions fall within the range $> 4.17 \leq 5.00$ – between often to always/always. 9/16 (56.3%) MSs are $> 3.33 \leq 4.17$ – between sometimes to often/often. The remaining 7/16 (62.5%) aspects' MSs are $> 2.50 \leq 3.33$ – between rarely to sometimes/sometimes.

Specification, method of fixing, and design (general) predominate. It is notable that type of structural frame is ranked tenth as it is the stage that impacts most on construction ergonomics (Smallwood 2002). Along with design (general) it provides the framework for a project in terms of construction ergonomics. Given that certain materials contain hazardous chemical substances it is notable that content of material achieved a ranking of thirteenth. Furthermore, given that materials handling, and more specifically the mass of materials contribute to manual materials handling, it is also notable that mass of materials has a MS of 3.43, higher than normal. Similarly, given that the surface area of many materials required for certain elements such as gypsum boards for ceilings and partitions, and glazing for shop fronts is large, the MS of 3.08 is notable. However, it should be noted that finishes and schedule, which encapsulate materials and processes, achieved rankings of fourteenth and fifteenth respectively.

Experience (22.7%) predominates in terms of respondents' source of ergonomics knowledge, followed by practice notes (13.6%), CPD seminars (11.4%), workshops (11.4%), conference papers (9.1%), magazine articles (9.1%), journal papers (6.8%), and other (6.8%). Post graduate qualifications and tertiary education were both identified by only 4.5% of respondents. Respondents' source of ergonomics knowledge is informal as opposed to formal.

Table 5 indicates the respondents' self-rating of their knowledge of ergonomics, construction ergonomics, and 'designing for ergonomics' skills in terms of percentage responses to a scale of 1 (limited) to 5 (extensive), and a MS ranging between 1.00 and 5.00. Given that the three MSs are $> 2.60 \leq 3.40$, the self-rating can be deemed to be between less than average to average/average. Furthermore, the ergonomics, and construction ergonomics MSs are both 3.36, which are close to the upper extremity of the range. The 7.1% 'unsure' response is notable.

Table 5. Respondents' self-rating of their knowledge with respect to ergonomics aspects.

Aspect	Response (%)						MS	R
	Unsure	Limited........................Extensive						
		1	2	3	4	5		
Ergonomics	0.0	0.0	0.0	64.3	35.7	0.0	3.36	1
Construction ergonomics	0.0	0.0	0.0	64.3	35.7	0.0	3.36	2
Designing for construction ergonomics	7.1	0.0	14.3	57.1	21.4	0.0	3.08	3

4 Conclusions

CHSAs understand and appreciate that construction ergonomics can be influenced during all the stages of projects, and especially design development.

CHSAs do consider/refer to construction ergonomics on various occasions, design included, and relative to various design related aspects, which indicates that it is important to CHSAs.

CHSAs do have an understanding and appreciation of the impact of design related aspects on construction ergonomics.

CHSAs are 'lacking' the requisite competencies, as they rate themselves as average in terms of their knowledge with respect to ergonomics. This in turn is likely to be attributable to their source of ergonomics knowledge being informal as opposed to formal.

5 Recommendations

CHSAs should be appointed at Stage 1 'Project initiation and briefing', Stage 2 'Concept and feasibility', and Stage 3 'Design development', in addition to the subsequent stages.

CHSAs should register for and complete formal tertiary education programmes that empower them in terms of construction economics, management, H&S, and science and technology as well as design management, procurement management, and project management.

Continuing professional development (CPD) courses should be evolved relative to the subject areas.

References

1. Construction Industry Development Board (cidb) (2009) Construction Health & Safety in South Africa Status & Recommendations. cidb, Pretoria
2. Republic of South Africa (2000) Project and construction management professions act 2000. No. 48 of 2000. Pretoria
3. Republic of South Africa (2014) No. R. 84 occupational health and safety act, 1993 construction regulations 2014. Government Gazette No. 37305. Pretoria
4. Republic of South Africa (2017) Draft ergonomics regulations No. 40578. Pretoria
5. Smallwood JJ (2004) The influence of engineering designers on health and safety during construction. J S Afr Inst Civ Eng 46(1):2–8

Environmental Design and Human Performance. A Literature Review

Erminia Attaianese(✉)

University of Naples Federico II, Via Toledo 402, Naples, Italy
erminia.attaianese@unina.it

Abstract. Ambient conditions inside the buildings may work as positive stimuli for occupants, or as stressors, given that the human body is regulated by physiological and psychological processes, that react to the physical factors. So, air, warmth or cold, lighting, daylight and views, sound and acoustic setting, layout and building details, must be seen today as important architecture elements to manage and control for designing places not only healthy and comfortable, but also effective and productive, because it may contribute to human performance and productivity. However, while references on occupant health and comfort in buildings results plentiful, literature on the effects of physical factors on human performance appears dated, scarce and inhomogeneous, and thus difficult to consult. The paper presents a literature review aimed at starting a state of the art about effects of physical factors of building interiors on occupant's performance. Particularly, the purpose is to overview environmental conditions and interior details of buildings, that have been experimented and resulted to have effect on occupant performance and ability, considering which category of occupant have been involved and which kind of ability have been observed; in which kind of buildings these observations have been made. As result, an updated selection of research about effects of buildings interiors feature on occupant performance is identified, based on which more confirmed research aspects emerge and those to be further developed detected.

Keywords: Environment · Physical factor · Cognitive performance
Building interiors

1 Introduction

A broad evidence-based literature, emerging from studies and experimentations carried out in the last fifty years in different research areas, confirms that the environment, particularly the built environment, plays a central role for occupant's health, comfort and well-being [1]. More recently, the increasing relevance of sustainability issues, is highlighting that a building should function productively, in a wider sense, contributing itself to the productivity of the hosted organization in different ways, some of them also including occupants [2]. Thus, the association of indoor environmental quality to occupants has been supposed not only in terms of health and comfort, but also as job performance, as ill-health and absenteeism costs are reported as relevant in decreasing productivity and relevant losses or gains are attributable to the design, management and

S. Bagnara et al. (Eds.): IEA 2018, AISC 825, pp. 486–495, 2019.
https://doi.org/10.1007/978-3-319-96068-5_55

use of the indoor environment. However, while references on occupant health and comfort in buildings results plentiful, literature on the effects of physical factors on human performance appears dated, scarce and inhomogeneous, and thus difficult to consult. The aim of this paper is to review a selection of existing studies that examine the effects of physical factor of built environment on human perceptions and behavior, particularly regarding outcomes on performance.

2 Conceptual Framework

2.1 Person-Environment Relationships Model

Defined within environmental behavior research since the 1960th, theories on person-environment relationships focus on the dynamics and effects that built environment factors may have on humans, directly and indirectly [3]. From an almost automatic stimulus-response model to more complex interdependent interaction between humans and its immediate environment, all theories agree that built environment is a set of human and physical interdependent elements, acting to support, obstacle, neutralize or negate conscious and unconscious behaviors. By integrating Smith and Barret et al. [3, 4], a conceptual framework of person-environment relationships is defined and developed in Fig. 1, describing how built environment may affect human perception, affect and cognition. Human behaviors are triggered off different circumstances. The starting performance of a person acting inside a designed space, derives from its personal features such as ability, gender, age, motivation and socioeconomic attributes, but is also influenced not only by the task demand that he or she must perform, but also by elements of the built environment. These factors arouse different type of outcomes, integrating physiological, cognitive and affective responses, by which human behaviors result. Effects are due to both short-term and long-term reactions [5], that may induce temporary or permanent states of stress/well-being, comfort/discomfort, illness/health and activate emotions, moods and cognitive responses, influencing, then, performance.

2.2 Human Performance and Task

Performance is a multi-dimensional concept, that includes process aspects referred to behaviors, and outcome aspects referred to results of these behaviors. By a goal-oriented approach, performance concerns those actions, which process is specifically defined in tasks, and whose effects are countable. Performance outcome are affected by ability, motivation and opportunity [6]. Effectiveness refers to the evaluations of the results of performance. Productivity is the ratio of effectiveness to the cost of attaining the outcome. A basic distinction between different type of performance has been made. *Task* performance refers to an in-role, predicted behavior, requiring expected abilities; *contextual* performance refers to extra-role behavior, discretional, sustained by motivation and personality. Moreover, *adaptive* performance has been defined, for indicating human capability to front changing and unpredictable situations [6]. Speed, accuracy and attentional demand in performing a task are considered elements to measure the quality of performance, assuming that creating a context to foster rapid,

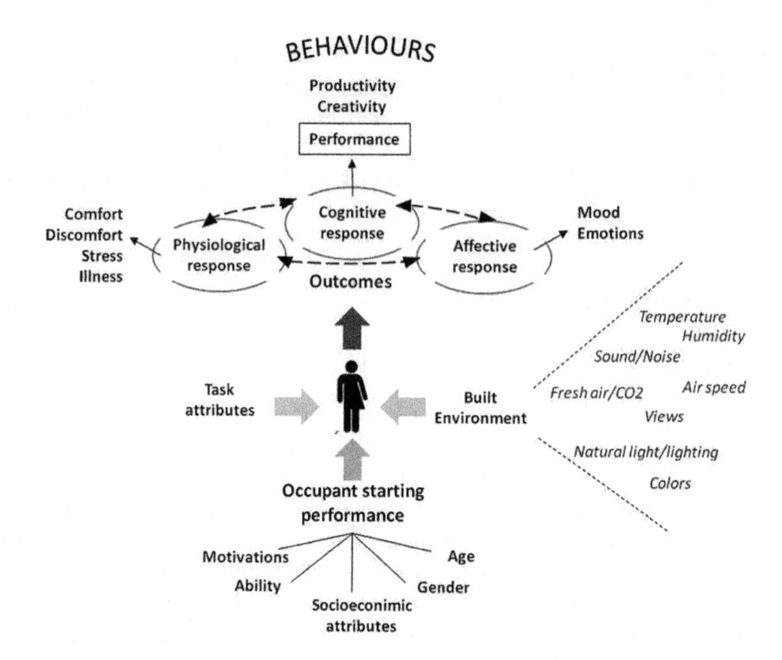

Fig. 1. Person-environment conceptual framework (adapted by Barret et al., 2015)

accurate and requiring less attentional demand tasks may be preferable in work environment [7], also taking into account human variability in job performance [8]. Some cognitive abilities are directly and indirectly reflected in performance, such as: perception, involving recognition and perception of sensory stimuli; attention, concerning ability of sustain concentration, including situational awareness; long-term and short-term memory; visual and spatial processing, also related to mapping and mental models; executive function, such as working memory, sequencing; flexibility; problem solving and decision-making, also including, obviously, motor skills [9].

2.3 Built Environment Factors

As it is the physical context where human tasks are performed, built environment has a crucial role in human performance quality. The elements in the physical space that can be identified as affecting fit or misfit between person and environment at work or in any other living situation, include both physical indoor parameters (i.c. temperature, humidity, air speed, CO_2. light, sound) and finishes, materials, space dimensions and interiors details [2]. In working environment, for example, performance is the result of job demands that require physical, mental or emotional effort [1]. Elements of the built environment may impact on job demands in terms of both physical and cognitive workload, and may act as job resources, creating condition supporting task performance. Noise and lighting in office, for instance, or close distance between workstations, affect attentional processes, increasing cognitive demand; dimensions and layout of workspace can enhance distances to reach and thus amplify required movements, both worsening workload. Likewise, diffused sound favors concentration, and adequate office furniture location may improve working condition lighting cognitive and physical efforts.

3 Research Method

3.1 Aims of the Literature Review

The purpose of the study is to overview researches about the effects of indoor environment factors of built environment on humans, with particular regard to the outcome on human performance. Particularly results were expected in terms of cognitive, physiological and affective response, influence the goal of performance. To this scope, the relations represented in conceptual framework (Fig. 1) oriented the search criteria. Formulation of the search terms were based on binomials "human performance AND single physical factor", and "single human performance AND single physical factor". Quantitative studies have been preferred to qualitative ones.

3.2 Literature Search and Selection

Literature selection was done within different search scopes such as Environmental Psychology, Interior Design, Environmental Design, Engineering Psychology, Human Factors/Ergonomics. Main inclusion criterion of literature has been papers that report evidence about the effect of an environment physical factor on a cognitive performance, independently to the type of the study (empirical or narrative), the context of investigation (working, learning or marketing) and the category of occupant involved (pupils, students; employees, customers). By applying a goal-oriented approach, evidences that prove effects of physical factors on the quality of a specific task performance, were preferred. For this reason, literature focuses only on comfort assessment, without its influence on performance, has been excluded (Table 1).

Table 1. Selected studies with regard to the effect of environment physical factors on human

Human performance	Number of selected paper	Narrative reviews
Cognitive performance	12	4
Generic cognitive task	5	2
Learning performance	6	–
Creativity	7	2
Communication and Socialization	8	2
Mood	21	4
Stress	3	2

4 Results

The greatest number of references resulted about mood (21) and cognitive performance (17), referred to individual cognitive skills (12) and generic cognitive load (5). Effects of physical factors have been proved on specific cognitive skills: memory, problem solving, decision making, motor skills and perception, attention and concentration, visual and spatial processing, processing speed (Tables 2 and 3).

Table 2. Results of the literature selection with regard to the effects on cognitive performance/1

Cognitive performance	Physical factor	Evidence
Memory	Light	Cool-white compared to warm-white lighting impairs the long-term memory recall of a novel text
		The combination of color temperature and illuminance that best preserved the positive mood enhanced performance in free recall tasks
	Noise	Learning from a text was disrupted by meaningful, irrelevant, conversational speech
	Colors	Warm colors and high brightness level can facilitate memory of spaces
	Heat	Higher degrees of temperature (32 °C–38 °C) led to slower acquisition and poor retention of the task as compared to the moderate level of temperature (25 °C). But lower degrees (7 °C–15 °C) of temperature did not have any significant negative effect on acquisition and retention
	Noise + Heat	Interactions were found between noise and heat effects on the long-term recall of a text
	Noise + Light	Interactions were found and between noise and light on the free recall of emotionally toned words
Problem solving	Colors	A color temperature which induced the least negative mood enhanced the performance in the problem-solving tasks
	Heat	The performance of two numerical and two language-based tests was significantly improved when the temperature was reduced from 25 °C to 20 °C
Decision making	Colors	Color improve decision making for first time customer in shops
Motor skills	Colors	Children strength decreased if they had been in the pink room, but increased after being in the blue room
Perception	Colors	When subjects perceived red as a warning cue in performance related task, drawing away attention of high task demand and moderating the perception of perceived task

Table 3. Results of the literature selection with regard to the effects on cognitive performance/2

Cognitive performance	Physical factor	Evidence
Attention and concentration	Light	Variable light could directly reduce restlessness
		With blue-enriched white light, in comparison to standard lighting conditions, students showed better concentration
	Noise	Attention worked faster in noise but with less accuracy
	Colors	A red office is perceived more distracting than a white office
		When performing a high-demanding task (e.g., managerial tasks) in a red environment, performance may decrease
		Red has the most arousal effects which has caused to less error made in certain task although it is highly rated as distracting
		Besides of arousal effect, red is found to cause avoidance behavior when subjects perceived it as a warning cue in performance related task, drawing away attention of high task demand and moderating the perception of perceived task
		Blue, gray and beige minimize attention or concentration
		Blue color has the most arousal effects and has the highest rating for performing environment
		Light beige evokes less covered area, less concentration and low in density of footstep in space circulation
	Heat	Concentration decrease if the office temperature is higher than the norm, but when it is too hot, they take up to 25% longer than usual to complete a task
	Fresh air	The mental attention of pupils is significantly slower when the level of CO_2 in classrooms is high and when the air exchange rate is low
Visual and spatial processing	Colors	Working in the white environment resulted in the most errors being made
	Space dimension	Rooms with higher ceilings activated visuospatial exploration structures in the dorsal stream. Open rooms activated structures underlying perceived visual motion
Processing speed	Colors	Blue-enriched white light on students′ performance has faster cognitive processing speed
	Heat	Increases with temperature up to 21 °C–22 °C and decreases with temperature above 23 °C–24 °C. Task performance decreases when the subjects feel warm. Performance of the tasks decreased at 30 °C compared with 22 °C. Optimum performance can be achieved slightly below neutral temperature

Results about mood involves studies on relationships between environmental indoor factors and affective response, regarding effects on both positive disposition, expressing feeling "up" and relaxation state, and negative mood, evoking anxiety and feeling "down" (Table 4). Some selected studies focus effects on learning performance (6), concerning pre-scholar, primary pupils, adolescent and academic students. Creativity resulted a cutting-edge issue, since on a total of selected studies (7) two are narrative studies, reporting systematically a lot of evidence about elements for environmental design of office for fostering creative performance (Table 5).

Table 4. Results of the literature selection with regard to the effects on mood

Mood	Physical factor	Evidence
Feeling up	Light	If the (white) light is warm, a higher level of illuminance will result in more positive mood, but when it became too bright the mood declined again
		The 3,000 K condition of illuminance was associated with better moods at 500 lx than other color temperatures
	Colors	Positive in the pink room and negative in the blue room
	Window	A window view may stimulate positive mood
	Plants	Plants can make people's mood more positive
Relaxion	Colors	Calming colors provide a relaxing experience are green, blue or blue/violet. Blue is also said to have a drowsy and sleepy effect. Cool colors tend to have a calming effect, but extreme areexciting
	Material	Less tension and fatigue were generated in the wooden rooms than in the non-wooden rooms when the participants did their work
Anxiety	Space	Enclosed rooms elicited exit decisions and activated a cingulate region connected with amygdala. Greater reactivity to stress when people placed in an enclosed rather than an open room has been demonstrated
	Colors	Anxiety is felt in a blue-green office
Feeling down	Light + windows	If the light is cool, a higher level of illuminance will result in lower positive mood. The workers' mood was at its lowest when the lighting was experienced as much too dark, but when the light became too bright the mood declined again. The relationship between mood and the distance to the nearest window was bimodal

Table 5. Results of the literature selection with regard to the human effects.

Human effect	Physical factors and synthesis of evidence
Creativity	Daylight, calming and inspiring colors, windows view to nature, space and layout, sound, plants, finishing/materials, privacy, indoor climate
Communication and socialization	Abundant and diffuse lighting, high ceilings exterior views, few elements of enclosure flexible furniture, adaptable equipment flexible physical barriers
Personal control	Negative effects derive by openness of work environments. If employees have non-assigned workstations, they have limited abilities to demonstrate psychological ownership within the office, negatively affecting well-being
Learning performance	The performance of two numerical and two language-based tests was significantly improved when the temperature was reduced from 25 °C to 20 °C. Natural light, space and furniture flexibility

5 Conclusion

The aim of this paper is to review a selection of existing studies that examine the effects of physical factor of built environment on human perceptions and behavior, particularly regarding outcomes on performance. The results, listed in the Tabs linking the human effects to the single element of built environment, may be seen as a starting point for the definition of a systematic frame of criteria for environmental design of buildings, especially interiors, compliant to physiologic, affective and cognitive responses, in order to create a really supportive architecture, where occupants can perform better their activities.

The study confirms that a limited number of selected research focus directly on effects of physical environment on specific task performance. Otherwise, research on comfort in working environment appears abundant, including how it affects stress. For this reason, qualitative studies prevail on quantitative ones.

Further research will be purposed toward more goals: to enlarge the study to other search engine, in order to have a wider research scope; to search evidence on the effects of single physical factor upon more specific cognitive skills, such as response time, shifting, hand-eye coordination, working memory; to include in selection literature about combined effect of more built environment factors on human performance.

References

1. Vischer CJ (2007) The effects of physical environment on job performance: a theoretical model of workspace stress. Stress Health 23:175–184
2. Clements-Croome D (2014) Sustainable Intelligent Buildings for Better Health, Comfort and Wee-Being. Report for Denzero Project supported by the TÁMOP- 4.2.2.A- 11/1/KONV-2012-0041 co-financed by the European Union and the European Social Fund
3. Smith D (2008) Color-person-environment relationships. Color Res Appl 33(4):312–319

4. Barrett P, Davies F, Zhang Y, Barrett L (2015) The impact of classroom design on pupils' learning: Final results of a holistic, multi-level analysis. Build Environ 89:118–133
5. De Croon E, Sluiter J, Kuijer P, Frings-Dresen M (2005) The effect of office concepts on worker health and performance: a systematic review of the literature. Ergonomics 48 (2):119–134
6. Sonnentag S, Volmer J, Spychala A (2008) Job performance. In: Barling J et al (eds) Micro Approaches. Los Angeles, Calif. [u.a]. SAGE, pp 427–447 (2008)
7. Wickens CD, Hollands JG, Banbury S, Parasuraman R (2016) Engineering psychology and human performance. 4th edn. Routeldge, London and New York, pp 2–3
8. Dalal RS, Bhave DP, Fiset J (2014) Within-person variability in job performance. Theor Rev Res Agenda 40(5):1396–1436 (2014)
9. Carrol JB (1993) Human Cognitive Abilities: A Survey of Factor-Analytic Studies. Cambridge University Press, Cambridge
10. Hidayetoglu LM, Yildirim K, Akalinc A (2012) The effects of color and light on indoor way finding and the evaluation of the perceived environment. J Environ Psychol 32(1):50–58
11. Ata S, Deniz A, Berrin A (2012) The physical environment factors in preschools in terms of environmental psychology: a review. Procedia – Soc Behav Sci 46:2034–2039
12. Barrett P, Davies F, Zhang Y, Barrett L (2015) The impact of classroom design on pupils' learning: final results of a holistic, multi-level analysis. Build Environ 89:118e133
13. Batra P, Garg R (2005) Effect of temperature on memory. J Indian Acad Appl Psychol 31(1–2):43–48
14. Dul J, Ceylan C, Aytac S (2008) Can the office environment stimulate a manager's creativity? Hum Fact Ergon Manuf 18(6):589–602
15. Dul J, Ceylan C (2011) Work environments for employee creativity. Ergonomics 54(1):12–20
16. Fich LB, Jönsson P, Kirtkegaard PH, Wallergård M, Garde AH, Hansen Å (2014) Can architectural design alter the physiological reaction to psychosocial stress? a virtual TSST experiment. Physiol Behav 135:91–97
17. Furnham A, Strbac L (2002) Music is as distracting as noise: the differential distraction of background music and noise on the cognitive test performance of introverts and extraverts. Ergonomics 45(3):203–217
18. Hygge S, Knez I (2001) Effects of noise, heat and indoor lighting on cognitive performance and self-reported affect. J Environ Psychol 21(3):291–299
19. Iyendo TO, Uwajeh PC, Ikenna ES (2016) The therapeutic impacts of environmental design interventions on wellness in clinical settings: a narrative review. Complement Ther Clin Pract 24:174e188
20. Jalil NA, Yunus RM, Said NS (2012) Environmental colour impact upon human behaviour: a review. Procedia – Soc Behav Sci 35:54–62
21. Keisa O, Helbig H, Streb J, Hill K (2014) Influence of blue-enriched classroom lighting on students' cognitive performance. Trends Neurosci Educ 3(3–4):86–92
22. Knez I (1995) Effects of indoor lighting on mood and cognition. J Environ Psychol 15 (1):39–51
23. Kowaltowski D, Mikami SA, Pina G, Barros R (2006) Architectural design analysis as a strategy for people environment studies: finding spaces that work. In: Environment, Health and Sustainable Development (IAPS 19 Conference Proceedings on CD-Rom). IAPS. Alexandria, Egypt (2006)
24. Küller R, Ballal S, Laike T, Mikellides B, Tonello G. (2006) The impact of light and colour on psychological mood: a cross-cultural study of indoor work environments. Ergonomics 49 (14):1496–1507

25. Lan L, Wargocki P, Liana Z (2011) Quantitative measurement of productivity loss due to thermal discomfort. Energy Build 43(5):1057–1062
26. Lan L, Wargocki P, Wyon DP, Lian Z (2011) Effects of thermal discomfort in an office on perceived air quality, SBS symptoms, physiological responses, and human performance, Indoor Air
27. Lee J, Moon JW, Kim S (2014) Analysis of occupants' visual perception to refine indoor lighting environment for office tasks. Energies 7:4116–4139
28. Ramlee N, Said I (2014) Review on atmospheric effects of commercial environment. Procedia – Soc Behav Sci 153:426–435
29. Savavibool N, Gatersleben B, Moorapun C (2016) The effects of colour in work environment: a systematic review. In: 7th AicE-Bs 2016, Edinburgh, UK, 27–30 July, E-BPJ, vol 1(4), pp. 262–270
30. Seppänen O, Fisk WJ, Lei Q (2006) ventilation and performance in office work. Indoor Air J 18:28–36
31. Shishegar N, Boubekri M (2016) Natural light and productivity: analyzing the impacts of daylighting on students' and workers' health and alertness, Int'l J Adv Chem Eng Biological Sci (IJACEBS) 3(1)
32. Suresh M, Dianne S, Jill F (2006) Person environment relationships to health and wellbeing: an integrated approach. IDEA J 87–102
33. Tantanatewin W, Inkarojrit V (2018) The influence of emotional response to interior color on restaurant entry decision. Int J Hospitality Manage 69:124–131
34. Taylor A, Engass K (2014) Linking Architecture and Education: Sustainable Design For Learning Environments. University of New Mexico Press, China
35. Vartanian O, Navarrete G, Chatterjee A., BrorsonFich L., Gonzalez-Mora JL, Leder H, Skov M (2015) Architectural design and the brain: effects of ceiling height and perceived enclosure on beauty judgments and approach-avoidance decisions. J Environ Psychol 41: 10–18
36. Wargocki P, Wyon DP (2007) The effects of moderately raised classroom temperature and classroom ventilation rate on the performance of schoolwork by children (1257-RP). HVAC&R Res 13(2):193e220)
37. Wessolowskia N, Koenig H, Lschulte-Markwort M, Barkmanna C (2014) The effect of variable light on the fidgetiness and social behavior of pupils in school. J Environ Psychol 39:101–108
38. Wohlers C, Hertel G (2017) Choosing where to work at work–towards a theoretical model of benefits and risks of activity-based flexible offices. Ergonomics 60(4):467–486
39. Zhanga X, Liana Z, Wub Y (2017) Human physiological responses to wooden indoor environment. Physiol Behav 174(15):27–34

Ergonomic Quality in Green Building Protocols

Ilaria Oberti(✉) and Francesca Plantamura

Politecnico di Milano, Milan, Italy
ilaria.oberti@polimi.it

Abstract. The control of living conditions in built spaces is a key aspect of design in order to guarantee the final user to live in healthy and safe spaces. This importance is also recognized in green building rating systems that, in fact, present criteria on wellbeing with particular link to physical issues. In this context, the adoption of an ergonomic approach would be useful in order to address the interaction between user and built environment as a whole, considering both factors that affect physical wellbeing (i.e. thermo-hygrometric, acoustic, lighting, air purity), and those that complete the complex sphere of wellbeing, that are psychological and social too. The aim of this study is to appraise how green building evaluation tools are able to bring out user's wellbeing in all its complexity. With respect to ergonomic quality, the study analyses LEED, a widely internationally widespread green building rating systems, and WELL, the first rating system with a specific focus on user wellbeing. A comparison between the two systems is carried out, with respect to their efforts in bringing out the ergonomic quality of built environment. The results show that both the analyzed tools give space to building features for user's wellbeing. Properly, the weight given to these aspects is less in LEED, because user's wellbeing is not its unique goal; however, some efforts would be useful to increase and reward the related criteria. In both cases, ergonomists within the project team can be a key element to pursue wellbeing in an effective user-centered vision.

Keywords: LEED · WELL · Wellbeing · Built environment · Ergonomics

1 Introduction

The control of living conditions in built spaces is a key aspect of design in order to guarantee the final user to live in healthy and safe spaces. This is even more important considering the long stay of the population in indoor environments, especially in urban areas [1], with the consequent impact of buildings on occupant health. In case of workplaces, this interest is increased by a growing awareness about correlated issues: the impact of indoor environment on employers' wellbeing, involving changes on their behavior and job performance [2]; the convenience in devoting resources to improve indoor quality of commercial buildings against benefits in terms of management of salary costs [3].

S. Bagnara et al. (Eds.): IEA 2018, AISC 825, pp. 496–503, 2019.
https://doi.org/10.1007/978-3-319-96068-5_56

This importance is also recognized in green building rating systems, tools addressed to control and promote buildings' sustainability during all their life cycle. Therefore, together with credits on the external impact of buildings, green rating systems also contain credits, or entire sections, dedicated to indoor environmental quality. An example is the Building Research Establishment Environmental Assessment Method (BREEAM), among the first systems for the environmental assessment of buildings, developed in UK and now widely internationally widespread [4]. This system covers ten categories of sustainability (Energy, Transport, Water, etc.), with one category dedicated to "Health & Wellbeing". Even the French Haute Qualité Environmental (HQE) system provides credits on indoor quality. This system, especially, offers a clear articulation around the two main goals: 1. control the impacts on the external environment; 2. create a healthy and comfortable indoor environment [5]. Attention to aspects of wellbeing is also present in the Leadership in Energy and Environmental Design (LEED) system, to date the most widespread green building rating system on an international level. This system is divided into several section (Energy & atmosphere, Material & resources, etc.) with one, "Indoor environmental quality", entirely dedicated to indoor features of buildings and their impacts on occupants' wellbeing [6].

However, buildings that achieve a high score in one of these rating systems do not always guarantee optimal conditions for occupant wellbeing. For example, a Post Occupancy Evaluation (POE) survey on 44 occupants of two LEED Platinum US college buildings shows, despite a generally positive assessment of health, performance and satisfaction, that in one of the two buildings remain critical issues on internal comfort, perceived health and occupant performance [7]. More generally, the correspondence between the achievements of specific IEQ credits, with the increase of satisfaction in the related area, is not proven, nor that between the overall achieved rating level and workplace satisfaction [8].

In order to maximize users' wellbeing through green rating systems, it can be useful to adopt an ergonomic approach. It will address the interaction between the user and the environment as a whole, considering both the factors that affect physical wellbeing (i.e. thermo-hygrometric, acoustic, lighting, air purity), and those that complete the complex sphere of wellbeing, that is psychological and social too.

In this context, the aim of this study is to appraise how green building evaluation tools are able to bring out the user wellbeing in all its complexity and how ergonomics can support this process with a user centered vision.

2 Method

The study is developed through the following three steps:

1. Analysis of LEED, one of the most relevant and widespread green rating systems, developed to assess and certify the overall environmental quality of buildings. The analysis is focused on the system's requirements related to ergonomic quality, including both physical and psychosocial factors of wellbeing.

2. Analysis of WELL Building Standard, the first rating system directly focused on health and wellbeing. The analysis is focused on the system's requirements related to building design, construction and management.
3. Comparison between the analyzed systems with respect to their efforts in bringing out ergonomic quality of built environment in all its complexity.

3 Results

3.1 LEED and Occupants' Wellbeing

Developed by United States Green Building Council (U.S.GBC) in collaboration with private companies and researchers, LEED is a green certification program for building design, construction, operations and maintenance. It has become one of the most widely international used standard. It includes a set of rating systems for the design, construction, operation, and maintenance of green buildings. Available for virtually all building, community and home project types, LEED provides a framework to create healthy, highly efficient and cost-saving green buildings.

The system is based on prerequisites and credits that a project has to meet to achieve a certification level: Certified, Silver, Gold and Platinum.

The environmental criteria are grouped into nine topics:

1. Integrative process credits,
2. Location & transportation,
3. Sustainable sites,
4. Water efficiency,
5. Energy & atmosphere,
6. Material & resources,
7. Indoor environmental quality,
8. Innovation,
9. Regional priority.

The "Indoor Environmental Quality (IEQ)" topic includes most of the credits related to the direct impact of building on occupant health and wellbeing. This section deals with the following aspects, divided between prerequisites (mandatory) and credits:

Prerequisite: Minimum Indoor Air Quality Performance
Prerequisite: Environmental Tobacco Smoke Control
Prerequisite: Minimum Acoustic Performance
Credit: Enhanced Indoor Air Quality Strategies
Credit: Low-Emitting Materials
Credit: Construction Indoor Air Quality Management Plan
Credit: Indoor Air Quality Assessment
Credit: Thermal Comfort
Credit: Interior Lighting
Credit: Daylight.
Credit: Quality Views
Credit: Acoustic Performance

Although, in this section (IEQ), LEED emphasizes physical aspects of wellbeing, some credits also cover aspects of a wider wellbeing. An example is the "Thermal Comfort" credit, which sets the goal of promoting occupants' productivity, comfort, and wellbeing by providing quality thermal comfort. In addition to thermal comfort design strategies, this credit requires to provide individual thermal comfort controls for at least 50% of individual occupant spaces and provide group thermal comfort controls for all shared multioccupant spaces. In addition to a climatically comfortable environment, this credit also recognizes the right of users to exercise control over their living or working environment. Credits related to occupant wellbeing are also present in other sections. An example is the "Open space" credit, in the "Site assessment" section. This credit involves psychosocial aspects of wellbeing: its goal is to create exterior open space that encourages interaction with the environment, social interaction and passive recreation, together with physical activities.

A credit, not mandatory, involved in user satisfaction is included in LEED O&M (Operation & Maintenance): the "Occupant comfort survey" credit, aimed to assess building occupants' comfort. Its request is to administer at least one occupant comfort survey (and then, a minimum of one new survey once every 2 years) to collect responses from a representative sample of building occupants regarding at least the following issues: acoustics; building cleanliness; indoor air quality; lighting; thermal comfort. If the results indicate that more than 20% of occupants are dissatisfied, it is request to develop and implement a corrective action plan to address comfort issues.

Credits aimed at wellbeing in a broader sense are included among the LEED "Pilot credits". One of these is the "Ergonomics approach for computer users" credit, aimed to improve occupant wellbeing (human health, sustainability and performance) through integration of ergonomic principles, specifically in the design of workspaces for computer users. During the design phases, among the requirements to meet this credit, there is the engagement of an Ergonomist or Health and Safety Specialist, to assist the team in the development of an ergonomics strategy, and the commitment to integrate ergonomics in general design. The ergonomics strategy has to start from occupant needs and get to the evaluation and maintenance of occupant wellbeing. Regarding both private and public health, there is the "Integrative Process for Health Promotion" pilot credit, aimed to facilitate a systematic consideration of the impact that project design and construction has on health and wellbeing, including physical, mental and social impacts. One of the pilot credits with a strong impact on the psychological wellbeing of users is the "Designing with Nature, Biophilic Design for the Indoor Environment" one, aimed to support and improve human health, wellbeing, and productivity by providing and incorporating elements of nature in the indoor environment.

3.2 WELL and Built Environment Quality

WELL is the first building standard focusing exclusively on building performances that affect health and wellbeing. Launched in 2014, the standard was developed by bringing together best practices in design and evidence-based medical and scientific research aimed at making the building a vehicle for occupant health and wellbeing. Administered by the International WELL Building Institute™ (IWBI™), this system is designed to work harmoniously with other international green building systems.

WELL, in particular, naturally interfaces with LEED, with many overlapping features. Moreover, the two systems are third party certified by the same organization, the Green Business Certification Inc. (GBCI).

WELL is made up of seven categories, "Concept", of wellness:

1. Air
2. Water
3. Nourishment
4. Light
5. Fitness
6. Comfort
7. Mind

Each concept is comprised of multiple features, which are intended to address specific aspects of occupant health, comfort or knowledge. As in LEED, features in WELL are categorized as either Preconditions, necessary for all levels of certification, and Optimizations, a certain percentage of which must be attained depending on the level of achievement that is pursued (Silver, Gold or Platinum).

"Air", "Light" and "Comfort" concepts are directly linked to the design and management aspects of built environment, with main reference to characteristics that affect users' physical wellbeing (Table 1).

Table 1. Air, Light and Comfort credits (WELL Standard)

Preconditions	Optimizations
AIR	
Air Quality Standards*	Air Flush
Smoking Ban*	Air Infiltration Management
Ventilation Effectiveness	Increased Ventilation
VOC Reduction	Humidity Control*
Air Filtration*	Direct Source Ventilation*
Microbe And Mold Control*	Air Quality Monitoring and Feedback*
Construction Pollution Management	Operable Windows*
Healthy Entrance*	Outdoor Air Systems
Cleaning Protocol	Displacement Ventilation
Pesticide Management	Pest Control*
Fundamental Material Safety	Advanced Air Purification*
Moisture Management	Combustion Minimization*
	Toxic Material Reduction
	Enhanced Material Safety
	Antimicrobial Activity for Surfaces
	Cleanable Environment*
	Cleaning Equipment*

(*continued*)

Table 1. (*continued*)

Preconditions	Optimizations
LIGHT	
Visual Lighting Design* Circadian Lighting Design* Electric Light Glare Control Solar Glare Control*	Low-Glare Workstation Design* Color Quality Surface Design Automated Shading And Dimming Controls Right To Light* Daylight Modeling Daylighting Fenestration*
COMFORT	
Accessible Design Ergonomics: Visual And Physical* Exterior Noise Intrusion* Internally Generated Noise* Thermal Comfort*	Olfactory Comfort Reverberation Time* Sound Masking* Sound Reducing Surfaces Sound Barriers Individual Thermal Control* Radiant Thermal Comfort

* Pending onsite post-occupancy Performance Verification testing.

Compared to the common green building rating systems, the "Mind" section offers an innovative vision on building requirements for occupant wellbeing (Table 2). This section contains numerous credits aimed to an overall wellbeing, also including all psychological and social aspects. One of the preconditions of "Mind" is the "Post-occupancy surveys" feature, for measuring the effectiveness in promoting and protecting the health and comfort needs of its occupants. Aggregate results from surveys have to be reported within 30 days both to the following building owners and

Table 2. Mind credits (WELL Standard)

Preconditions	Optimizations
MIND	
Health And Wellness Awareness* Integrative Design Post-Occupancy Surveys Beauty And Design I* Biophilia I - Qualitative*	Adaptable Spaces* Healthy Sleep Policy Business Travel Building Health Policy Workplace Family Support Self-Monitoring Stress And Addiction Treatment Altruism Material Transparency* Organizational Transparency* Beauty And Design II* Biophilia II - Quantitative* Innovation Feature (5 options)

* Pending onsite post-occupancy Performance Verification testing.

managers, occupants and to the International WELL Building Institute, in order to help further development in WELL Building Standard.

Appropriate design strategies are required also to achieve objectives apparently far from built environment aspects. For example, the "Interior Fitness Circulation" feature, within the "Fitness" section, requires attention to building aspects such as stairs that are accessible and easy to use by all occupants. In the same feature, moreover, the introduction of elements of aesthetic appeal is required, such as "View windows to the outdoors or building interior".

3.3 Built Environment and Ergonomic Quality – Systems' Comparison

The two systems have been developed to reach two different goals and so, obviously, there are deep differences between them both in weight given to users' wellbeing and in way to assess it. WELL is focused on health and wellness of building occupants, therefore the totality of obtainable points refers completely to this goal. LEED, on the contrary, is designed to achieve high performance in multiple areas of building environmental sustainability, so users' wellbeing is only one goal among a wider whole.

In the LEED system, credits for wellness are mainly concentrated in the Indoor Environmental Quality (IEQ) section. Thus, it is possible to compare, in general terms, the weight given to occupants' wellbeing in relation to the other objectives of the system. In the last version (v4) of LEED for New Construction and Major Renovations, the maximum number of points obtainable in IEQ is 16, out of 110 credits. This weight is comparable to sections as Location & Transportation (16) or Material & Resources (13), while it is about half of the Energy & Atmosphere Section (33 points) that is the most heavy within the system. An extra point in IEQ section can be obtained in the LEED O & M (Operation & Maintenance), which also contains the relevant "Occupant comfort survey" credit. An opening to the overall wellbeing of people is found in some pilot credit, as previously explained. One of these, "Ergonomics approach for computer users", explicitly refers to ergonomic quality but its requirements cover only some aspects of this subject and it is addressed only to computer users. Occupants that are not computer users are encouraged to be included in the ergonomics strategy but not required. Moreover, this credit leads, as well as the other credits related to health promotion and biophilia, to reach only one point despite the great effort required to achieve it, both as workload and as professionals to be integrated into the design team.

Substantial differences between LEED and WELL are also in the way of understanding wellbeing. The premise of WELL is the awareness of the impact that built environment can have on the many aspects of wellbeing (physical health, physiological comfort, cognitive and psychosocial wellbeing) and on user behavior. This awareness, partly shared by LEED, in WELL is the foundation of the system and it is the basis of numerous requirements, particularly in the Mind section. Moreover, the different way of conceiving wellbeing also implies a different way of developing its assessment. If LEED assesses mainly the design features of the building, in WELL the assessment of environments in use takes on great importance. In particular, in WELL the "Post-Occupancy surveys" credit is foreseen as a precondition (therefore, it is mandatory), while in LEED this credit, only foreseen in the O & M version, it is not a prerequisite but a simple credit (therefore not mandatory).

4 Conclusion

The results show that both the analyzed tools give space to user's wellbeing in their credits. However in LEED, users' wellbeing is only one goal among a wider whole and so the credits related to this objective in the cover a much smaller part than the one covered in WELL.

This difference in weight that the sphere of wellbeing has within the two systems is obvious, being LEED born to maximize all the environmental performance of a building, while WELL is built completely focused on the wellbeing of the occupants.

However, some improvements can be made in the way of understanding and measuring wellbeing features. In this regard LEED, and more generally all green building rating systems, can look to WELL for some useful insights regarding its comprehensive and interdisciplinary approaches, very close to ergonomics.

WELL meaningfully address the complex issues of human health, in a user-centered perspective. However, by not allowing an overall environmental assessment of buildings, this system can support green rating systems, but does not take their place in promoting environmental building quality.

It is therefore necessary to continue working to bring out more the ergonomic quality of the built in broad-spectrum systems (LEED, BREEAM, etc.) and to this goal ergonomics can play a key role.

References

1. Schweizer C, Edwards RD, Bayer-Oglesby L, Gauderman WJ, Ilacqua V, Jantunem MJ, Lai HK, Nieuwenhuijsen M, Kunzli N (2007) Indoor time–microenvironment–activity patterns in seven regions of Europe. J Eposure Sci Environ Epidemiol 17(2):170–181
2. Roelofsen P (2002) The impact of office environments on employee performance: The design of the workplace as a strategy for productivity enhancement. J Facilit Manage 1(3):247–264
3. Rocky Mountain Institute (2014) How to calculate and present deep retrofit value – A guide for owner-occupants
4. BREEAM Homepage. https://www.breeam.com/
5. HQE Homepage. http://www.behqe.com/fr
6. LEED Homepage. https://new.usgbc.org/leed. Accessed 23 Apr 2018
7. Hedge A, Dorsey JA (2013) Green buildings need good ergonomics. Ergonomics 56(3): 492–506
8. Altomonte S, Schiavon S, Kent MG, Brager G (2017) Indoor environmental quality and occupant satisfaction in green-certified buildings. Build Res Inf 1–20. https://doi.org/10.1080/09613218.2018.1383715

An Ergonomic Approach of IEQ Assessment: A Case Study

Erminia Attaianese[1], Francesca Romana d'Ambrosio Alfano[2(✉)], and Boris Igor Palella[3]

[1] DIARC - Dipartimento di Architettura, Università degli Studi di Napoli Federico II, Via Toledo, 402, 80135 Naples, Italy
[2] DIIn – Dipartimento di Ingegneria Industriale, Università degli Studi di Salerno, Via Giovanni Paolo II 132, 84084 Fisciano, Italy
fdambrosio@unisa.it
[3] Dipartimento di Ingegneria Industriale, Università degli Studi di Napoli Federico II, Piazzale V. Tecchio 80, 80125 Naples, Italy

Abstract. Only in the last fifteen years, the application of the ergonomics principles stated the need to achieve a good IEQ (indoor environmental quality) as a result of thermal, visual, acoustic comfort and indoor air quality (IAQ). The awareness increased that an adequate design of the indoor environment, where people work and live, requires a synergic approach to all facets involved in full compliance with sustainability. IEQ strictly affects the overall building energy performances and exhibits an antagonistic relationship with respect to the energy saving requirements. In addition, the effects of low IEQ levels on the health and the productivity at work could even greater than those related to the energy costs of building facilities.

The role played by IEQ is very important especially in school environments. Particularly, children are extra sensitive to a poor indoor environment as they are physically still developing and, in comparison to healthy adults, will suffer the consequences of a poor indoor environment with also negative effects in learning ability. Because of all mentioned issues an integrated approach in the design and in the assessment of school buildings is required where ergonomics plays a crucial role.

Since more than 20 years InEQualitES (Indoor Environmental Quality and Energy Saving) team, made by researchers and professors from Universities of Salerno and Naples, has focused the research on the environmental quality in schools. The experience gained in the field allowed to build a large database of subjective and objective data and helped us to find effective solution aimed to solve most common problems related to the application of assessment methods.

Based upon this experience, this discussion will be mainly focused to the main criticalities related to the thermal comfort assessment in schools with special reference to the integration of objective investigations (referred to a mean subject statistically significant) and subjective investigations which are the only able to show possible differences (age, gender and so on) in experienced perception.

Keywords: Indoor environmental quality · IEQ · Thermal comfort Schools · Sustainability

S. Bagnara et al. (Eds.): IEA 2018, AISC 825, pp. 504–513, 2019.
https://doi.org/10.1007/978-3-319-96068-5_57

1 Introduction

Only in recent years, the application of the ergonomics principles stated the need to achieve a good IEQ (Indoor Environmental Quality), as a result of thermal, visual, acoustic comfort and IAQ (Indoor Air Quality) [1, 2]. The awareness increased that an adequate design of the indoor environment – where people work and live 90% of their time – requires a synergic approach to all facets involved in full compliance with sustainability. IEQ strictly affects the overall building energy performances and exhibits an antagonistic relationship with respect to the energy saving requirements [3]. Therefore, project teams need to be multi-disciplinary and in an integrated design process be able to simulate energy performance of buildings, evaluate indoor environmental conditions. In addition, poor IEQ levels can promote increasing symptoms of SBS, acute respiratory illnesses sick leaves and a significant reduction of the people's performance [4].

This is even more crucial school buildings especially because these environments are characterized by a high density of persons who occupy the same environment. In addition, children are extra sensitive to poor thermal comfort levels with respect to healthy adults as they are physically still under development [4] exhibiting a different thermoregulation and reported thermal sensation due to their higher surfacearea-to-mass ratio, greater metabolic heat production per kg body mass and lower sweating rate [5].

Since more than 20 years the InEQualitES (Indoor Environmental Quality and Energy Saving), group - composed by professors of the Universities of Salerno and Naples, has focused the research on the environmental quality in schools [6]. The experience gained in the field allowed us to build a large database of subjective and objective data and helped us to find effective solution aimed to solve most common problems related to the application of assessment methods [6, 7].

Based upon this experience, in the present study will be discussed some results of a large experimental campaign survey (still in progress) about thermal comfort conditions in 110 classrooms of different level located in the Southern Italy. In particular, we will try to focus how the assessment of school building environments requires some adaptations and a full integration between objective and subjective investigations.

2 Thermal Comfort

According to the definition by ASHRAE [8], thermal comfort is the condition of mind that expresses satisfaction with the thermal environment. It can be distinguished into thermal comfort for the body as a whole and in local discomfort due to time and spatial inhomogeneities of the thermal environment (e.g. vertical air temperature differences, warm and cool floors, draughts and radiant asymmetries). Thermal comfort can be evaluated by means of objective and subjective investigations like other factors contributing to the indoor environmental quality.

The objective evaluation of overall comfort conditions is usually carried out by means of the PMV index which integrates the influence of the thermal comfort factors (air temperature, air velocity, mean radiant temperature, humidity, clothing and activity) into a value on a 7-points scale (Fig. 1A).

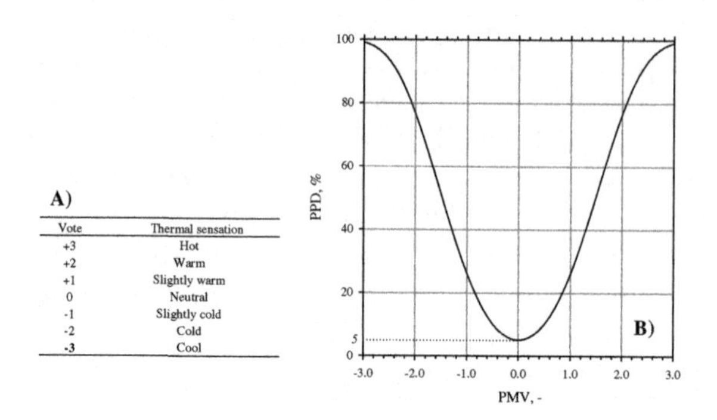
A)

Vote	Thermal sensation
+3	Hot
+2	Warm
+1	Slightly warm
0	Neutral
-1	Slightly cold
-2	Cold
-3	Cool

Fig. 1. 7-points ASHRAE thermal sensation scale [8] (A) and PMV/PPD relationship (B).

To express the quality of the thermal environment as a quantitative prediction of the percentage of thermally dissatisfied the PPD (Predicted Percentage of Dissatisfied) is also used. The PPD is correlated to the PMV value by means of a logistic relationship (Fig. 1B) whose mathematical structure reveals that a little percentage of dissatisfied (5%) can be expected under thermal neutrality conditions (i.e. PMV = 0).

Formulated by Fanger in the 70s [9] the PMV is an experimental index formulated for air-conditioned environments and based on an analysis of the heat balance equation for the human body together with the influence of the physical environment and expressed as a subjective sensation. This is the reason why in case of naturally ventilated environments it usually overestimates the thermal sensation felt by occupants [10]. This is the reason for the increasing success of the adaptive model proposed by Brager and de Dear [11] accepted by both ASHRAE [8] and CEN [2] in the absence of HVAC systems (or switched-off). Starting by this awareness, Fanger and Toftum [12] introduced two correction factors to extend PMV model to free-running buildings in warm climates. The first value, e (expectancy factor, which have to be multiplied for the PMV value to obtain the predicted thermal sensation vote), varies between 1 and 0.5 depending on the presence of HVAC system. For buildings provided with HVAC systems, the expectancy factor is equal to 1, whereas for naturally ventilated it is assumed to depend on the duration of warm climate and the possibility of compare these buildings with others in the region. The second factor introduced is a reduced activity level, as it is slowed down unconsciously by people feeling warm as a form of adaptation.

At a subjective level, thermal comfort conditions are usually carried out by administrating suitable questionnaires designed in compliance with ISO 10551 Standard [13].

From a standardization point of view, the thermal environments field is regulated by about 20 among CEN and ISO standards which can be divided into support standards, standards for evaluation of moderate and severe thermal environments (hot and cold), special applications (e.g. vehicles) and standards devoted to the contact with hot and cold surfaces [3].

3 Thermal Comfort in Schools

The assessment thermal comfort conditions in school buildings is a complex issue mainly due to the specificity of the users made up by children and boys with very different age groups. This is further complicated in those countries where HVAC systems are not common [10] or when the school building is protected by special laws for the safeguard of the cultural heritage (this is most common in Italy where about 20% of school building is built before 1940).

As recently reviewed by several authors [5, 6, 14] thermal comfort in schools his a topic widely investigated over the past years. It has been shown that, despite inter-individual variations, the perception of thermal comfort varies with age, gender and social status of the subjects [15]. In particular, young people suffer for high temperatures more than adults [16] and they show differentiated thermal sensations in relation to the activity schedule even among homogeneous groups. The pupils' perceived overall comfort is not always related to their thermal state as they feel hot but state that they are feeling comfortable. This is mainly because their discomfort is not necessarily related to the classroom microclimatic conditions but is often due to the tiredness (e.g. due to the number of hours spent at school and the present thermal environment). Other peculiarities concern the relationship between the predicted thermal sensation and the percentage of dissatisfied in naturally ventilated classrooms [8, 9]. Particularly, based on surveys in field, the estimated percentage of dissatisfied is not consistent with predicted, especially at higher or lower operative temperatures values. This is mainly due to the small number of investigated subjects (it is difficult finding subjective investigations carried out on a sample made by more than 1000 students) and seasonal or ventilation system effects.

The reasons of above discussed differences in the perception of thermal comfort in schools remain still debated issues. An explanation could be the limited availability of control of the thermal environment, generally carried out by operating on windows, doors and curtains (often decided by the teacher). In addition, the variation in activity levels and the strong relationship with outdoor microclimatic conditions make the school activities so diversified that the reference procedures reported in the technical standards are often ineffective.

In the present study will be discussed some results of a large experimental campaign survey (still in progress) about thermal comfort conditions in some school buildings located in Campania (Southern Italy) and carried out by InEQualitES team. The experience gained in the field allowed us to build a large database of subjective and objective data and helped us to find effective solution aimed to solve most common problems related to the application of assessment methods in so special environments as schools.

4 The Survey

This study is based on a large experimental campaign survey carried out during the period from 2004 to 2016 (and still in progress) in 110 classrooms of different level (age of the students in the range from 11 to 18 years) located in Campania (Southern Italy) only provided with heating boiler plants. The evaluations have been carried out in each classroom both in summer and in winter season. In Table 1 are reported specific details on the investigated sample.

Table 1. General details of investigated schools.

School Name, City and year	Number of classrooms	Age of the students
Scuola Media Statale S. Alfonso Maria de' Liguiri – Pagani (Salerno), 2016	22	From 11 to 13
Scuola Media Statale – Vietri sul Mare (Salerno), 2004	5	
Scuola Media Statale "Giovanni XXIII" – Cava de' Tirreni (Salerno), 2002	11	
Liceo Scientifico Statale "A. Genoino" – Cava de' Tirreni (Salerno), 2005	26	From 14 to 18
Istituto Professionale di Stato per l'Agricoltura e l'Ambiente – Castel San Giorgio (Salerno), 2000	7	
Liceo Scientifico Statale "B. Rescigno" – Roccapiemonte (Salerno), 2003	28	
Istituto Magistrale Statale "Virgilio" – Pozzuoli (Napoli), 2006	21	
Administrated questionnaires	**Male**	**Female**
Winter	1063	1211
Summer	940	1202

4.1 Subjective Investigation

This step of the survey has been carried out by administrating to each student a special questionnaire designed with the assistance of a team of engineers, psychologists and doctors [6]. The questionnaire is divided into five sections: personal information (in this section students have to describe their clothing at the moment of the survey), thermal comfort, visual and acoustic comfort, perceived air quality (PAQ), global sensations. The thermal comfort assessment has been carried out based on the answers given to the thermal comfort section of the questionnaire. The questions of this section have been formulated in compliance with the recommendations of ISO 10551 Standard [13] and deal with the acceptability of the environment (would you accept/reject this thermal environment (local climate) rather than reject/accept it?) and the thermal preference (please state how you would prefer to be now, warmer or cooler). A final question is devoted to the thermal perception; in this case students were asked a judgment on the temperature rather than on the sensation. The reason of this choice is mainly due to a linguistic matter as in Italian “to feel a hot sensation” and “to perceive a high temperature” are considered as synonyms. To make the subjective survey as truthful as possible, a post-processing of all the administrated questionnaires has been carried out. Contradictory answers, multiple choices have been ignored (about 6% of questionnaires has been rejected). Finally, statistical analyses (linear and non-linear regressions and the significance p-value test) have been carried out by means of commercial software [6].

4.2 Objective Investigation

This step has been carried out according to the prescriptions reported in ISO 7726 Standard [17]. In particular, a special Comfort Data Logger provided with sensors for the air temperature, the plane radiant temperatures, the air velocity and the dew point temperature has been used (all consistent with metrological requirements reported in [17]). The measurement protocol [6] is based upon the measurement of the air temperature, the mean radiant temperature, the air velocity and the relative humidity at three heights and the calculation of global comfort and local discomfort indices [6]. The metabolic rate has been evaluated according to ISO 8996 Standard and it was settled at 1.3 met [18]. Such a choice has been mainly due to take into account some adaptation [6] and to reduce the effect of the sensitivity of the PMV index with respect to this quantity. In fact, the discrepancies in the calculation of the PMV can easily reach 0.1–0.3 points or more [3]. The calculation of the static thermal insulation of the clothing has been carried out according to the ISO 9920 procedures [19] on the basis of the clothing ensembles that students declared to wear during the measurements and indicated in a special section of the questionnaire containing the pictures of the single garment [6, 7].

Before calculating PMV index, the intrinsic clothing insulation values of clothing ensembles have been finally corrected by the effect of body movements by means of special correlations as a function of the air velocity and the metabolic rate [17] as required by ISO 7730 [20]. Obtained values are reported in Table 2.

The calculation of the PMV and PPD indices has been carried out by means of the TEE package [21], a special software devoted to the assessment of the thermal environment in agreement with the whole International Standards devoted to the ergonomics of the thermal environment. Finally, the calculation of PMV and PPD indices of each environment has been carried out in compliance with the Fanger’s theory [7, 9] by averaging the values of PMV and PPD calculated in each position (averaging microclimatic parameters measured in the different measurement positions and then calculating PMV provide same results [7]). On the basis of the answers to the questionnaires some indicators of the subjective thermal comfort have been taken into account, particularly:

- TSV: Thermal Sensation Vote obtained by questionnaires on the typical 7-point scale and calculated as a mean value of the votes attributed to the environment;
- PD_F: percentage of dissatisfied obtained by the questionnaires and in compliance with the Fanger’s definition [9] (percentage of those who have voted ± 2 or ± 3 on the scale of the thermal perception);

Table 2. Resultant clothing insulation $I_{cl,r}$ distribution as a function of the predicted mean vote range collected in the investigation summarized in Table 1. Each value takes into account the effect of the body movements [6].

		$I_{cl,r}$,clo						
		PMV range						
		<−0.5	−0.5 ÷ −0.2	−0.2 ÷ 0	0 ÷ 0.2	0.2 ÷ 0.5	0.5 ÷0.7	>0.7
Winter	$I_{cl,max}$	–	1.4	1.44	1.44	1.49	1.4	1.4
	$I_{cl,min}$	–	1.06	1.23	1.20	1.2	1.28	1.28
	$I_{cl,mean}$	–	1.27	1.34	1.34	1.34	1.33	1.33
	SD	–	0.12	0.06	0.07	0.09	0.04	0.04
Summer	$I_{cl,max}$	–	–	–	0.69	0.7	0.72	0.72
	$I_{cl,min}$	–	–	–	0.57	0.56	0.54	0.54
	$I_{cl,mean}$	–	–	–	0.63	0.63	0.63	0.63
	SD	–	–	–	0.04	0.04	0.04	0.04

- PD_{acc}: percentage of dissatisfied in each classroom based on the acceptability criterion and calculated on the basis of occupants who felt the environment not acceptable from the thermal point of view.
- The values of the predicted percentage of dissatisfied PPD calculated by means of the equation proposed by Fanger [9] have been then compared with those obtained from the subjective survey.

5 Results

5.1 Comparison Between Subjective and Objective Investigations

To verify the agreement between subjective and objective investigations, in Fig. 2 the comparison between thermal sensation votes TSV obtained by questionnaires and PMV values from microclimatic surveys is depicted.

Based on objective and subjective investigations results, the microclimatic conditions observed in the different classrooms are typical of non-thermal neutrality with PMV > 0. Thereby we can assume that the investigated sample of students appeared more adapted to a warm condition than the Fanger's approach would suggest [6]. Anyway, it is noteworthy remind that early Fanger's study had been carried out under controlled conditions in climatic chamber on a sample of North-American students whereas the whole of the combined subjective and rational analysis here discussed dealt with naturally ventilated environments occupied by Mediterranean students.

Another matter to take into account is that the PMV values in Fig. 2 have been obtained by calculating the PMV index consistently with the real clothing insulation values of the interviewed sample (see Table 2) instead of reference values suggested by

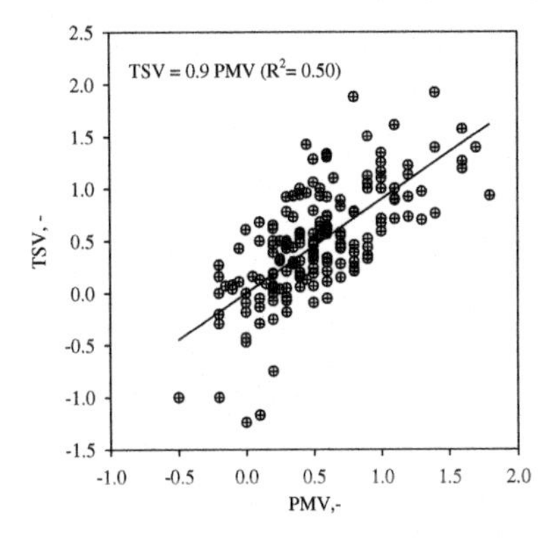

Fig. 2. Comparison between the thermal sensation vote (TSV) and the predicted thermal sensation in terms of TSV and PMV indices under winter and summer conditions [6].

ISO 7730 Standard [20]. Such a choice has been mainly due to the low correlation between the TSV observed values and PMV values calculated by means of reference summer and winter clothing insulation values (0.5 and 1.0 clo respectively) which were about 20–25% lower for the investigated sample of students [6].

The application of an expectancy factor equal to 0.9 results in a good agreement between predicted and subjective votes attributed to the environment in both seasons. This is consistent with Fanger and Toftum's findings who proposed a value in the range 0.9–1.0 [12, 22] for naturally ventilated buildings placed in climatic areas with short warm period (this the climate here investigated) during the summer and it is also surprising as it takes into account both summer and winter data.

5.2 Comparison Between Acceptability Criteria

A further analysis has been carried out to verify the relationship between the percentage of dissatisfied calculated based on the judgment of people who voted ±2 or ±3 (PD_F, according to the Fanger's theory) and the percentage of dissatisfied calculated by the subjective survey on the environment acceptability, PD_{acc}, based on the question: "how you would prefer to be now, warmer or cooler?".

Obtained PD_{acc} values, depicted in Fig. 3, have been correlated with TSV values by means of a logistic regression analysis [6].

Data in Fig. 3 clearly prove that under neutral sensation conditions (e.g. TSV = e·PMV = 0) the percentage of dissatisfied calculated according to the acceptability criterion is 15% instead of a 5% value consistent with the Fanger's criterion [9]. This result is consistent with almost the whole of data available in literature (also in residential buildings) because it seems almost accepted that the percentage of dissatisfied calculated on the basis of the acceptability criterion is greater than that obtained taking into account people who voted ±2 or ±3 [6].

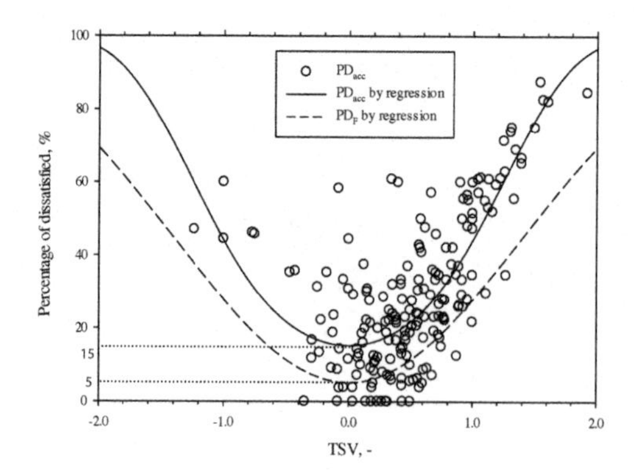

Fig. 3. Comparison between the percentage of dissatisfied calculated on the base of the acceptability criterion PD_{acc} and that predicted by means of Fanger's approach PD_F [6].

6 Conclusions

This investigation, based on a large experience in the field, has stressed how many aspects of the objective evaluation of thermal comfort in schools by means of the PMV/PPD indices require paying great attention. In particular, according to our findings, Fanger's basic approach for the assessment of thermal comfort conditions is effective also in naturally ventilated schools if a right expectancy factor and the real clothing insulation worn by students are known. Finally, our experience has demonstrated how important are the availability of special measurement protocols and the integration between subjective and objective analyses are crucial elements for reliable evaluations. Further studies addressed to verify the possible interaction between other IEQ facets (e.g. visual comfort and IAQ) and thermal comfort are presently in progress.

References

1. ASHRAE (2016) Interactions Affecting the Achievement of Acceptable Indoor Environments - ASHRAE Guideline 10-2016, American Society of Heating, Refrigerating and Air Conditioning, Atlanta
2. CEN (2007) Criteria for the Indoor Environment including thermal comfort indoor air quality, light and noise. EN Standard 15251. Comité Européen de Normalisation, Bruxelles
3. d'Ambrosio Alfano FR, Olesen BW, Palella BI, Riccio G (2014) Thermal comfort: design and assessment for energy saving. Energy Build 81:326–336
4. Bellia L, Boerstra A, van Dijken F, Ianniello E, Lopardo G, Minichiello F, et al (2010) Indoor environment and energy efficiency in schools part 1: principles. REHVA Guidebook 13
5. Teli D, Bourikas L, James PAB, Bahaj AS (2017) Thermal performance evaluation of school buildings using a children-based adaptive comfort model. Procedia Environ Sci 38:844–851
6. d'Ambrosio Alfano FR, Ianniello E, Palella BI (2013) PMV–PPD and acceptability in naturally ventilated schools. Build Environ 67:129–137

7. d'Ambrosio Alfano FR, Palella BI, Ranesi G, Riccio G: Parameters that affect PMV in schools. In: Heiselberg PK (Ed) CLIMA 2016 - Proceedings of the 12th REHVA World Congress, Aalborg, vol 10, pp 1–8
8. ASHRAE (2017) Thermal environmental conditions for human occupancy. ANSI/ASHRAE Standard 55. Atlanta: American Society of Heating, Refrigerating, and Air-Conditioning Engineers, Inc.
9. Fanger PO (1970) Thermal Comfort: Analysis and Applications in Environmental Engineering. Danish Technical Press, Copenhagen
10. Zhang G, Zheng C, Yang W, Zhang Q, Moschandreas DJ (2007) Thermal comfort investigation of naturally ventilated classrooms in a subtropical region. Indoor Built Environ, 148–158
11. Brager GS, de Dear RJ (1998) Thermal adaptation in the built environment: a literature review. Energy Build 27:83–98
12. Fanger PO, Toftum J (2002) Extension of the PMV model to non-air-conditioned buildings in warm climates. Energy Build 34:533–536
13. ISO (1995) Ergonomics of the thermal environment e assessment of the influence of the thermal environment using subjective judgement scales. ISO Standard 10551. International Standardization Organization, Geneva
14. Sadat Zomorodian Z, Tahsildoost M, Hafezi M (2016) Thermal comfort in educational buildings: a review article. Renew Sustain Energy Rev 59:895–906
15. Indraganti M, Daryani RK (2010) Effect of age, gender, economic group and tenure on thermal comfort: a field study in residential buildings in hot and dry climate with seasonalvariations. Energy Build 42(3):273–281
16. Teli D, Jentsch MF, James PAB (2014) The role of a building's thermal properties on pupils' thermal comfort in junior school classrooms as determined in field studies. Energy Build 82:640–654
17. ISO (1998) Ergonomics of the thermal environment – instruments for measuring physical quantities. ISO Standard 7726. International Standardization Organization, Geneva
18. ISO (1998) Ergonomics of the thermal environment e determination of metabolic rate. ISO Standard 8996. International Standardization Organization, Geneva
19. ISO (2007) Ergonomics of the thermal environment and estimation of the thermal insulation and evaporative resistance of a clothing ensemble. ISO Standard 9920. International Standardization Organization, Geneva
20. ISO (2005) Ergonomics of the thermal environment - analytical determination and interpretation of thermal comfort using calculation of the PMV and PPD indices and local thermal comfort criteria. International Standardization Organization, Geneva
21. d'Ambrosio Alfano FR, Palella BI, Riccio G (2016) Notes on the calculation of the PMV index by means of apps. Energy Procedia 101:249–256
22. d'Ambrosio Alfano FR, Olesen BW, Palella BI (2017) Povl Ole Fanger's impact ten years later. Energy Build 152:243–249

Human Factor and Energy Efficiency in Buildings: Motivating End-Users Behavioural Change

Verena Barthelmes[1], Valentina Fabi[1], Stefano Corgnati[1], and Valentina Serra[1,2(✉)]

[1] Energy Department, Politecnico di Torino, Corso Duca degli Abruzzi 24, Turin, Italy
valentina.serra@polito.it
[2] AiCARR, Via Melchiorre Gioia 168, Milan, Italy

Abstract. Energy efficiency in buildings does not only rely on efficient technical solutions and design of the building features, but is also highly dependent on how occupants decide to set their comfort criteria, as well as on their energy-related and environmental lifestyles. In this perspective, raising user awareness among occupants by training them to adopt a more "green" and energy-friendly behaviour has become a crucial aspect for reaching energy efficiency goals in buildings. Motivating occupants to change their behaviour can become a challenging task, especially if they are expected to internalize and adopt the new behaviour on a long term. This means that information and feedback provided to the occupants must be stimulating, easy to understand, and easy to adopt in the daily routine. In this context, first methodological progresses are here presented within an European project, designed to raise user awareness, reduce energy consumptions and improve health and IEQ conditions in different typologies of demonstration case studies by providing combined feedback on energy, indoor environmental quality, and health. In particular, this paper presents one out of five MOBISTYLE demonstration testbeds – a residence hotel - located in central Turin (IT). In detail, this paper describes the setup of a tailored engagement campaign for hotel apartments and the reception area. Based on selected monitored variables, user-friendly feedback was defined to provide the users with real-time information on energy use and environmental quality, as well as guidance on how to save energy and optimize consumption profiles while creating an acceptably comfortable and healthy indoor environment.

Keywords: Behavioural change · Energy use · Residence hotel

1 Introduction

Human interactions with the building systems and envelope have a significant impact on the energy efficiency and indoor environmental quality in buildings. It is well known that occupant's behaviour is a factor that affects building energy performances [1–4].

S. Bagnara et al. (Eds.): IEA 2018, AISC 825, pp. 514–525, 2019.
https://doi.org/10.1007/978-3-319-96068-5_58

The way building components and systems are used accounts for substantial uncertainty over a building's energy use and occupants' comfort. In the field of persuasive technology, various innovative systems have been realized to motivate people to change their behaviour in a more sustainable way [5–7]. For example, ambient displays can provide users with their real-time energy consumption, serious games might educate people on saving energy in an enjoyable manner, and mobile or web applications may provide virtual rewards based on user's sustainability performances.

Menezes [8] argued that the knowledge of the "unregulated energy use" could improve to reduce the performance gap between predicted and actual energy use.

ASHRAE Standard 90.1 [9] defines "regulated" components of energy use such as energy use for heating, cooling, ventilation, interior lighting, hot water, and a few other end uses. But there is a multitude of "unregulated" energy uses within the building that are not addressed by the standard. For example, in office environment, "regulated" energy use patterns and impact within the building stock is generally well understood, while the impact of "unregulated" energy use [10], such as small power and desktop equipment, on energy consumption is not clear. "Unregulated" energy use is highly affected by occupant behaviour. Kawamoto [11] reported that desktop equipment energy use could be reduced up to 43% in an average working day. Furthermore, Mulville [10] confirmed that many employees did not switch off the desktop equipment at the end of the day, reporting that in US only 44% of computers and 32% of monitors were switched off after working hours. In UK up to 60% of employees do not turn off monitors. The reported studies demonstrated then that there is further space of research on understanding the unregulated energy use offering a significant potential for saving energy. Since unregulated energy use accounts for 73–88% of total equipment energy use [11], behavioural change programs could help on reducing energy consumption.

As Staddon [12] stated, academic research aims at directly assessing the energy savings due to behavioural change initiatives.

Perceived behavioural control is a central item in the Theory of Planned Behaviour defined by Ajzen [13] and it was demonstrated to be effective in the commercial sector [8, 14]. In particular, the study from Menezes [8] reported this factor having a major impact with respect to social norms and attitudes. Other studies demonstrated that social norms and feedback containing social comparison are effective triggers for promoting behaviour change in the working environment [15–17]. Social norms could be defined as behaviours admitted as "the norm" within a certain group of people [14]. Furthermore, social comparison or peer comparison is constituted in the human being to assess opinions and capacities.

Investigations showed that feedback and feedforward interventions methods achieved up to 20% energy savings in buildings [18]. In their studies, Fischer [19] demonstrated that occupants could be influenced in making choices by feedback information learning about consequences of their past actions. Matthies [20] reported the same result applying feedforward information (i.e. educating users before they take actions). Receiving this kind of information, people have an idea of the consequences of their possible actions on comfort and on energy consumption; thus, they could choose to perform (or to not perform) them. In this way, occupants are empowered to interact with the built environment in a more conscious way. Moreover, users learn about the control opportunities in their working area, increasing their perception of control on the surrounding environment and by consequence their comfort perception.

Moreover, scientists and practitioners, experts not only in the health sector, recognized and demonstrated that holistic environmental conditions play a significant effect on building occupants' health, although the general public has only recently started to understand the effect that this relationship can have on their day-to-day lives and well-being [21, 22].

In this context, first methodological progresses within the HORIZON 2020 MOBISTYLE project "Motivating end-users behavioural change by combined ICT based tools and modular information services on energy use, indoor environment, health and lifestyle" (grant agreement No. 723032) [23] are presented. The main goal of this project is to increase user awareness in order to reduce energy consumptions and improve health and IEQ conditions. The project will be tested in different typologies of demonstration case studies (residences, university, offices, health care, hotel) by applying a holistic approach to the engagement campaign. In particular, users will be provided with combined feedback on energy, indoor environmental quality, and health.

The purpose of the demonstration case is to illustrate and demonstrate the approach and methodologies proposed by the project. From there, the developed and derived ICT tools and services can be assessed. Case studies will furthermore be aimed at providing answers to the following questions:

- parameters that should be monitored, how the information should be presented;
- drivers and motivators influencing user behaviour;
- effects of user behaviour on energy consumption and indoor environment;
- methods and approaches for creating awareness and a long term behavioural change in users.

In particular, this paper focuses on the application of the MOBISTYLE approach to one out of five MOBISTYLE demonstration testbeds – a residence hotel located in central Turin (IT). This residence hotel is an urban long-term residence hotel that aims at achieving the nZEH (nearly Zero Energy Hotel) target and therefore represents a suitable testbed for demonstrating how energy efficiency can be further improved by transferring knowledge and tailored advices to building occupants. In this case study, specific challenges include defining tailored feedback for different end users of the hotel, such as hotel guests, but also the hotel management, receptionists, and cleaning staff. A tailored monitoring and engagement campaign was set up in four residential hotel apartments and the reception area. The monitoring campaign is based on continuous measurements of indoor environmental variables (indoor air temperature, relative humidity, CO_2 concentration), electricity consumptions of domestic appliances (washing machine, dishwasher, microwave, TV), and behavioural patterns (e.g. window opening/closings, thermostat regulation, occupancy). Based on these selected monitored variables, user-friendly Key Performance Indicators (KPIs) were defined to provide the hotel guests and hotel staff/management with real-time information on energy use and environmental quality, as well as guidance on how to save energy and optimize consumption profiles while creating an acceptably comfortable and healthy indoor environment.

2 Behavioural Change Strategies

Changing behaviour of individuals as a strategy for saving energy has been on the agenda for many decades. Significant progress has been made in assessing behavioural strategies, but yet there is still much more to do.

Since the 1970s, many researchers from various fields have studied how feedback on energy use impacts residential consumer understanding and behaviour [24–26]. Studies involving informative billing and periodic feedback have realized energy savings between 10 and 20%. On average, these studies found that real-time energy feedback resulted in overall energy savings of 10–15% [24–26]. It is assumed, based on theory and field research, that if residential consumers had more detailed and/or frequent information about their consumption, they would both better understand their energy use patterns and be able to change them effectively.

Two main conclusions can be drawn from a broad group of studies.

Van Houwelingen and Van Raaij [27] outlined three main functions of feedback:

- Feedback has a learning function: users learn about the connection between the amount of energy they use and the energy consuming behaviour;
- Habit formation: users put the information they have learnt into practice and may develop a change in a routine habit;
- Internalisation of behaviour: when people develop new habits after a while they change their attitudes to suit that new behaviour.

The design and the exploitation of behavioural change programs related to the energy use reduction can differ greatly when applied to residential or to office occupants [20]. Households have a direct connection between their actions and their energy monthly cost. In office buildings, energy savings campaigns are generally defined at the top level of the organisation, and there is not a direct relationship with individual workers benefits. Workers' motivation to engage in energy efficiency behaviours at workplace is therefore complex and must rely on corporate and social responsibility objectives and reinforcing social norms. Moreover, the nature of competition between individual colleagues or different offices could be a more influential driver than financial gain. There are also some reasons suggested in the reviewed literature as to why people's attitudes are different at home compared to the workplace [1, 28]. For instance, at home users are isolated from others. In the workplace, different norms apply. In tertiary sector, behavioural-based strategies can become critical, since the effects of the energy saving is not directly paid back to the employees. For this reason, the financial motivation is not an effective trigger such as in the residential sector.

Further research is still needed to investigate motivations in the office environments and how to use them for promoting energy efficient behaviours in offices.

The approach used in MOBISTYLE is human- or people-centred. This approach focuses on actual needs of the users and attempts to include their habits, practices, ideas, desires in new products and services with the main goal to change their behaviour and save energy in buildings. Many scientists and public health practitioners recognize that environmental conditions have an effect on health, but the general public has only recently come to understand that this causal relationship affects their day-to-day lives.

Besides thermal conditions it is well known that indoor air quality is also strongly linked to human health.

This holistic approach is addressed to develop an effective methodology that combines together three main research areas related to the occupants (i.e. energy, indoor environmental quality and health) to gain new insights on the way people perform actions in the surrounding built environment. To stimulate users to perform more conscious actions, a combination of data collection, data analysis and elaboration, and tailored information should be provided.

As a first step, the proposed methodology relies on a data collection to investigate the operation of users on energy systems through behaviour-related data (including objective monitoring of the building and subjective data collection from the occupants). Then, data analysis and elaboration to depict human behaviour (related to energy, comfort, health) through user data (both wearable sensors and surveys, for example monitoring human presence and practices thanks to anthropological studies) allow for obtaining performance indicators. Finally, energy savings and the improvement of indoor environmental conditions and occupants' health and wellbeing are achieved by integrating behavioural communication strategies (awareness campaign).

The tailored information campaign should rely on simple and immediate information using different communication methods in order to be understood by the selected users and achieve an effective knowledge transfer. For this reason, different feedback typologies and strategies should developed to best suit the case studies. In the following, the methodology of MOBYSTYLE project is presented applied to a case study.

3 Case Study Description

A residence hotel, an urban hotel located in a central area of Turin, is the Italian demonstration case (Fig. 1). This case study presents a very traditional structure since the building was built at the beginning of 20th century with load bearing masonry walls. The building is heated by two condensing boilers powered by natural gas, also used for Domestic Hot Water (DHW) production. A chiller (cooling capacity 97 kW) is installed for the cooling system. Two-pipes fan coil units, placed in the false ceiling, are the terminals of the heating and cooling system (except radiators inside bathrooms). At present, the building does not have mechanical ventilation system (except for exhaust air systems in bathrooms and kitchens) and it does not a use any on-site renewable energy source.

3.1 Monitoring Campaign

The application of MOBISTYLE approach will be carried out in selected apartments and in the reception of the residence hotel. All the guestrooms are comparable to small apartments in terms of internal layout and equipment as the hotel business mainly relies on guests' long-term stays.

Fig. 1. Views of the case study.

The selected apartments are two-room or three-room flats but with similar size (between 37 m^2 and 39 m^2). The reception area has one entrance door and one window with northwest exposure.

Figure 2 depicts types and locations of the sensors installed in the two apartment typologies. In all apartments, measurements regarding energy consumption and indoor environmental quality (IAQ) will be taken. In particular, the following parameters will be gathered:

- indoor air temperature (T) [°C]: every 15 min;
- indoor relative humidity (RH) [%]: every 15 min;
- indoor CO_2 level (CO_2) [ppm]: every 10 min;
- electricity consumption [kWh]: every 10 min;
- electricity power [W]: every 10 min.

To understand the mechanisms of occupants' behaviours, it is necessary to monitor the behaviour itself but also to determine the cause and effect relationships that behaviours have with energy consumption, indoor and outdoor environment, and the occupants' health status.

The types of behaviour analysed in this case study can be listed as follows:

- Occupancy;
- Thermostat adjustment;
- Window opening;
- Whitegoods or other electrical devices;
- Door opening.

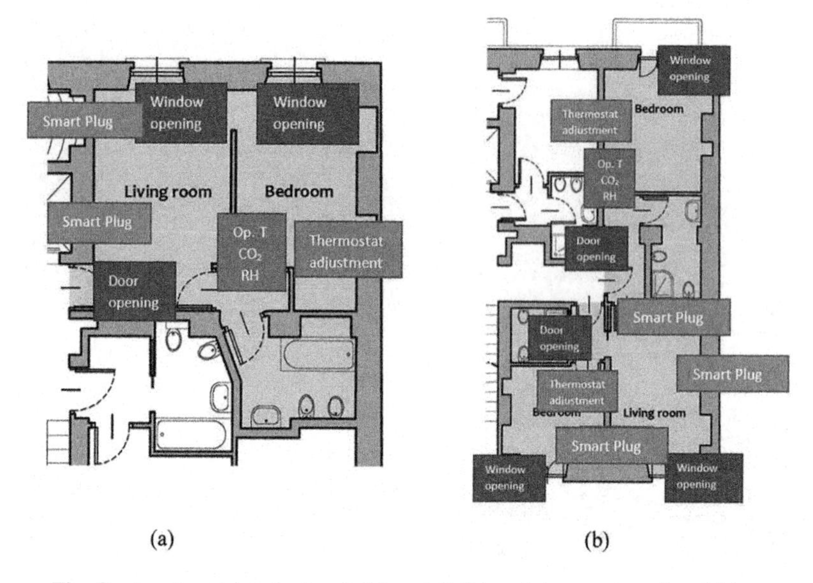

Fig. 2. Apartment typologies A (a) and B (b) and the monitored variables.

Furthermore, the monitoring campaign will also be aimed at measuring Physical activity (PA) in order to see if the PA pattern changes due to variability in the environment (such as climate).

Outdoor conditions during the campaign will be established by third-party outdoor data logging. Indeed, outdoor environmental variables, such as temperature, solar radiation, relative humidity, wind speed/direction, will be acquired through an online source. The monitoring campaign is currently ongoing and takes place for a period of 24 months, from March 2018 until March 2020 (both included).

3.2 Awareness Campaign

As MOBISTYLE demonstration in the residence hotel cannot be seen as a stand-alone measure, it will be supported with continuous awareness measures, information, and other feedback campaign activities for raising awareness around conscious energy usage, healthy lifestyle and the importance of living in a healthy indoor environment. Continuous interaction with all the building users of the residence hotel will be done through a mixture of communication channels to raise awareness of all end-users in the building (building manager, hotel guests, receptionists, cleaning staff). It is believed that their awareness will be increased even further when there is continuous provision of information and through different channels.

The actors involved in the project include selected guests of the monitored apartments and a focus group of users (6 participants). The focus group members will contribute with their interviews and feedback before, during, and after the measurement campaign as part of thick data collection, while the other hotel guests participating in the project will contribute with the data measured during the observation and monitoring period. It should be noted that the new and more aware behaviour encouraged at the residence hotel might lead to energy savings also in other situations the users will encounter in their everyday life. Furthermore, they may encourage other people to act according to the MOBISTYLE objectives and therefore more people are becoming aware (snow-ball effect).

Within the MOBISTYLE project, tailored information will be provided to the users of the residence hotel according to specific needs and characteristics of the hotel environment and occupants who interact with it (manager, guests, receptionists, cleaning staff). A tailored program will introduce to the building users the holistic approach of the project, its main objectives and the application at the case study. Furthermore, it will be clearly explained how data protection and user privacy is ensured.

Furthermore, awareness campaigns will promote "temperature trainings" not only as part of demonstrators but as a part of overall and healthy lifestyle and healthy ageing strategy. As example, users are educated that lowering down thermostat will not just bring energy savings but can also contribute to their better well-being and metabolic health. Moreover, it can be encouraged that users take stairs instead of an elevator as this does not only saves building's electricity consumption but also is healthier for them. Furthermore, the clever and conscious use of appliances (e.g. stand-by mode) and lighting might help to reduce energy consumptions and can be an interesting trigger for savings costs (manager), as well, next to reducing the environmental impact of the activity.

The project and its approach has been currently introduced to the hotel manager, two receptionists and the cleaning staff. The hotel manager engages in an aware energy-related management of the building and seems highly motivated to implement the MOBISTYLE strategies in the residence hotel to further improve the energy performance of the building and to raise awareness among the hotel guests and staff - while possibly optimizing their health and comfort conditions, as well. The hotel staff (receptionists and cleaning staff) generally seems to present a slightly lower level of knowledge regarding energy-related topics and achieving energy savings for them still seems to play a less important role, however, they appear very willing to learn and to be actively involved in the project. Hotel guests have not been investigated so far.

To benefit from the opportunity of collecting a large amount of data, the data interpretation and, subsequently, the feedback provided to the users as part of the monitoring campaign in the case study will be based on a combination of standards, measurements, and relative comparisons between an established baseline and the individual units in question.

Visualization of Feedback and KPIs. The feedback to the users will be given in a graphical form through an app on their mobile phones (see Fig. 3).

Information regarding energy use is based on the overall electricity use in the apartment and on the usage of single electrical appliances (depending on which appliances are used in the specific apartments). Depending on the user's choice, feedback can be based on data from the day before, the week before, the month before or the current year. The generic profile for electricity consumption and appliances usage will be developed using the energy consumption data collected during the initial

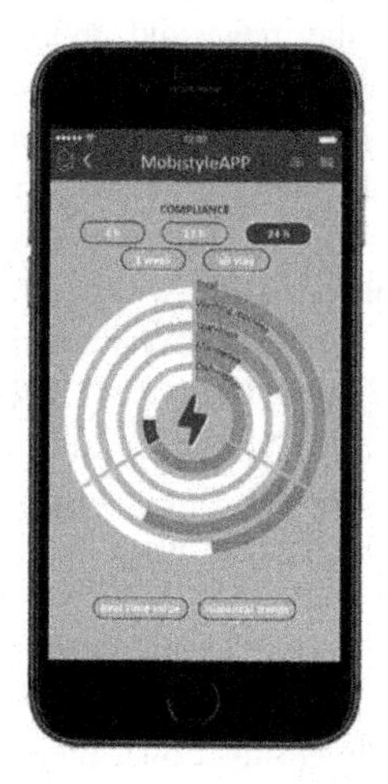

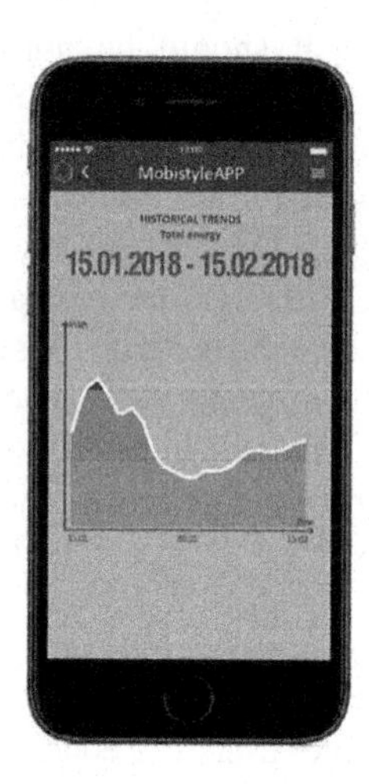

Fig. 3. Example of feedback visualization (energy use) on a hypothetical application.

monitoring. The energy consumption data will be displayed to the end user in a graphical form using three colour levels (green, orange and red). The relationship between colour levels and user consumption is based on the generic profile defined during benchmarking and energy saving targets for the case study (energy saving of 16%). Users must also have the possibility to read real-time consumption and historic records.

Similarly to the representation of energy usage, the IEQ conditions are translated to the user mainly through a graphical form that involves also in this case the use of three levels of colour, depending on their trend with respect to threshold values defined by the standards [29]. Historic records and real time values of temperature, CO_2 level and relative humidity of the apartment should be provided for the end users.

A challenge of the MOBISTYLE project is that, next to feedback on energy use and IEQ, the users will be able to visualize KPIs also related to health aspects, such as the heart rate. Heart rate is the amount of contractions of the heart muscle in a certain period of time and can be influenced easily by changes in environment or activity.

4 Future Steps

The experimentation is on-going and divided into three monitoring phases with provision of different levels of feedback, where each of them have their own objectives:

- Benchmark definition: Monitoring phase without feedback provision to establish reference scenarios;
- Feedback provision: Monitoring phase with feedback provision on energy, IEQ and health;
- Optimized feedback provision: Monitoring phase with optimized feedback provision - this optimization is based on an intermediate evaluation that measures the

impact on performance and changes in user practices by different feedback typologies.

Currently, the experimentation in the case study has started with the benchmark definition in March 2018. Feedback provision will start approximately in November 2018.

5 Conclusion

The aim of this paper was to describe first methodological progresses within the HORIZON 2020 MOBISTYLE project "Motivating end-users behavioural change by combined ICT based tools and modular information services on energy use, indoor environment, health and lifestyle" (grant agreement No. 723032). This paper is aimed at highlighting that for creating energetically sustainable and healthy environments at "human scale" it is necessary to raise user awareness through an holistic approach. In particular, this study underlines the necessity of applying a holistic approach to the engagement campaign by providing combined feedback on energy, indoor environmental quality, and health. In particular, this paper focused on the monitoring and awareness campaign in the Italian testbed, a residence hotel in Northern Italy. Indeed, this paper describes the setup and challenges of a tailored engagement campaign for hotel apartments and the reception area. Based on selected monitored variables, user-friendly feedback was defined to provide the users with real-time information on energy use and environmental quality, as well as guidance on how to save energy and optimize consumption profiles while creating an acceptably comfortable and healthy indoor environment.

Acknowledgements. This work is part of the research activities of an international project financed by European Community MOBISTYLE – Motivating end users Behavioural change by combined ICT based tools and modular Information services on energy use, indoor environment, health and lifestyle (Grant Agreement no: 723032). The Authors acknowledge the support of all the partners of the project for sharing their expertise for this activity.

References

1. Orland B, Ram N, Lang D, Houser K, Kling N, Coccia M (2014) Saving energy in an office environment: A serious game intervention. Energy Build 74:43–52
2. Andersen RK, Fabi V, Toftum J, Corgnati SP, Olesen BW (2013) Window opening behaviour modelled from measurements in Danish dwellings. Build Environ 69:101–113
3. D'Oca S, Fabi V, Corgnati SP, Andersen RK (2014) Effect of thermostat and window opening occupant behaviour models on energy use in homes. Building Simulation 7:683–694 ISSN 1996-3599
4. Fabi V, Andersen RV, Corgnati SP, Olesen BW (2012) Occupants' window opening behaviour: a literature review of factors influencing occupant behaviour and model. Build Environ 58:188–198

5. Gustafson C, Longland M (2008) Engaging employees in conservation leadership. In: Proceedings of 2008 ACEEE Summer Study Energy Efficiency in Buildings. http://www.aceee.org/files/proceedings/data/papers/7532.pdf
6. Adamson S (2010) Using TLC to reduce energy use. In: Proceeding ACEEE Summer Study Energy Efficiency in Buildings. http://www.aceee.org/files/proceedings/2010/data/papers/2124.pdf
7. Wood G, Newborough M (2003) Design and functionality of prospective of energy consumption displays. In: Proceeding of the 3rd International Conference on Energy Efficiency in Domestic Appliances and Lighting (EEDAL 2003)
8. Menezes AC, Tetlow R, Beaman CP, Cripps A, Bouchlaghem D, Buswell R (2012) Assessing the impact of occupant behaviour on electricity consumption for lighting and small power in office buildings. In: 7th International Conference on Innovation in Architecture, Engineering and Construction, Sao Paolo, Brazil
9. ANSI/ASHRAE/IES Standard 90.1-2016 – Energy Standard for Buildings Except Low-Rise Residential Buildings
10. Mulville M, Jones K, Huebner G (2014) The potential for energy reduction in UK commercial offices through effective management and behaviour change. Architectural Eng Des Manage 10(1–2):79–90
11. Kawamoto K, Shimoda Y, Mizuno M (2003) Energy saving potential of office equipment power management. Energy Build 36:915–923
12. Staddon SC, Cycil C, Goulden M, Leygue C, Spence A (2016) Intervening to change behaviour and save energy in the workplace: a systematic review of available evidence. Energy Res Soc Sci 17:30–51
13. Ajzen I (1991) The theory of planned behaviour. Organ Behav Human Decis Process 50:179–211
14. Mulville M, Jones, K, Huebner, G, Powell-Greig J (2016) Energy-saving occupant behaviours in offices: change strategies. Build Res Inf, 1–14
15. Chen H-M, Lin C-W, Hsieh S-H, Chao H-F, Chen C-S, Shiu R-S, Ye S-R, Deng Y-C (2012) Persuasive feedback model for inducing energy conservation behaviours of building users based on interaction with a virtual object. Energy Build 45:106–115
16. Fogg BJ (2003) Persuasive technology: using computers to change what we think and do. Morgan Kaufmann Publishers Inc., San Francisco
17. Emeakaroha A, Ang CS, Yan Y, Hopthrow T (2014) A persuasive feedback support system for energy conservation and carbon emission reduction in campus residential buildings. Energy Build 82:719–732
18. Meinke A, Hawighorst M, Wagner A, Trojan J, Schweiker M (2016) Comfort-related feedforward information: occupants' choice of cooling strategy and perceived comfort. Build Res Inf 45(1–2):222–238
19. Fischer C (2008) Feedback on household electricity consumption: A tool for saving energy? Energ Effi 1(1):79–104
20. Matthies E, Kastner I, Klesse A, Wagne H-J (2011) High reduction potentials for energy user behaviour in public buildings: how much can psychology-based interventions achieve? J Environ Stud Sci 1:241–255
21. Kingma B, van Marken Lichtenbelt W (2015) Energy consumption in buildings and female thermal demand. Nature Climate Change Epub
22. Malchaire J, d'Ambrosio FR, Palella BI (2017) Evaluation of the metabolic rate based on the recording of the heart rate. Ind Health 55:219–232
23. H2020-EE07-2016-IA - MOBISTYLE - MOtivating end-users Behavioural change by combined ICT based tools and modular Information services on energy use, indoor environment, health and lifestyle - Grant Agreement no: 723032

24. Faiers A, Cook M, Neame C (2007) Towards a contemporary approach for understanding consumer behaviour in the context of domestic use. Energy Policy 35:4381–4390
25. Miller S (1984) New essential psychology: experimental design and statistics. Routledge, London
26. Stern P (1992) What psychology knows about energy conservation. Am Psychol 47:1224–1231
27. Van Houwelingen JT, Van Raaij WF (1989) The effect of goal setting and daily electronic feedback on in-home energy use. J Consum Res 16:98–105
28. Gulbinas R, Jain RK, Taylor JE (2014) BizzWatts: a modular socio-technical energy management system for empowering commercial building occupants to conserve energy. Appl Energy 136:1076–1084
29. Standard EN 15251 (2008) In criteria for the indoor environment including thermal, indoor air quality, light and noise. European committee for standardization (CEN), Brussels, Belgium

An Application of Ergonomics in Workstation Design in Office

Ana Paula Lima Costa[1(✉)] and Vilma Villarouco[2]

[1] Ministério da Fazenda, Recife, PE, Brazil
aplimacosta@gmail.com
[2] Universidade Federal de Pernambuco, Recife, PE, Brazil

Abstract. This study sought to insert the systemic view provided by the ergonomic approach in the design practice of the office workspace, particularly in the aspect of delimiting the amount of area occupied by the user. It was verified that technical publications and regulatory norms provide data to mark office space, but it was identified in the specialized literature deficiency in the application of specific data of identification of the activities and the quantification of space. A methodology based on the Hypothetical-deductive research method was developed. As a case study, an office of a public office was proposed. From the analysis of the workspaces, it was noticed that, besides the dimension, the form of distribution influenced directly in the functionality of the same, concluding that not only the quantity of space determined its functionality, but that the form of distribution and use of work surfaces directly influences the functionality of the workspace. This article was extracted from the doctoral thesis of the author Costa.

Keywords: Ergonomia organizacional · Metodologia de projeto
Projeto de escritório

1 Introduction

Design methodologies of constructed environments rationally plan and impose space standards that often do not match people's cultural patterns. Assuming that a working environment is composed of a set of jobs, the quantification and configuration of this environment should be based on the set of needs of individuals using space. Therefore, without researching into the habits of users, designers will not be able to base their design parameters, needing to know the user in the performance of their duties to identify the activities performed and to be able to configure the environment according to their real needs. The help of a tool that introduced the use of ergonomic factors in the projects would contribute to identify essential elements of human activity, contributing to the project meeting the users' wishes.

The ergonomic approach to the design process of built environments can offer the possibility of new references for the configuration of the space, either by the architectural approach, or by the vision of the interior design, by the production environment designer, and by the company administrator, opening new possibilities for solutions that correspond to organizational needs and objectives.

S. Bagnara et al. (Eds.): IEA 2018, AISC 825, pp. 526–536, 2019.
https://doi.org/10.1007/978-3-319-96068-5_59

Thus, the application of ergonomic analysis to know the activities carried out by the user, in consonance with the systematic use of space dimensioning data, could provide an adequate space for the user to accommodate the activities.

This article show the aspect of how to conduct the configurations of workstation in the office, one of the subjects developed in the doctoral thesis of the author Costa [5] is presented.

2 The Office Space

The office exists for a company to accomplish its purposes with efficiency and effectiveness. Technical publications and regulatory standards focus on providing data to determine the shape and size of office desks, indicating the widths, depths and spaces of movement in the workplace, and organizing anthropometric data to be applied in the design of previous spaces [2, 12, 13]. Some authors focus on issues of organization and work space design, suggesting occupational areas to be assigned to each worker, following hierarchical and functional standards [3, 4, 6, 8, 10, 11, 14]. With this, the researchers intend to provide the community of designers and consultants with dimensional references that enable the development of projects in office environments that meet their expectations. The dimensioning of the space is also guided by technical norms that standardize anthropometric measures to be used, being documents of mandatory compliance by Brazilian companies [1].

3 Workstation Configuration Methodology

The present project script has resulted from the adaptation of the hypothetical-deductive scientific method [9], which, starting from a conflict among existing theories, offers a new theory that can be challenged in the intention to eliminate error. Applying to the space design case, existing theories could be characterized as the work space distribution patterns, identifying the deficiency in the application of specific data to identify the activities and the quantification of space to perform them.

In the adaptation of the Hypothetical-Deductive method, the steps were unfolded in phases that translate the result of the ergonomic investigation into guidelines of space configuration, where the data found result in the quantification of space.

The Methodology of job configuration is developed in the following steps: “Identification of the object of study”, to know the work situations; “Proposal of space organization”, to synthesize the activities carried out in the place; “Space use forecasting”, to define the patterns of actions and spaces; “Space formatting”, to assign patterns of activity and activity modules, resulting in the proposal of the job; and “Verification of the results found”, to verify the proposed solution (Diagram 1).

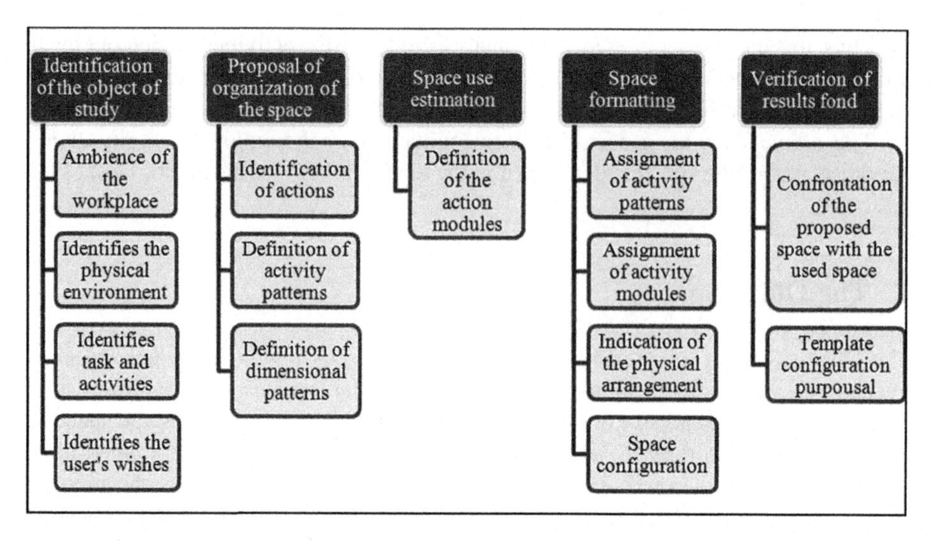

Diagram 1. Workstation Configuration Methodology Source: Author. 2014

4 Application of the Methodology of Configuration of Workstation

In the doctoral thesis was made a study in twelve kinds of workstation in a Brazilian public company. In this article, it will be demonstrated the application of the methodology in the design project of a one of the jobs of the public company.

4.1 Identification of the Study Object

Study carried out in environments investigating the planned activities to occur in the new space, verifying the area occupied in the execution of the activities.

In the workplace environment, a global approach is taken to know the productive unit and understand the operation of the company.

The company under study is part of the governmental and administrative structure of the Brazilian government. With a staff of two hundred and two employees, distributed in seven hierarchical levels, its duties are to technically support the areas of budgetary, financial and patrimonial execution and to carry out the activities of human and logistic resources.

In the Identification of the physical environment of the workplace, the activities with space are related in the following steps:

In the identification of the typology of space use it can be observed the distribution of the physical space and the relation with the administrative policy of the company.

The building is divided into two floors, occupying 1,364.60 m^2, divided into thirty-three work rooms and 16 service support rooms. The rooms have different configurations, having individual rooms and shared by people of different hierarchical levels. The jobs are fixed and individualized, with different sizes and formats.

In the analysis of the use of the space of the workplace in the horizontal plane, it is understood the use of the space of the workplace and the layout of the room.

The work room where this study was done there are three employees, occupies 15.22 m^2, allocating 5.07 m^2 of area to each employee. The area reserved for the specific workplace that the study was done offers room for user movement, but did not provide accommodation for an interlocutor. The workplace occupies 2.04 m^2, distributed in a rectangular space measuring 1.62 m $\times$ 1.26 m (Drawing 1).

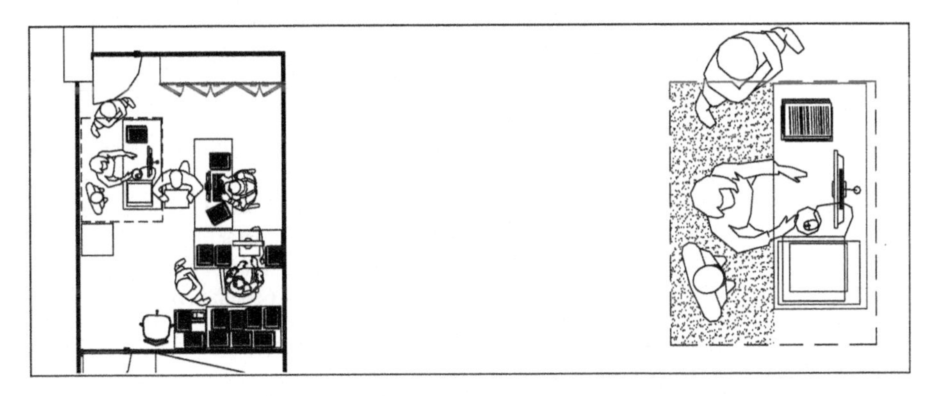

Drawing 1. Floor plan of the work room (left) and Floor plan of the workstation (right) Source: Author. 2014

In the Identification of the tasks and activities of the workplaces, the activities are related to each user and the position occupied.

In addition to collecting information with the performers themselves, the video recording is used as a means of recording the effective moment of the activity execution of these activities, through the iSEE program [7]. The software provides statistics indicating the postures adopted and the time spent in each of them. With this, a framework is assembled in which the information provided by the users, the prescribed tasks, the presumed activities and the activities actually carried out are confronted.

The work team of the user competed the analysis and instruction of the administrative processes, and executed the actions of fulfillment of judicial decisions. In order to relate the tasks and activities carried out by the workers, information was collected from the performers themselves.

The job of the user of the workplace has the function of support from the Inactive and Pensioners Service in the Human Recurs department. The Table 1 shows the tasks and activities performed by the user at the workplace.

The tasks assigned to the User were the analysis and execution of administrative processes and the provision of guidance. When counting the data through the iSEE program [7], it was observed that the user interacted with people, using 33.84% of their time. The interaction of the user with the computer occurred in 27.22%, and in 5.88% of that time, the user used the computer while interacting with other people. 9.57% of the time was spent on reading and writing papers and volumes, at 30.91% the user performed other actions and/or was absent from the job. Thus, it turns out that the user

Table 1. Tasks and activities

Tasks	Activities
Analysis of the processes for the granting of retirement and pensions	By receiving on a physical basis the processes performed by employees from other sectors, examines them, analyzing the content and the procedures adopted, in order to certify the compliance with established norms
Elaborate the annual report of processes	Control over number of published cases of the year
Review of processes	Analysis and study of procedural documents, in physical and digital media, to update the processes of retirement, through consultation of legislation. Handling of papers and searching for information on digital media
Orientation to servers	Face-to-face assistance to co-workers, with possible trips to third-party jobs, in which doubts about legislation are resolved, through consultation of legislation in physical and computerized environment

Source: Author. 2015

both uses his time to engage in bureaucratic activities and to interact with other people; it is concluded that the activities that he performs require interaction with other employees who are attached to his duties, which comprised 36.79% of the time he was in the job, which is consistent with the user's statements about his duties. However, it was not informed by the user whether these actions were carried out on scanned or paper documents, simultaneously or in isolation, which required different space for movement, and consequently space in the workplace. Likewise, guidance to colleagues was also not detailed whether it was conducted in digital, face-to-face or by telephone. It was verified in the filming the sharing of the screen of the computer with visitors, leading to understand that, at that moment the workers were analyzing together subjects of work, which also would demand work space that accommodated the users. Needs, feelings, and desires related to the environment are known through the "Poem of Desires" [5]. The user expressed that he would like more space of movement and storage, and that the work surface was bigger.

4.2 Proposal of Organization of Space

Using the data of the physical space and the actions performed in the work environments, the patterns of actions and dimensional patterns are defined.

Activities that occur in similar ways are a pattern that synthesizes events in place, generating a set of actions called the "action pattern". By setting up action standards, it becomes possible to allocate spaces that can support these actions. In view of the variety of actions identified in the workplace, through the analysis performed with the software, it was sought to group actions that presented similarity of execution to create groups of forms of execution of actions, obtaining a result which was more applied, to analyze the use of space, verifying the need for space. Thus, the activities were grouped, creating groups called "Action Pattern" (Table 2).

Table 2. Identification, description and standardization of actions in the workplace

Action identified	Description of the action	Action pattern
Reading of the computer screen	The user views the computer screen	Interaction with the computer
Entering data on the computer	The user interacts with the computer through the mouse and keyboard	
Attendance	The user talks to another individual, visitor to his post; receives from the visitor material such as: papers and documents; provides information to the visitor contained in physical and digital documents	Interaction with people
	The service is also carried out at a distance, by telephone	
Speak on the phone	The user interacts by talking on the phone	
Working together	The user performs with visitors reading the papers and the computer screen, writes on papers and enters data on the computer with the keyboard and mouse	
	The activities are carried out by both the user and the visitors, and sometimes the visitor assumes the position of the station owner to operate the computer, handle the mouse and the keyboard	
Using computer and talking	Introduction and reading of data on the computer simultaneously with interaction with people	Interaction with people and compute
Reading	The user reads papers and documents that are supported on the work surface.	Writing and reading
Writting	The user writes on paper using pens and pencils, and supports the material on the work surface	
Storage of work material	The user stores documents that are in your custody	Handling of paper and office supplies
Handling and storage of work and personal use material	The user manipulates and conditions office objects (stapler, calculator) and of personal use	

Source: Author. 2014

Based on the identification of activities that occurred most frequently in the workplace, a set of actions representative of the tasks performed in the office was defined. These actions are: "writing and reading", "computer interaction", "interaction with people", "interaction with people and computer simultaneously", and "handling and storage of papers and office supplies".

The definition of a dimensional standard to perform the activities was determined using the brasilian regulatory standard Associação Brasileira de Normas Técnicas - ABNT [1] and scientific research based on ergonomic principles of Neufert [12], Boueri [2] and Panero and Zelnik [13]. These areas of activity execution comprise the areas effectively used in the performance of activities. The deployment of complementary furniture for storage of work material or personal items is considered.

4.3 Estimate Use of Space

For each action that takes place in the space, a module with the necessary space for the activity to be developed, based on the analysis and selection carried out in the definition phase of dimensional patterns was assigned. The action modules are used as reference standards in the development of the project, involving the interface areas between the human body and the workplace, encompassing the physical dimension, scope and free space.

Definition of the Action Modules

The action modules were used as reference standards in the development of the project, involving the interface areas between the human body and the workplace, encompassing the physical dimension, scope and free space.

Action 1 - Writing and reading
Action 2 - Computer Interaction
Action 3 - Interaction with people
Action 4 - Interaction with people and computer
Action 5 - Handling and keeping of papers and office supplies

The action of "writing and reading" (Module 1) is performed in conjunction with "interaction with the computer" (Module 2). The fulfillment demands a work surface with compatible depth to accommodate two people sitting face to face (Module 3). When the consultation and reading of the computer screen are joint, the visitor can position himself next to the position holder, in the space he occupies when he will handle papers, while the other user interacts with the computer. There is also the demand for a compartment for the storage of personal objects (Module 4) and for storing work material (Module 5). The largest depth dimension of the surface is adopted for all modules in order to standardize the width of the table (Drawing 2).

4.4 Formatting the Space

The analyzes of the workplaces allowed the identification of the activities carried out and the spaces in which they are developed. The combination of results provided information to identify the spaces needed to meet the expectations of the company's administrative policy, the operational needs of the tasks and the demands of space to carry out the activities. As a result, the work space will be configured based on the

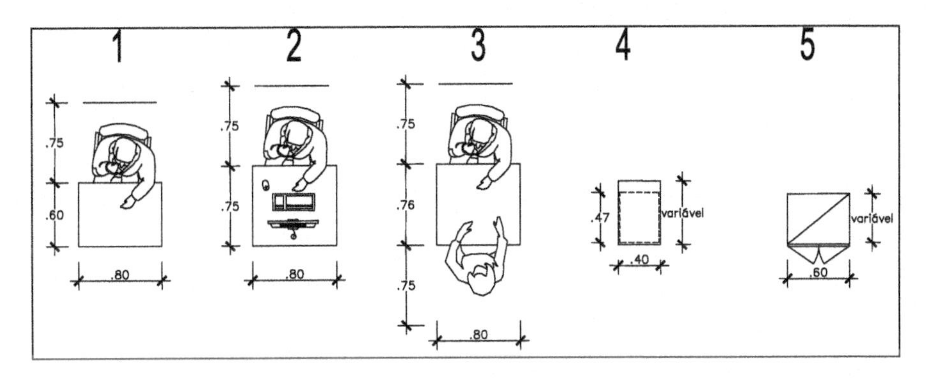

Drawing 2. Floor plan of Action Module

knowledge of the tasks assigned to the job and the activities carried out, to which "action patterns" and their "action modules" are related.

The combination of the modules will lead to the formatting of the physical space of the workplace.

The Process of "analysis and correction" activities are performed through "reading on the computer screen and on paper", "entering data with the mouse and keyboard". During the process, folders and papers are manipulated. During the orientation to the servers there are conversations between the user and visitors, in person and by telephone. It was observed the accumulation of material on the table, evidencing the need to provide storage compartments.

The Table 3 shows the indication of the physical arrangement and determination of the action modules according to the user's activities.

In this way, the workplace will have the frontal work surface measuring 1.75 m, the lateral surface with 2.12 m, the working side measuring 0.76 m, and the supporting side of the drawer and the cabinet measuring 0.60 m. The area occupied by the module will be five meters and two square centimeters ($1.75 \times 2.87 = 5.02$ m^2) (Drawing 3).

4.5 Verification of Results

In intention of verification of results, the studied object is confronted with the work model proposed in order to verify the efficiency of the proposed solution.

The graphical representation of the existing and proposal occupation is in the Drawing 4.

When comparing the existing workspaces and the workspaces proposed by the configuration methodology, it was verified that the areas occupied by the stations were smaller than those proposed by the configuration methodology; thus, the work space should be increased to meet the needs of the activities performed at the station. When comparing existing workspaces, it was noticed that the amount of area within the work room did not reflect the amount of workspace actually used by the user. It was found that the layout of the environment and the workstation was what actually determined the way the stations were used, since the arrangement of furniture had a direct influence on the user's movement spaces and on the ways of using the station. In this way, it was

Table 3. Activities, Actions, Modules and Layout

Activities	Actions	Actions standards	Actions modules	Layout - physical arrangement of the room
Process analysis. Correction of colleagues' processes.	Reading on computer screen and paper, entering data with the mouse and keyboard. Handles folders and papers.	Action 1 - Writing and reading Action 2 - Computer Interaction	Module 1 - work table dimension to perform activities with writing and reading roles and other office activities.	Shared room with occupants that carry out the same type of activity. Distancing jobs so that the conversation does not harm the room neighbors
Server orientation. It guides doubts about legislation, through consultation of legislation in physical and computerized environment.	Group conversation, with one person and on the phone. Digital consultation on legislation. Consultation in physical environment - papers and volumes.	Action 1 - Writing and reading Action 2 - Computer Interaction Action 3 - Interaction with people	Module 2 - work table dimension to carry out activities with the use of computer. Module 3 - work table dimension to perform service. Module 4 - Drawer for personal objects. Module 5 – Office cabinet	

Source: Author. 2015

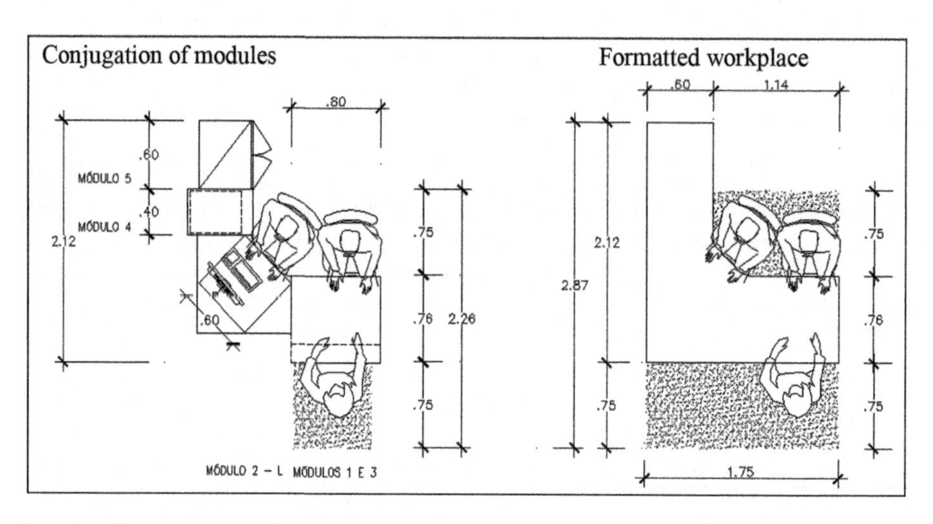

Drawing 3. Representation of the modules used and the space formatted with the use of the work surface in the "L" format of the User. Source: Author. 2015

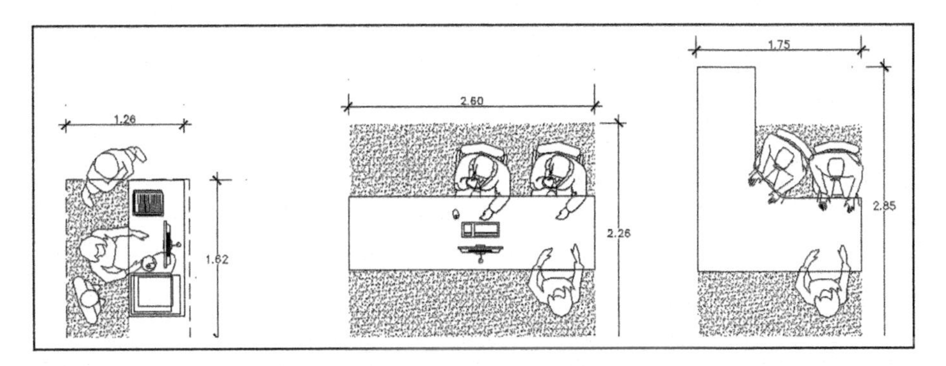

Drawing 4. Graphical representation of the existing configuration of the workplace (left), the Rectangular workplace occupation proposal of workplace (center), and the workplace purposal in "L" (right). Source: Author. 2015

concluded that it was not the amount of space, but the way the space was distributed and used that determined whether that space would help or hinder users from performing their functions.

When analyzing the layout that would be used in the rooms where the stations would be leased, it was verified that it could be proposed the layout of shared rooms, because the need for private rooms was not mentioned or detected in the analyzes.

5 Results

The decision about the physical arrangement of an office is related to organizational culture and functional need. The area of an environment that provides adequate conditions of use to the functions can be obtained by the spatial arrangement of these spaces of activities.

The structure of the proposed model deals with the adaptation of work spaces to human needs and limitations, which are the object of research in the scientific discipline of ergonomics, whose multidisciplinary application of knowledge can generate possibilities of hypotheses and contextualizations. From its application, it becomes possible to format the workstations for the activities carried out, size the physical spaces and list the typology of the spaces to meet the expectations of the company's administrative policy.

It was verified that the areas proposed by the methodology of space configuration were smaller than the areas indicated by the literature and by the standards of design of the company. However, the work surfaces proposed by the configuration methodology are larger than the work surfaces indicated in the literature [2–4, 6, 8, 11–14]. The reason is because the methodology conducts to planning of design to identify, fractionate and aggregate the activities planned to be performed at the station.

The results indicate that the amount of space does not determine the adequacy of space to the activity. It was detected in the analysis of the workstations that the configuration of the station, even with areas compatible with the areas proposed in the literature, did not always meet the needs of carrying out the activities.

In this way, it was noticed that in addition to the size and quantity of work surface, the way of distribution and use of work surfaces directly influences the functionality of the workspace.

References

1. ABNT Associação Brasileira de Normas Técnicas. http://www.abnt.org.br/abnt/conheca-a-abnt. Acesso em 21 May 2015
2. Filho B, Jorge J (2008) Projeto e dimensionamento dos espaços da habitação - Espaços de atividades. E - book- Livro II, Estação das Letras e Cores, São Paulo
3. Brill M, Margulis ST, Konar E, Bosti (1984) Using office design to increase productivity. Workplace Design and Productivity, Inc., vols 1 and 2, Buffalo, NY
4. Charles KE, Danforth AJ, Veitch JA, Zwierzchowski C, Johnson B, Pero K (2004) Workstation Design for Organizational Productivity. Practical advice based on scientific research findings for the design and management of open-plan offices. National Research Council of Canada and Public Works & Government Services Canada. ISBN 0-662-38514-4. Disponível em: http://irc.nrc-cnrc.gc.ca/ie/productivity/index. Acessado em 01 Apr 2012
5. Costa, APL (2016) Contribuições da egonomia para a composição de mobiliário e espaços de trabalho em escritório. Tese de Doutorado em Design na Universidade Federal de Pernambuco, Recife
6. Cury A (2005) Organização e métodos: uma visão holística. 8ª edição. Editora Atlas, São Paulo
7. Filgueiras EV (2012) Desenvolvimento de um método para avaliação da interacção homem/cadeira de um escritório numa perspectiva sistémica e ecológica. Tese de Doutorado de Motricidade Humana na especialidade de Ergonomia da Universidade Técnica de Lisboa
8. Harris DA, Palmer AE, Lewis MS, Munson DL, Meckler G, Gerdes R (1981) Planning and Designing the Office Environment. Van Nostrand Reinhold Company, New York
9. Marconi MdA, Lakatos EM (2009) Fundamentos de Metodologia Científica. 6ª Edição. Editora Atlas S. A. São Paulo
10. Marmot A, Eley J (2000) Office Space Planning – Designing For Tomorrow's Workplace. McGraw-Hill, New York
11. van Meel J, Martens, Y, van Ree HJ (2010) Planning Office Spaces - A Pratical Guide for Managers and Designers. Laurence King Publishing, London
12. Neufert E (2013) Neufert. Arte de projetar em arquitetura. 18ª edição, Editora Gustavo Gili, Barcelona
13. Panero J, Zelnik M (2008) Dimensionamento humano para espaços interiores. 1ª Edição, 4ª impressão. Editorial Gustavo Gili, SL, Barcelona
14. Pilc J (1984) Open Office Space (The Office Book Design Series). Facts On File, Inc, NY
15. Sanoff H (1991) Visual Research Methods in Design. Van Nostrand Reinhold, New York

Ergonomic Analysis of Secondary School Classrooms, a Qualitative Comparison of Schools in Naples and Recife

Thaisa Sampaio Sarmento[1(✉)], Vilma Villarouco[2], and Erminia Attaianese[3]

[1] Federal University of Alagoas, Maceió 57072900, Brazil
thaisa.sampaio@fau.ufal.br
[2] Federal University of Pernambuco, Recife 50740550, Brazil
villarouco@hotmail.com
[3] University of Naples Federico II, Naples 80134, Italy
erminia.attaianese@unina.it

Abstract. In contemporary educational systems, innovative learning environments require spatial rearrangements and the introduction of new tools such as cloud computing, tablets and smartphones, in addition to ergonomic spatial planning. Ergonomics of the Built Environment studies spatial adequacy for human activities and goes beyond technical issues by including users' needs and expectations. Because users interact with the environment, poorly built projects may affect cognitive perception and organizational ergonomics. Applying ergonomic criteria to interior design may contribute to a better environment performance in terms of wellbeing and user satisfaction, because ergonomics methods strongly recommend user participation as a principle. This research was carried out in Naples, Italy and Recife, Brazil. It aims to analyze two secondary school environments, considering user characteristics, means of socialization and classrooms conditions. Guidelines were elaborated for new learning environment design, considering ergonomic recommendations, technological adequacies and user satisfaction.

Keywords: Ergonomics for built environment · Learning environment
Users' perception

1 Introduction

School environments must be prepared for the necessary upgrades required by new behavior in contemporary youth, new activities with technological resources and recognition of the importance of blended-learning. A school that does not prepare citizens for the reception and critical use of Media, Digital Information and Communication Technologies (TMDICs) is the greatest contradiction in the educational system, since new generations of students leave school without any preparation for the contemporary way of life [1].

The challenge in school environment design is the need to rethink and readjust its functions and configurations based on activities that require technological resources

S. Bagnara et al. (Eds.): IEA 2018, AISC 825, pp. 537–546, 2019.
https://doi.org/10.1007/978-3-319-96068-5_60

and space for blended-learning activities. Upgrading classroom environments involves the inclusion of computers and the improvement of electrical and internet networks. Nevertheless, the spatial configuration remains unchanged and only the layout for expository activities is considered.

2 Research Method

This research was conducted in two existing high school environments which were selected based on accessibility and because they offer environments in which computational resources are used for learning purposes with teacher orientation. The selected schools were Liceo Statale Don Lorenzo Milani, in Naples, Italy, and Escola Técnica Cícero Dias, in Recife, Brazil. Data collection took place between April and June 2017. The techniques and instruments used were:

Quick Ethnographic Observation – researchers' observations and recording of their perception about positive and negative aspects;

Interviews and Questionnaires – applied to students and teachers in order to obtain data about social and family profiles and preferences for the subjective aspects of learning environments;

Identification of environmental configuration – measurement of physical-environmental conditions that affect the environment adequacy (temperature, ventilation, noise, lighting, surface cladding, colors, materials, etc.), which is a technique recommended by MEAC [2].

Analysis of environment in use and Task analysis – monitoring the position and movement of users within a given space. It can register paths and occupation patterns, generating a usage map for each evaluated environment. The analysis of the environment in use is a technique recommended by MEAC [2], the elaboration of usage maps [3–6]. Task analysis in ergonomic assessment of built environments is also recommended [7].

User Perception Analysis – data collection about how users perceive the quality and satisfaction of the environment they use, a technique recommended by MEAC [2]. The brainstorming tool was applied, in which user groups discussed and filled out panels with questions about what they like/do not like and how they would improve the environments they use.

The analysis of the data obtained generated an ergonomic diagnosis and the elaboration of guidelines for technology enriched learning environments.

3 Trends in Learning Environments

Student performance in school is related to building quality [8–10]. An older school building, today, hinders the introduction of 21st century education, especially in relation to the flexibility of spaces and adoption of Information and Communication Technologies (ICTs). However, as stated in a groundbreaking contemporary project, a well-designed school building can act as a catalyst for pedagogical changes [11].

Regarding teachers' satisfaction about environmental conditions of classrooms, they demonstrate a preference for natural lighting, the possibility to control light levels (such as curtains) and thermal conditions (by thermostats or windows). Operable windows are most desirable for cooling and ventilation of environments [8].

Good design for learning environment should allow students to experience four ways of learning: Campfires (learn with an expert); Watering hole (learn with peers); Cave cavern by introspection – individual study); Life (learn by doing) [12], in way to enable students the complete benefit of education [13]. Principal elements to reconfigure learning environments are highlighted as follows:

- Learning studio and learning suite (two or more learning studios combined) to increase the impact on learning. Teachers are not alone with a single class but joining two or even four classes in a single multidisciplinary space where the work is fully collaborative [14];
- An engaging learning model which reinforces the idea of versatility of internal layouts. Adaptable furniture increases the opportunities for spatial reconfiguration depending on daily dynamic necessities [15];
- The layout model for learning environments developed by the Centre for Excellence in Teaching and Learning (CETL) [16] from the Universities of Sussex and Brighton in the UK;
- The award-winning, innovative solutions for Vittra School design in Denmark by Rosan Bosch [17], which create autonomous learning spaces and design of constructive elements compatible to the collaborative educational method adopted;
- Smart classroom that proposes a configuration for presentation optimization, free access to learning resources and interactivity provided by the continuous use of ICT [18].

A classroom project should be designed considering these criteria [13]: more flexible, wider classrooms, accessible spatial conditions, layouts designed to accommodate the pedagogical activities proposed, different furniture inside classrooms to allow a variation of arrangements, the option to choose furniture based on comfort, well trained teachers in the correct use of equipment, clear instructions in each room on the use of equipment and finally, more attractive and comfortable interior spaces.

4 Analyzed Schools

4.1 Escola Técnica Cícero Dias (ETE-CD) in Recife, Brazil

The Cicero Dias State Technical School (ETE-CD) located in Recife, Brazil offers full time technical high school education. In 2017, the school counted 452 students who attended for free. Annually, 180 new vacancies are offered through a public admission test for teenagers between 14 and 17 years old, most of them coming from public schools. Two professional courses are offered: game programming and multimedia.

4.2 Liceo Statale Don Lorenzo Milani (LS-DLM) in Naples, Italy

The Liceo Statale Don Lorenzo Milani is located in the district of San Giovanni a Teduccio in Naples, Italy. Its facilities consist of two two-story buildings connected by an internal corridor. The educational method associates traditional teaching with digital resources; thus, classrooms are wide and internet-ready. The school is public and serves approximately 790 students distributed in 39 classes, with 20 students on average per class. Their pedagogical project offers language training, visual arts and architecture.

5 Data Analysis

In both schools, most classrooms are set for expository activities. For group activities, the reorganization of the layout is problematic because large and heavy furniture and limited space hinder dynamic activities. Both computer rooms are adaptations because they offer ample space, but their layout is set for generic learning use. Neither specific nor digital classes were considered when environments were designed. Thus, inadequacies related to comfort conditions, user flow, ergonomics and user satisfaction were identified.

5.1 Environmental Configuration Analysis

Dimensional and functional data observed in the classrooms were registered using photographs and annotations.

School 1: ETE-CD

Spatial dimensions: 120,61 m^2 considering the maximum occupancy capacity (28 people using computers) - 4.30 m^2/student. It is excessive because the recommended ratio is 1.5 m^2/student. There is unused space even in a class with 33 students. The lack of concentration is a constant problem.

Furniture distribution: There are 19 sets of tables and chairs on wheels, next to the windows, 04 round tables with 03 chairs each for computers, 02 rectangular tables with 06 chairs each for group activities; facing the blackboard, 01 table for the teacher, 01 digital board with a suspended projector, 02 cabinets, Wi-Fi equipment and power outlets, 29 computers provided by the school (01 used by the teacher).

School 2: LS-DLM

Spatial dimensions: 78,80 m^2 considering the maximum occupancy capacity (20 people), with a 3.94 m^2/student ratio. It is over dimensioned since the common recommendation is 1.5 m^2/student. There is unused space and the size and weight of furniture are critical.

Furniture distribution: There are 12 sets of high tables and chairs on wheels in the center of the room, 04 sets of double tables and 08 regular chairs for computer use next to the wall, 02 support tables for the speaker (teacher) facing the blackboard; 05 cabinets; 01 projection screen, 01 suspended projector, Wi-Fi equipment and power outlets.

Table 1. Physic data obtained in analyzed environments. (Source: the authors).

Analysed factor	Recommended value	Obtained value school 1 ETE-CD	Obtained value school 2 LS-DLM
Temperature (°C)	23 °C [19]	23	22
Noise (dB)	25–35 dB [20] 40–50 [21]	46–76	55–85
Lighting (Lux)	300–500 [22] | 300–average, 500 at blackboard [23]	–	–
Natural	300–500	200–205	455–485
Artificial	300–500	200–266	476–496
Surfaces materials	School 1 ETE-CD	School 2 LS-DLM	
Floor	White ceramic	Beige ceramic	
Walls	Brick, 14 cm thickness, no thermal insulation, white and orange colored painted surfaces	Brick, 20 cm thickness, interior thermal insulation, White colored painted surfaces	
Ceiling	Corrugated concrete Slab gray colored	White painted plywood panels	
Furniture	School 1 ETE-CD	School 2 LS-DLM	
Chairs	Chair on wheels - gray metallic structure, white PVC seat. Seat size 42 × 42 cm, seat height 45 cm. armrests size 22 cm	High chair, seat size 40 × 46 cm, seat height 68 cm Black metallic structure, dark green upholstered seat	
	Chair on wheels - black metallic structure, dark green upholstered seat, seat size 45 × 48 cm, seat height 47 cm, armrests size 22 cm	Regular chair, seat size 53 × 56 cm, seat height 48 cm, black metallic structure, blue upholstered seat	
Tables	Round 4 places table, 150 cm diameter, height 72 cm, plywood structure, light gray melamine covered surface	High table. Size 120 × 80 cm, height 90 cm black metallic structure, plywood top, white melamine covered surface	
	Individual table, top size 80 × 60 cm, height 73 cm light plywood structure and top surface	Regular table, size top 150 × 60 cm, height 76 cm black metallic structure, plywood top, white melamine covered surface	
	Rectangular 6 places table, size 200 × 100 cm, height 76 cm plywood structure with light gray melamine covered surface	–	
Bookshelves	Plywood bookshelves, light gray melamine covered surfaces	Plywood bookshelves, light gray melamine covered surfaces	
Doors	Size 168 × 220 cm, width panel 80 cm. Plywood structure, white painted	Size 138 × 218 cm, width panel 90 cm. Plywood structure, blue melamine covered surface	
Windows	Aluminum structure, frosted glass, size 150 × 280 cm	Size 380 × 137/105 cm. White PVC structure, translucent glass. Polyester blackouts and PVC shutter	

Measurement of performance indices for both environments: Digital tools used: thermal performance - DS Thermometer (DS Software), acoustic performance - Decibel (Sound Meter, ABC Apps), and light performance - O2 Led Light Meter (www.o2led.com.br) (see Table 1).

5.2 Analysis of Environment in Use and Task Analysis

Classroom at School 1: ETE-CD

This room presented spatial inadequacies for computational resources, mainly regarding the size and organization of seats when switching between expository to individual activity on computers. To listen to the lesson, the students had to drag their chairs (with casters) to the front of the room, later returning to their desks. Seats are not fixed all year round, so students choose where to sit based on social affinity. The ambience is very relaxed, and the students show interest in the content. The yellow area highlights the most used areas for learning activities: (1) expository - students bring their chairs forward to face the blackboard; (2) desktop work - individual chairs and tables lined up on front and side walls and circular four-place tables on one side of the room. Highlighted in purple, the less used areas in the back of the room or laterally by the blackboard. The user flow is displayed in green for entrance and orange for exit (see Fig. 1).

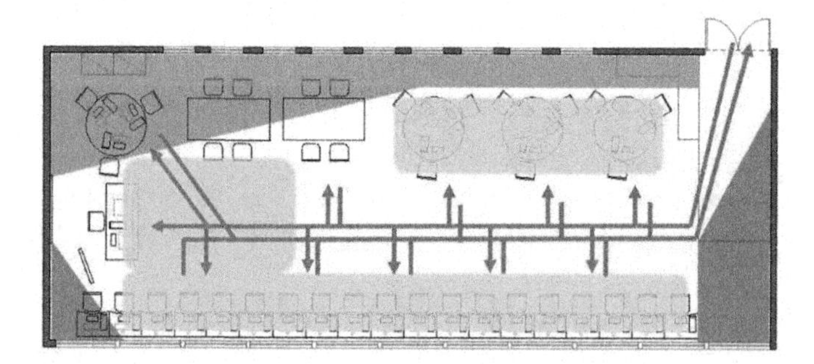

Fig. 1. Computing room plan (not scaled) demonstrates the use and flow – school 1: ETE-CD. (Source: the authors)

Classroom at School 2: LS-DLM

The teacher proposed an individual activity using desktops computers and authorized the students to listen to music, giving individual orientations from desk to desk while they performed the exercise. Some students frequently left the room, showed apathy or engaged in informal conversations with classmates. The lack of computers for all students contributed to the dispersion. The interaction between students and teachers seemed calm and open, although dispersion increased at the end of the lesson. The areas of greater use are highlighted in yellow: room center - individual or group activities, next to the blackboard - oral presentations and next to the left wall - use of desktop computers. In purple, the areas of least use are highlighted: at the back of the room, from where one cannot see nor hear clearly, and on the right side of the blackboard, forming a small area by the window with a curtain. The areas in purple accumulate papers and disposal material. User flow is marked in green for entrance and orange for exit (see Fig. 2).

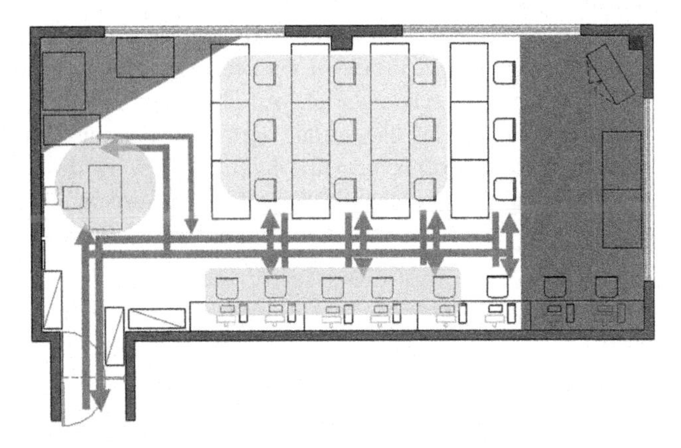

Fig. 2. The computer room plan (not scaled) shows use and flow – School 2: LS- DLM. (Source: the authors)

5.3 Analysis of User Perception

A brainstorming technique was applied to collect environment perception data from users. The results show students' impressions based on their personal and cultural preferences (see Table 2).

Table 2. User perception synthesis from both schools analyzed. (Source: the authors).

School 1 ETE-CD	School 2 LS-LM
Which aspects do you like in this classroom?	
Wide space, use of colors, computers always available, different forms of spatial organization, air conditioning and multimedia installed, good lighting; good relationship between people	The width and lighting of the classroom, seat organization, wide circulation halls, multimedia resources, excellent relationship between students and teachers
Which aspects do you not like in this classroom?	
Uncomfortably small chairs, slow internet, inconvenient talks during classes, digital board does not work, closed rooms without openings to the outside, low natural light, damaged tables, inefficient air-conditioning, small blackboard, water infiltration on walls	Uncomfortable internal temperature all year round, precarious cleaning service, bad materials for practical activities and exercise results poorly showed to the student community, bad distribution of space, bad acoustic and thermal insulation, unused outside areas, water infiltration on walls, insufficient computer/student ratio, poor maintenance, insufficient bookshelves and work material, limited internet
If you could design a new classroom/school, which aspects would you improve?	
About classroom: Upholstered chairs, larger boards, boards to write formula tips on the walls, more efficient and well-maintained air-conditioning, use of digital boards, room and time for resting	*About classroom:* Allow the use of cell phones for learning purposes, install more power outlets, acquire new furniture and equipment, allow colored walls, install more window curtains, offer more materials for practical work, allow outdoor classes, increased digital interactivity

5.4 Ergonomic Diagnosis

Inadequacies were found in both analyzed environments regarding: construction quality and maintenance, cleaning conditions and electrical installations. The main problem was that the environment layouts are adaptations and were not planned to accommodate technology enriched activities, nor do they offer the necessary flexibility for multiple learning activities - lectures, practical classes with teacher assistance, individual and group activities. The elongated, rectangular shaped classroom and the furniture organization cause user dispersion during classes and high concentration areas. Furniture is ergonomically inadequate and worn out, while lighting on work desks is poor. The desktops screen position in relation to the blackboard is awkward and the opposite wall windows cause glare on them. There is little protection against direct sunlight.

Table 3. Ergonomics guidelines proposed for technology enriched learning environments. (Source: the authors).

Built environment
Install doors at the back of the room, avoiding the obstruction of the blackboard; provide well-distributed electrical outlets to charge students' devices at their desks, install doors in positions that do not disturb the attention or blackboard visualization; install two restrooms nearby for both genders, avoiding long walks to reach them, as well as open-air resting areas; position windows for both outside view (large and suitable for daylight use), and inside corridor, allow students to customize, color and exhibit their results on school and classroom walls, design cozy and welcoming areas, provide dynamic and flexible conditions for multiples activities, favoring concentration and visual comfort for screen use
Comfort conditions
Provide adequate visual and listening conditions for the content/speech [24], paint the front wall in a darker tone to allow a better focus on blackboard, ensure good comfort conditions, a sense of identity and security, adapt the environment for seasonal conditions of comfort, provide adequate thermal sensation all year round, observing international standard recommendations
Layout and furniture
Organize tables and chairs, including desktops facing the blackboard, avoid overshadow the screens and favor attention to the blackboard or speaker, provide adjustable, wheeled seats if possible, footrest and armrests, make two types of blackboards available for writing and digital resources, propose a flexible layout for both general and specific activities. Adapt furniture for multiple learning activities, collaborative work, personal space and a sense of coexistence and communication [13], provide sufficient space for the speaker to organize personal belongings at the front and walk freely during speech, prevent disorganized equipment wiring, hindering the view or access to the blackboard, use light tables and chairs, if possible on wheels, allowing layout changes for different activities, provide permanent storage for students' personal belongings, organize space to exhibit students' creations
Technology
Manage information technology resources democratically, maintain information and knowledge networks in and out of school, produce and share content, maintain digital and secure internet access, keep classrooms connected to online teaching modules to load and provide content, keep learning methods aligned with both technology resources and physical infrastructure [25], allow students to access the internet on laptops and smartphones, install environmental control system (digital and online) for illumination, shading, air conditioning and audiovisual resources

5.5 Ergonomic Proposition

Based on the analyzes performed in the environments, ergonomic guidelines were elaborated for technology-enhanced learning environments. These guidelines (see Table 3) are categorized into four groups: Built Environment, Comfort Conditions, Layout and Furniture, and Technology.

6 Conclusion

This research demonstrates ergonomic inadequacies in the learning environments of secondary schools located in different countries and cultures. In both situations, a need for further studies and insights was observed, which could better investigate ergonomic issues in educational built environments, in order to adapt spatial configurations for technology enriched educational methods.

References

1. Sibilia P (2012) Redes ou Paredes. A escola em tempos de dispersão. Contraponto, Rio de Janeiro
2. Villarouco V (2009) An ergonomic look at the work environment. In: Proceedings of IEA 2009, 17th World Congress on Ergonomics, Springs, Beijing
3. Sanoff H (1991) Visual research methods in design. Van Nostrand Reinhold, New York
4. Sommer R, Sommer B (1997) A practical guide to behavioral research, tools and techniques. Oxford University Press, New York
5. Fascioni, L Métodos de pesquisa etnográfica. http://www.faberludens.com.br. Accessed 02 Aug 2017
6. Rheingantz PA et al (2009) Observando a qualidade do lugar: procedimentos para a avaliação pós-ocupação. Universidade Federal do Rio de Janeiro, Rio de Janeiro
7. Attaianese E, Duca G (2012) Human factors and ergonomic principles in building design for life and work activities: an applied methodology. Theor Issues Ergon Sci 13(2):187–202. https://doi.org/10.1080/1463922X.2010.504286
8. Powell M (2015) Reacting to classroom design: a case study of how corrective actions impact undergraduate teaching and learning. Doctoral thesis, Lesley University
9. Kaup M, Kim H, Dudek M (2013) Planning to learn: the role of interior design in educational settings. Int J Des Learn 4(2):41–55. https://doi.org/10.14434/ijdl.v4i2.3658
10. Baker L (2011) What school buildings can teach us: post-occupancy evaluation surveys in K-12 learning environments. Master Thesis, University of California, Berkeley
11. Oblinger D (2006) Learning Spaces. [s.l.] Educause e-book
12. Thornburg D (1999) Campfire in Cyberspace. Starsong, Lake Barrington
13. Nair P (2014) Blueprint for tomorrow: redesigning schools for student-centered learning. Harvard Education Press, Cambridge
14. Nair P, Fielding R, Lackney J (2013) The language of school design: design patterns for 21st century schools, 3rd edn. Designshare.com, Minneapolis
15. Cannon Design. vs furniture, Bruce Mau Design (2010) The Third Teacher: A Collaborative Project. Abrams, New York

16. Higher Education Funding Council for England - HEFCE. Designing Spaces for Effective Learning. JISC Development Group, University of Bristol, UK. http://www.jisc.ac.uk/eli_learningspaces.html Accessed 01 Mar 2016
17. Archdaily VittraTelefonplan/ Rosan Bosch. http://www.archdaily.com/202358/vittra-telefonplan-rosan-bosch. Accessed 03 Feb 2016
18. Huang R et al (2012) The functions of smart classroom in smart learning age. In: Open education research, vol 2012, pp 22–27
19. Kowaltoski D (2011) Arquitetura Escolar, o projeto do ambiente de ensino. São Paulo, Oficina de textos
20. Berglung B, Lindvall T, Schwela DH (eds) (1999) Guidelines for community noise, World Healthy Organization
21. Associação Brasileira De Normas Técnicas NBR 10152 (2000) Avaliação de ruído ambiente em recintos de edificações visando o conforto dos usuários – Procedimentos, Rio de Janeiro, ABNT
22. European Committee for Standardization EN 12464-1 (2002) Light and Lighting, Light of workplace. Part 1: indoor workplace
23. Associação Brasileira de Normas Técnicas. NBR ISO/CIE 8995-1 (2013) Iluminação de ambientes de trabalho. Parte 1: interior. Rio de Janeiro, ABNT
24. Gee L (2006) Human-centered design guidelines. In: Oblinger D (ed) Learning Spaces, pp 10.1–10.13. http://www.educause.edu/learningspacesch10. Accessed 05 Apr 2017
25. Nou DS (2016) Ergonomic Concerns in Universities and Colleges. In: Hedge A (ed) Ergonomic workplace design for health, wellness, and productivity. Taylor & Francis, Boca Raton, CRC, p 662, pp 342–376

Prototyping a Learning Environment, an Application of the Techniques of Design Science Research and Ergonomics of the Built Environment

Thaisa Sampaio Sarmento[1(✉)], Vilma Villarouco[2], and Alex Sandro Gomes[3]

[1] Federal University of Alagoas, Maceió 57072-900, Brazil
thaisa.sampaio@fau.ufal.br
[2] Federal University of Pernambuco, Recife 57740-550, Brazil
villarouco@hotmail.com
[3] Federal University of Pernambuco, Recife 57740-560, Brazil
asg@cin.ufpe.br

Abstract. School environments and infrastructure are fundamental elements to the success of the educational system. The purpose of this article is to describe the triangulation of three methods used to develop a conceptual model of learning environment appropriate to technology-based learning. An ergonomic approach to the built environment should consider constructive aspects, psychological effects and users' expectations about their learning environment. The application of ergonomic methods is a strategic advantage to obtain innovative ideas to be applied in different types of environments. One of the highlights of this research is its participative nature, involving high school students and teachers. Motivating these people to get involved in a future perspective thinking exercise is one of the principles of Science Design [1]. The central method of this research is Design Science Research [2], which was supplemented with ergonomic techniques for the analysis and design of built environments - The Ergonomic Methodology for the Built Environment - EMBE (MEAC in Portuguese) [3] and the studies on ergonomics for design the built environment [4]. A conceptual model of learning environment was obtained and presented in four different layouts: for activity rotation, group activities, laboratory activities and expository activities. Additionally, technical specifications were also presented regarding: constructive elements, comfort conditions, environmental systems, furniture and equipment criteria. Finally, the developed method is presented as a contribution to the field of Design and Ergonomics of the Built Environment.

Keywords: Design science research · Ergonomics of the built environment
Participatory techniques

S. Bagnara et al. (Eds.): IEA 2018, AISC 825, pp. 547–556, 2019.
https://doi.org/10.1007/978-3-319-96068-5_61

1 Introduction

Space is the environmental matrix for human relations in its complexity [5, 6]. It is simultaneously the result of cultural, social and institutional factors. The relationship of exchange between the person and the physical environment is dialectical. The physical environment influences human practices and at the same time they act on the spatial factors it is made of. Therefore, the environment must be planned to offer fully functional conditions and satisfaction to its users.

The epistemological bases underlying this research are supported by the ideas of Constructive Design [7], Participatory Design, Design Science Research [2], ergonomic evaluation methods - Ergonomic Methodology for the Built Environment – EMBE [3, 8] and the stages of ergonomic for building design [4]. This article discusses the triangulation of methods to guide the design of a conceptual model of learning environment appropriate to practices with Blended Learning in public high schools. This research aimed to develop a conceptual model of the physical learning environment, in which hybrid educational practices are applied and the built environment acts as a learning facilitator.

2 Development of Models and Artifacts

Participatory design gives the user the same value as a project specialist [9, 10], participating from planning to prototyping. One of the principles of Participatory Design is the need for people to engage in the design of the future. This brings us closer to the principles of Design Science [1] and Constructive Design Research [7], which refer to Design research with the objective of constructing products, models, systems and spaces.

The focus is the ideation, the construction of new things, the description and explanation of the processes of these constructions. The elaboration of models is a material hypothesis that expresses the relationships between constructs, which are used to describe and think about human practices [11]. Models can also be understood as a description, in which the component elements are clearly defined.

2.1 Design Method - Design Science Research

Design Science Research is an epistemological paradigm of quantitative and/or qualitative approach used to build knowledge on how to design, carry out a project, seek to develop and design solutions to improve existing systems, solve problems or create artifacts. It is applied in twelve methodological steps of a procedural nature. The steps generate products that feed the next steps or can also go back to the methodological sequence in order to refine the design process. By adopting DSR to design models, the relationships between its utility and the conditions for capturing the general structure of reality are continuously observed [2].

3 Ergonomics of the Built Environment - Evaluation and Proposition of Environments

For this research two ergonomic approaches were combined, the first one of analytical nature - Ergonomic Methodology for the Built Environment – EMBE [3, 8] and another of a design nature - Ergonomic Design [4], which complement each other in the evaluation and proposition of built environments.

3.1 Evaluation Method - EMBE (Ergonomic Methodology for the Built Environment)

EMBE [3, 8] consists of an ergonomic evaluation that seeks to identify conflicts caused by missing or inadequate elements in an environment, based on the opinions and suggestions of its users and identified through the research tools of ergonomics, architecture and environmental psychology. It has the Ergonomic Workplace Analysis (EWA) as its starting point and tries to establish an analogy between the phases of the traditional analysis and those necessary for space evaluation focusing on the work performed in it, verifying possible interactions that are detrimental to productivity or that could improve working conditions.

The stages of EMBE are: global analysis of the environment, identification of the environmental configuration, evaluation of user's perception, ergonomic diagnosis and ergonomic propositions. The researcher is required to use tools of perception and environmental psychology in any environment to be evaluated when using EMBE. The result of the EMBE should be the preparation of a list of duly justified recommendations or the proposition of a project that brings solutions to the problems identified in the analyzes.

3.2 Projective Method for Human-Center Building Design

Ergonomic principles should be applicated in the design of buildings as a methodology applied to a project [4]. This methodology essentially involves users, thought the ISO 13407/1999 international standard, which determines the user-centered design of systems with a clear understanding of the characterization of users and all the tasks involved in the functioning of a system of activities. The methodology follows a sequence of seven steps: design briefing, user profile elaboration, task analysis, elaboration of requirements for service to users, architectural detailing, validation of design solutions, monitoring of user performance in the use of the building.

4 Methodological Triangulation

As a way of achieving the research objectives, a triangulation of the three methods mentioned above was elaborated (see Fig. 1.): the left column describes the ergonomic methods that fed the main DSR steps (central column), the right column describes the steps in carrying out the research to reach the proposed objectives.

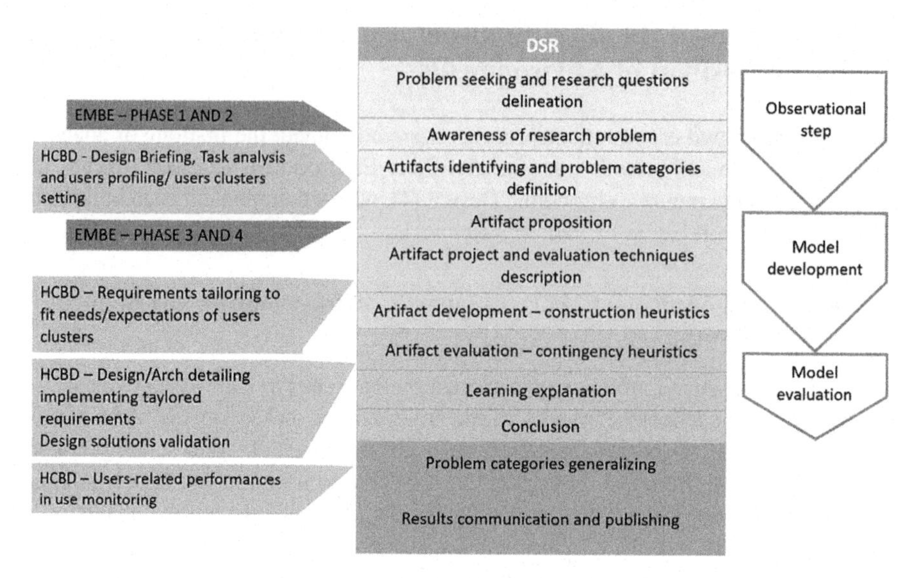

Fig. 1. Synthesis of the triangulation of methods used. Source: the authors

5 Description of Methodological Steps

The application of the method occurred in three main stages: (1) Observational Step, (2) Model Conception and Development and (3) Model Evaluation.

5.1 Observational Step

The observational step occurred in high school environments to analyze the users carrying out their learning activities and the built environment. The techniques and instruments employed aimed to: (1) characterize and understand the users, their social and cultural profiles and tasks in the environments; (2) characterize and understand how environments are configured and used; and (3) collect data on user's perception of these environments. The data obtained were recorded by means of notes, audios, photographs and forms.

The environment observation focused on collecting data on: Identification of the Environmental Configuration (survey of physical-environmental conditions that affect the environment adequacy - temperature, ventilation, noise level, insolation, artificial lighting, surface coatings, colors, materials, etc.) and Analysis of the Environment in Use and Task Analysis (monitoring the positioning and movement of people within the space in a period).

Prototyping techniques were used throughout the research process, beginning with brainstorming, through which user groups discussed and filled out panels with questions about what they like or dislike and how they would improve the environments used (see Fig. 2).

Fig. 2. Students performing brainstorming activity on preferences for learning environments. Source: the authors

5.2 Design and Development of the Conceptual Model

The techniques and instruments used for the Development of the Conceptual Model sought to: (1) generate educational and spatial concepts for the Learning Environment Conceptual Model, (2) provide user participation through walkthrough desktop techniques (see Fig. 3) – the use of miniatures for people and environment elements to simulate activities, improvements and innovative scenarios [12], (3) generate design solutions for questions on shape, design, layout, circulation, materials, colors, furniture, openings, equipment, among others (see Fig. 4), meet ergonomics parameters and user needs/expectations, and (4) design and represent solutions generated in 2D and Virtual Reality (see Fig. 3) for interaction with evaluators.

Fig. 3. Students performing the desktop walkthrough technique to elaborate ideas for innovative learning environments. Source: the authors

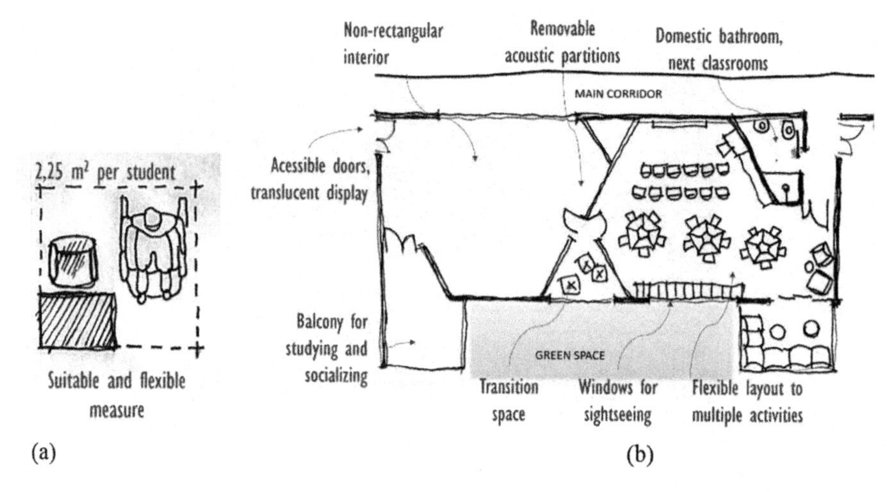

Fig. 4. Sketches for prototype solutions (a) Individual space measurement, (b) Space and layout solutions. Source: the authors

The shape started as a rectangle, typically adopted to design public school buildings. It may both simplify the construction of new buildings and the renovation of existing ones. On the inside, the spatial solution prioritized non-parallel walls, a layout for a versatile environment focused on multiactivity. To improve acoustic conditions, the ceiling is higher towards the back of the classroom. Each environment was dimensioned to offer a maximum occupancy of 30 students resulting in a 2.25 m2/person ratio, which offers comfort, accessibility and flexibility to an individual workstation area.

The results are environments with changeable layouts combining four different forms of spatial organization. They are suitable for blended learning practices, especially for rotation models.

- Layout for laboratory activities (1) and layout for activity rotation (2);
- Layout for expositive activities (3) and layout for group activities (4).

Interior areas (67.50 m^2) are identical for each layout. This solution allows multiple combinations of layouts in order to address spatial needs or just to provide better accommodation by allowing students to change furniture position during lessons. It is possible to obtain a new layout organization whenever necessary, instead of moving the students to a different classroom. To divide contiguous rooms, removable acoustic panels are proposed. Regarding furniture type and distribution, the solutions prioritized items such as comfort, flexibility for organizational modifications – slim wheeled chairs and tables, individual desks can be joined into a circular table for 5 students, soft and comfortable reading armchairs for indoor and outdoor environments.

5.3 Conceptual Model Evaluation

The evaluation of the Conceptual Model was mostly qualitative but included a quantitative evaluation stage. The entire elaboration of the model is described above (see Fig. 5).

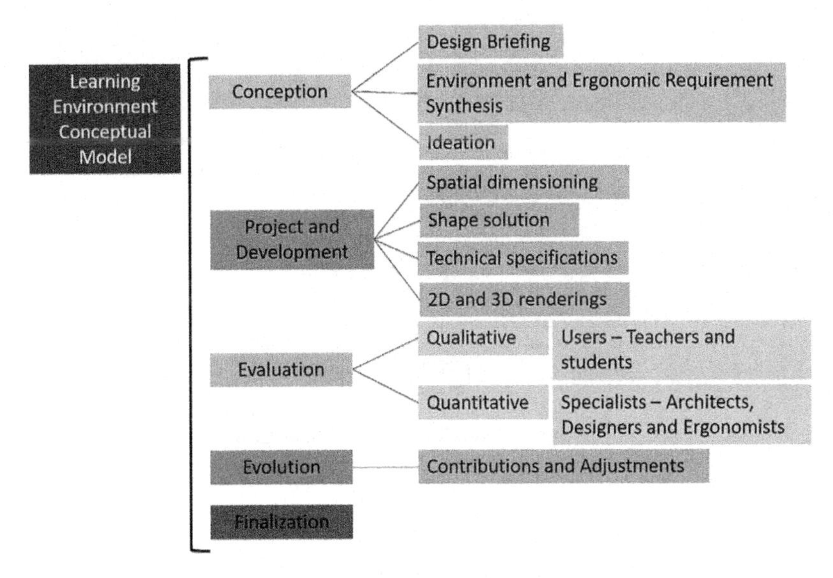

Fig. 5. Model elaboration and evaluation diagram. Source: the authors

The evaluation techniques used were participatory and applied in two steps: (1) Qualitative assessment in focus groups (users answered a 30 min focal group interview on the perceived aspects of the learning environment conceptual model) and (2) Quantitative evaluation with specialists (designers and architects answered a calibrated questionnaire to measure satisfaction levels regarding technical aspects of the learning environment conceptual model). Examples are given by the evaluators (see Fig. 6).

Evaluation results were organized in synthetic tables containing: analysis origin, description of the analyzed item and changes, either applied or pending for the evolution of the Conceptual Model.

6 Learning Environment Conceptual Model Evaluation Results

The modifications carried out in the Conceptual Model generated reconfigured floor plans (see Figs. 7 and 8), where one can notice the changes in the location and quantity of restrooms, the position and dimension of the main doors and access to the transition space, as well as the adoption of a larger area to access the learning environments. With these changes, users on the balconies can access the main corridors or bathrooms without necessarily entering the main learning environments.

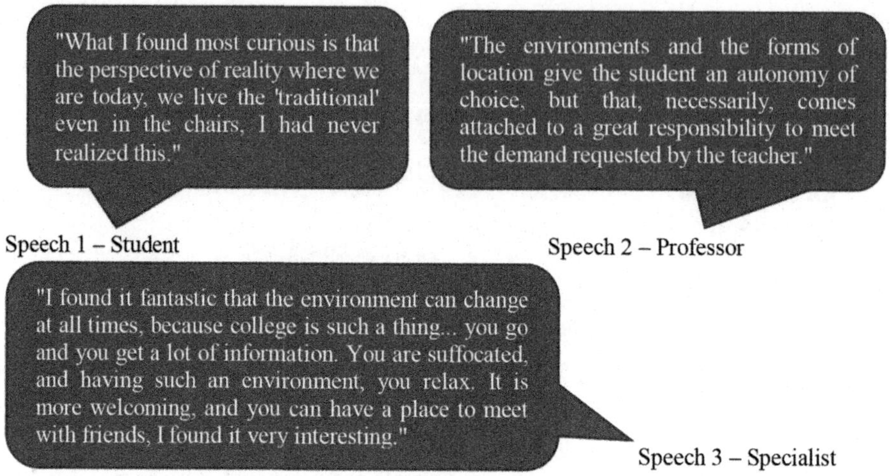

"What I found most curious is that the perspective of reality where we are today, we live the 'traditional' even in the chairs, I had never realized this."

Speech 1 – Student

"The environments and the forms of location give the student an autonomy of choice, but that, necessarily, comes attached to a great responsibility to meet the demand requested by the teacher."

Speech 2 – Professor

"I found it fantastic that the environment can change at all times, because college is such a thing... you go and you get a lot of information. You are suffocated, and having such an environment, you relax. It is more welcoming, and you can have a place to meet with friends, I found it very interesting."

Speech 3 – Specialist

Fig. 6. Speech samples from evaluators obtained in focus groups. Source: the authors

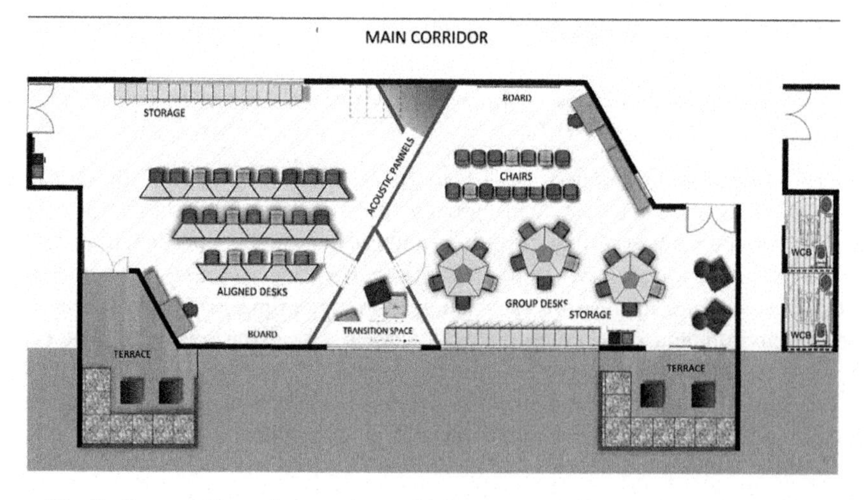

Fig. 7. Layouts for station rotation and laboratory activities. Source: the authors

The useable areas (67.50 m^2) were maintained and the internal configuration of the environments did not change significantly. The largest modifications were performed in restrooms, which were resized and duplicated to serve both genders and meet accessibility conditions.

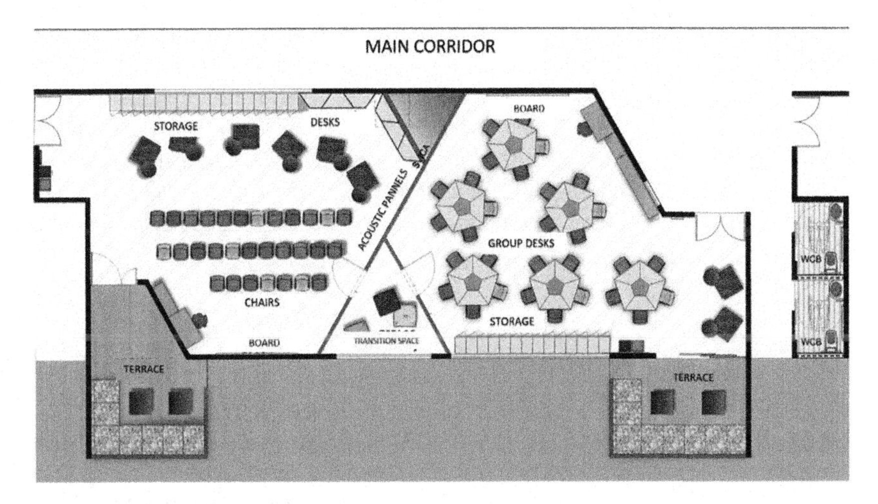

Fig. 8. Layouts for exhibit activities and group activities. Source: the authors

7 Final Considerations

The concepts of flexibility, autonomy for users to customize and adapt the environment to activities were pointed out as necessary solutions for the evolution of school environments. The results of this research demonstrated the importance of user participation in the processes of environment design, considering their experiences in the spaces they use. The trends in learning environment innovation were tested and spatial solutions were elaborated and approved by the participants of the research.

The triangulation between the DSR and the methods of evaluation and ergonomic design for the built environment brought more depth to the theme studied and allowed to acquire (see Fig. 9).

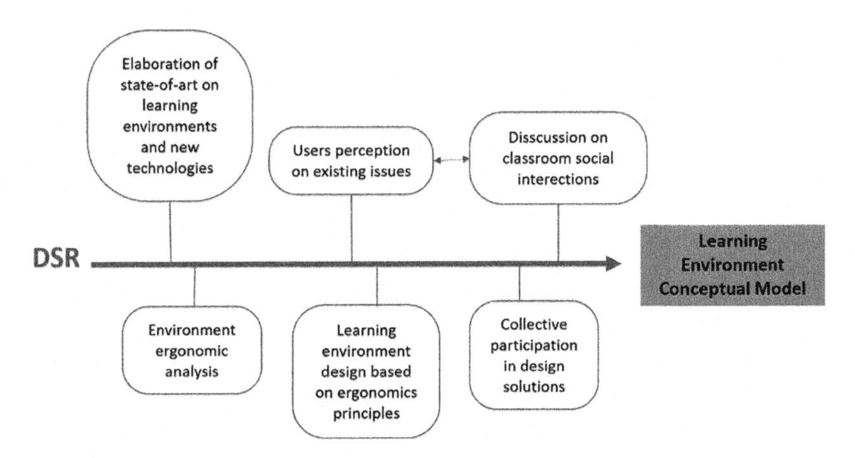

Fig. 9. Scheme of perceived learning with methodological triangulation. Source: the authors

The qualitative nature of the methodological approaches paved the way for the elaboration of an innovative, in-depth method on how to treat a constructed object. This strategy represents an important gain in the scientific field of Ergonomics of the Built Environment, since the method developed in this thesis encompasses methods and techniques of analysis and proposition for the design of environments according to ergonomic parameters. Although this method has been applied only once, focusing on school buildings, it allows improvements so that it can be applied to studies on different constructive typologies.

References

1. Simon HA (1996) The Sciences of the Artificial, 3rd edn. MIT Press, Cambridge
2. Dresch A, Lacerda DP, Antunes JAV Jr (2015) Design Science Research: Método de Pesquisa para Avanço da Ciência e Tecnologia. Bookman, Porto Alegre
3. Villarouco V (2009) An ergonomic look at the work environment. In: Proceeding IEA 09: 17th World Congress on Ergonomics, Beijing, China
4. Attaianese E, Duca G (2012) Human factors and ergonomic principles in building design for life and work activities: an applied methodology. Theor Issues Ergon Sci 13(2):187–202. https://doi.org/10.1080/1463922X.2010.504286
5. Fischer GN (1994) Psicologia Social do Ambiente, Instituto Piaget, Ed. SIG, Lisboa
6. Oliveira E (2011) Por uma arquitetura socioeducativa para adolescentes em conflito com a lei: uma abordagem simbólica da relação pessoa-ambiente. In: 11° Ergodesign Usihc - Congresso Internacional de Ergonomia e Usabilidade de Interface Humano-tecnologia: Produto, Informações, Ambiente Construído e Transporte, Manaus
7. Koskinen I, Zimmerman J, Binder T, Redström J, Wensveen S (2011) Design Research Through Practice, From the Lab, Field, and Showroom. Elsevier, Waltham
8. Villarouco V (2011) Tratando de ambientes ergonomicamente adequados: seriam ergoambientes? In: Mont'alvão C, Villarouco V Um novo olhar para o projeto: a ergonomia no ambiente construído, pp 25–46, Faperj, 2AB, Rio de Janeiro
9. Asaro P (2000) Transforming society by transforming technology: the science and politics of participatory design. Account Manag Inf Technol 10:257–290
10. Muller MJ (2003) Participatory design: the third space in HCI. In: Mahway NJ (ed) Handbook of HCI. Lawrence Erlbaum, New York
11. March ST, Smith, GF (1995) Design and natural science research in Information Technology. Decis Support Syst 15:251–266 https://doi.org/10.1016/0167-9236(94)00041-2 Accessed 03 Sept 2016
12. Stickdorn M, Schneider J (orgs.) (2014) Isto é Design Thinking de Serviços. Fundamentos, ferramentas, casos. Bookman, Porto Alegre

Architectural Risk of Buildings and Occupant Safety: An Assessment Protocol

Erminia Attaianese[1(✉)] and Raffaele d'Angelor[2]

[1] University of Naples Federico II, Via Toledo 402, Naples, Italy
erminia.attaianese@unina.it
[2] Contarp, Inail, Direzione Regionale Campania, Via Nuova Poggioreale, Naples, Italy

Abstract. Architectural risk of buildings relates to the possibility that technical and environmental elements of buildings interiors and outdoor spaces, may create dangerous situations for health and safety of occupants due to their engineering properties and their state of preservation, maintenance and use. Despite dangerous situations arising from architectural features of buildings are mentioned but undervalued in safety regulations, and a limited number of built environment aspects are currently analyzed in standard assessments of health and safety on work, many evidences demonstrate the strong relation between injuries or diseases of occupants and technical and environmental features of life and work environments. From this background, the study presents a Protocol for the Assessment of Architectural Risk (ARAP) for working environment proposed by Laboratory of Applied and Experimental Ergonomics of University of Naples Federico II (LEAS), with the Campania Chapter of INAIL, the Italian National Institute for Insurance against Accidents at Work. Main results of an application of the ARAP Protocol to an office building are also presented.

Keywords: Injuries prevention in buildings · Work place · Work environment

1 Introduction

Relationship between built environment and human health was clear since ancient times, but only recently the evidence of the potential effects of specific technical aspects of the building on the physical, social and mental health of the occupants highlighted the strong influence of the architectural features of the building on risk. The Environmental Burden of Disease Associated with Inadequate Housing Report, published in 2011 by the World Health Organization, demonstrates the relevance of health risks associated with unhealthy and insecure buildings [1]. Moreover, in working environments, the evidence that occupational buildings influence the health of workers is increasing today, since a huge literature gradually confirms not only that there is a link between the way in which buildings are designed, constructed and maintained and safety of occupants, but also that the physical and constructive characteristics of buildings affect well-being and productivity, in a direct and quantifiable way [2, 13].

S. Bagnara et al. (Eds.): IEA 2018, AISC 825, pp. 557–566, 2019.
https://doi.org/10.1007/978-3-319-96068-5_62

Safety is a complex condition since it involves different aspects of the sociotechnical systems, since it is characterized by the interaction between people and means that, in acting to pursue goals, operate in a physical environment and in given contexts of rules and relationships. Every type of organized activity requires the use of external or internal built spaces, which technological and environmental setting are evidently related to the nature of the activities, to their size and organization. Thus, occupational buildings are production infrastructure. Building infrastructures play an important role in the safety and health of workers, since buildings and work spaces affect the activities and behaviors of the occupants, like systems and equipment, representing not only the physical context in which the production is implemented, but also a resource for the quality and efficiency of production. The EN ISO 9001 standard, for example, dedicates a specific paragraph to the management of company infrastructures, highlining that to achieve the compliance with product requirements, the organization must define, prepare and maintain the necessary infrastructures, such as buildings, work spaces and related services [3]. Building performance affect workers performance, anda safe working environment is a pre-requisite for employee's health, well-being and productivity [4].

2 The Need of a Holistic Concept of Architectural Risk

For considering, in a systematic and global way, the different technical and environmental elements of buildings that, in a direct and indirect way, may affect the overall safety performance of the working environment for occupants, the concept of architectural risk has been defined, and a protocol for the risk assessment based on a dynamic and performance approach has been elaborated [6].

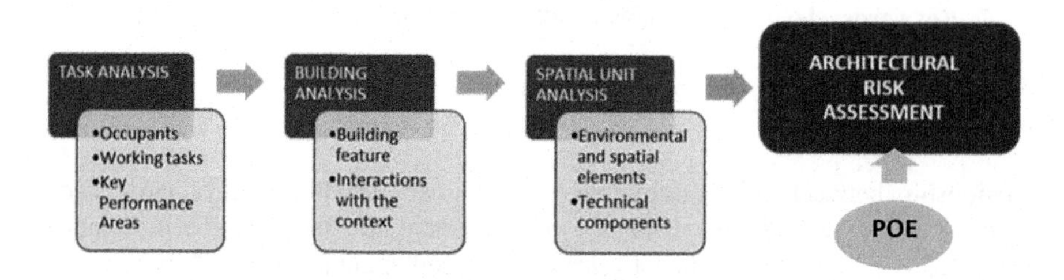

Fig. 1. Architectural risk protocol steps

The risk deriving from the building architectural components pertains the probability that technical and environmental elements of building systems and of the internal and external spaces, in which the work activities are carried out, can determine dangerous conditions for the health and safety of the operators because of their technical and constructive characteristics and their state of conservation, maintenance and use. Since according to WHO, health is a state of complete physical, psychological and social wellbeing [5] safety requirements need to be associated to comfort. A safe and comfortable built environment ensures occupants the most adequate conditions to meet their personal and social prerogatives, thus protecting health. Discomfort, after all, produces a lowering of the physical and mental capacities of individuals, with the consequent increase, among other things, of risk exposure levels [7].

3 The Assessment Protocol of Architectural Risk

The architectural risk assessment protocol (ARAP) is aimed at detecting the conditions of safety and comfort offered by the work environments and is based on the evaluation of a series of key performance areas (KPA), in relation to which the set of connotating requisites and their relative indicators are indicated, referring to both technical standards and practical experiences from the international literature [7]. Key performance areas are: Injury safety, Fire safety, Stability and Comfort, as detailed in Table 1. Control of the risk conditions, in each functional space of the building (rooms, corridors, atria, entrance hall, stairs, etc.), is based on the systematic observation of technical elements and environmental conditions from which potential dangers can arise for the user in the interaction with the physical environment. The protocol steps are fixed in Fig. 1. The protocol is structured into items, related to two scales: the scale of the building, and the scale of the single environment (spatial unit). At both scales, the identification of the source of potential risk deriving from architectural factors is carried out through a set of different survey sheets, relating to both technical elements (walls, roofs, floors, ceilings, stairs), and environmental and spatial aspects (dimensional, material, acoustics, thermal, luminous and chromatic features).

Table 1. Key performance areas

Key performance	Corresponding set of requirements
Injury safety	Protection from falls on the same level
Protection against falls from a height	
Impact protection	
Protection against entrapment	
Protection against falling objects from above	
Impact protection with vehicles	
Burn protection	
Electrocution protection	
Impact protection with vehicles	
Fire safety	Rescue operation efficacy
User safety in the evacuation	
Fire stability	
Stability	
Comfort	Hygro-thermal comfort
	Visual comfort
	Acoustic comfort
	Indoor air quality
	Psychological comfort

The protocol is applied through three types of survey sheets: one for detecting the architectural risk referring to the scale of the building, and two for detecting the architectural risk referring to the scale of the spatial unit, made by technical elements and environmental-spatial elements, as showed in Table 2. Particularly, the survey is aimed at reporting and analyzing those factors that affect each key performance areas, generating risk situations for occupants due to details and finishing of the building. Building elements, that may be risky factors for occupant's safety, are then broken down into ever more elementary components, as detailed in Tables 3 and 4 [8].

Each selected element is linked to three types of causes: the intrinsic properties of the building element; the human alterations suffered by the element; the level of usury and/or degradation of the element, as specified in Tables 5 and 6.

Table 2. Survey sheets

Type of survey sheet
Architectural risk survey referred to the building scale
Architectural risk survey referred to the spatial unit scale/technical elements
Architectural risk survey referred to the spatial unit scale/environmental elements

Table 3. Architectural risk survey referred to the building scale

Building categories of elements to observe
Building layout and volumes
Orientation
Façade configuration
Construction materials and techniques
Installations
Green areas and natural elements of the site
Additional functional areas of the buildings

Table 4. Architectural risk survey referred to spatial unit scale

Technical elements to observe	Environmental elements to observe
Walls	Morpho-dimensional features
Roofs	Hygro-thermal features
Floors	Acoustic features
Ceilings	Light, lighting and colors
Stairs	Haptic features
HVAC	Electromagnetics
Electrical system	
Water system	
Elevators	

This is a distinctive aspect of the protocol, since it allows to connect the architectonic risk, not only to the specific features of the building components and details that interact with the occupants, and to the technical choices made by the designers, but also to the level of conservation and efficiency of work environment, directly by associating safety of the built environment with their maintenance and maintainability.

Table 5. Possible causes of the architectural risk

Cause	What it means
Intrinsic properties of the building element	How the building element is done, in terms of material, dimensional consistency and functionality
Human alterations suffered by the building	How the element is used, tampered and element modified consequently to improper actions
Usury of the building element	How the element appears consequently to its aging and as expression of the lack or incorrect maintenance

Moreover, the possibility of obtaining a general picture of the risk factors linked to the different type of causes, allows a direct assessment of the possible protection measures for their mitigation, since may be evidences in advance whether "hard" interventions are required, based on the replacement of the element or its maintenance, or if "soft" actions can be foreseen, based on organization actions for preventing their misuse by occupants.

Table 6. Example of architectural risks referred to walls in spatial units

Walls details	Intrinsic properties	Human alterations	Usury
Finishing	• surface texture creates visible shadow effects • the decorative pattern of the component has a strong visual impact • the reflection factor does not fall within the range between 0.3 and 0.8 • the reflection factor of the wall is smaller than the flooring one • the finishing material is glossy and/or reflective	• presence of furniture and decorations with shiny and mirrored surfaces (glass, paintings, mirrors, glazed doors) • repairs carried out with materials with different textures/colors/consistencies • no removal of dirt and stains	• presence of opacifying deposits on the surfaces (e.g. dust, moisture spots, ecc.)

4 Application of the Protocol to an Office Building

The assessment protocol was used to identify the architectural risk in the building that houses the regional headquarters of the INAIL, the Italian National Institute for Insurance against Accidents at Work.

This is a tower-type building built in the '80s, with a reinforced concrete structure and a central core, a curtain wall with reflecting insulating glass, internal mobile partitions and a centralized two-pipe air-conditioning system. The evaluation was carried out by analyzing the ground floor, which has the function of entrance with guard room, and the fourteenth floor, considered as a type from a technical and distribution point of view. This floor is designed, like the others, to house offices and archives, connected by a perimeter corridor to the central core, and with a space from which access to the elevators and the stairwell, located in the central core. The rooms have a mineral fiber countertop and a linoleum floor. The set of the used survey sheets was that relating to the space unit, which led to the "objective" identification of the individual technical and environmental-spatial aspects that could create risk to the safety and comfort of the occupants.

Moreover, a Post Occupational Evaluation survey was defined to detect "subjective" assessment of safety and comfort of the building, based on a questionnaire of 27 questions, divided in 5 sections. To the occupants were asked to express judgments about their perception about the office, expressed on a scale of five points of satisfaction: Excellent, Good, Just Sufficient, Mediocre and Poor.

The results of the questionnaire, administered to 35 occupants, were compared and integrated to the objective results detected through protocol sheets.

5 Results

The assessment, conducted between June and September 2014, highlighted the detailed framework of architectural risk [10].

Risk situations mainly concern comfort and, to a much lesser degree, user safety, while, in terms of causes, the intrinsic properties of the technical elements and finishes, were found to be the most frequent risk factors, demonstrating that several unsuitable technical choices during the design and construction of the building were made [9].

Visual comfort results the most critical performance area, mostly due to high reflection factors of walls, floors and ceilings, and to the presence of fixed, non-maintainable, external transparent enclosures (curtain wall), which alters color perception and generate glare and visual fatigue, as confirmed by subjective evaluation made by occupants. It must be noticed that the glass curtain wall building, are fixed façade, which washing requires access from the outside, implying complex and expensive maintenance operations [11]. Thus, the low frequency of the windows cleaning (usually no more than two time a year) implies that dirty and opaque glass walls alter outside light accessin the rooms, giving the effect of an inhomogeneous natural lighting causing a sensible visual discomfort for occupants. This visual discomfort also influences occupant's psychological well-being, causing a general sense of annoyance towards the environment [12].

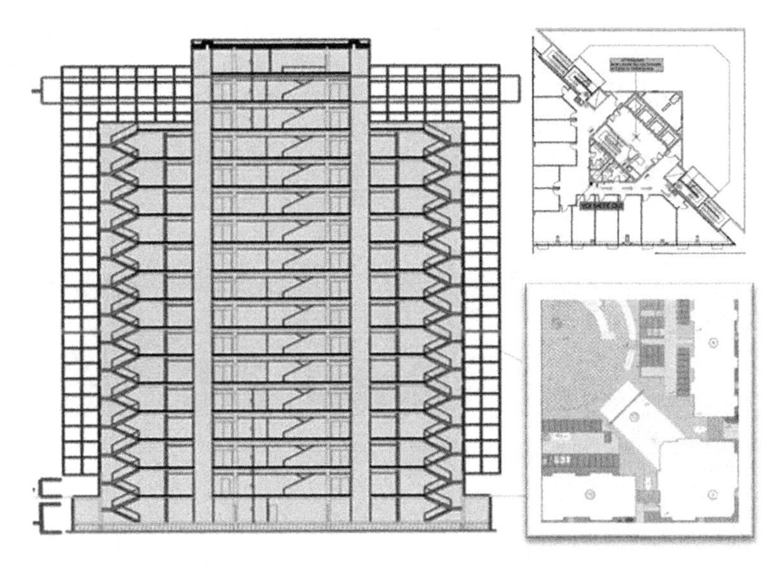

Fig. 2. Longitudinal section, localization and type floor of the office building

Other criticalities, related to thermal-hygrometric comfort, is connected both to the fixed transparent external walls, and therefore to the lack of an adjustable outward opening system, which does not allow to optimize the entry and distribution of the air flow coming from outside, both to the lack, in many environments, of adequate systems of darkening and shielding to solar radiation. About user safety, the risk of slip, trip and fall (fall in same level) emerges, due to a series of factors related to the wear of the linoleum flooring, and to the efficacy of lighting system (in the corridors) (Figs. 2 and 3) and (Table 7).

Table 7. Results about the key performance area of comfort. Frequency of causes referring to technical elements of the building

Requirements	Intrinsic properties	Human alterations	Usury and/or degradation
Visual comfort	57%	29%	14%
Hygro-thermal comfort	40%	13%	47%
Acoustic comfort	25%	–	75%
Indoor air quality	62%	33%	5%
Psychological comfort	60%	14%	26%

The POE reported that occupants suffer mainly about comfort. Particularly they consider as inadequate: shading systems, that are perceived as not fulfilling their functions, affecting negatively visual comfort; the sound-absorbing walls, that are old and not replaceable because no more available on the market, with consequent problems of acoustic comfort; window openings, that are not adjustable; the air conditioning system, that does not usable and adjustable, take into a scarce account the need to regulate thermal parameters during mediums seasons. Moreover, occupants are resulted to suffer an excessive humidity degree of the indoor air, confirmed by the presence of moisture spots probably due to condensation phenomena on the walls. The lighting system presents too much lamps and does not provide for the possibility of adjusting the lighting intensity to comply with the request for personal preference.

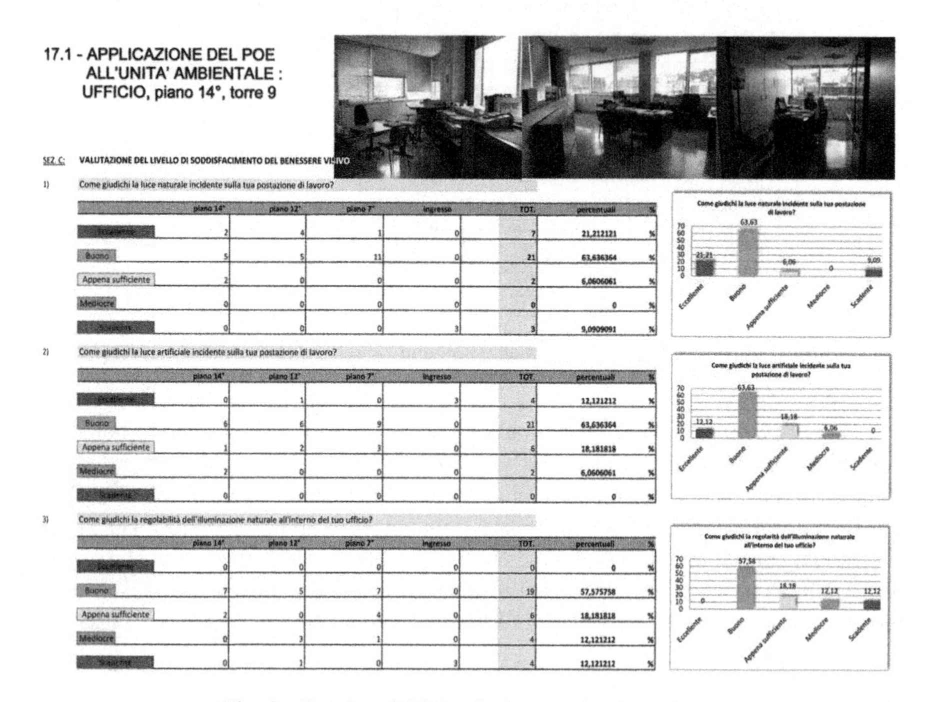

Fig. 3. Results of POE referring to visual comfort

6 Conclusions

The ARAP protocol was developed to systematically and globally identify and evaluate the risk conditions for occupants deriving from how buildings, internal and external areas have been designed, used and maintained. In the working contexts, buildings may function for supporting workers in their jobs, but also may be potential source of dangerous situations, due to building elements and finishing that create environmental and technical conditions affecting safety and comfort. The ARAP protocol allow to improve the risk assessment at work integrating it with architectural components of risks for workers, since Italian legislation is lacking about approaches focusing to analyze globally potential human related concerns related to buildings in which working activities are performed, so much that prevention and protection measures result difficult to identify and apply.

For this purpose, the identification of the recurring causes that the ARAP protocol allows to carry out, associating technological and environmental elements of the building that may pose a risk to occupants' health and safety, to their different sources, integrates the assessment with the possibility of defining and estimating the improving interventions for work environment.

In fact, the situations that are at risk due to the intrinsic properties of elements of the construction, being the effect of a bad design are more complex and expensive to change because they involve the need to replace constitutive parts of the building (i.e. structure, envelope, etc.). Differently, risk situations deriving from elements of the building that have undergone changes over time by the occupants (i.e. misuse, voluntary adaptations, etc.) and those due to their wear, can be mitigated and improved through interventions affecting changes in behavior and a more appropriate maintenance process of the building.

By the application of the ARAP protocol to the presented building, it is confirmed that the architectural and technological choices made upstream by the designer are crucial for the safety of the occupants, since most of the resulted risks are related to the fundamental components, such as the curtain wall, the windows topology and building systems. Moreover, it need to be highlighted that the scarce building maintainability, due to an indifferent architecture design to the ergonomic maintenance concerns, is another relevant cause of risk for occupant's safety and comfort.

References

1. WHO (2011) Environmental burden of disease associated with inadequate housing (EDB), World Health Organization, New York
2. Clements-Croome D (2006) Creating Productive Workplaces, 2nd edn. Routledge, Oxford
3. ISO 9001 (2015) Quality management systems. Requirements
4. Roelofsen P (2002) The impact of office environments on employee performance: the design of the workplace as a strategy for productivity enhancement. J Facil Manage 1(3):247–264
5. Dilani A (2004) Design and Health III: Health Promotion Through Environmental Design. International Academy for Design and Health, Stockholm
6. Attaianese E, D'Angelo R, Duca G (2013) Architectural risk and workplace safety: proposal for an assessment method. Ital J Occup Environ Hyg 4(2):81–88
7. Attaianese E (ed) (2015) Valutare il rischio architettonico negli ambienti di lavoro, Fascicolo 2 della Collana RAS. INAIL Campania e Università di Napoli Federico II, Edicampus, Ricercare e Applicare la Sicurezza, Roma
8. UNI 8290-1: 1981- Ediliziaresidenziale. Sistema tecnologico. Classificazione e terminologia
9. Manuele FA (2008) Prevention through design (PtD). Hist Future J Saf Res 39:127–130
10. Attaianese E, d'Angelo R (2016) La, valutazione del rischio di natura architettonica per la sicurezza dei luoghi di lavoro: applicazione del Protocollo di Valutazione del Rischio Architettonico ad un edificio per uffici. RivistaItaliana di Ergonomia. Special Issue 1:129–133

11. Chew MYL, De Silva N (2003) Maintainability problems of wet areas in high-rise residential buildings. Building Research and Information 31(1):60–69 ScholarBank@NUS Repository
12. Boyce PR, Wilkins A (2018) Visual discomfort indoors. Light Res Technol 50:98–114
13. Vischer J, Wifi M (2015) The effect of workplace design on quality of life at work. In: Fleury Bahi G, Pol E, Navarro O (eds) Handbook of Environmental Psychology and Quality of Life Research. Springer, London

The Particular View: The User's Environmental Perception in Architectural Design

Rodrigo Mendes Pinto(✉)

Brasília University, Brasília 70675-180, Brazil
rodimendes@gmail.com

Abstract. From the analyzes on correctional facilities, the ergonomic approach is still relatively strict. Thus, although this work is not innovative, the application to the physical space of these institutions reveals a gap. In this way, this research seeks to increase the knowledge of the architecture and to reduce the existent gap on the subject in the brazilian ergonomics literature. The analysis aims the perception of the users as a determining factor in the elaboration of correctional facilities for adolescents in sanction compliance. For the purpose, we use the theoretical and methodological apparatus of the Ergonomic Analysis of Work and Cognitive Ergonomic, with the support of the technique idealized by Abraham Moles for the extraction of the user's perception, called the "Constellation of Attributes". This technique collects data about the idealized space and the occupied space, explaining its (in)compatibilities. The research also explored the analysis of perception based on characteristics such as: gender, time in the public service, seeking to identify the relationship between these characteristics and the interviewees' perception. By examining the collected data, the distance between the space project and its respective user was ratified, generating conflicts in the activity dynamics, damaging the mental health and the results of the public policy, besides the financial impact after the occupation of the building due to adaptations and reforms. We also identify a distance between the users' imagery and the socio-educational determination of public policy, which allows us to question how much architecture contributes to the effectiveness of adolescent resocialization.

Keywords: Socio-architecture · Perception · Attributes constellation

1 Introduction

Even though the discussion about the relationship between the user and the space (whether or not it is built) is more and more popular, some themes are seldom regarded from the ergonomics point of view. Among them, is the Socio-educative Architecture.

When it comes to the physical space, the theme to be addressed in this research for this architecture aspect, what stands out throughout Brazilian history is the adaptation of places for the development of adolescents' socio-educational activities. These spaces were not designed to shelter this public, nor are they suited to resocializing public policies either. Thus, socio-educative practices were made much more by individual initiative than guided by general directions.

S. Bagnara et al. (Eds.): IEA 2018, AISC 825, pp. 567–578, 2019.
https://doi.org/10.1007/978-3-319-96068-5_63

In this context, the understanding of the needs of the professionals who work in juvenile correctional facilities during the stage of the architectural project development, that is, the translation of the subjectivity into the language of the professional designer, is of crucial importance to propose suitable spaces that allow the effective rehabilitation of the adolescent.

As a research object, we used a juvenile correctional facility[1], inaugurated after 2006, the year of publication of a document called the National System of Socio-Educative Care - SINASE[2] in Portuguese abbreviation. The main objective of this document is to beacon public policy intended to reinstate to society the adolescents serving a punitive measure. Thus, in principle this building is in accordance with the stated by the document, considering the project was approved by the then Secretary of Human Rights of the Presidency of the Republic.

Based on studies in the area of Ergonomics, and its main objective which is the transformation of the work, one of its purposes is to presume the participation of the subjects, since an ergonomic action is, ultimately, a process of collective construction between the team of ergonomists and the body of social actors involved (Abrahão *et al.* 2009).

Rouilleault (2001 apud Guérin *et al.* 2001 p. XIV) also emphasizes the need for effective participation of users, where they are invited to '*observe as closely as possible* what connects the material and organizational conditions of the work to their results, *actual work activity*, and to take it into account from the outset, broadening the collective involved in the conception and its objectives.'

Hence, the analysis of user perception[3] is essential for a well-oriented ergonomic action. Of cognitive essence, the analysis approaches the instruments used in environmental psychology. The perspective of this type of analysis seeks to "evaluate the adaptability of these spaces to the developed activities" (Vasconcelos et al. 2010), not the opposite. For this purpose, it will be used the technique conceived by Moles (1974), called "Constellation of Attributes".

2 Methodology

Villarouco (2002) claims that there is no possibility of studying the constructed space dismissing the aspects related to the environmental perception of those who use it. Elali also stresses the need for users to participate in the evaluation of built spaces, "because their direct and daily contact with the object turns them into a qualified critic" Elali (2002).

With this understanding stated, we will use the technique called "Attributes Constellation" for the development of this study. It consists of an experimental technique for the analysis of free and spontaneous associations induced on a particular site.

---

[1] Due to legal restrictions, this work will not make any reference as to which facility was used as the object of study.

[2] Free translation.

[3] For this study: socio-educational agents, psychologists, pedagogues, social workers and lawyers.

The experiment is divided in two parts: a spontaneous one, where it is inquired "what attributes do you relate to [the imagined environment]?"; and another one induced, where it is questioned "what attributes do you relate to [the occupied environment]?". Due to the freedom of answers, we can determine the desired number of responses (Ekambi-Schmidt 1972).

After this phase, the attributes are separated in a decreasing order of citation frequency (Table 1) and a mathematical calculation, utilizing Fechner's Law, establishes the "psychological distance" that separates the central object from its qualifiers [or attributes]. Afterwards, a graphic that resembles a constellation is designed (see Fig. 1).

Table 1. Example tabulation of spontaneous qualifiers for a room.

Item	Position	Spontaneous qualifiers	Number of mentions
1	1	Brightness	10
2	2	Silence	8
3	3	Clean	7
4		Wide	7
5	4	With television	5
6	5	Outlets in every wall	2
7	6	With connected bathroom	1

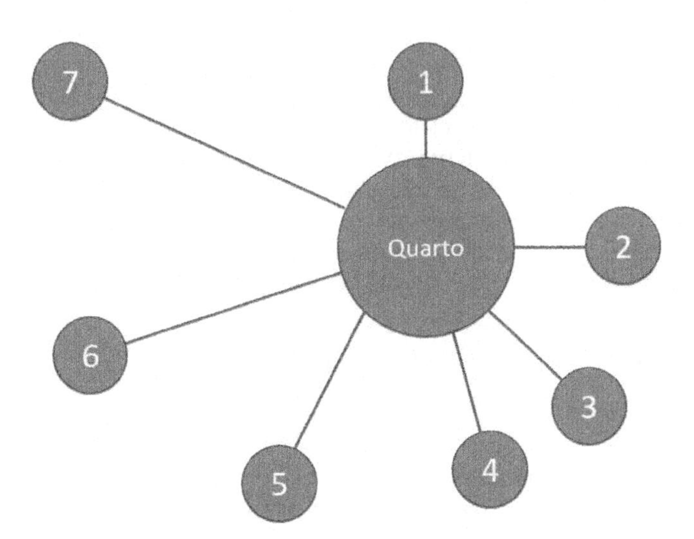

Fig. 1. Resulting graph Table 1

3 Correctional Facilities Under Focus

3.1 The Juvenile Correctional Facilities – JCFs and the Socio-Educative Policy

The history of Brazilian specific institutions that receive adolescents who are serving a punitive measure started with the creation of the Two Rivers Correctional Colony (Colônia Correcional Dois Rios) in 1894 in Rio de Janeiro and the Disciplinary Institute (Instituto Disciplinar) in 1902 in São Paulo.

Since then, there have been many changes in various aspects related to socio-educative policy. These changes were funded on the paradigm change in regards to the recognition of children and adolescents as subjects of rights.

At present, since the institution of the current Federal Constitution (1988), of the Child and Adolescent Statute - CAS (1990) and of the National System of Socio-Educative Care - SINASE (2006), there has been distinct attention to the physical area as a facilitator of the socio-educative process.

In this matter, SINASE is essential due to its detailing of the guidelines. It pinpoints the physical structure and the relationship between it and the user, clarifying that it must be pedagogically appropriate to the development of socio-educative action. It transmits messages to people in occurring a symbiotic relationship between spaces and people. Thus, the physical space embodies an element that promotes the personal, relational, affective and social development of adolescents in compliance with socio-educational measures (BRASIL, 2006).

It also emphasizes that the organization of the physical space should predict and enable the change of phases of the adolescent's care through the change of environments (spaces) according to the goals established and achieved in the individual plan of care (IPC), favoring greater concreteness in regards to their advances and/or setbacks of the socio-educative process. Hence, there are three phases of socio-educative care: (a) initial phase of care: period of reception, recognition and preparation by the adolescent of the process of individual and group coexistence, based on the goals established in the IPC; (b) intermediate phase: period of sharing in which the adolescent presents improvements related to the goals agreed in the IPC; and (c) conclusive phase: period in which the adolescent presents clarity and awareness about the goals achieved in his socio-educative process. Regardless of the socio-educative phase in which the adolescent currently is, there is a need to have a physical space reserved for those who are threatened in their physical and psychological integrity, called protective coexistence in SINASE.

However, even with the guidelines having been improved, the physical structure of socio-educative correctional facilities did not advance at the same speed, still presenting characteristics similar to prison facilities, something prohibited by SINASE[4].

[4] In spite of the criticism and prohibition in the similarity between the architecture of socio-educative juvenile detention and the penitentiary architecture, SINASE itself alleges the construction of walls with a minimum height of 5,00 m and the presence of watchtowers (lookouts) in the second, which, in our view, contradicts the one advocated by the policy.

On the one hand, we attribute the maintenance of these characteristics to the fact that old buildings that housed adolescents, therefore, designed and built under the light of other orientations, continued in the operation. On the other hand, the distance between the real users[5] of the units and their construction (we understand that this takes place since the design of the architectural project) causes the project and reforms to be disconnected from the real needs of the socio-educative policy, apart from discarding the relevant interdisciplinary, individual and experiential contribution.

For this research, we used only the public servants that were working in the correctional facilities, not accounting for the adolescents, due to the need of attending to legal and security protocols that, due to time, were not possible to bring to this work.

3.2 Attributes Constellation

Of a cognitive nature, this analysis seeks to extract the expectations and impressions the users have about the environments investigated. Here the focus is only the user, looking inside; not looking at the action, nor at task performance.

As a tool that will help capture the perception of the ambient by the user, it will be used the method called Attributes Constellation.

Attributes Constellation is a technique for extracting the user's perception of their ambient, allowing the identification of real attributes of the spaces, as well as expressing the ideal environments aspirations. (Paiva et al. 2016)

In order to quantitatively limit responses, we asked the interviewees to indicate until five words that would answer each of the following questions:

- What are the attributes (not just adjectives) that come to your mind when you think of an IDEAL Juvenile Correctional Facility?
- What are the attributes (not just adjectives) that come to your mind when you think about the Juvenile Correctional Facility THAT YOU WORK FOR?

In addition to the two questions that compose the mentioned technique, the users were asked information about their gender, what servant position they occupy, length of work in the facility, whether they ever were in a commissioned position and, if so, for how long. These questions seek to identify if these variables relate to the nature of the perception of the focused place.

3.3 Results of the Applied Technique

The application of the technique in juvenile correctional facilities is an initial and still ongoing part of the research carried out by the author for his masters' dissertation, also in progress, under the light of Ergonomics.

Until the closing of this work a total of 8 public servants answered the aforementioned questions, being them 1 social worker, 1 pedagogue and 6 social-educative agents.

After the personal information, we chose to question the participant about the individual's perception of the imaginary environment and then question him about the

[5] Mainly servants working in the facility and adolescents.

Table 2. Synthesis of raw data collected

Which public servant position you occupy?	How many years of placement working in Juvenile Correctional Facility?	Have you held any commissioned positions during your placement in Juvenile Correctional Facility?	In case the answer for previous question was 'YES', for how long did you occupy the placement (s)?	What are the attributes (not just adjectives) that come to your mind when you think of an IDEAL Juvenile Correctional Facility?	What are the attributes (not just adjectives) that come to your mind when you think about the Juvenile Correctional Facility THAT YOU WORK FOR?
Social-educative agent	More than 15 years	Yes	More than 10 years	Wide; adequate lighting, emergency exit, fire system, modules with better ventilation	Adequate school area, larger patio, inspection system inadequate for facility, equipped room for technical service, dormitory for servants
Social-educative agent	Between 5 and 10 years	No		Discipline; Satisfied servants; Education; Family participation; Crime related treatment	Indiscipline; Inadequate space; Unsatisfied servants; Negative family influence
Social-educative agent	Between 5 and 10 years	Yes	Less than 1 year	Schooling; Professionalization; Culture; Sport; discipline	Lack of dialogue; Tiring; Lack of adequate tools; Adequate physical structure; Ineffective schooling
Social-educative agent	Between 5 and 10 years	No		Must have communication; must be a welcoming place; Resocialization as a priority over punishment	Very hot; Lack of convenience; it is a prison (for all); Lack of communication; Lack of concrete projects for resocialization
Pedagogue	Between 5 and 10 years	Yes	Between 2 and 4 years	Dialogue; Emotional and psychic weariness; Intellectual work; Conflicts; Discomfort	Change - management; Team work; Conflict; Resistance; Advance; Culture

(*continued*)

Table 2. (*continued*)

Which public servant position you occupy?	How many years of placement working in Juvenile Correctional Facility?	Have you held any commissioned positions during your placement in Juvenile Correctional Facility?	In case the answer for previous question was 'YES', for how long did you occupy the placement (s)?	What are the attributes (not just adjectives) that come to your mind when you think of an IDEAL Juvenile Correctional Facility?	What are the attributes (not just adjectives) that come to your mind when you think about the Juvenile Correctional Facility THAT YOU WORK FOR?
Social-educative agent	More than 15 years	No		Professionalization; Individualization; Professionalism; Discipline; Political interest or commitment	Lack of: maintenance, professional training, engagement, personal use material
Social-educative agent	Between 2 and 4 years	No		Respect the maximum number of adolescents per accommodation; Daily and all-day activities; Use of sports courts; Effective multidisciplinary work with defined roles; Managers who are committed and are from social-educative careers	Ample and clean space; Reasonable resting place for servants; Scarce work material; Communication difficulty between departments; Delay in maintenance of the modules
Social worker	Up to 1 year	No		Socio-education; Accountability; Precarious social policy	Work; Belief in the human being; Political management

occupied environment. This approach sought to produce as minimal as possible interference on the answers, seeing that when the first question is carried out on the occupied environment, the following question is strongly limited to the antonyms of the first one (Tables 2, 3 and 4).

Table 3. Spontaneous qualifiers for juvenile correctional facility

Categories	Position	Spontaneous qualifiers	Number of mentions	Psychological distance
Organizational	1	Social-education	5	0,82
Organizational	2	Discipline	3	1,00
Organizational	2	Professionalization	3	1,00
Organizational	3	Accountability	2	1,21
Organizational	3	Culture/Sport	2	1,21
Organizational	3	Communication	2	1,21
Architectonic	4	Ample	1	1,91
Architectonic	4	Adequate lighting	1	1,91
Security	4	Fire system	1	1,91
Architectonic	4	Better ventilation	1	1,91
Organizational	4	Precarious social policy	1	1,91
Organizational	4	Satisfied public servants	1	1,91
Security	4	Emergency exit	1	1,91
Organizational	4	Family participation	1	1,91
Organizational	4	Crime related treatment	1	1,91
Architectonic	4	Welcoming place	1	1,91
Human	4	Physical and psychic weariness	1	1,91
Human	4	Conflicts	1	1,91
Organizational	4	Individualization	1	1,91

Thus, we have the following chart that gathers the collected data:

After that, the answers were divided into 4 groups, according to their characteristics, being them:

- Organizational: relates to the form of operationalization and the approach of public policy, which varies when management changes;
- Human: linked to the physical and mental perception of the operator;
- Security: Although it has architectural repercussions, the essence of the category is security.

- What are the attributes (not just adjectives) that come to your mind when you think of an IDEAL Juvenile Detention Center?

In the analysis of the attributes constellation graph for spontaneous qualifiers, it is noticeable that the items with the highest incidence do not refer to architectural aspects.

From the 6 first attributes in the ranking, 5 are related to how the socio-educative policy is operated and managed, with the most frequent one being the attribute "Socio-education", which, although in a generic way, reinforces the perception about the educative character in detriment to the sanctioning one of the correctional measure.

As for architectural attributes, they all appear with the same frequency. Of these, we highlight the attribute "Welcoming", pointed out by a Socio-Educative Agent, as characteristic of an ideal correctional facility, demonstrating a possible change of perspective, contrary to the perception of restrictive tendency of the mentioned position for the application of the correctional measure.

It is worth highlighting the recurrence of attributes aimed at the offer of available activities to the adolescent and the approach during their time in the facility. They are: socio-education, discipline, professionalization, culture/sport, family participation, crime related treatment and individualization. There were 16 appearances of the 30 cited, revealing which perception and direction the servants desire of the socio-educative policy (Figs. 2 and 3).

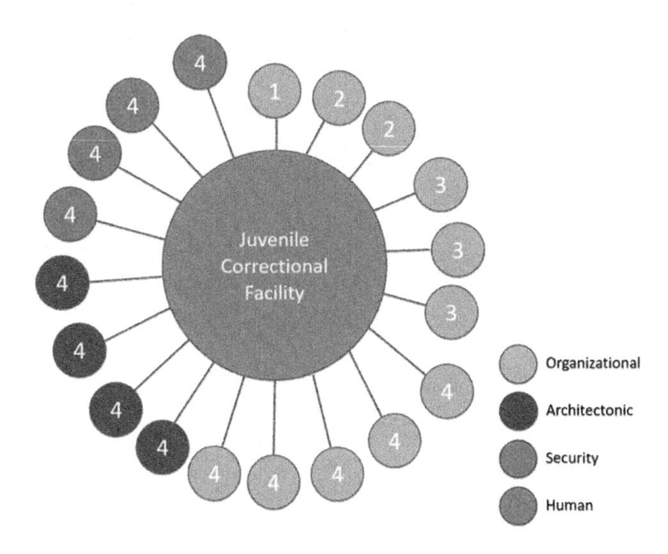

Fig. 2. Spontaneous attributes

When taking in consideration a segmentation by means of the position occupied, a greater incidence of operational aspects is observed, that is, more objective ones, present in the citations of the socio-educative agents, whereas greater subjectivity is verified in the expression of the pedagogue and the social worker.

Table 4. Induced qualifiers for a JCF

Categories	Position	Induced qualifiers	Number of mentions	Psychological distance
Architectonic	1	Inadequate space	3	1,10
Organizational	1	Lack of communication	3	1,10
Architectonic	2	Absence of an equipped technical treatment room	2	1,36
Architectonic	2	Absence of a resting room for the socio-educative agents	2	1,36
Organizational	2	Work	2	1,36
Organizational	2	Political management	2	1,36
Organizational	2	Dissatisfied servants	2	1,36
Architectonic	2	Adequate space	2	1,36
Organizational	2	Lack of maintenance	2	1,36
Organizational	2	Lack of material	2	1,36
Architectonic	3	Adequate school	1	2,32
Architectonic	3	Small patio	1	2,32
Architectonic	3	Inspection room in inadequate place in the facility	1	2,32
Human	3	Belief in the human being	1	2,32
Organizational	3	Indiscipline	1	2,32
Organizational	3	Negative family influence	1	2,32
Human	3	Tiresome	1	2,32
Organizational	3	Ineffective schooling	1	2,32
Architectonic	3	Heat	1	2,32
Organizational	3	Ineffective resocialization	1	2,32
Human	3	Conflict	1	2,32
Human	3	Resistance	1	2,32
Organizational	3	Progress	1	2,32
Organizational	3	Culture	1	2,32
Organizational	3	Lack of capacitation	1	2,32

Advancing the analysis to the induced qualifiers, the strong expansion of the architectural components is highlighted, and of they show among the first ones, reflecting how much the inadequacy or absence of spaces can compromise the activity.

Still, there is little reference to the environments strictly attended by adolescents, such as bedrooms, living areas in the modules, patio, guest building. There are also no notes on the appearance of the facilities and elements that refer to the prisons - walls surrounding the facility, at least 5 meters high, equipped with razor wire and lookouts. This way, it becomes apparent the disconnection between the idealized (welcoming places and effective socio-educative activity) and that which is perceived as architectural elements that may facilitate the social reintegration of the adolescent.

It is also possible to see a certain reflection of the answers onto the ideal environment, for example lack of communication, dissatisfied servants, negative influence of family members, heat, and ineffective resocialization.

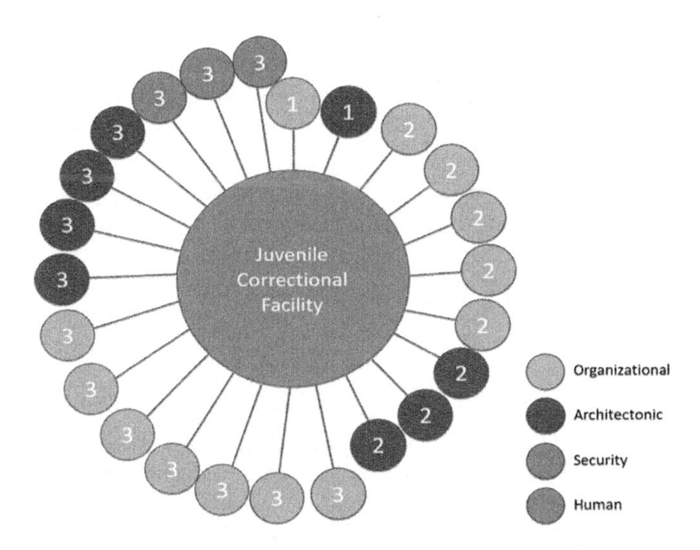

Fig. 3. Induced attributes

Another relevant fact is the presence of positive aspects, albeit in a subtle way (contrary to others for the same question), that consider the physical space adequate, besides the offer of activities (culture), apparently sufficient to meet the resocializing policy.

When centralizing the analysis on the time worked in juvenile correctional facilities, it is evident the directly proportional relationship between time worked in that specific place and the number of negative attributes cited, reinforcing the probable greater critical knowledge about the policy implemented and its directions. From another angle, one should not overlook the refreshing view of those who have recently arrived, said to tend to be an "utopian" look by the older servants.

4 Final Considerations

The limited incidence of architectural features in the list of spontaneous qualifiers revealed low understanding of the role of architecture as a relevant contributor to the socio-educative process.

Nonetheless, there was a substantial increase when approaching the induced qualifiers. Therefore, the influence of spaces on the performance of activities is clear. However, such relevance is only evident during the usage, denying the design phase. Fonseca and Rheingantz (2009) emphasize that it is necessary to study more carefully how the different types of environments stimulate the thought and its subsequent transformation into actions. These omissions end up manifesting itself, usually related to negative aspects, during the use and occupation of the environments.

We clarify that we do not advocate that architecture is a modifying element of "the individual's personality, but it can influence the perception and cognition of spaces and thus it provides satisfaction in use" (Kowaltowski et al. 2000).

Kowaltowski (2000) also emphasizes it is necessary to use methodologies that allow to adequately filter information and that assist in the interpretation of observations and in the actual determination of user satisfaction. This is the purpose of this research using the Attributes Constellation.

In this sense, the aim is to add elements of reference for new projects to the existing dimensional parameters, since the simple fulfillment of these requirements does not ensure an environment qualified for usage, nor does it supply all the needs of the relation between person and environment.

References

Abrahão J et al (2009) Introdução à Ergonomia: da prática à teoria. Blucher, São Paulo

Ekambi-Schimidt J (1972) La perception de l'habitat. Paris: édition universitaires

Elali GA (2002) Ambientes para educação infantil: um quebra-cabeça? São Paulo: [s.n.]

Fonseca JF, Rheingantz, PA (2009) O ambiente está adequado? Prosseguindo com a discussão. In: Produção, Florianópolis, Agosto, pp 502–513

Guerín F et al (2001) Compreender o trabalho para transformá-lo: a prática da ergonomia, 1ª edn. Blücher, São Paulo

Kowaltowski D et al (2000) Ambiente Construído e Comportamento Humano necessidade de uma metodologia. ENTAC 2000 Modernidade e Sustentabilidade, Salvador

Moles AA (1974) Sociodinâmica da cultura. Universidade de São Paulo, São Paulo

Oliveira GRD, Mont'alvão C (2015) Metodologias utilizadas nos estudos de ergonomia do ambiente construído e uma proposta de modelagem para projetos de design de interiores. In: 15º Congresso Internacional de Ergonomia e Usabilidade de Interfaces Humano-Tecnologia, Produto, Informações, Ambientes Construídos e Transporte. Blucher Proceedings, Recife

Paiva MMB et al (2016) Discussão acerca da percepção ambiental, suas ferramentas e cognição. In: 15º Ergodesign. Recife: Edgard Blücher Ltda

Vasconcelos CFE, Villarouco V, Soares MM (2010) Contribuição da Psicologia Ambiental na Análise Ergonômica do Ambiente Construído. Ação Ergonômica 5(3):14–20 ISSN 1519-7859

The Environmental Contribution to Wayfinding in Museums: Enhancement and Usage by Controlling Flows and Paths

Federica Romagnoli[1], Teresa Villani[1], and Angelo Oddi[2(✉)]

[1] Department of Planning, Design and Technology of Architecture, Sapienza University of Rome, Rome, Italy
romagnolifederica@outlook.it,
teresa.villani@uniroma1.it
[2] CNR, National Research Council of Italy, Rome, Italy
angelo.oddi@istc.cnr.it

Abstract. The field of research in which wayfinding is situated refers to the way people move in reaction to environmental stimulation. It therefore fully concerns not just signage but also space designing, its geometric configuration, technical solutions and their material characterization. The focus is consequently on environmental factors that facilitate wayfinding in a museum (accessibility, visibility, etc.) and on other elements such as spatial configuration, architectural features and functional aspects. These factors influence relational phenomena and therefore visitors' satisfaction. Methods and tools for designing and managing spaces have been studied in the research. The configurational analysis method of space has been used to objectify syntactic features of space. In particular, the outcomes of an experimental project, which have been analyzed in a master's thesis on the re-functionalization of the museum of Palazzo dei Diamanti in Ferrara, are presented. Permeability, proximity, connections of spaces, namely meaningful features to ensure wayfinding have been examined. Space parameters resulting from the geometry of the layout, from the visual connections and from the changes of direction were then evaluated. The outcomes have been used as inputs for designing a unitary tour route circuit, that also reconnects the museum's second floor, and for planning three independent alternative routes for a differentiated use of the museum.

Keywords: Wayfinding · Museum · Configurational analysis · Space syntax

1 Introduction

1.1 The Environmental Contribution to Wayfinding

Decisive factors that make the wayfinding project effective concern the ease with which space can be decoded and "read" by users, also thanks to the use of environmental stimulations addressed to various sensory channels, and which provide in a coherent and organized way all the necessary information for the museum to be usable, welcoming and safe even for those who enter for the first time. In fact, thanks to a well-done design that maximizes the level of environmental communicability, it is possible

S. Bagnara et al. (Eds.): IEA 2018, AISC 825, pp. 579–588, 2019.
https://doi.org/10.1007/978-3-319-96068-5_64

to improve cultural and environmental accessibility and to reduce the "visitor's stress", namely the gradual loss of attention towards the exhibition due to the mental fatigue of the user who is too busy making decisions on where to go inside the museum.

The field of research of which wayfinding is part of therefore concerns the way people move in reaction to environmental stimulations [1]. Hence, it fully concerns the designing of space [2], technical solutions and artifacts. It is not only about communicative systems, but also spatial configuration and material characterization [3]. Environmental factors that simplify wayfinding in a museum (accessibility, visibility, etc.) along with other elements such as spatial configuration, architectural features and functional aspects, all influence relational phenomena [4] and therefore visitors' satisfaction [5]. The aim of the proposed contribution is to deepen the knowledge of the spatial layout of museums and qualify, by measuring it, the integration between its various functional areas in order to improve wayfinding and fruition [6]. For this reason, methods and tools for monitoring, evaluating and verifying flows are proposed, with the aim of managing and designing museum spaces. The study focuses on the proximity relations between spaces, on hierarchy in circulation, on the differentiation of routes, on the legibility of accesses, on journey times, and on the control of personalized exhibition routes. Methods and tools in support of space design and management have been studied in order to improve visitors' flows. One of the most used methods is the analysis of spatial configurations, also employed in the research by using the software deptmapX [7–12]. It is a method that makes the syntactic characteristics of space objectifiable and that has been applied in several museums (British Museum [13], Victoria & Albert Museum [14]) with the aim of improving their use. The results of the configurational tests carried out in the mentioned examples led to the replanning of wayfinding strategies, with consequently an optimization of the different museum areas and a relaunch of the image of the museums.

This paper, in line with what has already been experimented at a methodological level, reports the outcomes of an experimental project that has been analyzed in a master's thesis on an emblematic case study: the re-functionalization of the flows and routes of the Palazzo dei Diamanti museum in Ferrara, which aimed to optimizing ways of use and journey times in relation to the possibility of dividing the museum and the exhibition spaces depending on the features of consistency and interest of temporary exhibitions (functional lots) [15]. By using depthmapX (configurational analysis) some significant properties of space were identified in the interest of way-finding (permeability, proximity, connections). It was possible to give a value to each element of space according to its connection with other spaces, the permeability with adjacent spaces and the proximity to the entrance (information point) [16]. Space parameters resulting from the geometry of space, from the visual relations, and from changes of direction of the layout were then evaluated. The results have been used as inputs for re-designing a unitary tour route circuit (sequential structure), that also reconnects the museum's second floor which is now separated, and for planning three autonomous sub-rings that represent alternative uses of the museum.

2 Case Study: Enlargement and New Functionality of Palazzo dei Diamanti in Ferrara

2.1 The Topic of the Project

The general aim of the proposed experimentation concerns the potentials of wayfinding to enhance museum spaces, as a discipline integrated to designing, in order to influence the quality of the visit and the overall satisfaction of the visitors.

The experimental project focused on the Palazzo dei Diamanti of Ferrara, the most important work of Ferrara's Renaissance architecture and for which the city announced an international design competition [17]. Currently there are three different museums in the Palazzo: the National Art Gallery, located on the main floor, the Galleries of Modern and Contemporary Art, within the two historical wings of the ground floor, and the Museum of the Risorgimento and the Resistance, in the southern area of the complex (Fig. 1). The latter will be transferred within a few years to a new location (designed ad hoc) in order to free spaces that will be requalified and allow the enlargement of the Galleries of Modern and Contemporary Art in terms of exhibitions and provided services. The invitation to tender was based on the necessity to provide the structure with supplementary services that are currently missing and to adapt the visit route, which is currently not very functional, to contemporary museum standards. In fact, the exhibition route of the Galleries of Modern and Contemporary Art is within two wings of the Palazzo (called "ala Biagio Rossetti" and "ala Benvenuto Tisi") that are separated and not contiguous and which conjoin with an outdoor "platform roof". Consequently, one of the main issues in the request of proposal was the design of a new air-conditioned connecting unit between the two wings of the building that could also be used separately and independently from the exhibition route to be able to host big events and conferences and to exhibit large format works of art.

2.2 Analysis of the Current State of Affairs

To develop such an articulated Construction Plan, in compliance with the building's tutelage requirements for its historical and cultural importance, wayfinding problems were studied as a strategic aspect for the re-functionalization of the entire complex from a point of view of the reorganization and hierarchization of spaces and the necessary usability. Which are the tools that the architectural plan can use to make buildings more hospitable and easily decipherable for someone that has never been there?

Configurational analysis represents a supportive tool for planning decisions. Starting from a careful analysis of the current state of affairs, both from an environmental and technological point of view to understand the potentialities of the spaces and the feasible intervention levels, the planimetric and spatial configuration was checked by considering and prioritizing the systems of flows inside the building. Subsequently, different interventions were suggested to characterize the areas from a material point of view by working on the perceptual and sensorial properties of the architectural elements (finishing materials, furnishing, devices). Also used was the application of methodologies of numerous international examples of improvements and upgrades of very complex museum spaces based on scientific and objective support for

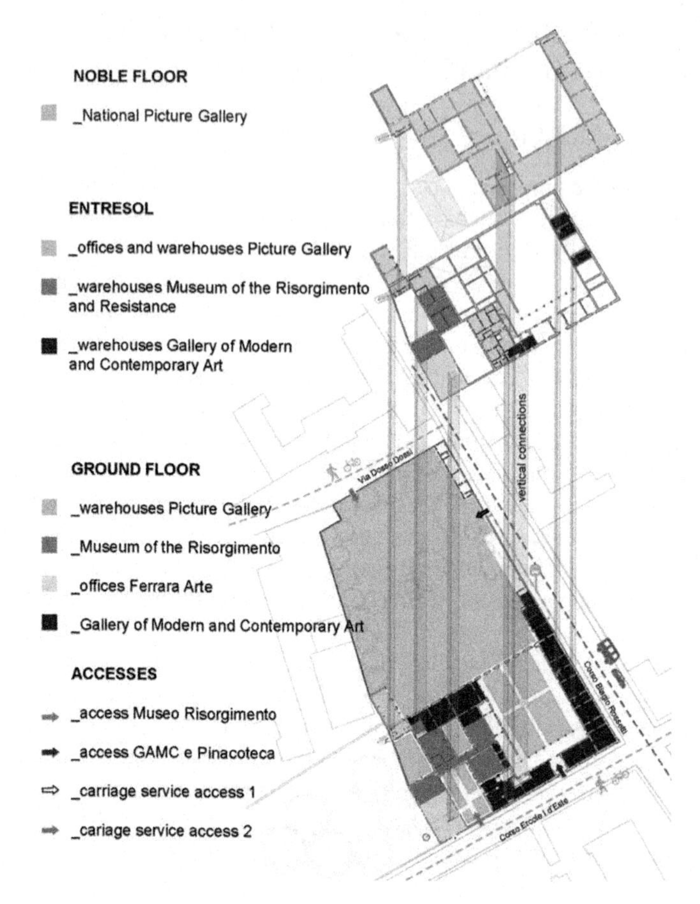

Fig. 1. Functional subdivision of Palazzo dei Diamanti, current state.

the verification of project choices. The software used, made available by the University College, was dephtmapX which can analyze the spatial configuration and was useful for carrying out an initial verification of the project choices. The verification provided a first representation of the building through a graph that identifies convex spaces (environmental units) as nodes and their direct relationships as connecting lines. In this case, the construction of the graph highlighted the predominantly sequential connections of the wing environments destined for exhibition areas, which therefore would require the user to follow a progressive-linear tour. Starting from the graph the successive levels of analysis led to the measurement of the level of connectivity of each environment, that is, for a given space, the maximum number of spaces adjacent to it. Repeating this calculation for each space it was possible to measure the degree of integration of the convex space by adding the distances from the latter to all the other nodes. The software made it possible to qualitatively appreciate the results by coloring in warm colors the best integrated spaces and in blue those that could be reached only with a greater number of passages, see (Fig. 2). It should be noted how, in the present state, with the distribution elements most connected to the different environmental units, there is therefore no central "core" inside the building, rather two main routes, one of which leads outdoors, which is not very convenient especially in the winter. Even through the analysis of the "HH" spatial integration, it can be seen how the distribution axes are more integrated than the rest of the complex, nevertheless access to the rooms to the south of the building (where the request for the proposal requires the new services for the public to be installed) is easier than the north exhibition wing. This means that the required reception areas, cloakrooms and toilets end up being poorly integrated. Continuing with the application of the verification tools, we abstract the space by means of a point grid from which it is possible to calculate the visual integration i.e. all the points of the grid that are mutually visible to each other. Again, the most visible spaces are in red and the most hidden in blue. From this, it is clear how the new services areas, given their position on the lay-out, are visible only from specific positions, so their presence would not be perceptible to visitors without specific signage. Where the warmer colors are, on the other hand, the portico has maximum visibility compared to the rest of the building, a feature that makes it appropriate to be the first reception point, where the visitor could find the first signs on possible routes to take.

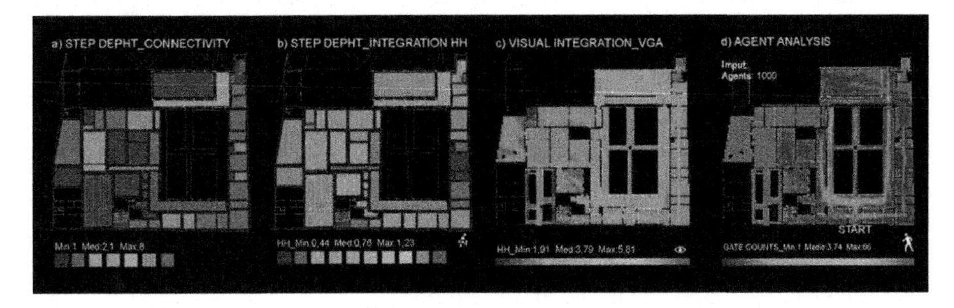

Fig. 2. Space syntax analysis (with depthmapX) of ground floor of Palazzo dei Diamanti, current state. (a) Connectivity (b) Integration HH (c) Visual integration (d) Agent analysis. (Color figure online)

The visibility graph (Fig. 2c) is also fundamental for another type of analysis because this is related to the movement of people in space. In fact, the results of this type of calculation simulates the flows within the space through the "agents" or entities capable of perceiving the space surrounding them to make the predicted decisions based on statistically expected human behavior. In the simulation the agents started from the main entrances of the building in numbers proportional to the flows of people observed in reality; each agent can choose a direction of movement (action) based on his visual field (perception) derived from the analysis of visual integration. The layout is broken down by a uniform grid and for each agent occupying a given square of the grid, the next destination is randomly selected among all the visible squares within a visual field view cone (the angle is generally large, i.e. 170°), giving the agent the opportunity to change direction. In the case study of Palazzo dei Diamanti the process was repeated 10,000 times. From the results obtained it is observed that different areas of the ground floor are less frequented by the agents, as they are less visible. In particular, it is possible to notice how, using the current access to the Modern Art Galleries, as indicated by the invitation to tender, agents are not "spontaneously" able to reach the premises where the new services will be installed, thus such results do not support the original idea (given in the invitation to tender) for a convenient and integrated usage of all areas of the building.

To summarize, by applying these analyses to the current planimetric configuration of Palazzo dei Diamanti, several critical issues have emerged: the most integrated and connected spaces are those of distribution, one of which is the uncovered portico, and the other a connecting space, which is quite narrow. The analysis of the agents also shows how, by accessing the historical entrance, due to the lack of visibility of the areas destined to new services, it would be statistically unlikely, for an individual who has never been in the building, to "spontaneously" find the museum visitor support services. These results have been decisive for setting up a project that proposes an improved alternative in response to the invitation to tender, starting from the study of the individual environmental units present inside the building and assessing the environmental characteristics of usability, visual wellbeing, transformability, connectivity and visibility, to the in-depth analysis of the technological system and the heterogeneity of the technical solutions present in the building. These analyzes have further highlighted how within the museum an intervention was needed to even out the accessibility of the various parts of the building so that they are accessible for all types of users.

As far as the aspects the wayfinding are concerned, the imbalance between the two entrances is evident: the Modern Art Galleries' entrance, already stands out for its characteristic sculptural façade, while the Museum of the Risorgimento and the Pinacoteca are much less noticeable from the outside.

2.3 Meta-Project and Proposed Interventions

A meta-project proposal was thus established for the main entrance of the museum to be the functional reception area. Moreover, this area, taking advantage of the courtyard as a full height space, would represent a new main entrance point for the Pinacoteca, leaving however the possibility of using the previous access through the historic staircase (Fig. 3). In this way it will also be possible to organize around the new access/InfoPoint the spaces for supplementary services available to users at the beginning and end of the exhibition itinerary. To avoid the inconvenience of passing through the open portico, the proposal foresees its closure through a system of continuous glazed facades.

Following the indications of the invitation to tender that required the construction of a multi-purpose pavilion that connected the two wings of the building, some alternatives for positioning the new pavilion were tested. The proposal recommended placing and organizing it to be contingent to the rest of the building, while allowing an air-conditioned and covered passage between the two exhibition wings.

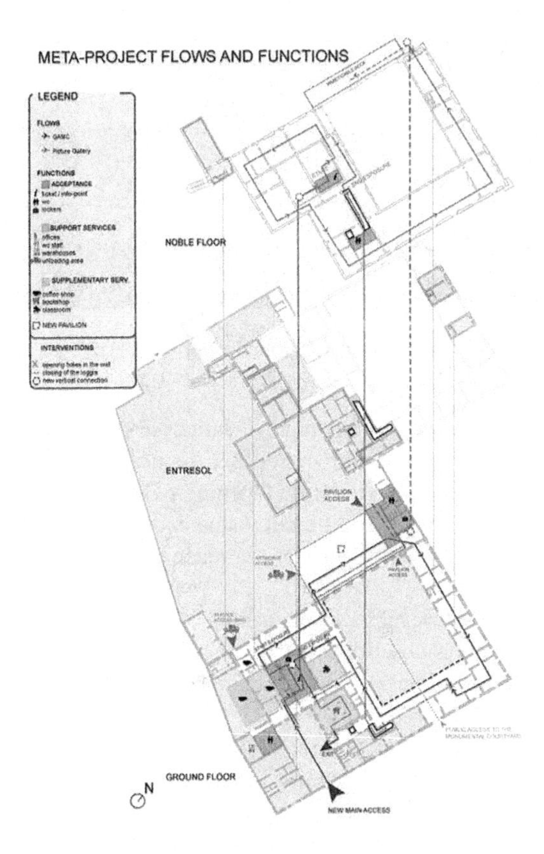

Fig. 3. Meta-project of flows and functions.

The exhibition path will thus be able to end where it started: inside the courtyard that will be covered, as required by the invitation to tender. Finally, the exit from the museum will finally be placed at the bookshop.

2.4 Verifications

Repeating the configurational analysis (depthmapX) has confirmed that these solutions maximize the integration of the new "core" designed to welcome visitors. Additionally, having agents start from the new proposed entrance, it allows them to complete the itinerary "autonomously" (Fig. 4d), as opposed to the first simulation.

The analysis of the integration (Fig. 4b) shows how it is necessary to make it possible to use the open gallery or main walkway also in the winter, given that it is necessarily part of the museum itinerary and a high level connection between the parts. By moving the entrance to the support service area the museum will also have a new center of gravity that will allow for a simpler distribution of the different sectors. This operation will make the area intended for the new services more integrated: therefore it is here that it is proposed to insert a new vertical connection to the picture galley at the

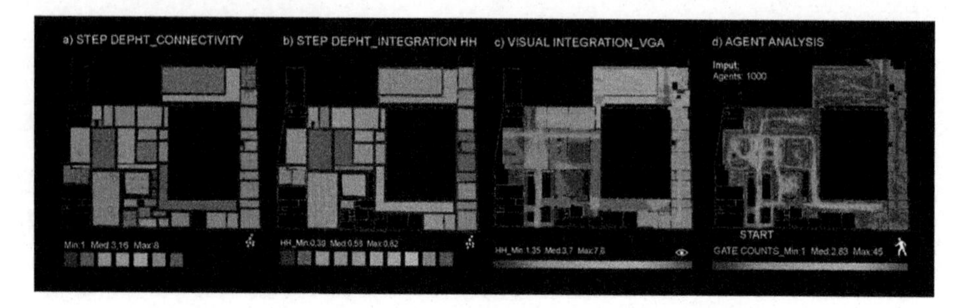

Fig. 4. Space syntax analysis of ground floor of Palazzo dei Diamanti.

noble floor, using the court to its full high. Qualitative observation of the graphical analysis of visual integration (Fig. 4c) shows that the new entrance to the newly built pavilion will reach a high level of visibility. Opening new spaces that are presently still closed inside the walls will generate different visual axes that will allow visitors to find their way around more easily, allowing them to reach various areas that today are less visible.

The last simulation, which refers to the agents' use of planimetric space (Fig. 4d), shows that, thanks to the proposed layout, the distribution spaces will end up being a lot less crowded, while the highest concentration of people will be in the welcome areas, as it should be.

Later the possibility of the exhibition rooms being used in a fractioned way by organizing a division into "functional spaces" was looked into, since the ground floor is intended for temporary exhibition: they can be used in different ways according to the nature and size of exhibitions. Every functional space is independent, and potentially could has its own welcome area by using the various possible ways into the building (Fig. 5). As a maximum complexity configuration, it was supposed the simultaneous use of several routes, such that all start from the same access point and end in the same spot. More specifically, the type A scenario schematically represents the situation in which access to educational, dining, and commercial services, is considered. This scenario takes into account the potential lack of temporary exhibits. The type B scenario would instead occur when an exhibit is held only in the Biagio Rossetti Wing, but integrated services could still be accessed at the end of the visit. The third layout illustrates how it would be possible, thanks to the multiple access points, to use the Gallery even when the ground floor is not used. Even the new pavilion would be accessible through its own entrance, separately from the rest of the building.

The last layout represents the simultaneous use of the different routes inside the museum. Thanks to the barycentric position of the new welcome area, it will be possible to host at least three different temporary exhibits without compromising the accessibility of the noble floor.

Environmental and technological requirements were defined in accordance with the organization established in the meta-project on flows, so that the building might be adapted and transformed for the project's requirements. Since the pavilion is a new construction, several possibilities of aggregation of the environmental units necessary for its functionality were explored and considered upgrading the building's historic

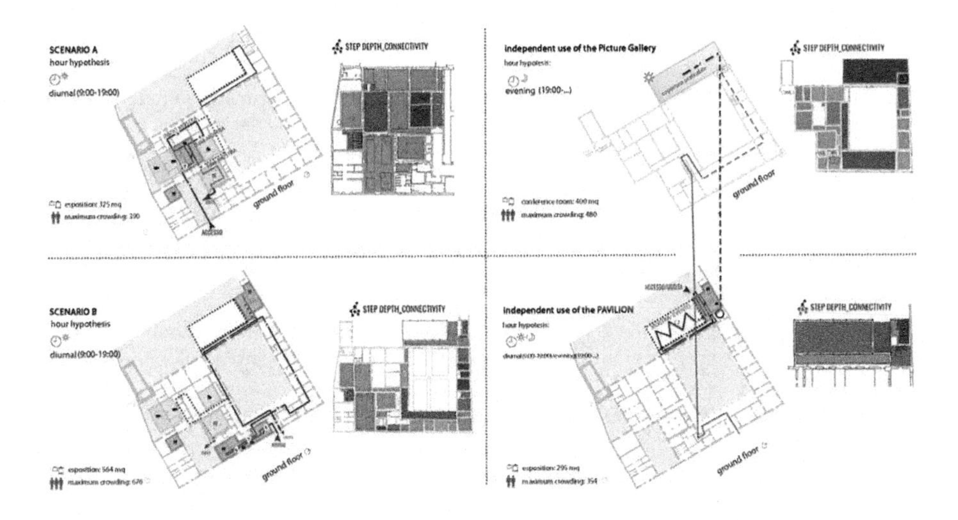

Fig. 5. Meta-project of alternative sub-lots and checks with depthmapX.

wings. The chosen configuration will allow for a flexible use of the inside space, meeting the grant's requirements of hosting large-scale art-work, as well as widely attended events and conferences. The renovation works were listed in order of priority, from those necessary for maximizing the museum's accessibility to those that would guarantee its greatest functionality.

3 Conclusions

The case study of Palazzo dei Diamanti of Ferrara showed how the configurational analysis has made objective a series of critical issues found in the museum's usage and supported the formulation and the hierarchization of the project objectives. By using the same kind of analysis once the proposed layout was verified, we were able to qualitatively assess our choices. This process supports the notion that solutions that facilitate wayfinding inside complex spaces have positive impacts on the accessibility of the entire building, as long as they are conceived in a holistic way from the start of the planning process. The proposed analyses and verifications aided the design of the functional layout and guided the technical choices to satisfy the invitation to tender requirements. However, this is only an initial but fundamental step in actualizing wayfinding. Not enough work has yet been carried out for the space to be communicative and easily read by its users. For this reason, the next steps of the project focused on the material characterization of the different functional areas. Each area's distinct characteristics were made clearer for anyone experiencing the museum by developing specific superficial coverings (especially flooring), and by choosing furnishing that, through color codes and matrices, organically increased the space's ability to communicate with all other present elements.

References

1. Thorndyke PW, Hayes-Roth B (1992) Difference in spatial knowledge acquired from maps and navigation. Cogn Psychol 12:560–589
2. Lynch K (1960) The Image of the City. MIT Press, Cambridge
3. Passini R (1984) Wayfinding in Architecture. Van Nostrand Reinhold, New York
4. Villani T, Silvestri A (2016) Wayfinding and environmental communication in museums for the promotion of cultural heritage and sustainability. In: Gambardella C (ed) World Heritage and Degradation. Smart Design, Planning and Technologies. La scuola di Pitagora Editrice, Napoli, pp 32–40
5. Montella M, Dragoni P (2010) Musei e valorizzazione dei beni culturali. Atti della Commissione per la definizione dei livelli minimi di qualità delle attività di valorizzazione. Clueb, Bologna
6. Salgamcioglu ME, Cabadak D (2017) Permanent and temporary museum spaces: a study on human behavior and spatial organization relationship in refunciond warehouse spaces of Karakoy, Istanbul. In: 11th International Proceedings on International Space Syntax Symposium. Instituto Superior Tecnico, Departamento de Engenharia Civil, Arquitetura e Georrecursos, Lisboa, pp 22-1–22-18
7. Hillier B, Raford N (2010) Description and discovery in socio-spatial analysis: the case of space syntax. In: Walford G et al (eds) The Sage Handbook of Measurement. SAGE, London
8. Hillier B, Hanson J (1984) The Social Logic of Space. Cambridge University Press, Cambridge
9. Hillier B (1996) Space is the Machine: A Configurational Theory of Architecture. Cambridge University Press, Cambridge
10. Hiller B, Tzortzi K (2011) Space syntax: the language of museum space. In: Macdonald S (ed) A Companion to Museum Studies. Wiley-Blackwell, London, pp 282–301
11. Tzortzi K (2015) Spatial concepts in museum theory and practice. In: 10th International Proceedings of Space Syntax Symposium, London
12. Penn A (2008) Architectural research. In: Knight A, Ruddock L (eds) Advanced Re-search Methods in the Built Environment. Wiley-Blackwell, Oxford, pp 14–27
13. Dursun P (2007) Space syntax in architectural design. In: 6th International Proceedings of the Space Syntax Symposium, Instanbul
14. Kwon SJ, Sailer K (2015) Seeing and being seen inside a museum and a department store - a comparison study in visibility and co-presence patterns. In: 10th International Proceedings of Space Syntax Symposium, London
15. Choi YK (1999) The morphology of exploration and encounter in museum layouts. Environ Plan 26:241–250
16. Tzortzi K (2007) Museum building design and layout: patterns of interaction. In: 6th International Proceedings of Space Syntax Symposium. ITU Faculty of Architecture, Istanbul, pp. 072-01–072-15
17. Comune di Ferrara. http://servizi.comune.fe.it/376/bandi-scaduti-ed-aggiudicazioni. Accessed 20 Dec 2017

Printed by Printforce, the Netherlands